Probability Theory and Stochastic Modelling

Volume 104

Probability Theory and Stochastic Modelling publishes cutting-edge research monographs in probability and its applications, as well as postgraduate-level textbooks that either introduce the reader to new developments in the field, or present a fresh perspective on fundamental topics.

Books in this series are expected to follow rigorous mathematical standards, and all titles will be thoroughly peer-reviewed before being considered for publication.

Probability Theory and Stochastic Modelling covers all aspects of modern probability theory including:

- Gaussian processes
- Markov processes
- Random fields, point processes, and random sets
- Random matrices
- Statistical mechanics, and random media
- Stochastic analysis
- High-dimensional probability

as well as applications that include (but are not restricted to):

- Branching processes, and other models of population growth
- Communications, and processing networks
- Computational methods in probability theory and stochastic processes, including simulation
- Genetics and other stochastic models in biology and the life sciences
- Information theory, signal processing, and image synthesis
- Mathematical economics and finance
- Statistical methods (e.g. empirical processes, MCMC)
- Statistics for stochastic processes
- Stochastic control, and stochastic differential games
- Stochastic models in operations research and stochastic optimization
- Stochastic models in the physical sciences

Probability Theory and Stochastic Modelling is a merger and continuation of Springer's Stochastic Modelling and Applied Probability and Probability and Its Applications series.

Sergey Bobkov • Gennadiy Chistyakov
Friedrich Götze

Concentration and Gaussian Approximation for Randomized Sums

Springer

Sergey Bobkov
School of Mathematics
University of Minnesota
Minneapolis, MN, USA

Gennadiy Chistyakov
Fakultät für Mathematik
Universität Bielefeld
Bielefeld, Germany

Friedrich Götze
Fakultät für Mathematik
Universität Bielefeld
Bielefeld, Germany

ISSN 2199-3130 ISSN 2199-3149 (electronic)
Probability Theory and Stochastic Modelling
ISBN 978-3-031-31148-2 ISBN 978-3-031-31149-9 (eBook)
https://doi.org/10.1007/978-3-031-31149-9

This Springer imprint is published by the registered company Springer Nature Switzerland AG
The registered company address is: Gewerbestrasse 11, 6330 Cham, Switzerland

Preface

Given a random vector $X = (X_1, \dots, X_n)$ on a probability space $(\Omega, \mathcal{F}, \mathbb{P})$ with values in the Euclidean space $\mathbb{R}^n$, $n \geq 2$, define the weighted sums

$$\langle X, \theta \rangle = \sum_{k=1}^{n} \theta_k X_k,$$

parameterized by points $\theta = (\theta_1, \dots, \theta_n)$ of the unit sphere $\mathbb{S}^{n-1}$ in $\mathbb{R}^n$ (with center at the origin and radius one). In general, the distribution functions of weighted sums $\langle X, \theta \rangle$, say

$$F_\theta(x) = \mathbb{P}\{\langle X, \theta \rangle \leq x\}, \quad x \in \mathbb{R},$$

essentially depend on the parameter θ. On the other hand, a striking observation made by V. N. Sudakov [169] in 1978 indicates that, under mild correlation-type conditions on the distribution of X, and when n is large, most of the F_θ's are concentrated around a certain "typical" distribution function F. Here "most" should be understood in the sense of the normalized Lebesgue measure $\mathfrak{s}_{n-1}$ on $\mathbb{S}^{n-1}$. A more precise statement can be given, for example, under the isotropy condition

$$\mathbb{E}\,\langle X, \theta \rangle^2 = 1, \quad \theta \in \mathbb{S}^{n-1},$$

which frequently appears in many applications. Similar to the classical central limit theorem, Sudakov's result thus represents a rather general principle of convergence, with various interesting aspects. A related phenomenon was discovered later by Diaconis and Freedman [79] in terms of low-dimensional projections of non-random data (cf. also von Weizsäcker [176]).

The phenomenon of concentration of the family $\{F_\theta\}_{\theta \in \mathbb{S}^{n-1}}$ naturally begs the question of closeness of F_θ to F for all θ from a large part of the sphere in terms of standard distances d in the space of probability distributions on the real line. A canonical choice would be the Kolmogorov (uniform) distance

$$\rho(F_\theta, F) = \sup_x |F_\theta(x) - F(x)|.$$

Less sensitive alternatives would be the Lévy distance

$$L(F_\theta, F) = \inf\Big\{h \geq 0 : F(x-h) - h \leq F_\theta(x) \leq F(x+h) + h \ \text{ for all } x \in \mathbb{R}\Big\},$$

as well as the distances in L^p-norms

$$d_p(F_\theta, F) = \Big(\int_{-\infty}^{\infty} |F_\theta(x) - F(x)|^p \, \mathrm{d}x\Big)^{1/p}, \quad p \geq 1,$$

among which $W = d_1$ and $\omega = d_2$ are most natural. For a given distance d, the behavior of the average value $m = \mathbb{E}_\theta \, d(F_\theta, F)$, as well as the deviation from the mean in spherical probability $\mathfrak{s}_{n-1}\{d(F_\theta, F) \geq m + r\}$, is of interest as a function of n and $r > 0$.

In this context the model of independent random variables X_k has been intensively studied in the literature. When X_k are independent and identically distributed (the i.i.d. case) and have mean zero and variance one, the distribution functions F_θ are known to be close to the standard normal distribution function

$$\Phi(x) = \frac{1}{\sqrt{2\pi}} \int_{-\infty}^{x} \mathrm{e}^{-y^2/2} \, \mathrm{d}y,$$

as long as $\max_k |\theta_k|$ is small. If the 3-rd absolute moment $\beta_3 = \mathbb{E}\,|X_1|^3$ is finite, the Berry–Esseen theorem allows us to quantify this closeness by virtue of the bound

$$\rho(F_\theta, \Phi) \leq c\beta_3 \sum_{k=1}^{n} |\theta_k|^3,$$

which holds for some absolute constant $c > 0$. Although the right-hand side is greater than or equal to $c\beta_3/\sqrt{n}$ for any $\theta \in \mathbb{S}^{n-1}$, the bound above implies a similar upper bound on average: $\mathbb{E}_\theta \, \rho(F_\theta, \Phi) \leq c'\beta_3/\sqrt{n}$.

The i.i.d. case inspires the idea that, under some natural moment and correlation-type assumptions, most of the F_θ might also be close to the standard normal law. But, in light of Sudakov's theorem, this is equivalent to a similar assertion about the typical distribution – a property which is determined by the distribution of the Euclidean norm

$$|X| = \big(X_1^2 + \cdots + X_n^2\big)^{1/2}.$$

Indeed, in general, the typical distribution can be identified as the spherical average

$$F(x) = \int_{\mathbb{S}^{n-1}} F_\theta(x) \, \mathrm{d}\mathfrak{s}_{n-1}(\theta) \equiv \mathbb{E}_\theta F_\theta(x),$$

which may be alternatively described as the distribution of $|X|\,\theta_1$, where the first coordinate of a point on the sphere is treated as a random variable independent of X. (In the sequel, $\mathbb{E}_\theta$ is always understood as the integral with respect to the measure $\mathfrak{s}_{n-1}$.) Since $\theta_1\sqrt{n}$ is nearly standard normal, F will be close to Φ if and only if the random variable $R^2 = \frac{1}{n}\,|X|^2$ is approximately 1 in the sense of the weak topology. In many situations, this can be verified directly by computing, for example, the variance of R^2, while in some others it represents a non-trivial "thin shell" type concentration problem.

This book aims to describe the current state of the art concerning Sudakov's theorem. In particular, using the metrics d mentioned above, we will focus on the derivation of various bounds for $\mathbb{E}_\theta\, d(F_\theta, F)$ and $\mathbb{E}_\theta\, d(F_\theta, \Phi)$, as well as on large deviation bounds. Our investigations rely on several basic tools. Besides classical techniques of Fourier Analysis (such as Berry–Esseen-type bounds), many arguments rely upon the spherical concentration phenomenon, that is, concentration properties of the measures $\mathfrak{s}_{n-1}$ for growing dimensions n, including the associated Sobolev-type and infimum-convolution inequalities. Concentration tools are also used for various classes of distributions of X when trying to approximate the typical distribution F by the standard normal law.

In order to facilitate the readability of the presentation of the results related to the Sudakov phenomenon, we decided to make the presentation more self-contained by including these auxiliary techniques in the first three chapters. Thus we describe in a separate part (Part II) some general results on concentration in the setting of Euclidean and abstract metric spaces. Most of this material can also be found in other publications, including the excellent survey and monograph by M. Ledoux [129], [130], and the recent book by D. Bakry, I. Gentil, and M. Ledoux [8].

The spherical concentration is discussed separately in Part III. It is a classical well-known fact (whose importance was first emphasized by V. Milman in the early 1970s) that any mean zero smooth (say, Lipschitz) function f on the unit sphere $\mathbb{S}^{n-1}$ has deviations at most of the order $1/\sqrt{n}$ with respect to the growing dimension n. Moreover, as a random variable, $\sqrt{n}\, f$ has Gaussian tails under the measure $\mathfrak{s}_{n-1}$. In addition to this spherical phenomenon, we present recent developments on the so-called second order concentration, which was pushed forward by the authors as an advanced tool in the theory of randomized summation. Roughly speaking, the second order concentration phenomenon indicates that, under proper normalization hypotheses in terms of the Hessian, any smooth f on $\mathbb{S}^{n-1}$ orthogonal to all affine functions in $L^2(\mathfrak{s}_{n-1})$ actually has deviations at most of the order $1/n$. Moreover, as a random variable, nf has exponential tails under the measure $\mathfrak{s}_{n-1}$. Part III also contains various bounds on deviations of elementary polynomials under $\mathfrak{s}_{n-1}$ and collects asymptotic results on special functions related to the distribution of the first coordinate on the sphere.

These tools are needed to quantify Sudakov's theorem in terms of several moment and correlation-type conditions, and for various classes of distributions of X. With this aim, we shall introduce and discuss the following moment type quantities for a parameter $p \geq 1$,

$$M_p = \sup_{\theta \in \mathbb{S}^{n-1}} \big(\mathbb{E}\,|\,\langle X, \theta\rangle\,|^p\big)^{1/p}, \quad m_p = \frac{1}{\sqrt{n}}\,\big(\mathbb{E}\,|\,\langle X, Y\rangle\,|^p\big)^{1/p},$$

as well as the variance-type functionals

$$\sigma_{2p} = \sqrt{n}\,\bigg(\mathbb{E}\,\Big|\frac{|X|^2}{n} - 1\Big|^p\bigg)^{1/p}, \quad \Lambda = \sup_{\sum a_{ij}^2 = 1} \mathrm{Var}\bigg(\sum_{i,j=1}^n a_{ij} X_i X_j\bigg),$$

where Y is an independent copy of X. For example, $M_2 = m_2 = 1$ in the isotropic case, and $\sigma_4^2 = \frac{1}{n}\,\mathrm{Var}(|X|^2)$, which can often be estimated via evaluation of the covariances of X_i^2 and X_j^2. The relevance of these functionals will be clarified in various examples; they are also connected with analytic properties of the distribution μ of X expressed in terms of isoperimetric or Poincaré-type inequalities. For example, there is a simple bound $\Lambda \leq 4/\lambda_1^2$ via the spectral gap λ_1 associated to μ.

We shall now outline several results on upper bounds for $\mathbb{E}_\theta\, d(F_\theta, F)$ and $\mathbb{E}_\theta\, d(F_\theta, \Phi)$ involving these functionals for various distances d. They are discussed in detail in the remaining Parts IV–VI of this monograph.

• *Lévy distance*. Here the moments M_1 and M_2 will control quantitative bounds on fluctuations of F_θ around the typical distribution F in the metric L providing polynomial rates with respect to n. Namely, for some absolute constant $c > 0$ we have

$$\mathbb{E}_\theta\, L(F_\theta, F) \leq c\, \frac{M_1 + \log n}{n^{1/4}}, \quad \mathbb{E}_\theta\, L(F_\theta, F) \leq c\, \Big(\frac{\log n}{n}\Big)^{1/3} M_2^{2/3}.$$

• *Kantorovich L^1 transport distance*. Here rates can be improved in terms of the moments M_p of higher order. In particular, we have

$$\mathbb{E}_\theta\, W(F_\theta, F) \leq c_p M_p\, n^{-\frac{p-1}{2p}} \quad (p > 1),$$

where the constants c_p depend on p only. However, a classical rate of $1/\sqrt{n}$ from other contexts will not be achievable via these bounds.

• *Kolmogorov distance*. Using the variance-type functionals σ_p, it is possible not only to replace the typical distribution F with the normal distribution function Φ, thus proving a law of attraction for F_θ, but also to show a standard rate as well, assuming a finite third moment. Analogously to the classical Berry–Esseen theorem, it is shown that, if $\mathbb{E}\,|X|^2 = n$ and $\mathbb{E}X = a$, then

$$\mathbb{E}_\theta\, \rho(F_\theta, \Phi) \leq \frac{A}{\sqrt{n}}$$

with $A = c\,(m_3^{3/2} + \sigma_3^{3/2} + |a|)$ up to some absolute constant c. Here, one may eliminate the parameter a, by using elementary bounds $m_3 \leq M_3^2$ and $\sigma_3 \leq \sigma_4$ (the latter requires, however, the finiteness of the 4-th moment). A slightly worse estimate can also be derived under less restrictive moment assumptions. For example,

$$\mathbb{E}_\theta\, \rho(F_\theta, \Phi) \leq c\,(M_2^2 + \sigma_2)\, \frac{\log n}{\sqrt{n}}.$$

• *Trigonometric and other functional models of random variables.* Modulo a logarithmic factor, the upper bounds such as

$$\mathbb{E}_\theta\, \rho(F_\theta, \Phi) \le c\, \frac{\log n}{\sqrt{n}}$$

turn out to be optimal with respect to n in many examples of orthonormal systems $X = (X_1, \dots, X_n)$ of functions in L^2. These include in particular the trigonometric system of size n with components

$$\begin{aligned} X_{2k-1}(t) &= \sqrt{2}\, \cos(kt), \\ X_{2k}(t) &= \sqrt{2}\, \sin(kt), \qquad -\pi < t < \pi, \;\; k = 1, \dots, n/2 \;\; (n \text{ even}), \end{aligned}$$

with respect to the normalized Lebesgue measure $\mathbb{P}$ on $\Omega = (-\pi, \pi)$. More precisely, we derive lower bounds such as

$$\mathbb{E}_\theta\, \rho(F_\theta, \Phi) \ge \frac{c}{\sqrt{n}\, (\log n)^s}$$

for some positive c and s independent of n. A similar bound also holds for the sequence of the first n Chebyshev polynomials on the interval $\Omega = (-1, 1)$, for the Walsh system on the Boolean cube $\{-1, 1\}^p$ (with $n = 2^p - 1$), for systems of functions $X_k(t_1, t_2) = f(kt_1 + t_2)$ with 1-periodic f (such functions X_k form a strictly stationary sequence of pairwise independent random variables on the square $\Omega = (0, 1) \times (0, 1)$ under the restricted Lebesgue measure), and some others.

• *L^2 distance.* In order to develop lower bounds as above, similar upper and lower bounds will be needed for the L^2-distance ω, in combination with upper bounds for the Kantorovich-distance W. A number of general results in this direction will be obtained under moment and correlation-type assumptions, as in the case of Kolmogorov distance ρ. In fact, in the case of ω, the correct asymptotic behavior of $\mathbb{E}_\theta\, \omega^2(F_\theta, F)$ will be derived up to the order $1/n^2$. For instance, when the random vector X has an isotropic symmetric distribution and satisfies $|X| = \sqrt{n}$ a.s. (and thus all $\sigma_p = 0$), one has

$$\mathbb{E}_\theta\, \omega^2(F_\theta, F) \sim \frac{1}{n^4}\, \mathbb{E}\, \langle X, Y \rangle^4$$

with an error term of order $1/n^2$, and a similar result holds for the Gaussian limit Φ instead of the typical distribution F. Here, as before, Y denotes an independent copy of X.

• *Improved rates in the i.i.d. case.* Returning to the classical i.i.d. model with $\mathbb{E}X_1 = 0$, $\mathbb{E}X_1^2 = 1$, a remarkable result due to Klartag and Sodin [125] which we include in this monograph improves the pointwise Berry–Esseen bound as follows

$$\mathbb{E}_\theta\, \rho(F_\theta, \Phi) \le \frac{c\beta_4}{n}, \qquad \beta_4 = \mathbb{E}X_1^4.$$

In fact, this result holds in the non-i.i.d. situation as well with replacement of β_4 with the arithmetic means of the 4-th moments of X_k. This bound can be complemented by corresponding large deviation bounds. Thus, for typical coefficients, the distances $\rho(F_\theta, \Phi)$ are at most of order $1/n$, which is not true in general when the coefficients are equal to each other!

Moreover, we show that, if the distribution of X_1 is symmetric, and the next moment $\beta_5 = \mathbb{E}\,|X_1|^5$ is finite, it is possible to slightly correct the normal distribution Φ to obtain a better approximation such as

$$\mathbb{E}_\theta\, \rho(F_\theta, G) \leq \frac{c\beta_5}{n^{3/2}}.$$

Here, G is a certain function of bounded variation which is determined by β_4 and depends on n (but not on θ).

- *The second order correlation condition.* Certainly Sudakov's theorem begs the question whether or not similar results continue to hold for dependent components X_k. This is often the case, although the orthonormal systems mentioned above may serve as counter examples. More precisely, the variance functional $\Lambda = \Lambda(X)$ turns out to be responsible for improved rates of normal approximation for F_θ's on average and actually for most θ's. When X has an isotropic symmetric distribution, it will be shown by virtue of the second order spherical concentration that

$$\mathbb{E}_\theta\, \rho(F_\theta, \Phi) \leq \frac{c \log n}{n}\, \Lambda,$$

which thus extends the i.i.d. case modulo a logarithmic factor. The symmetry assumption can be removed at the expense of additional terms reflecting higher order correlations. In particular, in the presence of the Poincaré-type inequality, we have

$$\mathbb{E}_\theta\, \rho(F_\theta, \Phi) \leq \frac{c_1 \log n}{n}\, \lambda_1^{-1},$$

which may be complemented by corresponding deviation bounds.

- *Distributions with many symmetries.* Special attention will be devoted to the case where the distribution of X is symmetric about all coordinate axes and isotropic (which reduces to the normalization condition $\mathbb{E}X_k^2 = 1$). The Λ-functional then simplifies, and under the 4-th moment condition, the Berry–Esseen bound "on average" takes the form

$$\mathbb{E}_\theta\, \rho(F_\theta, \Phi) \leq \frac{c \log n}{n} \Big(\max_{k \leq n} \mathbb{E}X_k^4 + V_2\Big),$$

where

$$V_2 = \sup_{\theta \in \mathbb{S}^{n-1}} \mathrm{Var}(\theta_1 X_1^2 + \dots + \theta_n X_n^2).$$

If additionally the distribution of X is invariant under permutations of coordinates, it yields a simpler bound

$$\mathbb{E}_\theta\, \rho(F_\theta, \Phi) \le \frac{c \log n}{n}\, \big(\mathbb{E}X_1^4 + \sigma_4^2\big).$$

Here the last term σ_4^2 may be removed in some cases, e.g. when $\mathrm{cov}(X_1^2, X_2^2) \le 0$.

These results can be sharpened under some additional assumptions on the shape of the distribution of X. We include the proof of the following important variant of the central limit theorem due to Klartag [123]: If the random vector X has an isotropic, coordinatewise symmetric log-concave distribution, then, for all $\theta \in \mathbb{S}^{n-1}$,

$$\rho(F_\theta, \Phi) \le c \sum_{k=1}^n \theta_k^4$$

up to some absolute constant c. Here, the average value of the right-hand side is of order $1/n$. Although the class of log-concave probability distributions is studied in many investigations, their basic properties are discussed in this text as well. In particular, we include the proof of the Brascamp–Lieb inequality, which serves as a main tool in Klartag's theorem.

Finally, in the last chapter we conclude with brief historical remarks on results about randomized variants of the central limit theorem, in which coefficients have a special structure.

Acknowledgements. This work started in 2015 during the visit of the first author to the Bielefeld University, Germany, and he is grateful for their hospitality. The authors were supported by the SFB 701 and the SFB 1283/2 2021 – 317210226 at Bielefeld University. The work of the first author was also supported by the Humboldt Foundation, the Simons Foundation, and NSF grants DMS-1855575, DMS-2154001.

It is our great pleasure to thank Michel Ledoux for valuable comments on the draft version of the monograph.

In Memoriam. Shortly after completion of this book, Gennadiy Chistyakov passed away after a prolonged illness in December 2022. We mourn the loss of our friend and colleague.

Contents

Part VI Distributions and Coefficients of Special Type

Part I
Generalities

Chapter 1
Moments and Correlation Conditions

In this chapter, we introduce general functionals of an algebraic character associated to probability distributions on $\mathbb{R}^n$ with sufficiently many finite moments. These include maximal moments of linear functionals of a given order and the characteristics defined via correlation-type conditions. Such functionals appear to be natural tools in the study of Khinchine-type inequalities, as well as in the estimation of "small ball" probabilities.

As usual, $\mathbb{R}^n$ denotes the real Euclidean n-space, endowed with the Euclidean norm $|x| = (x_1^2 + \dots + x_n^2)^{1/2}$ and the inner product $\langle x, y\rangle = x_1 y_1 + \dots + x_n y_n$, where $x = (x_1, \dots, x_n),\ y = (y_1, \dots, y_n) \in \mathbb{R}^n$.

1.1 Isotropy

A random vector $X = (X_1, \dots, X_n)$ in $\mathbb{R}^n$ defined on a probability space $(\Omega, \mathcal{F}, \mathbb{P})$ is said to have finite second moment if the expectation $\mathbb{E}\,|X|^2$ is finite (that is, when all components X_i have finite second moments $\mathbb{E}X_i^2$). Let us start with the following classical notion.

Definition 1.1.1 A random vector X in $\mathbb{R}^n$ with finite second moment is called isotropic, or one says that X has an isotropic distribution, if

$$\mathbb{E}\,\langle X, a\rangle^2 = |a|^2 \quad \text{for all } a \in \mathbb{R}^n.$$

This definition is frequently used in Convex Geometry, especially for random vectors which are uniformly distributed over a convex body (in which case the body is called isotropic, cf. [144]). In the probabilistic context, the above condition may be rewritten in terms of the components of X as

$$\mathbb{E}X_i X_j = \delta_{ij}, \quad i, j = 1, \dots, n,$$

S. Bobkov et al., *Concentration and Gaussian Approximation for Randomized Sums*, Probability Theory and Stochastic Modelling 104,
https://doi.org/10.1007/978-3-031-31149-9_1

where $\delta_{ij} = 1_{\{i=j\}}$ is the Kronecker symbol. Often, one additionally assumes that all X_i have mean zero, and then the above means that these random variables are non-correlated and have variances one.

The property of being isotropic is stable under orthogonal transformations: the random vector $Y = UX$ is isotropic, whenever X is isotropic, and $U : \mathbb{R}^n \to \mathbb{R}^n$ is an orthogonal linear operator (or an orthogonal matrix).

With a random vector X in $\mathbb{R}^n$ having finite second moment, we associate a linear operator R, or $n \times n$ matrix, defined by

$$\mathbb{E}\,\langle X, a\rangle^2 = \langle Ra, a\rangle\,, \quad a \in \mathbb{R}^n.$$

It is a covariance matrix of X, when X has mean zero, but we call it so in the general case as well. Then, the isotropy condition means that R is an identity matrix (or, an identity linear operator on $\mathbb{R}^n$).

In general, if R is non-degenerate, there always exists a linear map $A : \mathbb{R}^n \to \mathbb{R}^n$ such that $Y = AX$ is an isotropic random vector. Indeed, one may take $A = R^{-1/2}$. In many problems, the isotropy condition thus simply represents a normalization only, and often does not lead to the loss of generality.

Let us mention a few immediate consequences from isotropy.

Proposition 1.1.2 *Assume that a random vector $X = (X_1, \dots, X_n)$ in $\mathbb{R}^n$ is isotropic. Then*

a) $\mathbb{E}X_k^2 = 1$ *for all* $k = 1, \dots, n$.
b) Hence $\mathbb{E}\,|X|^2 = n$.
c) We have $\sum_{k=1}^n (\mathbb{E}X_k)^2 \le 1$.
d) If Y is an independent copy of X, then $\mathbb{E}\,\langle X, Y\rangle^2 = n$.

The property *c)* follows from Bessel's inequality, using orthogonality of X_i as elements in the space $L^2(\Omega, \mathcal{F}, \mathbb{P})$. The last assertion is also obvious:

$$\mathbb{E}\,\langle X, Y\rangle^2 = \mathbb{E}\bigg(\sum_{i=1}^n X_i Y_i\bigg)^2 = \sum_{i,j=1}^n (\mathbb{E}\,X_i X_j)^2 = n.$$

Thus, although isotropy does not mean that the X_k have mean zero, the expectations $\mathbb{E}X_k$ must be bounded by 1 and must be small on average.

The assertion *d)*, which may be taken as a natural weakening of isotropy, is equivalent to the property that

$$\frac{1}{n}\sum_{k=1}^n \lambda_k^2 = 1,$$

where the λ_k are eigenvalues of the covariance operator R of X. Indeed, if U is an orthogonal linear operator on $\mathbb{R}^n$, then the correlation operator of the random vector $\widetilde{X} = UX$ is equal to $\widetilde{R} = URU^{-1}$. Moreover, since R is symmetric, one can always find U such that the matrix $\widetilde{R}$ is diagonal, with diagonal elements λ_k being the eigenvalues of R. In particular, $\mathbb{E}\widetilde{X}_k^2 = \lambda_k$. Since

$$\mathbb{E}\left\langle \widetilde{X}, a \right\rangle^2 = \langle Ra, a \rangle = \sum_{k=1}^{n} \lambda_k a_k^2, \quad a \in \mathbb{R}^n,$$

we conclude that

$$\mathbb{E}\,\langle X, Y \rangle^2 = \mathbb{E}\,\langle UX, UY \rangle^2 = \mathbb{E}\left\langle \widetilde{X}, \widetilde{Y} \right\rangle^2 = \mathbb{E} \sum_{k=1}^{n} \lambda_k \widetilde{X}_k^2 = \sum_{k=1}^{n} \lambda_k^2,$$

where Y is an independent copy of X and $\widetilde{Y} = UY$.

On the other hand,

$$\mathbb{E}\,|X|^2 = \mathbb{E}\,|\widetilde{X}|^2 = \sum_{k=1}^{n} \mathbb{E}\widetilde{X}_k^2 = \sum_{k=1}^{n} \lambda_k.$$

By Cauchy's inequality, $\frac{1}{n} \sum_{k=1}^{n} \lambda_k^2 \geq \frac{1}{n} (\sum_{k=1}^{n} \lambda_k)^2$ with equality if and only if all eigenvalues λ_k are equal to each other. Hence, we have the following general relation, which also allows one to compare *b*) with property *d*) from Proposition 1.1.2.

Proposition 1.1.3 *For any random vector X in $\mathbb{R}^n$ with finite second moment,*

$$\mathbb{E}\,\langle X, Y \rangle^2 \geq \frac{1}{n} \left(\mathbb{E}\,|X|^2\right)^2,$$

where Y is an independent copy of X. Moreover, assuming that $\mathbb{E}\,|X|^2 = n$, we have $\mathbb{E}\,\langle X, Y \rangle^2 = n$ if and only if X is isotropic.

1.2 First Order Correlation Condition

A natural weakening of the isotropy is the following property requiring that the components X_k of X are almost non-correlated.

Definition 1.2.1 We say that a random vector X in $\mathbb{R}^n$ satisfies a first order correlation condition with constant M_2 if

$$\left(\mathbb{E}\,\langle X, a \rangle^2\right)^{1/2} \leq M_2 |a| \quad \text{for all } a \in \mathbb{R}^n.$$

The optimal value of this constant is given by

$$M_2(X) = \sup_{\theta \in \mathbb{S}^{n-1}} (\mathbb{E}\,\langle X, \theta \rangle^2)^{1/2},$$

where the supremum extends over the unit sphere $\mathbb{S}^{n-1} = \{\theta \in \mathbb{R}^n : |\theta| = 1\}$. Representing the maximal eigenvalue of the covariance matrix R associated with X,

$$M_2(X) = \max_k \lambda_k,$$

it is invariant under orthogonal transformation of the space and therefore does not depend on the choice of the coordinate system in $\mathbb{R}^n$.

Similarly to Proposition 1.1.2, we have:

Proposition 1.2.2 *Assume that a random vector $X = (X_1, \dots, X_n)$ in $\mathbb{R}^n$ satisfies a first order correlation condition with constant M_2. Then*

a) $\mathbb{E}X_k^2 \le M_2^2$ *for all* $k = 1, \dots, n$.
b) Hence $\mathbb{E}\,|X|^2 \le M_2^2\, n$.
c) We have $\sum_{k=1}^n (\mathbb{E}X_k)^2 \le M_2^2$, *that is,* $|\mathbb{E}X|^2 \le M_2^2$.
d) $M_2(X - \mathbb{E}X) \le M_2(X)$.

To derive c), note that the vector $\mathbb{E}X = (\mathbb{E}X_1, \dots, \mathbb{E}X_n)$ satisfies

$$\langle \mathbb{E}X, \theta\rangle^2 \le \mathbb{E}\,\langle X, \theta\rangle^2 \le M_2^2|\theta|^2 = M_2^2$$

for all $\theta \in \mathbb{S}^{n-1}$. Taking the maximum over all θ yields the desired bound $|\,\mathbb{E}X| \le M_2$.

For the inequality in d), just note that, for all $a \in \mathbb{R}^n$,

$$\mathbb{E}\,\langle X, a\rangle^2 \ge \mathrm{Var}(\langle X, a\rangle) = \mathbb{E}\,\langle X - \mathbb{E}X, a\rangle^2.$$

Proposition 1.2.2 is proved. Another useful observation is:

Proposition 1.2.3 *If the random variables $X_1, \dots, X_n$ are non-correlated, with means $\mathbb{E}X_k = 0$ and variances $v_k^2 = \mathrm{Var}(X_k)$, $v_k \ge 0$, then $M_2(X) = \max_k v_k$.*

1.3 Moments and Khinchine-type Inequalities

The functional $M_2(X)$ has a natural generalization in terms of the p-th absolute moments of linear functionals of X.

Definition 1.3.1 Given a random vector X in $\mathbb{R}^n$ and a number $p \ge 1$, put

$$M_p(X) = \sup_{\theta \in \mathbb{S}^{n-1}} \Big(\mathbb{E}\,|\,\langle X, \theta\rangle\,|^p\Big)^{1/p}.$$

These quantities are not decreasing in p. An inequality of the form $M_p(X) \le M$ means that X satisfies a p-th moment condition (with constant M), in full analogy with the terminology accepted for i.i.d. (that is, independent, identically distributed) random variables. In the case $p = 2$, we are reduced to Definition 1.2.1. However, in the sequel we will be especially interested in $p = 4$ and larger values.

Recall that, given a random variable ξ, by Hölder's inequality, the L^p-norms $\|\xi\|_p = (\mathbb{E}\,|\xi|^p)^{1/p}$ do not decrease. The reverse relations of the form

$$\|\xi\|_p \le C_{p,q}\,\|\xi\|_q, \quad 0 < q < p,$$

when they are stated for classes of random variables with constants $C_{p,q}$ independent of members of a given class, are commonly called Khinchine-type inequalities.

For example, if the random vector X is isotropic, we get by the definition that

$$\| \langle X, a \rangle \|_p \le M_p(X) \| \langle X, a \rangle \|_2 \quad \text{for any } a \in \mathbb{R}^n,$$

and the constant here is best possible. This relation represents a Khinchine-type inequality with $p > 2$ and $q = 2$.

The quantity $M_p(X)$ may be used to control the p-th moment of the Euclidean length $|X|$, according to the following extension of Proposition 1.2.2 in part b).

Proposition 1.3.2 *Given $p \ge 2$, for any random vector X in $\mathbb{R}^n$,*

$$(\mathbb{E}\,|X|^p)^{1/p} \le M_p(X)\sqrt{n}.$$

Note that, if X is isotropic, there is an opposite inequality

$$(\mathbb{E}\,|X|^p)^{1/p} \ge (\mathbb{E}\,|X|^2)^{1/2} = \sqrt{n}.$$

Hence, the rate $\sqrt{n}$ for the L^p-norm of $|X|$ with respect to n is correct as long as $M_p(X)$ is of order 1.

Proof If $\theta = (\theta_1, \dots, \theta_n)$ is a random vector uniformly distributed in the unit sphere $\mathbb{S}^{n-1}$, the distribution of the linear functional $\langle a, \theta \rangle$ depends on the Euclidean length $|a|$, only. In particular, taking $a = |a| e_1 = (|a|, 0, \dots, 0)$, we have

$$\mathbb{E}_\theta\,|\langle a, \theta \rangle|^p = |a|^p\,\mathbb{E}_\theta\,|\theta_1|^p, \quad a \in \mathbb{R}^n,$$

where $\mathbb{E}_\theta$ denotes the integral over the uniform probability measure $\mathfrak{s}_{n-1}$ on $\mathbb{S}^{n-1}$. Inserting here $a = X$, we get

$$\mathbb{E}_\theta\,|\langle X, \theta \rangle|^p = |X|^p\,\mathbb{E}_\theta\,|\theta_1|^p.$$

Next, take the expectation with respect to X and use $\mathbb{E}\,|\langle X, \theta \rangle|^p \le M_p(X)$ to arrive at the upper bound

$$\mathbb{E}\,|X|^p \le \frac{M_p^p(X)}{\mathbb{E}_\theta\,|\theta_1|^p}.$$

Moreover, since $\mathbb{E}_\theta\,\theta_1^2 = \frac{1}{n}$, we get

$$(\mathbb{E}_\theta\,|\theta_1|^p)^{1/p} \ge (\mathbb{E}_\theta\,|\theta_1|^2)^{1/2} = \frac{1}{\sqrt{n}}.$$

This proves the proposition. □

The growth of the functionals M_p can be controlled by means of the Orlicz norms, which may be stronger in comparison with L^p-norms. In particular, the use of the exponential moments leads to the following:

Proposition 1.3.3 *For a number $\lambda > 0$, assume that $\mathbb{E}\, e^{|\langle X,\theta\rangle|/\lambda} \le 2$ for all $\theta \in \mathbb{S}^{n-1}$. Then for all $p \ge 1$,*

$$M_p(X) \le \lambda p.$$

Proof Put $\xi = |\langle X,\theta\rangle|$. Using the inequality $x^p e^{-x} \le (\frac{p}{e})^p$, $x \ge 0$, we have

$$\mathbb{E}\,\xi^p = \lambda^p\, \mathbb{E}\Big(\frac{\xi}{\lambda}\Big)^p e^{-\xi/\lambda}\, e^{\xi/\lambda} \le \Big(\frac{\lambda p}{e}\Big)^p\, \mathbb{E}\, e^{\xi/\lambda} \le 2\Big(\frac{\lambda p}{e}\Big)^p < (\lambda p)^p.$$

It remains to take the maximum over all θ. □

1.4 Moment Functionals Using Independent Copies

As closely related moment-type functionals, let us also introduce the following.

Definition 1.4.1 Given a random vector X in $\mathbb{R}^n$ and a number $p \ge 1$, put

$$m_p(X) = \frac{1}{\sqrt{n}}\Big(\mathbb{E}\,|\langle X,Y\rangle|^p\Big)^{1/p},$$

where Y is an independent copy of X.

In particular, as we discussed in Section 1.1, we have

$$m_2(X) = \frac{1}{\sqrt{n}}\Big(\sum_{k=1}^n \lambda_k^2\Big)^{1/2}$$

in terms of the eigenvalues λ_k of the covariance matrix associated with X.

These quantities M_p are also non-decreasing in p and are invariant under linear orthogonal transformations of Euclidean space.

The next elementary observation will play an important role in our study of the randomized central limit theorem.

Proposition 1.4.2 *Given independent random vectors X and Y in $\mathbb{R}^n$, for any $p \ge 2$,*

$$\frac{1}{\sqrt{n}}\Big(\mathbb{E}\,|\langle X,Y\rangle|^p\Big)^{1/p} \le M_p(X)M_p(Y).$$

In particular,

$$m_p(X) \le M_p^2(X).$$

Indeed, by Definition 1.3.1, for any particular value of Y,

$$\mathbb{E}_X|\langle X,Y\rangle|^p \le M_p^p(X)\,|Y|^p.$$

Here one should take the expectation with respect to Y and apply Proposition 1.3.2.

A possible rate of growth of m_p may be controlled, for example, by the exponential moment assumption

$$\sup_{\theta\in\mathbb{S}^{n-1}} \mathbb{E}\, e^{|\langle X,\theta\rangle|/\lambda} \leq 2 \quad (\lambda > 0).$$

Here one may combine Proposition 1.4.2 with Proposition 1.3.3 to conclude that $m_p \leq (\lambda p)^2$. This bound implies that for some absolute constant $c > 0$,

$$\mathbb{E}\, \exp\left\{\frac{c}{\lambda n^{1/4}}\sqrt{|\langle X,Y\rangle|}\right\} \leq 2.$$

Another natural, although rough, bound for the functional m_p may be given for certain classes of interest like in the following elementary statement.

Proposition 1.4.3 *If the random vector X is isotropic, and $|X| \leq b\sqrt{n}$ a.s. for some $b > 0$, then*

$$m_3 \leq b^{2/3}\, n^{1/6}, \quad m_4 \leq b\, n^{1/4}.$$

To show this, let Y be an independent copy of X. Then $|\langle X,Y\rangle| \leq b^2 n$ a.s., while $\mathbb{E}\,\langle X,Y\rangle^2 = n$. Hence

$$m_3 = \frac{1}{\sqrt{n}}\Big(\mathbb{E}\,|\langle X,Y\rangle|^3\Big)^{1/3} \leq \frac{1}{\sqrt{n}}\Big(\mathbb{E}\,\langle X,Y\rangle^2\cdot b^2 n\Big)^{1/3} = b^{2/3}\, n^{1/6}$$

and

$$m_4 = \frac{1}{\sqrt{n}}\Big(\mathbb{E}\,\langle X,Y\rangle^4\Big)^{1/4} \leq \frac{1}{\sqrt{n}}\Big(\mathbb{E}\,\langle X,Y\rangle^2\cdot b^4 n^2\Big)^{1/4} = b\, n^{1/4}.$$

In addition to $m_p(X)$, for integers $p \geq 1$ one may also consider related functionals

$$\alpha_p(X) = \frac{1}{\sqrt{n}}\Big(\mathbb{E}\,\langle X,Y\rangle^p\Big)^{1/p},$$

where Y is an independent copy of X (of course, $\alpha_p = m_p$ for even p). Once X has a finite moment of order p, $\alpha_p(X)$ is well-defined, finite and non-negative. Indeed, writing $X = (X_1,\dots,X_n)$, $Y = (Y_1,\dots,Y_n)$, we have

$$\mathbb{E}\,\langle X,Y\rangle^p = \sum_{i_1,\dots,i_p=1}^{n}\Big(\mathbb{E}\, X_{i_1}\dots X_{i_p}\Big)^2 \geq 0.$$

The argument can be extended to involve more general functionals.

Proposition 1.4.4 *Let X be a random vector in $\mathbb{R}^n$ with finite moment of an integer order $p \geq 1$, and Y be its independent copy. For any real number $0 \leq \alpha \leq p$,*

$$\mathbb{E}\,\frac{\langle X,Y\rangle^p}{(|X|^2+|Y|^2)^\alpha} \geq 0,$$

where the ratio is defined to be zero if $X = Y = 0$. In addition, for $\alpha \in [0, 2]$,

$$\mathbb{E}\,\frac{\langle X, Y\rangle^2}{(|X|^2 + |Y|^2)^\alpha} \geq \frac{1}{n}\,\mathbb{E}\,\frac{|X|^2\,|Y|^2}{(|X|^2 + |Y|^2)^\alpha}.$$

Proof First, let us note that

$$\mathbb{E}\,\frac{|\langle X, Y\rangle|^p}{(|X|^2 + |Y|^2)^\alpha} \leq \mathbb{E}\,\frac{(|X|\,|Y|)^p}{(|X|\,|Y|)^\alpha} = (\mathbb{E}\,|X|^{p-\alpha})^2,$$

so, the expectation is finite. Moreover, without loss of generality, we may assume that $0 < \alpha \leq p$ and $\xi = |X|^2 + |Y|^2 > 0$ with probability 1. We use the identity

$$\int_0^\infty \exp\big\{-\xi t^{1/\alpha}\big\}\,\mathrm{d}t = c_\alpha\,\xi^{-\alpha} \quad \text{where } c_\alpha = \int_0^\infty \exp\big\{-s^{1/\alpha}\big\}\,\mathrm{d}s,$$

which gives

$$c_\alpha\,\mathbb{E}\,\langle X, Y\rangle^p\,\xi^{-\alpha} = \int_0^\infty \mathbb{E}\,\langle X, Y\rangle^p\,\exp\big\{-\xi t^{1/\alpha}\big\}\,\mathrm{d}t.$$

Writing $X = (X_1, \dots, X_n)$, $Y = (Y_1, \dots, Y_n)$, we have

$$\begin{aligned}\mathbb{E}\,\langle X, Y\rangle^p\,\exp\big\{-\xi t^{1/\alpha}\big\} &= \mathbb{E}\,\langle X, Y\rangle^p\,\exp\big\{-t^{1/\alpha}(|X|^2 + |Y|^2)\big\} \\ &= \sum_{i_1,\dots,i_p=1}^n \Big(\mathbb{E}\,X_{i_1}\dots X_{i_p}\,\exp\big\{-t^{1/\alpha}\,|X|^2\big\}\Big)^2,\end{aligned}$$

which shows that the left expectation is always non-negative. Integrating over $t > 0$, this proves the first assertion of the proposition.

For the second assertion, write

$$\begin{aligned}c_\alpha\,\mathbb{E}\,\langle X, Y\rangle^2\,\xi^{-\alpha} &= \int_0^\infty \mathbb{E}\,\langle X, Y\rangle^2\,\exp\big\{-t^{1/\alpha}(|X|^2 + |Y|^2)\big\}\,\mathrm{d}t \\ &= \int_0^\infty \mathbb{E}\,\langle X(t), Y(t)\rangle^2\,\mathrm{d}t,\end{aligned}$$

where

$$X(t) = \exp\big\{-t^{1/\alpha}|X|^2/2\big\}\,X, \quad Y(t) = \exp\big\{-t^{1/\alpha}|Y|^2/2\big\}\,Y.$$

Since $Y(t)$ represents an independent copy of $X(t)$, one may apply Proposition 1.1.3 to these random vectors, which gives

$$\mathbb{E}\,\langle X(t), Y(t)\rangle^2 \geq \frac{1}{n}\,\mathbb{E}\,|X(t)|^2\,|Y(t)|^2.$$

Hence

$$\int_0^\infty \mathbb{E}\,\langle X(t),Y(t)\rangle^2\,dt \geq \frac{1}{n}\int_0^\infty \mathbb{E}\,|X(t)|^2\,|Y(t)|^2\,dt$$
$$= \frac{1}{n}\int_0^\infty \mathbb{E}\,|X|^2\,|Y|^2\,\exp\big\{-t^{1/\alpha}(|X|^2+|Y|^2)\big\}\,dt$$
$$= \frac{c_\alpha}{n}\,\mathbb{E}\,|X|^2\,|Y|^2\,\xi^{-\alpha}.$$

This proves the second inequality of the proposition. □

1.5 Variance of the Euclidean Norm

For various reasons, one often needs to use correlations of products X_iX_j. In fact, already the information about the mutual covariances of X_i^2 (i.e., when $i = j$) has applications to randomized variants of the central limit theorem. A particular functional of significant interest is

$$\sigma_4^2(X) = \frac{\mathrm{Var}(|X|^2)}{n} = \frac{\mathrm{Var}(X_1^2+\dots+X_n^2)}{n} \qquad (\sigma_4 \geq 0).$$

If $\mathbb{E}\,|X|^2 = n$, the boundedness of this quantity means that the values of X are concentrated around the Euclidean sphere of radius $\mathbb{E}\,|X|$ (which is typically large and of order $\sqrt{n}$). This can be illustrated by the following observation.

Proposition 1.5.1 *For any random vector X in $\mathbb{R}^n$ such that $\mathbb{E}\,|X|^2 > 0$, we have*

$$\mathrm{Var}(|X|) \leq \frac{\mathrm{Var}(|X|^2)}{\mathbb{E}\,|X|^2}.$$

Thus, if $\mathbb{E}\,|X|^2 = n$ (in particular, if X is isotropic), then $\mathrm{Var}(|X|) \leq \sigma_4^2(X)$.

The first inequality follows from the general relation $\mathbb{E}\xi^2\,\mathrm{Var}(\xi) \leq \mathrm{Var}(\xi^2)$, which is valid for any non-negative random variable ξ. Indeed, assuming that the second moment $a^2 = \mathbb{E}\xi^2$ is finite ($a > 0$), we have

$$\mathrm{Var}(\xi^2) = \mathbb{E}\,(\xi^2-a^2)^2 = \mathbb{E}\,(\xi-a)^2(\xi+a)^2$$
$$\geq \mathbb{E}\,(\xi-a)^2\cdot a^2 \geq \mathrm{Var}(\xi)\cdot a^2.$$

The advantage of $\mathrm{Var}(|X|)$ compared with σ_4^2 is that it is finite under the second moment condition (i.e., when $\mathbb{E}\,|X|^2 < \infty$), while the finiteness of σ_4^2 requires the 4-th moment condition. On the other hand, often the latter functional is easier to handle. In addition, similarly to $M_4(X)$, $\sigma_4^2(X)$ allows one to control the L^4-norm of $|X|$, since, by the very definition and using Proposition 1.2.2,

$$\mathbb{E}\,|X|^4 = (\mathbb{E}\,|X|^2)^2 + \sigma_4^2(X)n \le M_2^4(X)\,n^2 + \sigma_4^2(X)n.$$

This bound can be an alternative to the one given in Proposition 1.3.2, i.e., to

$$\mathbb{E}\,|X|^4 \le M_4^4(X)\,n^2.$$

Let us also mention a simple observation about the behavior of σ_4 under shifts.

Proposition 1.5.2 *For any random vector X in $\mathbb{R}^n$,*

$$\sigma_4(X - \mathbb{E}X) \le \sigma_4(X) + \frac{2}{\sqrt{n}}\,M_2^2(X).$$

Proof Let us write $M_2 = M_2(X)$, $\sigma_4 = \sigma_4(X)$, and $a = \mathbb{E}X$. Since $|X - a|^2 = |X|^2 - 2\,\langle X, a\rangle + |a|^2$, we have

$$\mathrm{Var}(|X - a|^2) = \mathrm{Var}(|X|^2) + 4\,\mathrm{Var}(\langle X, a\rangle) - 4\,\mathrm{cov}(|X|^2, \langle X, a\rangle).$$

By Proposition 1.2.2, $|a| \le M_2$, so,

$$\mathrm{Var}(\langle X, a\rangle) \le \mathbb{E}\,\langle X, a\rangle^2 \le M_2^2|a|^2 \le M_2^4,$$

and, by Cauchy's inequality,

$$\mathrm{cov}(|X|^2, \langle X, a\rangle) \le \sqrt{\mathrm{Var}(|X|^2)}\,\sqrt{\mathbb{E}\,\langle X, a\rangle^2} \le \sigma_4\sqrt{n}\,M_2^2.$$

Collecting these bounds, we get that

$$\mathrm{Var}(|X - a|^2) \le \sigma_4^2 n + 4M_2^4 + 4\sigma_4\sqrt{n}\,M_2^2 = \big(\sigma_4\sqrt{n} + 2M_2^2\big)^2,$$

so that

$$\sigma_4^2(X - a) \le \Big(\sigma_4 + \frac{2}{\sqrt{n}}\,M_2^2\Big)^2.$$

Hence, Proposition 1.5.2 is proved. □

In the light of Proposition 1.5.1, it is natural to look for conditions that ensure an equivalence of $\mathrm{Var}(|X|)$ and $\sigma_4^2(X)$.

Proposition 1.5.3 *Suppose that $\mathbb{E}\,|X|^2 = n$.*

a) If $|X|^2 \le 2n$ a.s., then

$$\sigma_4^2(X) \le 16\,\mathrm{Var}(|X|).$$

b) If $M_6(X)$ is finite, then

$$\sigma_4^2(X) \le 16\,\mathrm{Var}(|X|) + \frac{1}{4}M_6^6(X).$$

Thus, the finiteness/boundedness of M_6 allows us in some sense to reverse the inequality of Proposition 1.5.1. The argument is based on the following lemma from calculus (which will also be needed later on).

Lemma 1.5.4 *For any random variable $\xi \geq 0$ with finite 4-th moment,*

$$1 - \mathbb{E}\xi \leq \frac{1}{2}\,\mathbb{E}\,(1-\xi^2) + \mathbb{E}\,(1-\xi^2)^2.$$

If $\xi^2 \leq 2$ a.s., then we also have a lower bound

$$1 - \mathbb{E}\xi \geq \frac{1}{2}\,\mathbb{E}\,(1-\xi^2) + \frac{1}{16}\,\mathbb{E}\,(1-\xi^2)^2.$$

Proof By Taylor's formula for the function $w(\varepsilon) = 1 - \sqrt{1-\varepsilon}$ around zero on the half axis $\varepsilon < 1$, we have

$$w(\varepsilon) = \frac{1}{2}\,\varepsilon + \frac{1}{8}\,\varepsilon^2 + \frac{w'''(\varepsilon_1)}{6}\,\varepsilon^3$$

for some point ε_1 between zero and ε. Since $w'''(\varepsilon) = \frac{3}{8}\,(1-\varepsilon)^{-5/2} \geq 0$, we have an upper bound

$$w(\varepsilon) \leq \frac{1}{2}\,\varepsilon + \frac{1}{8}\,\varepsilon^2, \quad \varepsilon \leq 0.$$

Also, $w'''(\varepsilon) \leq \frac{3}{8}\,3^{5/2} < 6$ for $0 \leq \varepsilon \leq \frac{2}{3}$, so,

$$\frac{1}{8}\,\varepsilon^2 + \frac{w'''(\varepsilon_1)}{6}\,\varepsilon^3 \leq \varepsilon^2\left(\frac{1}{8} + \varepsilon\right) \leq \varepsilon^2.$$

Thus, in both cases,

$$w(\varepsilon) \leq \frac{1}{2}\,\varepsilon + \varepsilon^2, \quad \varepsilon \leq \frac{2}{3}.$$

Let us now prove this inequality for the remaining values $\frac{2}{3} \leq \varepsilon \leq 1$. Since then $\frac{1}{2}\,\varepsilon + \varepsilon^2 \geq \frac{7}{6}\,\varepsilon$, while $w(\varepsilon) \leq 1$, we may assume that $\varepsilon \in [\frac{2}{3}, \frac{6}{7}]$. In this interval it is sufficient to verify that $w(\varepsilon) \leq \frac{7}{6}\,\varepsilon$, i.e., $1 - \varepsilon \geq (1 - \frac{7}{6}\,\varepsilon)^2$. But the latter inequality is fulfilled at the endpoints of the interval and therefore at all inner points $\frac{2}{3} < \varepsilon < \frac{6}{7}$.

Hence, for all $\varepsilon \leq 1$, we have $w(\varepsilon) \leq \frac{1}{2}\,\varepsilon + \varepsilon^2$. Applying this inequality with $\varepsilon = 1 - \xi^2$, we get a pointwise bound

$$1 - \xi \leq \frac{1}{2}\,(1-\xi^2) + (1-\xi^2)^2.$$

Taking the expectation of both sides, the first bound of the lemma follows.

Let us now turn to the lower bound. From Taylor's formula for w, using again the property $w'''(\varepsilon) \geq 0$, we have $w(\varepsilon) \geq \frac{1}{2}\,\varepsilon + \frac{1}{8}\,\varepsilon^2$ for all $\varepsilon \geq 0$. In the case $\varepsilon \leq 0$, we have $w'''(\varepsilon) \leq \frac{3}{8}$. Hence, for some point ε_1 between zero and $\varepsilon \in [-1, 0]$,

$$w(\varepsilon) = \frac{1}{2}\,\varepsilon + \varepsilon^2\left(\frac{1}{8} + \frac{w'''(\varepsilon_1)}{6}\,\varepsilon\right) \geq \frac{1}{2}\,\varepsilon + \varepsilon^2\left(\frac{1}{8} - \frac{3}{8}\frac{|\varepsilon|}{6}\right) \geq \frac{1}{2}\,\varepsilon + \frac{1}{16}\,\varepsilon^2.$$

Applying the resulting inequality with $\varepsilon = 1 - \xi^2$ and taking the expectation, we obtain the second bound of the lemma. □

Proof (of Proposition 1.5.3) In $a)$, the random variable $\xi = \frac{|X|}{\sqrt{n}}$ satisfies the condition of the second part in Lemma 1.5.4. Hence, applying its lower bound, we get

$$\begin{aligned}\mathrm{Var}(|X|) &= n\,(1 - \mathbb{E}\xi)\,(1 + \mathbb{E}\xi)\\ &\geq n\,(1 - \mathbb{E}\xi) \geq \frac{n}{16}\,\mathbb{E}\,(1 - \xi^2)^2 = \frac{1}{16}\,\sigma_4^2(X),\end{aligned}$$

thus proving the first assertion.

To prove $b)$, introduce the events $A = \{|X|^2 \leq 2n\}$ and $B = \{|X|^2 > 2n\}$. In the end of the proof of Lemma 1.5.4 we obtained the lower bound

$$1 - \sqrt{1-\varepsilon} \geq \frac{1}{2}\,\varepsilon + \frac{1}{16}\,\varepsilon^2, \quad -1 \leq \varepsilon \leq 1.$$

Applying it with $\varepsilon = 1 - \xi^2$ on the set A and using $\mathbb{E}\,\xi^2 = 1$, we get

$$\begin{aligned}\frac{1}{16}\,\mathbb{E}\,(1 - \xi^2)^2\,1_A &\leq \mathbb{E}\,(1 - \xi)\,1_A - \frac{1}{2}\,\mathbb{E}\,(1 - \xi^2)\,1_A\\ &= \mathbb{E}\,(1 - \xi) - \mathbb{E}\,(1 - \xi)\,1_B + \frac{1}{2}\,\mathbb{E}\,(1 - \xi^2)\,1_B.\end{aligned}$$

Here $-(1 - \xi) + \frac{1}{2}\,(1 - \xi^2) = -\frac{1}{2}\,(1 - \xi)^2 \leq 0$. Hence

$$\frac{1}{16}\,\mathbb{E}\,(1 - \xi^2)^2\,1_A \leq \mathbb{E}\,(1 - \xi) = \frac{1}{n}\,\frac{\mathrm{Var}(|X|)}{1 + \mathbb{E}\xi} \leq \frac{1}{n}\,\mathrm{Var}(|X|).$$

Thus, we have obtained a more general statement in comparison with $a)$, namely

$$\frac{1}{n}\,\mathbb{E}\,(|X|^2 - n)^2\,1_A \leq 16\,\mathrm{Var}(|X|),$$

which becomes the desired claim when $|X|^2 \leq 2n$.

As for the expectation over the complementary set B, one may involve the moment functionals $M_p = M_p(X)$ with suitable values $p \geq 4$. Recall that, by Proposition 1.3.2, $\mathbb{E}\,|X|^p \leq M_p^p\,n^{p/2}$. Hence, by Markov's inequality,

$$\mathbb{P}(B) \leq \frac{1}{(2n)^p}\,\mathbb{E}\,|X|^p \leq \frac{1}{2^p\,n^{p/2}}\,M_p^p.$$

Applying Hölder's inequality with exponents $q > 1$ and $q' = \frac{q}{q-1}$, we then get that

$$\begin{aligned}\frac{1}{n}\,\mathbb{E}\,(|X|^2-n)^2\,1_B &\le \frac{1}{n}\,\mathbb{E}\,|X|^4\,1_B\\ &\le \frac{1}{n}\,(\mathbb{E}\,|X|^{4q})^{1/q}\,\mathbb{P}(B)^{1/q'} \le \frac{1}{n}\,M_{4q}^4\,n^2\Big(\frac{M_p^p}{2^p\,n^{p/2}}\Big)^{1/q'}\\ &= M_{4q}^4\Big(\frac{M_p^p}{2^p}\Big)^{1/q'}\,n^{1-\frac{p}{2q'}}.\end{aligned}$$

The power of n is vanishing for $p = 2q'$. On the other hand, if we want to minimize the maximal index of the moment functionals, one should choose $p = 4q$. The two constraints are fulfilled for $q = 3/2$, $q' = 3$, and thus $p = 6$, in which case

$$\frac{1}{n}\,\mathbb{E}\,(|X|^2-n)^2\,1_B \le M_6^4\Big(\frac{M_6^6}{2^6}\Big)^{1/3}.$$

Thus Proposition 1.5.3 is proved. □

The normalized variance σ_4^2 is a member of the family of the following variance-type functionals which allow us to control the concentration of $|X|$ about its mean in the case $\mathbb{E}\,|X|^2 = n$.

Definition 1.5.5 Given a random vector X in $\mathbb{R}^n$ with $\mathbb{E}\,|X|^2 = n$, put

$$\sigma_{2p}(X) = \sqrt{n}\,\Big(\mathbb{E}\,\Big|\frac{|X|^2}{n}-1\Big|^p\Big)^{1/p}, \qquad p \ge 1.$$

The finiteness of $\sigma_{2p}(X)$ is equivalent to the finiteness of the moment of $|X|$ of order $2p$ (thus explaining the appearance of the index $2p$).

Note that σ_{2p} represents a non-decreasing function of p, which attains a minimum at $p = 1$ with value

$$\sigma_2(X) = \frac{1}{\sqrt{n}}\,\mathbb{E}\,\big|\,|X|^2-n\big|.$$

At the same time, all these functionals dominate the L^1-deviation of $|X|$ from $\sqrt{n}$.

Proposition 1.5.6 *If* $\mathbb{E}\,|X|^2 = n$, *then*

$$\sigma_{2p}(X) \ge \sigma_2(X) \ge \mathbb{E}\,\big|\,|X|-\sqrt{n}\,\big|$$

for all $p \ge 1$. *Moreover,*

$$\frac{1}{4}\,\sigma_2^2(X) \le \mathrm{Var}(|X|) \le 2\sigma_2(X)\sqrt{n}.$$

The first line of inequalities is obvious, and we only need to explain the last line. In terms of the random variable $\xi = \frac{|X|}{\sqrt{n}}$, we have $\xi \ge 0$, $\mathbb{E}\xi^2 = 1$, and

$$\mathrm{Var}(|X|) = n\,\mathrm{Var}(\xi) = n\,(1-(\mathbb{E}\xi)^2) = n\,(1-\mathbb{E}\xi)\,(1+\mathbb{E}\xi),$$

while $\sigma_2 = \sqrt{n}\,\mathbb{E}\,|1-\xi^2|$. By Cauchy's inequality,

$$(\mathbb{E}\,|1-\xi^2|)^2 \le \mathbb{E}\,(1-\xi)^2\,\mathbb{E}\,(1+\xi)^2 = 4\,\mathbb{E}\,(1-\xi)\,\mathbb{E}\,(1+\xi),$$

implying that $\sigma_2^2 \le 4\,\mathrm{Var}(|X|)$.

For the last bound, just note that $\mathbb{E}\xi \le 1$, so that $\mathrm{Var}(|X|) \le 2n\,\mathbb{E}\,(1-\xi)$, while

$$\sigma_2 = \sqrt{n}\,\mathbb{E}\,|1-\xi|\,(1+\xi) \ge \sqrt{n}\,\mathbb{E}\,|1-\xi|.$$

1.6 Small Ball Probabilities

The functionals

$$\sigma_{2p} = \sigma_{2p}(X), \quad m_p = m_p(X), \quad \text{and} \quad M_p = M_p(X)$$

can be used in the problem of bounding probabilities of “small"” balls. We can already see this from the following elementary claim.

Proposition 1.6.1 *Given a random vector X in $\mathbb{R}^n$ such that $\mathbb{E}\,|X|^2 = n$, we have, for all $p \ge 1$ and $0 < \lambda < 1$,*

$$\mathbb{P}\big\{|X|^2 \le \lambda n\big\} \le \frac{1}{(1-\lambda)^p}\,\frac{\sigma_{2p}^p}{n^{p/2}}.$$

In addition,

$$\mathbb{P}\big\{|X|^2 \le \lambda n\big\} \le \frac{4}{(1-\lambda)^2}\,\frac{\mathrm{Var}(|X|)}{n}.$$

Here, the second inequality sharpens the first one in the range $1 \le p \le 2$ up to a λ-dependent factor (when $\mathrm{Var}(|X|)$ is of order σ_{2p}^p).

Proof According to Definition 1.5.5,

$$\sigma_{2p}^p = n^{-p/2}\,\mathbb{E}\,\big|n - |X|^2\big|^p.$$

Hence, by Chebyshev's inequality,

$$\begin{aligned}\mathbb{P}\big\{|X|^2 \le \lambda n\big\} &= \mathbb{P}\Big\{n - |X|^2 \ge (1-\lambda)\,\mathbb{E}\,|X|^2\Big\}\\ &\le \frac{\sigma_{2p}^p\,n^{p/2}}{(1-\lambda)^p\,(\mathbb{E}\,|X|^2)^p} = \frac{\sigma_{2p}^p}{(1-\lambda)^p\,n^{p/2}},\end{aligned}$$

thus proving the first inequality of the proposition.

To prove the second inequality, put

$$v^2 = \mathrm{Var}(|X|) = n - (\mathbb{E}\,|X|)^2 \quad (v \ge 0),$$

so that $v^2 \le n$ and $\mathbb{E}\,|X| = \sqrt{n - v^2}$. In the case $v^2 \le (1-c^2)n$ with $\sqrt{\lambda} < c < 1$, we have $\mathbb{E}\,|X| \ge c\sqrt{n}$, and by Chebyshev's inequality,

$$\begin{aligned}\mathbb{P}\Big\{|X|^2 \le \lambda n\Big\} &= \mathbb{P}\Big\{|X| \le \sqrt{\lambda n}\Big\} \\ &= \mathbb{P}\Big\{\mathbb{E}\,|X| - |X| \ge \sqrt{n - v^2} - \sqrt{\lambda n}\Big\} \\ &\le \mathbb{P}\Big\{\mathbb{E}\,|X| - |X| \ge c\sqrt{n} - \sqrt{\lambda n}\Big\} \le \frac{v^2}{(c - \sqrt{\lambda})^2\, n}.\end{aligned}$$

The last ratio is greater than or equal to 1, as long as $v^2 \ge (c - \sqrt{\lambda})^2\, n$, and then the resulting inequality holds automatically. Therefore, for all $c \in (\sqrt{\lambda}, 1)$,

$$\Big(v^2 \le (1 - c^2)\, n \ \text{ or } \ v^2 \ge (c - \sqrt{\lambda})^2\, n\Big) \Longrightarrow \mathbb{P}\{|X|^2 \le \lambda n\} \le \frac{v^2}{(c - \sqrt{\lambda})^2\, n}.$$

In particular, the latter bound holds true without any constraint on v, as long as $(1 - c^2)n \ge (c - \sqrt{\lambda})^2\, n$, that is, $2c^2 - 2c\sqrt{\lambda} + \lambda - 1 \le 0$. This quadratic inequality in the variable c takes place in the interval $[c_0, c_1]$ with the right endpoint $c_1 = \frac{1}{2}\,(\sqrt{\lambda} + \sqrt{2 - \lambda})$, which satisfies $\sqrt{\lambda} < c_1 < 1$. Hence $c = c_1$ is the best admissible value minimizing the ratio $\frac{v^2}{(c - \sqrt{\lambda})^2\, n}$. And for this value we have

$$\mathbb{P}\{|X|^2 \le \lambda n\} \le \frac{v^2}{(c_1 - \sqrt{\lambda})^2\, n} = \frac{v^2}{n}\,\frac{(\sqrt{\lambda} + \sqrt{2 - \lambda})^2}{(1 - \lambda)^2} \le \frac{v^2}{n}\,\frac{4}{(1 - \lambda)^2},$$

which was required. The proposition is proved. □

Thus, if the variance functionals σ_{2p} are of order 1, the resulting bounds are of order $n^{-p/2}$. However, they may be considerably sharpened under the convolution operation. Given an independent copy Y of X, write

$$|X - Y|^2 = |X|^2 + |Y|^2 - 2\,\langle X, Y\rangle\,.$$

By the first inequality of Proposition 1.6.1 with $\lambda = \frac{3}{4}$,

$$\mathbb{P}\Big\{|X|^2 + |Y|^2 \le \frac{3}{4}\,n\Big\} \le \mathbb{P}\Big\{|X|^2 \le \frac{3}{4}\,n\Big\}\,\mathbb{P}\Big\{|Y|^2 \le \frac{3}{4}\,n\Big\} \le \frac{4^{2p}\,\sigma_{2p}^{2p}}{n^p}.$$

Alternatively for the case $p = 2$, the second inequality of the proposition leads to the improvement

$$\mathbb{P}\Big\{|X|^2 + |Y|^2 \le \frac{3}{4}\,n\Big\} \le \frac{4^6}{n^2}\,\mathrm{Var}^2(|X|).$$

On the other hand, by Markov's inequality, for any $q \ge 1$,

$$\mathbb{P}\Big\{|\,\langle X, Y\rangle\,| \ge \frac{1}{4}\,n\Big\} \le \frac{4^q\,\mathbb{E}\,|\,\langle X, Y\rangle\,|^q}{n^q} = \frac{4^q\, m_q^q}{n^{q/2}}.$$

Splitting the event $|X - Y|^2 \le \frac{1}{4}\,n$ to the case where $|\,\langle X, Y\rangle\,| \ge \frac{1}{4}\,n$ and to the case where an opposite inequality holds, we have a set inclusion

$$\Big\{|X-Y|^2 \le \frac{1}{4}\,n\Big\} \subset \Big\{|\left<X,Y\right>| \ge \frac{1}{4}\,n\Big\} \cup \Big\{|X|^2 + |Y|^2 \le \frac{3}{4}\,n\Big\}.$$

Hence, we arrive at:

Proposition 1.6.2 *Let Y be an independent copy of a random vector X in $\mathbb{R}^n$ such that $\mathbb{E}\,|X|^2 = n$. Then for all $p, q \ge 1$,*

$$\mathbb{P}\Big\{|X-Y|^2 \le \frac{1}{4}\,n\Big\} \le \frac{4^q}{n^{q/2}}\,m_q^q + \frac{4^{2p}}{n^p}\,\sigma_{2p}^{2p}.$$

In particular, choosing $q = 2p$,

$$\mathbb{P}\Big\{|X-Y|^2 \le \frac{1}{4}\,n\Big\} \le \frac{C}{n^p}$$

with constant $C = 4^{2p}\,(m_{2p}^{2p} + \sigma_{2p}^{2p})$. We also have the bound

$$\mathbb{P}\Big\{|X-Y|^2 \le \frac{1}{4}\,n\Big\} \le \frac{4^q}{n^{q/2}}\,m_q^q + \frac{4^6}{n^2}\,\mathrm{Var}^2(|X|).$$

Recall that, by Proposition 1.4.2, one may replace m_q with M_q^2 in these bounds.

In the most interesting cases $p = 1$, $p = \frac{3}{2}$, $p = 2$ with $q = 2p$, we thus get respectively

$$\mathbb{P}\Big\{|X-Y|^2 \le \frac{1}{4}\,n\Big\} \le \frac{16}{n}\,(m_2^2 + \sigma_2^2),$$
$$\mathbb{P}\Big\{|X-Y|^2 \le \frac{1}{4}\,n\Big\} \le \frac{64}{n^{3/2}}\,(m_3^3 + \sigma_3^3),$$
$$\mathbb{P}\Big\{|X-Y|^2 \le \frac{1}{4}\,n\Big\} \le \frac{256}{n^2}\,(m_4^4 + \sigma_4^4).$$

Choosing $p = q = 2$, we also have

$$\mathbb{P}\Big\{|X-Y|^2 \le \frac{1}{4}\,n\Big\} \le \frac{16}{n}\,m_2^2 + \frac{256}{n^2}\,\sigma_4^4.$$

Remark 1.6.3 If $|X| = \sqrt{n}$ a.s. and thus all $\sigma_{2p} = 0$, the latter bounds yield

$$\mathbb{P}\Big\{|X-Y|^2 \le \frac{n}{4}\Big\} \le c_1\,\frac{m_2^2}{n}, \qquad c_1 = 16,$$
$$\mathbb{P}\Big\{|X-Y|^2 \le \frac{n}{4}\Big\} \le c_2\,\frac{m_4^4}{n^2}, \qquad c_2 = 256.$$

In the isotropic case, the second inequality implies the first one, up to an absolute factor. Indeed, since $|X| = \sqrt{n}$ a.s., we have $|\left<X,Y\right>| \le n^2$, so,

$$m_4^4 = \frac{1}{n^2}\,\mathbb{E}\left<X,Y\right>^4 \le \mathbb{E}\left<X,Y\right>^2 = m_2^2\,n.$$

In fact, both inequalities are optimal in the sense that, up to absolute factors, they are attained for the random vector X with the isotropic distribution

$$\mathcal{L}(X) = \frac{1}{n} \sum_{k=1}^{n} \delta_{\sqrt{n}\, e_k},$$

where $e_1, \dots, e_n$ is an orthonormal basis in $\mathbb{R}^n$. In this example, $m_2 = 1$, while $m_4^4 = n$. Hence $\frac{m_2^2}{n} = \frac{m_4^4}{n^2} = \frac{1}{n}$. Similarly,

$$\mathbb{P}\Big\{|X - Y|^2 \le \frac{n}{4}\Big\} = \mathbb{P}\big\{X = Y\big\} = \frac{1}{n}.$$

Note that the distributions F_θ of linear functionals $\langle X, \theta \rangle$ in this particular example represent empirical measures

$$F_\theta = \mathcal{L}(\langle X, \theta \rangle) = \frac{1}{n} \sum_{k=1}^{n} \delta_{\sqrt{n}\, \langle \theta, e_k \rangle},$$

based on n "observations" $\sqrt{n}\, \langle \theta, e_1 \rangle, \dots, \sqrt{n}\, \langle \theta, e_n \rangle$.

Let us now extend the bounds of Proposition 1.6.2 for a growing number of convolutions, using the variance functional $\sigma_4 = \sigma_4(X)$ and the moment functionals $m_{2p} = m_{2p}(X)$ with an integer value of $p \ge 1$. Given independent copies $\{X^{(k)}\}_{1 \le k \le p}$ of the random vector X in $\mathbb{R}^n$, consider the sum of the form

$$S_p = \sum_{k=1}^{p} \varepsilon_k X^{(k)}$$

with arbitrary coefficients $\varepsilon_k = \pm 1$ and write $|S_p|^2 = \Sigma_1 + \Sigma_2$ with

$$\Sigma_1 = \sum_{k=1}^{p} |X^{(k)}|^2, \quad \Sigma_2 = \sum_{1 \le i \ne j \le p} \varepsilon_i \varepsilon_j \left\langle X^{(i)}, X^{(j)} \right\rangle.$$

The second sum contains $p(p-1) < p^2$ terms, and each of them has L^{2p}-norm

$$\| \left\langle X^{(i)}, X^{(j)} \right\rangle \|_{2p} = \Big(\mathbb{E} \left\langle X^{(i)}, X^{(j)} \right\rangle^{2p} \Big)^{1/2p} = m_{2p} \sqrt{n}.$$

Hence, by the triangle inequality, $\|\Sigma_2\|_{2p} \le p^2\, m_{2p} \sqrt{n}$, and by Markov's inequality,

$$\mathbb{P}\Big\{|\Sigma_2| \ge \frac{1}{4}\, n\Big\} \le \frac{\mathbb{E}\, |\Sigma_2|^{2p}}{(\frac{1}{4}\, n)^{2p}} \le (4p^2)^{2p}\, \frac{m_{2p}^{2p}}{n^p}.$$

On the other hand, by Proposition 1.5.1 with $p = 2$ and $\lambda = \frac{1}{2}$, we have

$$\mathbb{P}\Big\{\Sigma_1 \leq \frac{1}{2}\,n\Big\} \leq \mathbb{P}\Big\{|X|^2 \leq \frac{1}{2}\,n\Big\}^p \leq \Big(\frac{4\sigma_4^2}{n}\Big)^p.$$

Hence

$$\mathbb{P}\Big\{|S_p|^2 \leq \frac{1}{4}\,n\Big\} \leq \mathbb{P}\Big\{|\Sigma_2| \geq \frac{1}{4}\,n\Big\} + \mathbb{P}\Big\{|\Sigma_1| \leq \frac{1}{2}\,n\Big\} \leq \frac{C}{n^p}$$

with $C = (4p^2\, m_{2p})^{2p} + (4\sigma_4^2)^p$. Actually, we need such a bound for the sum of $2p$ random vectors.

Proposition 1.6.4 *Let $X^{(k)}$, $Y^{(k)}$ $(1 \leq k \leq p)$ be independent copies of the random vector X in $\mathbb{R}^n$ such that $\mathbb{E}\,|X|^2 = n$. Then*

$$\mathbb{P}\Big\{\Big|\sum_{k=1}^p (X^{(k)} - Y^{(k)})\Big|^2 \leq \frac{1}{4}\,n\Big\} \leq \frac{C}{n^{2p}}$$

with constant

$$C = \big(16\,p^2 m_{4p}\big)^{4p} + \big(4\sigma_4^2\big)^{2p}.$$

1.7 Second Order Correlation Condition

In the context of higher order concentration (see Chapters 13 and 17), a more powerful natural correlation condition emerged which led to the following requirement.

Definition 1.7.1 We say that a random vector $X = (X_1, \dots, X_n)$ in $\mathbb{R}^n$ satisfies a second order correlation condition with constant Λ if, for any collection $a_{ij} \in \mathbb{R}$,

$$\mathrm{Var}\bigg(\sum_{i,j=1}^n a_{ij} X_i X_j\bigg) \leq \Lambda \sum_{i,j=1}^n a_{ij}^2.$$

In the matrix form, using $A = (a_{ij})_{i,j=1}^n$ with the Hilbert–Schmidt norm $\|A\|_{\mathrm{HS}} = \big(\sum a_{ij}^2\big)^{1/2}$, the definition becomes

$$\mathrm{Var}(\,\langle AX, X\rangle\,) \leq \Lambda\, \|A\|_{\mathrm{HS}}^2.$$

It is sufficient to consider in this inequality only symmetric matrices A. Indeed, the matrix $B = \frac{1}{2}\,(A + A')$ is symmetric, where A' is the transpose of A, and $\langle AX, X\rangle = \langle BX, X\rangle$. On the other hand, by the triangle inequality, $\|B\|_{\mathrm{HS}} \leq \|A\|_{\mathrm{HS}}$.

Any random vector X in $\mathbb{R}^n$ with finite 4-th moment satisfies a second order correlation condition for some finite constant, and otherwise $\Lambda = \infty$. By the definition, the optimal value $\Lambda = \Lambda(X)$ represents the maximal eigenvalue of the covariance matrix associated with the n^2-dimensional random vector

$$X^{(2)} = \big(X_i X_j - \mathbb{E} X_i X_j\big)_{i,j=1}^n.$$

That is, in terms of the M_2-functional, $\Lambda(X) = M_2^2(X^{(2)})$.

For short, let us write $M_p = M_p(X)$, $m_p = m_p(X)$, $\sigma_4^2 = \sigma_4^2(X)$, $\Lambda = \Lambda(X)$.

Putting $a_{ij} = \delta_{ij}$ in Definition 1.7.1, we obtain an elementary relation between the two variance functionals.

Proposition 1.7.2 *For any random vector X in $\mathbb{R}^n$, we have $\sigma_4^2 \le \Lambda$.*

There can be a large difference between the values of these two functionals. For example, when $|X| = \text{const}$, we necessarily have $\sigma_4^2 = 0$.

The moment functional M_4 can also be controlled in terms of Λ and M_2. Putting $a_{ij} = \theta_i\theta_j$, $\theta = (\theta_1, \dots, \theta_n) \in \mathbb{S}^{n-1}$, Definition 1.7.1 leads to $\mathrm{Var}(\langle X, \theta\rangle^2) \le \Lambda$. Using $\mathbb{E}\,\langle X, \theta\rangle^2 \le M_2^2$, we obtain:

Proposition 1.7.3 *For any random vector X in $\mathbb{R}^n$, we have $M_4^4 \le M_2^4 + \Lambda$. In particular, if X is isotropic, then $M_4^4 \le 1 + \Lambda$.*

Recalling Proposition 1.4.2, we also have:

Proposition 1.7.4 *If Y is an independent copy of the random vector X in $\mathbb{R}^n$, then*

$$\mathbb{E}\,\langle X, Y\rangle^4 \le \big(M_2^4 + \Lambda\big)^2\, n^2.$$

Equivalently, $m_4^2 \le M_2^4 + \Lambda$.

Similarly to $M_p(X)$, $m_p(X)$ and $\sigma_4^2(X)$, the functional $\Lambda(X)$ is invariant under linear orthogonal transformations of the Euclidean space $\mathbb{R}^n$. Indeed, if $Y = UX$, where $U : \mathbb{R}^n \to \mathbb{R}^n$ is orthogonal, then

$$\begin{aligned}\mathrm{Var}(\langle AY, Y\rangle) &= \mathrm{Var}(\langle U^{-1}AUX, X\rangle) \\ &\le \Lambda\,\|U^{-1}AU\|_{\mathrm{HS}}^2 = \Lambda \|A\|_{\mathrm{HS}}^2.\end{aligned}$$

Hence, one may always assume that $\mathbb{E}X_iX_j = \lambda_i\delta_{ij}$, where the λ_i are eigenvalues of the covariance operator of X. In that case, Definition 1.7.1 becomes

$$\mathbb{E}\bigg(\sum_{i,j=1}^n a_{ij}\,(X_iX_j - \lambda_i\delta_{ij})\bigg)^2 \le \Lambda \sum_{i,j=1}^n a_{ij}^2.$$

In the isotropic case, all $\lambda_i = 1$, and the definition of $\Lambda(X)$ is further simplified to

$$\mathbb{E}\bigg(\sum_{i,j=1}^n a_{ij}\,(X_iX_j - \delta_{ij})\bigg)^2 \le \Lambda \sum_{i,j=1}^n a_{ij}^2.$$

All the above definitions extend to complex-valued random variables X_i with complex numbers a_i and a_{ij} (in which case, a_i^2 and a_{ij}^2 should be replaced with $|a_i|^2$ and $|a_{ij}|^2$ respectively). Note that, if ξ is a complex-valued random variable, its variance is defined by

$$\mathrm{Var}(\xi) = \mathbb{E}\,|\xi - \mathbb{E}\xi|^2 = \mathbb{E}\,|\xi|^2 - |\mathbb{E}\xi|^2.$$

Proposition 1.7.5 *For any isotropic random vector X in $\mathbb{R}^n$, we have $\Lambda \geq \frac{n-1}{n}$.*

Proof Applying the inequality from Definition 1.7.1 to the matrix A with only one non-zero entry on the (i, j)-place, we get

$$\mathrm{Var}(X_i X_j) = \mathbb{E} X_i^2 X_j^2 - \delta_{ij} \leq \Lambda.$$

Summing this over all i, j leads to $\mathbb{E}\,|X|^4 - n \leq n^2 \Lambda$. But $\mathbb{E}\,|X|^4 \geq (\mathbb{E}\,|X|^2)^2 = n^2$, thus proving the proposition. □

When $n = 1$, it is possible that $\Lambda = 0$ (which is attained for the ± 1 Bernoulli random variables). However, if $n \geq 2$, we have a universal bound $\Lambda(X) \geq \frac{1}{2}$ in the whole class of isotropic probability distributions.

Chapter 2
Some Classes of Probability Distributions

The relevance of the previously introduced moment and correlation-type functionals can be illustrated by examples of some classical classes of probability distributions. In this chapter, these functionals are discussed for product measures (in which case one can also refine upper bounds on "small ball" probabilities), for joint distributions of pairwise independent random variables, and for coordinate-symmetric distributions. We also discuss the class of logarithmically concave measures and include some additional background material which will be needed later on.

2.1 Independence

One basic example, which may be used for illustrations of the previously introduced moment-like functionals, corresponds to independent components of X. Let us clarify the meaning of these functionals in this important case.

As before, let $X = (X_1, \dots, X_n)$. From the very definition it follows that

$$\sigma_4^2(X) = \frac{1}{n} \sum_{k=1}^{n} \mathrm{Var}(X_k^2) \tag{2.1}$$

as long as $X_1, \dots, X_n$ are independent. We also have:

Proposition 2.1.1 *Let $n \geq 2$. If the random variables $X_1, \dots, X_n$ are independent and have mean zero, then*

$$M_2(X) = \max_k \big(\mathbb{E}X_k^2\big)^{1/2}, \quad \Lambda(X) = \max_k \big\{\mathrm{Var}(X_k^2), 2ab\big\},$$

where a and b are the first two maximal numbers in the sequence $\mathbb{E}X_1^2, \dots, \mathbb{E}X_n^2$. In particular, $\Lambda(X) \leq 2 \max_k \mathbb{E}X_k^4$.

S. Bobkov et al., *Concentration and Gaussian Approximation for Randomized Sums*, Probability Theory and Stochastic Modelling 104,
https://doi.org/10.1007/978-3-031-31149-9_2

Proof The first equality is obvious. As for the second assertion, put $v_i^2 = \mathbb{E}X_i^2$ and note that the random variables

$$X_{ij}^{(2)} = X_i X_j - \mathbb{E}X_i X_j = X_i X_j - \delta_{ij} v_i^2$$

have mean zero. Their covariances are given by

$$\begin{aligned} \mathbb{E}\,X_{ij}^{(2)} X_{kl}^{(2)} &= \mathbb{E}\,(X_i X_j - \delta_{ij} v_i^2)\, X_k X_l \\ &= \mathbb{E}\,X_i X_j X_k X_l - \delta_{ij}\delta_{kl}\, v_i^2 v_k^2. \end{aligned} \tag{2.2}$$

Case 1: $i \neq j$. The right-hand side of (2.2) is vanishing unless $(i, j) = (k, l)$ or $(i, j) = (l, k)$. In both these cases it is equal to

$$\mathbb{E}\,X_{ij}^{(2)} X_{ij}^{(2)} = \mathbb{E}\,X_{ij}^{(2)} X_{ji}^{(2)} = v_i^2 v_j^2.$$

Case 2: $i = j$. The right-hand side of (2.2) may be non-zero only when $k = l$.
Case 2a): $i = j$, $k = l$, $i \neq k$. Then the right-hand side of (2.2) is equal to

$$\mathbb{E}\,X_{ii}^{(2)} X_{kk}^{(2)} = \mathbb{E}X_i^2\, \mathbb{E}X_k^2 - \mathbb{E}X_i^2\, \mathbb{E}X_k^2 = 0.$$

Case 2b): $i = j = k = l$. The right-hand side is equal to

$$\mathbb{E}\,X_{ii}^{(2)} X_{ii}^{(2)} = \mathbb{E}X_i^4 - \mathbb{E}X_i^2\, \mathbb{E}X_i^2 = \mathrm{Var}(X_i^2).$$

Summarizing, we see that the covariance $\mathbb{E}\,X_{ij}^{(2)} X_{kl}^{(2)}$ may be non-zero only when $(i, j) = (k, l)$ or $(i, j) = (l, k)$ for $i \neq j$ and when $i = j = k = l$, in which case it is equal to

$$\mathbb{E}\,X_{ij}^{(2)} X_{ij}^{(2)} = \mathbb{E}\,X_{ij}^{(2)} X_{ji}^{(2)} = (1 - \delta_{ij})\, v_i^2 v_j^2 + \delta_{ij} \mathrm{Var}(X_i^2).$$

Therefore, for any real symmetric matrix $(a_{ij})_{i,j=1}^n$,

$$\begin{aligned} \mathrm{Var}\bigg(\sum_{i,j=1}^n a_{ij} X_i X_j\bigg) &= \sum_{i,j=1}^n \sum_{k,l=1}^n a_{ij} a_{kl}\, \mathbb{E}\,X_{ij}^{(2)} X_{kl}^{(2)} \\ &= 2 \sum_{i \neq j} a_{ij}^2\, v_i^2 v_j^2 + \sum_i a_{ii}^2\, \mathrm{Var}(X_i^2). \end{aligned}$$

Clearly, the maximum of this variance subject to $\sum_{i,j=1}^n a_{ij}^2 \leq 1$ is equal to

$$\max\Big\{2 \max_{i \neq j}\, v_i^2 v_j^2,\ \max_i\, \mathrm{Var}(X_i^2)\Big\},$$

and the desired assertion follows. □

Corollary 2.1.2 *For i.i.d. random variables $X_1, \dots, X_n$ with mean zero,*

$$M_2(X) = \left(\mathbb{E}X_1^2\right)^{1/2}, \quad \sigma_4^2(X) = \mathrm{Var}(X_1^2),$$

and if $n \geq 2$,

$$\Lambda(X) = \max\left\{\mathrm{Var}(X_1^2), 2\left(\mathbb{E}X_1^2\right)^2\right\} \leq 2\,\mathbb{E}X_1^4.$$

Let us now turn to the functionals M_p with arbitrary p.

Proposition 2.1.3 *For independent random variables $X_1, \dots, X_n$ with mean zero,*

$$\max_k \left(\mathbb{E}\,|X_k|^p\right)^{1/p} \leq M_p(X) \leq C_p \max_k \left(\mathbb{E}\,|X_k|^p\right)^{1/p}$$

for any $p \geq 2$, where one may put $C_p = 2\sqrt{p}$.

In the i.i.d. situation, $M_p(X)$ is thus equivalent to the L^p-norm $(\mathbb{E}\,|X_1|^p)^{1/p}$.

Proof The lower bound follows from $M_p \geq M_2$ and Proposition 2.1.1.

For the upper bound, we use the symmetrization argument. By Jensen's inequality, if η is an independent copy of a random variable ξ with mean zero and finite p-th absolute moment, then $\mathbb{E}\,|\xi|^p \leq \mathbb{E}\,|\xi - \eta|^p$. Let $Y = (Y_1, \dots, Y_n)$ be an independent copy of the random vector X and put $X' = X - Y$. Its components $X'_k = X_k - Y_k$ are independent and have symmetric distributions. Hence

$$\mathbb{E}\,|\langle X, \theta\rangle|^p \leq \mathbb{E}\,|\langle X', \theta\rangle|^p = \mathbb{E}\left|\sum_{k=1}^n \theta_k \varepsilon_k X'_k\right|^p,$$

where the $\varepsilon_k = \pm 1$ are arbitrary. Here one can take the average over all ε_k using Khinchine's inequality

$$\mathbb{E}_\varepsilon\left|\sum_{k=1}^n a_k \varepsilon_k\right|^p \leq B_p\left(\sum_{k=1}^n a_k^2\right)^{p/2},$$

which holds with a constant $B_p > 0$ depending on p only. This gives

$$\mathbb{E}\,|\langle X, \theta\rangle|^p \leq B_p^p\,\mathbb{E}\left(\sum_{k=1}^n \theta_k^2 X_k'^2\right)^{p/2} \leq B_p^p\,\mathbb{E}\sum_{k=1}^n \theta_k^2\,|X'_k|^p,$$

where we also used the convexity of the power function $x \to x^{p/2}$ together with the assumption $\theta_1^2 + \dots + \theta_n^2 = 1$. Maximizing over θ and using $\mathbb{E}\,|X'_k|^p \leq 2^{p-1}\,\mathbb{E}\,|X_k|^p$ (again, by Jensen's inequality), we arrive at

$$M_p^p(X) \leq 2^{p-1} B_p^p \max_k \mathbb{E}\,|X_k|^p.$$

Finally, let us recall a standard argument providing an explicit bound on B_p. Let $Z_n = a_1\varepsilon_1 + \cdots + a_n\varepsilon_n$ with $a_1^2 + \cdots + a_n^2 = 1$. For every $t > 0$, we have

$$\mathbb{E}_\varepsilon\, e^{t|Z_n|} \le 2\,\mathbb{E}_\varepsilon\, e^{tZ_n} = 2\prod_{k=1}^n \cosh(a_k t) \le 2\prod_{k=1}^n e^{(a_k t)^2/2} = 2\,e^{t^2/2}.$$

Using $x^p e^{-x} \le (\frac{p}{e})^p$, $x \ge 0$, this gives

$$\mathbb{E}_\varepsilon|Z_n|^p = t^{-p}\,\mathbb{E}_\varepsilon\,|tZ_n|^p\, e^{-t|Z_n|}\, e^{t|Z_n|} \le \Big(\frac{p}{et}\Big)^p\, \mathbb{E}_\varepsilon\, e^{t|Z_n|} \le 2\Big(\frac{p}{et}\Big)^p\, e^{t^2/2}.$$

Here, the (optimal) choice $t = \sqrt{p}$ leads to $\mathbb{E}_\varepsilon\,|Z_n|^p \le 2\,(\frac{p}{e})^{p/2}$. Hence, a similar bound holds true for B_p, which yields $C_p \le (\frac{4p}{e})^{p/2} < (2\sqrt{p})^p$. We thus arrive at the desired upper bound. □

Next, let us also stress that small ball bounds like the one of Proposition 1.6.1 with $p = 2$,

$$\mathbb{P}\Big\{|X|^2 \le \lambda n\Big\} \le \frac{\sigma_4^2(X)}{(1-\lambda)^2}\,\frac{1}{n}, \qquad 0 < \lambda < 1 \ \ (\text{where } \mathbb{E}\,|X|^2 = n),$$

and even the bounds of Propositions 1.6.2 and 1.6.4 may be considerably sharpened for independent summands with respect to the growing parameter n.

To recall the standard argument, consider the sum $S_n = \xi_1 + \cdots + \xi_n$ of independent random variables $\xi_k \ge 0$ such that $\mathbb{E}S_n = n$ and $\mathrm{Var}(S_n) = \sigma^2 n$. Given a parameter $t > 0$ (to be chosen later on), we have, for any $0 < \lambda < 1$,

$$\mathbb{E}\, e^{-tS_n} \ge e^{-\lambda t n}\, \mathbb{P}\{S_n \le \lambda n\}.$$

On the other hand, every function $u_k(t) = \mathbb{E}\, e^{-t\xi_k}$ is positive, convex, and admits Taylor's expansion near zero up to the quadratic form, which implies that

$$u_k(t) \le 1 - t\,\mathbb{E}\xi_k + \frac{t^2}{2}\,\mathbb{E}\xi_k^2 \le \exp\Big\{-t\,\mathbb{E}\xi_k + \frac{t^2}{2}\,\mathbb{E}\xi_k^2\Big\}.$$

Multiplying these inequalities, we get

$$\mathbb{E}\, e^{-tS_n} \le \exp\Big\{-tn + \frac{bt^2}{2}\Big\}, \qquad b = \sum_{k=1}^n \mathbb{E}\xi_k^2.$$

The two bounds yield

$$\mathbb{P}\{S_n \le \lambda n\} \le \exp\Big\{-(1-\lambda)nt + bt^2/2\Big\},$$

and after optimization over t (in fact, $t = \frac{1-\lambda}{b}\,n$), we arrive at the exponential bound

$$\mathbb{P}\{S_n \le \lambda n\} \le \exp\Big\{-\frac{(1-\lambda)^2}{2b}\,n^2\Big\}.$$

Note that

$$b = \mathrm{Var}(S_n) + \sum_{k=1}^{n} (\mathbb{E}\xi_k)^2 \le \Big(\sigma^2 + \max_k\,(\mathbb{E}\xi_k)^2\Big)\,n,$$

so

$$\mathbb{P}\{S_n \le \lambda n\} \le \exp\Big\{-\frac{(1-\lambda)^2}{2\left[\sigma^2 + \max_k\,(\mathbb{E}\xi_k)^2\right]}\,n\Big\}.$$

In the case $S_n = |X|^2 = X_1^2 + \cdots + X_n^2$, thus with $\xi_k = X_k^2$, we have $\sigma^2 = \sigma_4^2(X)$, and according to Proposition 2.1.1,

$$M_2^4(X) = \max_k\,(\mathbb{E}X_k^2)^2 = \max_k\,(\mathbb{E}\xi_k)^2.$$

Hence, we get:

Proposition 2.1.4 *If the random variables* $X_1, \dots, X_n$ *are independent and* $\mathbb{E}\,|X|^2 = n$*, then for all* $0 < \lambda < 1$,

$$\mathbb{P}\{|X|^2 \le \lambda n\} \le \exp\Big\{-\frac{(1-\lambda)^2}{2\,(\sigma_4^2 + M_2^4)}\,n\Big\},$$

where $\sigma_4^2 = \sigma_4^2(X)$ *and* $M_2 = M_2(X)$*. In particular, in the i.i.d. case,*

$$\mathbb{P}\{|X|^2 \le \lambda n\} \le \exp\Big\{-\frac{(1-\lambda)^2}{2\,\mathbb{E}X_1^4}\,n\Big\}.$$

In fact, the property that the probabilities $\mathbb{P}\{|X|^2 \le \lambda n\}$ decrease exponentially fast does not require the finiteness of the 4-th moment. To show this, let us return to the sum $S_n = \xi_1 + \cdots + \xi_n$ of independent random variables $\xi_k \ge 0$, assuming that the summands are identically distributed with $\mathbb{E}\xi_1 = 1$, so that $\mathbb{E}S_n = n$. As before, given $t > 0$ and $0 < \lambda < 1$, we have a lower bound on the Laplace transform

$$\mathbb{E}\,\mathrm{e}^{-tS_n} \ge \mathrm{e}^{-\lambda t n}\,\mathbb{P}\{S_n \le \lambda n\}.$$

Let V denote the common distribution of ξ_k which is supported on the positive half-axis $[0,\infty)$. The function

$$u(t) = \mathbb{E}\,\mathrm{e}^{-t\xi_1} = \int_0^\infty \mathrm{e}^{-tx}\,\mathrm{d}V(x), \quad t \ge 0,$$

is positive, convex, non-increasing, and has a continuous, non-decreasing derivative

$$u'(t) = -\mathbb{E}\,\xi_1 \mathrm{e}^{-t\xi_1} = -\int_0^\infty x\mathrm{e}^{-tx}\,\mathrm{d}V(x),$$

with $u(0) = 1$, $u'(0) = -1$.

Now, given $p \in (0, 1)$, let $\kappa = \kappa_p$ denote the maximal quantile for the probability measure $x\mathrm{d}V(x)$ on $(0, \infty)$ of order p, i.e., the minimal number $\kappa > 0$ such that

$$\int_{(\kappa,\infty)} x\,\mathrm{d}V(x) \le 1 - p.$$

Using $1 - \mathrm{e}^{-y} \le y$ $(y \ge 0)$, we have, for all $s > 0$,

$$\begin{aligned} 1 + u'(s) &= \int_0^\infty x(1 - \mathrm{e}^{-sx})\,\mathrm{d}V(x) \\ &= \int_{0<x\le\kappa} x(1 - \mathrm{e}^{-sx})\,\mathrm{d}V(x) + \int_{x>\kappa} x(1 - \mathrm{e}^{-sx})\,\mathrm{d}V(x) \\ &\le s\int_{0<x\le\kappa} x^2\,\mathrm{d}V(x) + 1 - p \le 1 - p + \kappa^2 s, \end{aligned}$$

that is, $u'(s) \le -p + \kappa^2 s$. This gives

$$u(t) = 1 + \int_0^t u'(s)\,\mathrm{d}s \le 1 - pt + \kappa^2\,\frac{t^2}{2} \le \exp\Big\{-pt + \kappa^2\,\frac{t^2}{2}\Big\},$$

and therefore

$$\mathbb{P}\{S_n \le \lambda n\} \le \mathrm{e}^{\lambda nt}\,\mathbb{E}\,\mathrm{e}^{-tS_n} = \mathrm{e}^{\lambda nt}\,u(t)^n \le \exp\Big\{-n\Big((p-\lambda)t - \kappa^2\,\frac{t^2}{2}\Big)\Big\}.$$

If $p > \lambda$, the right-hand side is minimized at $t = (p - \lambda)/\kappa^2$, and we get

$$\mathbb{P}\{S_n \le \lambda n\} \le \mathrm{e}^{-n\,(p-\lambda)^2/2\kappa^2}.$$

One may take, for example, $p = (1 + \lambda)/2$, and then we arrive at

$$\mathbb{P}\{S_n \le \lambda n\} \le \mathrm{e}^{-n(1-\lambda)^2/8\kappa^2}.$$

Let us summarize this for the case $\xi_k = X_k^2$.

Proposition 2.1.5 *Let the random variables $X_1, \dots, X_n$ be independent and identically distributed, with $\mathbb{E}X_1^2 = 1$. Given $0 < \lambda < 1$, let a number $\kappa > 0$ be chosen to satisfy*

$$\mathbb{E}\,X_1^2\,1_{\{X_1^2>\kappa\}} \le \frac{1+\lambda}{2}.$$

Then

$$\mathbb{P}\{|X|^2 \le \lambda n\} \le \exp\Big\{-\frac{(1-\lambda)^2}{8\kappa^2}\,n\Big\}.$$

2.2 Pairwise Independence

A certain weakening of the independence property is the case where the random variables $X_1, \dots, X_n$ are pairwise independent. In that case we still have the important identity

$$\sigma_4^2(X) = \frac{1}{n} \sum_{k=1}^{n} \mathrm{Var}(X_k^2).$$

If additionally these random variables are identically distributed, then $\sigma_4^2(X) = \mathrm{Var}(X_1^2)$, which is thus dimension-free.

Proposition 2.1.1 is also partially saved. If $X_1, \dots, X_n$ are pairwise independent and have mean zero, then

$$M_2(X) = \max_k \left(\mathbb{E}X_k^2\right)^{1/2}, \quad X = (X_1, \dots, X_n).$$

Moreover, if all $\mathbb{E}X_k^2 = 1$, then we deal with an isotropic random vector X.

Example. Given a real-valued, 1-periodic, Borel measurable function f on the real line, an interesting example of pairwise independent random variables is described by the model

$$X_k(t, s) = f(kt + s) \quad (0 < t, s < 1, \ \ k = 1, 2, \dots) \tag{2.3}$$

on the probability space $\Omega = (0, 1)\times(0, 1)$, which we equip with the Borel σ-algebra $\mathcal{F}$ and the Lebesgue measure $\mathbb{P}$.

This system is related to the well-studied systems $\{f(kt)\}$ (like in the case of trigonometric systems), but an additional "mixing" argument s adds a number of remarkable properties. In particular, the following statement holds.

Proposition 2.2.1 *$\{X_k\}_{k=1}^{\infty}$ is a strictly stationary sequence of pairwise independent random variables on Ω.*

Proof Put $\eta_k(t, s) = kt + s$ (mod 1). Since $X_k = f(\eta_k)$, it is sufficient to show that the η_k form a strictly stationary sequence of pairwise independent random variables. Moreover, instead of η_k, one may consider the sequence

$$\xi_k = \zeta z^k, \quad k = 1, 2 \dots,$$

where ζ and z are independent, complex-valued random variables uniformly distributed on the unit circle S^1 of the complex plane. Note that

$$\mathbb{E}\,\zeta^m = \begin{cases} 1, & \text{if } m = 0, \\ 0, & \text{if } m \neq 0, \end{cases} \quad \text{for all } m \in \mathbb{Z}.$$

By independence of ζ and z, we have, for all integers $m_1, \dots, m_N$ and $h \geq 0$,

$$\mathbb{E}\,\xi_{1+h}^{m_1} \cdots \xi_{N+h}^{m_N} = \mathbb{E}\,\zeta^{m_1+\dots+m_N}\,\mathbb{E}\,z^{(1+h)m_1+\dots+(N+h)m_N}.$$

Here, the right-hand side is either equal to 0 or 1, and is equal to 1 if and only if

$$m_1 + \cdots + m_N = 0 \quad \text{and} \quad (1+h)m_1 + \cdots + (N+h)m_N = 0,$$

that is, if and only if $m_1 + \cdots + m_N = 0$ and $1 \cdot m_1 + \cdots + N \cdot m_N = 0$. The latter description does not depend on h, which implies the strict stationarity by applying Weierstrass' density theorem. Similarly, for indexes $k > l \geq 1$ and all $n, m \in \mathbb{Z}$, one easily verifies that

$$\mathbb{E}\, \xi_k^n \xi_l^m = \mathbb{E}\, \xi_k^n \, \mathbb{E}\, \xi_l^m,$$

which is an equivalent form of independence of ξ_k and ξ_l (for S^1-valued random variables). Proposition 2.2.1 is proved. □

Being pairwise independent, ξ_k form an almost deterministic sequence in the sense that $\xi_k = g_k(\xi_1, \xi_2)$ for certain measurable functions g_k on $S^1 \times S^1$. Moreover, there is a relation $\xi_k = g_{i,j,k}(\xi_i, \xi_j)$, $i \neq j$, as long as $\frac{k-j}{i-j}$ is an integer.

Since the function $(t, s) \longrightarrow kt + s \bmod(1)$ is uniformly distributed in $(0, 1)$ as a random variable on Ω, we have

$$\mathbb{E}X_k = \int_0^1 f(x)\,\mathrm{d}x, \quad \mathbb{E}\,|X_k|^p = \int_0^1 |f(x)|^p\,\mathrm{d}x.$$

Hence, when $\int_0^1 f(x)\,\mathrm{d}x = 0$ and $\int_0^1 f(x)^2\,\mathrm{d}x = 1$, the vector $X = (X_1, \dots, X_n)$ will be isotropic, with components being pairwise independent. The latter insures that, if f has finite 4-th moment on $(0, 1)$, then the variance functional

$$\sigma_4^2(X) = \frac{1}{n}\,\mathrm{Var}(|X|^2) = \int_0^1 f(x)^4\,\mathrm{d}x - \Big(\int_0^1 f(x)^2\,\mathrm{d}x\Big)^2$$

is finite and does not depend on n.

2.3 Coordinatewise Symmetric Distributions

Another illustrative example is the class of random vectors $X = (X_1, \dots, X_n)$ such that the distributions of $(\varepsilon_1 X_1, \dots, \varepsilon X_n)$ with arbitrary $\varepsilon_k = \pm 1$ do not depend on the choice of ε_k. Equivalently, the distribution of X is invariant under reflections $(x_1, \dots, x_n) \rightarrow (\varepsilon_1 x_1, \dots, \varepsilon_n x_n)$ of the space $\mathbb{R}^n$ about coordinate axes. In particular, X and $-X$ should have the same distribution, that is, the distribution of X should be symmetric about the origin. We call such probability distributions coordinatewise symmetric, although in the literature they are also called distributions with an unconditional basis.

This class includes all symmetric product measures on $\mathbb{R}^n$, which corresponds to the case where the components X_k are i.i.d. random variables with symmetric distributions on the real line. It is therefore not surprising that many formulas from the previous sections extend to the coordinatewise symmetric distributions. In particular, we have the following elementary assertion, which is a full analog of Proposition 2.1.3 about mean zero independent components.

Proposition 2.3.1 *If $X = (X_1, \dots, X_n)$ has a coordinatewise symmetric distribution, then*

$$M_2(X) = \max_k \big(\mathbb{E}X_k^2\big)^{1/2}.$$

Moreover, for any $p \geq 2$,

$$\max_k \big(\mathbb{E}\,|X_k|^p\big)^{1/p} \leq M_p(X) \leq C_p \max_k \big(\mathbb{E}\,|X_k|^p\big)^{1/p},$$

where one may take $C_p = 2\sqrt{p}$.

The first assertion is obvious, as well as the lower bound in the second one. For an upper bound, using the coordinatewise symmetry, we may write

$$\mathbb{E}\,|\langle X, \theta\rangle|^p = \mathbb{E}\,\bigg|\sum_{k=1}^n \theta_k \varepsilon_k X_k\bigg|^p,$$

where the $\varepsilon_k = \pm 1$ are arbitrary. The remaining part of the argument is identical to the one from the proof of Proposition 2.1.3.

Recalling Proposition 1.4.2, we obtain as a consequence that

$$m_p(X) = \frac{1}{\sqrt{n}}\Big(\mathbb{E}\,|\langle X, Y\rangle|^p\Big)^{1/p} \leq C_p^2 \max_k \big(\mathbb{E}\,|X_k|^p\big)^{2/p}$$

(where Y is an independent copy of X). In fact, this inequality can be refined in terms of the mean moments. Let us restrict ourselves to the mean 4-th moment

$$\bar{\beta}_4 = \frac{1}{n}\sum_{k=1}^n \mathbb{E}X_k^4,$$

which may also be treated as a normalized Lyapunov coefficient (in the central limit theorem).

Proposition 2.3.2 *If the random vector X has a coordinatewise symmetric distribution, then $m_4^4(X) \leq 3\bar{\beta}_4^2$.*

Indeed, we have a general identity

$$\mathbb{E}\,\langle X, Y\rangle^4 = \sum_{i_1,\dots,i_4=1}^n \Big(\mathbb{E}\,X_{i_1}X_{i_2}X_{i_3}X_{i_4}\Big)^2.$$

The last expectation is non-vanishing if and only if in the sequence i_1, i_2, i_3, i_4 there are either two distinct numbers which are listed twice, or all of them coincide. Hence, by Cauchy's inequality,

$$\begin{aligned}\mathbb{E}\,\langle X,Y\rangle^4 &= 3\sum_{i\neq j}(\mathbb{E}X_i^2X_j^2)^2+\sum_i(\mathbb{E}X_i^4)^2\\ &\le 3\sum_{i\neq j}\mathbb{E}X_i^4X_j^4+\sum_i(\mathbb{E}X_i^4)^2 = 3\Big(\sum_i\mathbb{E}X_i^4\Big)^2-2\sum_i(\mathbb{E}X_i^4)^2.\end{aligned}$$

It follows that $\mathbb{E}\,\langle X,Y\rangle^4 \le 3n^2\bar{\beta}_4^2$.

As an important subclass, one may consider random vectors $X=(X_1,\dots,X_n)$ with spherically invariant distributions. The distribution of X is then uniquely determined by the distribution of its Euclidean norm $|X|$, and in polar coordinates one may write $X=r\theta$, assuming that θ is a random vector uniformly distributed in the unit sphere $\mathbb{S}^{n-1}$ and $r\ge 0$ is a random variable independent of θ. For example, when X is standard normal, r^2 has a χ^2-distribution with n degrees of freedom.

In this subclass, the distributions of linear functionals $\langle X,a\rangle$ depend on $|a|$ only, and similarly to the i.i.d. case, many assertions about X can be stated in terms of X_1 or r. In particular,

$$M_2^2(X)=\mathbb{E}X_1^2,$$

so that X is isotropic if and only if $\mathbb{E}X_1^2=1$, or equivalently $\mathbb{E}\,|X|^2=n$.

More generally, one may consider coordinatewise symmetric distributions that are invariant under permutations of coordinates, i.e., under mappings $(x_1,\dots,x_n)\to(x_{\pi(1)},\dots,x_{\pi(n)})$ for arbitrary permutations π of $\{1,\dots,n\}$. Concerning the functional $\sigma_4^2(X)$, the situation is not similar to the i.i.d. case, since the covariances $\mathrm{cov}(X_i^2,X_j^2)$, $i\neq j$, may not vanish. More precisely, all such covariances are equal to each other, and we have

$$\sigma_4^2(X)=\frac{1}{n}\,\mathrm{Var}(|X|^2)=\mathrm{Var}(X_1^2)+(n-1)\,\mathrm{cov}(X_1^2,X_2^2). \tag{2.4}$$

Hence, in order to bound $\sigma_4^2(X)$ from above by a dimension-free quantity, one needs a property like

$$\mathrm{cov}(X_1^2,X_2^2)\le\frac{c}{n}.$$

The same remark applies to a more complicated Λ-functional, which is often of the same order as $\sigma_4^2(X)$. To be more precise, in the coordinatewise symmetric case, $\Lambda(X)$ may be in essence reduced to the moment-type functional

$$V_2(X)\;=\;\sup_{\theta\in\mathbb{S}^{n-1}}\mathrm{Var}(\theta_1X_1^2+\dots+\theta_nX_n^2)=M_2^2(Y),$$

where Y is a random vector with components $Y_k=X_k^2-\mathbb{E}X_k^2$. That is, $V_2(X)$ represents the maximal eigenvalue of the covariance matrix $\{\mathrm{cov}(X_i^2,X_j^2)\}_{i,j=1}^n$.

Proposition 2.3.3 *Given a random vector $X=(X_1,\dots,X_n)$ with a coordinatewise symmetric distribution, we have*

$$V_2(X)\le\Lambda(X)\le 2\max_k\,\mathbb{E}X_k^4+V_2(X). \tag{2.5}$$

If additionally the distribution of X is invariant under permutations of coordinates, then

$$\sigma_4^2(X) \le \Lambda(X) \le 2\,\mathbb{E}X_1^4 + \sigma_4^2(X), \tag{2.6}$$

where the last term $\sigma_4^2(X)$ may be removed in the case $\mathrm{cov}(X_1^2, X_2^2) \le 0$ $(n \ge 2)$.

These assumptions are met, in particular, when the components are independent and have a common symmetric distribution on the line. In this case, (2.6) is consistent with representation for $\Lambda(X)$ from Proposition 2.1.1.

Proof The lower bound on $\Lambda(X)$ in (2.5) follows from Definition 1.7.1 by choosing the coefficients of the form $a_{ij} = \theta_i\delta_{ij}$.

In the proof of the second inequality (the upper bound), the argument is similar to the one used in the proof of Proposition 2.1.1. Put $v_i^2 = \mathbb{E}X_i^2$ and define

$$X_{ij}^{(2)} = X_iX_j - \mathbb{E}X_iX_j = X_iX_j - \delta_{ij}v_i^2.$$

As before, the covariances of these mean zero random variables are given by

$$\mathbb{E}\,X_{ij}^{(2)}X_{kl}^{(2)} = \mathbb{E}\,X_iX_jX_kX_l - \delta_{ij}\delta_{kl}\,v_i^2v_k^2. \tag{2.7}$$

Case 1: $i \neq j$. By the symmetry about the coordinate axes, the right-hand side of (2.7) is vanishing unless $(i,j) = (k,l)$ or $(i,j) = (l,k)$. In both these cases, it is equal to

$$\mathbb{E}\,X_{ij}^{(2)}X_{ij}^{(2)} = \mathbb{E}\,X_{ij}^{(2)}X_{ji}^{(2)} = \mathbb{E}X_i^2X_j^2.$$

Case 2: $i = j$. Again, the right-hand side in (2.7) is non-zero only when $k = l$.

Case 2a): $i = j$, $k = l$, $i \neq k$. The right-hand side in (2.7) is equal to

$$\mathbb{E}\,X_{ii}^{(2)}X_{kk}^{(2)} = \mathbb{E}X_i^2X_k^2 - \mathbb{E}X_i^2\,\mathbb{E}X_k^2 = \mathrm{cov}(X_i^2, X_k^2).$$

Case 2b): $i = j = k = l$. The right-hand side is equal to

$$\mathbb{E}\,X_{ii}^{(2)}X_{ii}^{(2)} = \mathbb{E}X_i^4 - \mathbb{E}X_i^2\,\mathbb{E}X_i^2 = \mathrm{Var}(X_i^2).$$

In both subcases, $\mathbb{E}\,X_{ii}^{(2)}X_{kk}^{(2)} = \mathrm{cov}(X_i^2, X_k^2)$. Therefore, for any collection of real numbers a_{ij} such that $a_{ij} = a_{ji}$, $i,j = 1,\dots,n$, and with $\sum_{i,j=1}^n a_{ij}^2 = 1$,

$$\begin{aligned}\mathrm{Var}\Big(\sum_{i,j=1}^n a_{ij}X_iX_j\Big) &= \sum_{i,j=1}^n\sum_{k,l=1}^n a_{ij}a_{kl}\,\mathbb{E}\,X_{ij}^{(2)}X_{kl}^{(2)} \\ &= 2\sum_{i\neq j} a_{ij}^2\,\mathbb{E}X_i^2X_j^2 + \sum_{i,k} a_{ii}a_{kk}\,\mathrm{cov}(X_i^2, X_k^2).\end{aligned}$$

Here, the first sum on the right-hand side does not exceed

$$\max_{i\neq j}\,\mathbb{E}X_i^2X_j^2\sum_{i\neq j}a_{ij}^2 \le \max_i\,\mathbb{E}X_i^4\sum_{i\neq j}a_{ij}^2 \le \max_i\,\mathbb{E}X_i^4$$

(by applying Cauchy's inequality). As for the second sum, it does not exceed $V_2(X)$, and we obtain

$$\Lambda(X) \le 2 \max_{i \neq j} \mathbb{E}X_i^2 X_j^2 + V_2(X), \tag{2.8}$$

from which (2.5) follows immediately.

Turning to the second assertion, we may assume that $n \ge 2$. The first inequality in (2.6) is general, cf. Proposition 1.7.2. For the upper bound, we note that, for any $\theta = (\theta_1, \dots, \theta_n) \in \mathbb{S}^{n-1}$,

$$\begin{aligned} \mathrm{Var}(\theta_1 X_1^2 + \dots + \theta_n X_n^2) &= \sum_{i,j=1}^n \theta_i \theta_j \,\mathrm{cov}(X_i^2, X_j^2) \\ &= \sum_{i \neq j} \theta_i \theta_j \,\mathrm{cov}(X_1^2, X_2^2) + \mathrm{Var}(X_1^2) \\ &= \Big(\sum_{i=1}^n \theta_i\Big)^2 \mathrm{cov}(X_1^2, X_2^2) - \mathrm{cov}(X_1^2, X_2^2) + \mathrm{Var}(X_1^2). \end{aligned}$$

Here, in the case $\mathrm{cov}(X_1^2, X_2^2) \ge 0$, the last sum is maximized for equal coefficients, and using (2.4), we then get

$$\mathrm{Var}(\theta_1 X_1^2 + \dots + \theta_n X_n^2) \le (n-1)\mathrm{cov}(X_1^2, X_2^2) + \mathrm{Var}(X_1^2) = \sigma_4^2(X).$$

Hence, (2.8) implies (2.6). In the case where $\mathrm{cov}(X_1^2, X_2^2) \le 0$, we have similarly

$$\mathrm{Var}(\theta_1 X_1^2 + \dots + \theta_n X_n^2) \le -\mathrm{cov}(X_1^2, X_2^2) + \mathrm{Var}(X_1^2) = \mathbb{E}X_1^4 - \mathbb{E}X_1^2 X_2^2,$$

which means that $V_2(X) \le \mathbb{E}X_1^4 - \mathbb{E}X_1^2 X_2^2$. Hence, by (2.8),

$$\Lambda(X) \le 2\,\mathbb{E}X_1^2 X_2^2 + V_2(X) \le \mathbb{E}X_1^2 X_2^2 + \mathbb{E}X_1^4 \le 2\,\mathbb{E}X_1^4.$$

Hence, (2.6) follows in this case as well, without the $\sigma_4^2(X)$-functional.

Proposition 2.3.3 is proved. □

2.4 Logarithmically Concave Measures

A positive Borel measure μ on $\mathbb{R}^n$ is called logarithmically concave or log-concave, if for all $t, s > 0$, $t + s = 1$, and all non-empty compact subsets A, B of $\mathbb{R}^n$,

$$\mu(tA + sB) \ge \mu(A)^t \,\mu(B)^s. \tag{2.9}$$

Here,

$$tA + sB = \{tx + sy : x \in A,\ y \in B\}$$

denotes the Minkowski average of sets, which is compact for compact A and B. In this case, (2.9) extends to all non-empty Borel sets in $\mathbb{R}^n$ (and then it is safer to use the lower measure μ_* on the left-hand side). If a random vector X in $\mathbb{R}^n$ has a log-concave distribution, one also says that X is log-concave.

Log-concave measures form an important class of a great interest, especially in Convex Geometry. They possess a number of remarkable properties resembling the behavior of product measures, and here we briefly mention a few of them.

a) The property of being log-concave is dimension free: If μ is log-concave on $\mathbb{R}^n$ viewed as a linear subset of $\mathbb{R}^m$ ($m \geq n$), then μ is log-concave on $\mathbb{R}^m$.
b) Stability under linear transformations: If $\nu = \mu T^{-1}$ is the image of a log-concave measure μ on $\mathbb{R}^n$ under a linear map $T : \mathbb{R}^n \to \mathbb{R}^m$, then the inequality (2.9) continues to hold for ν on $\mathbb{R}^m$. In particular, if a random vector $X = (X_1, \dots, X_n)$ is log-concave, then all components X_i, and moreover – all linear functionals $\langle X, v \rangle$ are log-concave on the real line.
c) The product $\mu = \mu_1 \otimes \cdots \otimes \mu_k$ of log-concave measures μ_i on $\mathbb{R}^{n_i}$ is log-concave on $\mathbb{R}^n$ of dimension $n = n_1 + \cdots + n_k$. Combining this property with *b*), we also obtain that the class of log-concave measures is preserved under convolutions.
d) The support $K = \mathrm{supp}(\mu)$ of a log-concave probability measure μ on $\mathbb{R}^n$ is a closed convex set. In particular, any affine subspace of $\mathbb{R}^n$ has μ-measure 0 or 1 (zero-one law). By the dimension of μ one means the dimension of K.

Proposition 2.4.1 (C. Borell) *Any log-concave probability measure μ on $\mathbb{R}^n$ is absolutely continuous with respect to the Lebesgue measure λ_K on the support set K and has density $f = \mathrm{e}^{-V}$ for some convex function $V : K \to (-\infty, \infty]$. Conversely, if V is convex, the equality $\mathrm{d}\mu = \mathrm{e}^{-V}\,\mathrm{d}\lambda_K$ defines a log-concave measure.*

The proof of the existence of densities for log-concave probability measures may be found in [63], [64]. In dimension one, it is however simple. Assume that the distribution of X is not a delta-measure, so that $a = \operatorname{ess\,inf} X < b = \operatorname{ess\,sup} X$, and introduce the distribution function $F(x) = \mathbb{P}\{X \leq x\}$, $x \in \mathbb{R}$. Applying (2.9) to the half-axes, we obtain that $\log F$ and $\log(1 - F)$ are concave functions on (a, b). This implies that these functions are locally absolutely continuous, and thus F has a density on (a, b). Moreover, the distribution of X may not have a positive mass at a, or at b.

Note that the density f in Proposition 2.4.1 may be defined according to the Lebesgue differentiation theorem by

$$f(x) = \liminf_{\varepsilon \downarrow 0} \frac{\mu(B(x,\varepsilon))}{\lambda_K(B(x,\varepsilon))}, \quad x \in \mathbb{R}^n,$$

where $B(x, \varepsilon)$ denotes the Euclidean ball of radius ε centered at x. Hence, by (2.9) applied to the balls, we have, for all $t, s > 0$ ($t + s = 1$) and $x, y \in \mathbb{R}^n$,

$$f(tx + sy) \geq f(x)^t f(y)^s.$$

That is, $\log f$ must be a concave function, necessarily finite in the relative interior $\mathrm{int}(K)$. In other words, f must be log-concave. In particular, a probability measure

μ on $\mathbb{R}^n$ which is absolutely continuous with respect to the n-dimensional Lebesgue measure is log-concave if and only if it has a log-concave density.

For example, the Lebesgue measure $\mu = \lambda$ on $\mathbb{R}^n$ and its restriction $\mu = \lambda_K$ to any convex body K in $\mathbb{R}^n$ are log-concave. This also follows from the definition (2.9) by applying the Brunn–Minkowski inequality

$$\lambda(tA + sB) \geq \left(t\lambda(A)^{1/n} + s\lambda(B)^{1/n}\right)^n, \tag{2.10}$$

where A and B may be arbitrary Borel sets in $\mathbb{R}^n$. Other interesting examples include Gaussian measures, exponential distributions, and their normalized restrictions to arbitrary convex bodies.

The sufficiency part in the characterization of log-concave measures may be established by several methods. One of them is the classical bisection method going back to Hadwiger ([73], [63]). A simpler approach relies upon the Prékopa–Leindler theorem, which is natural to include here (cf. also [157], [68]).

Proposition 2.4.2 *Let $t, s > 0$, $t + s = 1$, and let $u, v, w : \mathbb{R}^n \to [0, \infty)$ be Borel measurable functions satisfying*

$$w(tx + sy) \geq u(x)^t v(y)^s \quad \textit{for all } x, y \in \mathbb{R}^n. \tag{2.11}$$

Then

$$\int w(z)\,\mathrm{d}z \geq \left(\int u(x)\,\mathrm{d}x\right)^t \left(\int v(y)\,\mathrm{d}y\right)^s. \tag{2.12}$$

Proof An immediate induction argument reduces the assertion to dimension one. If $n \geq 2$, fix $x, y, z \in \mathbb{R}^{n-1}$ such that $tx + sy = z$ and consider the functions $u_1(x') = u(x, x')$, $v_1(y') = v(y, y')$, $w_1(z') = w(z, z')$ in variables $x', y', z' \in \mathbb{R}$. Since they satisfy (2.11) on the real line, we obtain (2.12), that is, the relation

$$\tilde{w}(tx + sy) \geq \tilde{u}(x)^t\, \tilde{v}(y)^s, \quad x, y \in \mathbb{R}^{n-1}, \tag{2.13}$$

for the functions

$$\tilde{u}(x) = \int_{-\infty}^{\infty} u(x, x')\,\mathrm{d}x', \quad \tilde{v}(y) = \int_{-\infty}^{\infty} v(y, y')\,\mathrm{d}y',$$
$$\tilde{w}(z) = \int_{-\infty}^{\infty} w(z, z')\,\mathrm{d}z'.$$

The relation (2.13) is of the form (2.11), so that we may apply the induction hypothesis in dimension $n - 1$ to obtain (2.12) for $(\tilde{u}, \tilde{v}, \tilde{w})$, which, by Fubini's theorem, is the same as (2.12) in dimension n for the original functions u, v, w.

Now, let $n = 1$. By a simple truncation, one may assume that u and v are bounded. Moreover, since both (2.11) and (2.12) do not change when the involved functions are multiplied by positive constants, let $\operatorname{ess\,sup} u = \operatorname{ess\,sup} v = 1$. In this case, the sets

$$A(\alpha) = \{x \in \mathbb{R} : u(x) > \alpha\}, \quad B(\alpha) = \{y \in \mathbb{R} : v(y) > \alpha\},$$
$$C(\alpha) = \{z \in \mathbb{R} : w(z) > \alpha\}$$

are non-empty for all $0 < \alpha < 1$. The hypothesis (2.11) yields the set inclusion $tA(\alpha) + sB(\alpha) \subset C(\alpha)$, so, we may apply the one dimensional Brunn–Minkowski inequality (2.10). It gives

$$\begin{aligned}\int_{-\infty}^{\infty} w(z)\,dz &= \int_0^{\infty} \lambda(C(\alpha))\,d\alpha \geq \int_0^1 \lambda(C(\alpha))\,d\alpha \\ &\geq t\int_0^1 \lambda(A(\alpha))\,d\alpha + s\int_0^1 \lambda(B(\alpha))\,d\alpha \\ &= t\int_{-\infty}^{\infty} u(x)\,dx + s\int_{-\infty}^{\infty} v(y)\,dy \geq \left(\int_{-\infty}^{\infty} u(x)\,dx\right)^t \left(\int_{-\infty}^{\infty} v(y)\,dy\right)^s,\end{aligned}$$

which is the desired conclusion (2.12) for $n = 1$. □

Let us also note that the inequality (2.10) in dimension one is elementary. Indeed, let A and B be compact sets such that $\sup A = \inf B = 0$ (otherwise, one may properly shift these sets, since (2.10) is shift-invariant). Then $(tA) \cup (sB) \subset tA + sB$, $(tA) \cap (sB) = \{0\}$, and thus, by the additivity, $\lambda(tA + sB) \geq t\lambda(A) + s\lambda(B)$.

As for higher dimensions, the Prékopa–Leindler theorem actually extends the Brunn–Minkowski inequality. The hypothesis (2.11) is fulfilled for the indicator functions $u = 1_A$, $v = 1_B$, $w = 1_{tA+sB}$, and (2.12) then leads to the multiplicative (log-concave) variant of (2.10),

$$\lambda(tA + sB) \geq \lambda(A)^t\, \lambda(B)^s.$$

Being applied with weights $t = \frac{a}{a+b}$, $s = \frac{b}{a+b}$, where $a = \lambda(A)^{1/n}$, $b = \lambda(B)^{1/n}$, and with the sets $A' = \frac{1}{a}A$, $B' = \frac{1}{b}B$, the above relation yields $\lambda(A + B) \geq (a + b)^n$, which is an equivalent variant of (2.10).

More generally, given a convex function V on an open convex domain in $\mathbb{R}^n$ and non-empty Borel sets $A, B \subset \mathbb{R}^n$, the hypothesis (2.11) is fulfilled for the functions $u = e^{-V} 1_A$, $v = e^{-V} 1_B$, $w = e^{-V} 1_{tA+sB}$, and then (2.12) yields the inequality (2.9) for the measure $d\mu(x) = e^{-V(x)}\,dx$. This means that μ is log-concave.

One immediate consequence of the Prékopa–Leindler theorem (Proposition 2.4.2) is the following useful observation due to Prékopa [160].

Corollary 2.4.3 *If $f(x, y)$ is an integrable log-concave function on $\mathbb{R}^n \times \mathbb{R}^m$, then the function*

$$g(x) = \int_{\mathbb{R}^m} f(x, y)\,dy$$

is log-concave on $\mathbb{R}^n$.

Indeed, the measure μ with density f is log-concave on $\mathbb{R}^n \times \mathbb{R}^m$. Hence, by property b), its image μ' under the projection $T(x, y) = x$ represents a log-concave measure on $\mathbb{R}^n$. It is absolutely continuous with respect to the Lebesgue measure, and the function g appears as the density of μ'. Therefore, g is log-concave.

One remarkable consequence of Corollary 2.4.3 is that the convolution of any two integrable log-concave functions on $\mathbb{R}^n$ is log-concave as well. This property was discovered by Davidovich, Korenbljum and Khacet, with a rather straightforward elementary argument, cf. [78].

Now, let us briefly comment on the one-dimensional case (to which we will return later on). According to Borell's characterization, a probability measure μ on the real line is log-concave if and only if it is a delta-measure (mass point) or if it has a log-concave density f supported on some open interval (a, b), finite or not. In particular, it must be unimodal: f is non-increasing on $(a, x_0]$ and is non-decreasing on $[x_0, b)$ for some $x_0 \in [a, b]$. The latter can be seen from another useful characterization.

With any probability measure μ which is supported on (a, b) and has there an almost everywhere positive density f, one may associate the function

$$I_\mu(p) = f(F^{-1}(p)), \quad 0 < p < 1, \tag{2.14}$$

where $F^{-1} : (0, 1) \to (a, b)$ is the inverse to the distribution function $F(x) = \mu((a, x))$, $a < x < b$, which is obviously increasing on (a, b). Note that, up to a shift, any positive continuous function I on $(0, 1)$ defines a unique probability measure μ with the associated function $I = I_\mu$, for example, via the identity

$$F^{-1}(p) = m + \int_{1/2}^{p} \frac{dt}{I(t)},$$

where m is a median for μ.

For example, for the two-sided exponential distribution ν with density $\frac{1}{2}\,e^{-|x|}$, the associated function is given by $I_\nu(p) = \min\{p, 1 - p\}$.

Borell's characterization yields the following useful description.

Proposition 2.4.4 *A non-degenerate probability measure μ on the real line is log-concave if and only if it is supported on an open interval, finite or not, and has there a positive continuous density, such that the function I_μ is concave on* $(0, 1)$.

2.5 Khinchine-type Inequalities for Norms and Polynomials

The density f of an absolutely continuous log-concave probability measure μ on $\mathbb{R}^n$ must decay exponentially fast at infinity, and one may easily derive an upper bound $f(x) \leq C\,e^{-c\,|x|}$ with positive constants c and C which do not depend on x. Therefore, linear functionals over log-concave probability measures have a finite exponential moment. Moreover, the L^p-moments turn out to be equivalent to each other, so that Khinchine-type inequalities hold true. In fact, there is a more general observation due to Borell [63] about the L^p-norms

$$\|X\|_p = (\mathbb{E}\,\|X\|^p)^{1/p}$$

for an arbitrary norm $\|\cdot\|$ on $\mathbb{R}^n$.

Proposition 2.5.1 *If a random vector X in $\mathbb{R}^n$ has a log-concave distribution, then*

$$\|X\|_p \le Cp\,\mathbb{E}\,\|X\|, \quad p \ge 1, \tag{2.15}$$

for some positive absolute constant C.

Proof We follow a simple argument by Borell. Putting $A = \{x \in \mathbb{R}^n : \|x\| < r\}$ and $B = \mathbb{R}^n \setminus \lambda A$ ($\lambda > 1$), we have a set inclusion

$$\frac{\lambda - 1}{\lambda + 1}\,A + \frac{2}{\lambda + 1}\,B \subset \mathbb{R}^n \setminus A.$$

Hence, by (2.9),

$$1 - \mu(A) \ge \mu(A)^{\frac{\lambda-1}{\lambda+1}}\,(1 - \mu(\lambda A))^{\frac{2}{\lambda+1}},$$

or equivalently,

$$\mathbb{P}\{\|X\| \ge \lambda r\} \le \alpha\left(\frac{1-\alpha}{\alpha}\right)^{\frac{\lambda+1}{2}},$$

where $\alpha = \mathbb{P}\{\|X\| < r\}$. Since (2.15) is homogeneous with respect to X, we may assume that $r = 1$ is a quantile of $\|X\|$ of order $\alpha = 2/3$. This leads to

$$\mathbb{P}\{\|X\| \ge \lambda\} \le \frac{1}{3}\,2^{-\frac{\lambda-1}{2}}, \quad \lambda \ge 1.$$

By direct integration,

$$\begin{aligned}\mathbb{E}\,\|X\|^p\,1_{\{\|X\| \ge 1\}} &= \frac{1}{3} + \int_1^\infty \mathbb{P}\{\|X\| \ge \lambda\}\,d\lambda^p \le \frac{1}{3} + \frac{1}{3}\int_1^\infty 2^{-\frac{\lambda-1}{2}}\,d\lambda^p \\ &\le \frac{1}{3}\int_0^\infty 2^{-\frac{\lambda-1}{2}}\,d\lambda^p = \frac{\sqrt{2}}{3}\left(\frac{2}{\log 2}\right)^p \Gamma(p+1).\end{aligned}$$

Using $\Gamma(p+1) \le p^p$, we get that

$$\mathbb{E}\,\|X\|^p \le 1 + \mathbb{E}\,\|X\|^p\,1_{\{\|X\| \ge 1\}} \le 1 + \left(\frac{2p}{\log 2}\right)^p \le \left(1 + \frac{2}{\log 2}\right)^p p^p,$$

so that $(\mathbb{E}\,\|X\|^p)^{1/p} \le (1 + \frac{2}{\log 2})\,p$. Finally, since $\frac{1}{3} \le \mathbb{P}\{\|X\| \ge 1\} \le \mathbb{E}\,\|X\|$, we arrive at the conclusion of the proposition with constant $C = 3\,(1 + \frac{2}{\log 2}) < 12$. $\square$

Corollary 2.5.2 *If a random variable ξ has a log-concave distribution, then*

$$(\mathbb{E}\,|\xi|^p)^{1/p} \le Cp\,\mathbb{E}\,|\xi|, \quad p \ge 1, \tag{2.16}$$

for some absolute constant C. Hence, if a random vector X in $\mathbb{R}^n$ has a log-concave distribution, then

$$M_p(X) \le Cp\,M_1(X).$$

The second assertion uses the property that all linear functionals of X represent log-concave random variables.

In fact, to derive (2.16), elementary convexity-like arguments may be sufficient, and here we mention one of them. First assume that the distribution of ξ is symmetric about zero and is not a mass point. Suppose for normalization that the quantile κ_α for the distribution function $F(x) = \mathbb{P}\{|\xi| \le x\}$ of order $\alpha \in (0,1)$ satisfies $\kappa_\alpha = -\log(1-\alpha)$. At this step, we only use the weaker property that $S(x) = -\log(1-F(x))$ is a convex function on $[0,\infty)$.

Let $b = \sup\{x \ge 0 : F(x) < 1\}$ and let $T = S^{-1} : [0,\infty) \to [0,b)$ be the inverse function, which is thus increasing and concave. Note that $T(\kappa_\alpha) = \kappa_\alpha$ and $|\xi| = T(\eta)$, where the random variable η has a standard exponential distribution with density e^{-x} $(x \ge 0)$. Let $l = l(x)$ be an affine function majorizing T and such that $l(\kappa_\alpha) = T(\kappa_\alpha)$. Equivalently, for some $c \in \mathbb{R}$,

$$l(x) = T(\kappa_\alpha) + c(x - \kappa_\alpha) = cx + (1-c)\kappa_\alpha, \qquad x \ge 0.$$

Since $l(0) \ge 0$, we necessarily have $0 \le c \le 1$, and by the convexity of the power function, $l(x)^p \le cx^p + (1-c)\kappa_\alpha^p$. Hence

$$\begin{aligned}\mathbb{E}\,|\xi|^p &= \mathbb{E}\,T(\eta)^p \le \mathbb{E}\,l(\eta)^p \le c\,\mathbb{E}\,\eta^p + (1-c)\kappa_\alpha^p \\ &= c\,\Gamma(p+1) + (1-c)\kappa_\alpha^p \le \max\{\Gamma(p+1), \kappa_\alpha^p\}.\end{aligned}$$

It remains to relate the quantile to the first absolute moment of ξ. Applying Markov's inequality, we have $1 - \alpha = \mathbb{P}\{|\xi| \ge \kappa_\alpha\} \le \frac{1}{\kappa_\alpha}\,\mathbb{E}\,|\xi|$, hence

$$\mathbb{E}\,|\xi|^p \le \frac{\max\{\Gamma(p+1), \kappa_\alpha^p\}}{((1-\alpha)\kappa_\alpha)^p}\,(\mathbb{E}\,|\xi|)^p.$$

This inequality is homogeneous with respect to ξ of order p, and therefore one may drop the normalization condition. Choosing $\alpha = 1 - 1/e$, we arrive at

$$\mathbb{E}\,|\xi|^p \le \mathrm{e}^p\,\Gamma(p+1)\,(\mathbb{E}\,|\xi|)^p.$$

Since $\Gamma(p+1)^{1/p} \le p$, we obtain the inequality (2.16) with constant $C = \mathrm{e}$.

To remove the symmetry assumption, assume that $\mathbb{E}\xi = 0$ and apply the previous step to $\xi - \xi'$, where ξ' is an independent copy of ξ. By Jensen' inequality, we get

$$\begin{aligned}\mathbb{E}\,|\xi|^p &\le \mathbb{E}\,|\xi - \xi'|^p \\ &\le \mathrm{e}^p\,\Gamma(p+1)\,(\mathbb{E}\,|\xi - \xi'|)^p \le (2e)^p\,\Gamma(p+1)\,(\mathbb{E}\,|\xi|)^p.\end{aligned}$$

Finally, in the general one-dimensional case, write $\xi = \eta + \mathbb{E}\xi$. Then $\|\eta\|_1 \le 2\,\|\xi\|_1$, and, by the previous step applied to η,

$$\begin{aligned}\|\xi\|_p &\le \|\eta\|_p + \|\xi\|_1 \\ &\le 2e\,\Gamma(p+1)^{1/p}\,\|\eta\|_1 + \|\xi\|_1 \le (4e\,\Gamma(p+1)^{1/p} + 1)\,\|\xi\|_1,\end{aligned}$$

which yields the inequality (2.16) with constant $C = 4\mathrm{e} + 1$.

Another important feature of the log-concavity is that the L^1-norm on the right-hand side of (2.15) may be replaced with a smaller quantity, the so-called "L^0-norm" defined by

$$\|X\|_0 = \lim_{p \to 0} \|X\|_p = \exp\big\{\mathbb{E}\,\log\|X\|\big\}.$$

Introduce the distribution function

$$F(r) = \mathbb{P}\{\|X\| \le r\}, \quad r \ge 0.$$

The next Khinchine-type inequality is due to Latala [127]. We will obtain it as a consequence of a stronger property following an argument of [20].

Proposition 2.5.3 *If the random vector X has a log-concave distribution on $\mathbb{R}^n$, then the function $F(r)\,r^{-1/(2\mathrm{e})}$ is non-decreasing in the interval $0 < F(r) \le \frac{\mathrm{e}}{\mathrm{e}+1}$. In particular,*

$$\mathbb{E}\,\|X\| \le C\,\|X\|_0 \tag{2.17}$$

for some absolute constant C.

Proof Without loss of generality, assume that the distribution of X is not a mass point, so that the interval $\Delta = \{r : 0 < F(r) < 1\}$ is not empty. By the definition (2.9), the function $\log F(r)$ is concave in Δ, so $F(r)$ has a left continuous Radon–Nikodym derivative $f(r)$ on Δ. Moreover, using a simple approximation argument, we may assume that the function f is continuous on Δ.

For $r, s \in \Delta$, $s > r$, consider the set $A = \{x \in \mathbb{R}^n : r < \|x\| \le s\}$, and denote by $B = \{x \in \mathbb{R}^n : \|x\| \le 1\}$ the unit ball for the given norm. Since for $\lambda \in (0, \frac{1}{2}]$,

$$(1-\lambda)A + \lambda\,(rB) \subset \{x \in \mathbb{R}^n : (1-2\lambda)\,r < \|x\| \le (1-\lambda)\,s + \lambda r\},$$

we obtain from (2.9) that

$$F((1-\lambda)s + \lambda r) - F((1-2\lambda)r) \ge (F(s) - F(r))^{1-\lambda}\,F(r)^{\lambda}.$$

Here, there is an equality for $\lambda = 0$, and comparing the derivatives of both sides at $\lambda = 0$, we arrive at

$$f(s)(r-s) + 2f(r)r \ge (F(s) - F(r))\log\frac{F(r)}{F(s) - F(r)}.$$

Hence, since $r < s$,

$$2f(r)r \geq (F(s) - F(r)) \log \frac{F(r)}{F(s) - F(r)}.$$

Now, given $0 < \alpha < 1$ and $r \in \Delta$ such that $F(r) < \frac{1}{1+\alpha}$, one can choose $s \in \Delta$ so that $F(s) - F(r) = \alpha F(r)$, and then the above inequality yields

$$2f(r)r \geq \frac{\alpha}{\log(1/\alpha)}\, F(r),$$

that is, $\frac{f(r)}{F(r)} \geq \frac{p}{r}$ with constant $p = \frac{\alpha}{2\log(1/\alpha)}$. But this is equivalent to the property that the function $\log F(r) - \log(r^p)$ is non-decreasing on the interval $0 < F(r) < \frac{1}{1+\alpha}$. Choosing $\alpha = 1/\mathrm{e}$, which maximizes the power $p = p(\alpha)$, the first assertion of the proposition is thus proved.

Assuming for normalization that $\|X\|$ has a quantile of order $2/3$ at $\kappa = 1$, then $F(r) \leq F(1)\, r^{1/(2\mathrm{e})} = \frac{2}{3}\, r^{1/(2\mathrm{e})}$ for all $0 < r \leq 1$ (since $\frac{2}{3} \leq \frac{\mathrm{e}}{\mathrm{e}+1}$). Hence

$$\begin{aligned}\mathbb{E} \log \|X\| &= \int_0^\infty \log r \, \mathrm{d}F(r) \geq \int_0^1 \log r \, \mathrm{d}F(r)\\ &= -\int_0^1 \frac{F(r)}{r}\, \mathrm{d}r \geq -\frac{2}{3}\int_0^1 \frac{r^{1/(2\mathrm{e})}}{r}\, \mathrm{d}r = -\frac{4\mathrm{e}}{3}.\end{aligned}$$

Dropping the normalization condition on the quantile, we arrive at the relation $\|X\|_0 \geq \exp\{-\frac{4\mathrm{e}}{3}\}\,\kappa$. But, by Proposition 2.5.1, $\mathbb{E}\,\|X\| \leq c\kappa$, which has been derived with $c = 1 + \frac{2}{\log 2}$. Hence, the inequality (2.17) follows with constant $C = c\, \exp\{\frac{4\mathrm{e}}{3}\}$. Proposition 2.5.3 is thus proved. □

Similar dimension-free Khinchine-type inequalities continue to hold for polynomials of bounded degree in place of the norms. Let us mention without proof the following important result (which generalizes Corollary 2.5.2).

Proposition 2.5.4 *Suppose that the random vector X has a log-concave distribution on $\mathbb{R}^n$. For any polynomial Q in n real variables of degree d, and any $p \geq 1$,*

$$(\mathbb{E}\,|Q(X)|^p)^{1/p} \leq C_{p,d}\, \mathbb{E}\,|Q(X)|$$

with some constant $C_{p,d}$ depending on p and d only.

One immediate consequence of this result concerns the variance-type functionals

$$\sigma_{2p}(X) = \sqrt{n}\,\Big(\mathbb{E}\,\Big|\frac{|X|^2}{n} - 1\Big|^p\Big)^{1/p}, \quad p \geq 1.$$

Since $Q(x) = |x|^2$ is a quadratic polynomial on $\mathbb{R}^n$, Proposition 2.5.4 yields:

Corollary 2.5.5 *If a random vector X in $\mathbb{R}^n$ has a log-concave distribution, then*

$$\sigma_{2p}(X) \leq C_p\, \sigma_2(X)$$

for some constant C_p depending on p only.

This relation holds regardless of the condition $\mathbb{E}\,|X|^2 = n$ which was used in the definition of $\sigma_{2p}(X)$.

If X is log-concave and isotropic, then $M_2(X) = 1$ and all moment functionals $M_p(X)$ are bounded by constants depending on p only (Corollary 2.5.2). Hence, by Propositions 1.5.1 and 1.5.3, we also have the relations

$$\mathrm{Var}(|X|) \leq \sigma_4^2(X) \leq 16\,\mathrm{Var}(|X|) + C$$

for some positive absolute constant C.

2.6 One-dimensional Log-concave Distributions

The description of log-concavity in dimension one, given in Proposition 2.4.4, may be used to establish a number of relations between various moment-like functionals and the value of the density f at the median m. In particular, we have:

Proposition 2.6.1 *Given a random variable ξ with a log-concave density f,*

$$\frac{1}{12\,\mathrm{Var}(\xi)} \leq f(m)^2 \leq \frac{1}{2\,\mathrm{Var}(\xi)}.$$

An equality in the first inequality is achieved for the uniform distribution on any finite interval, while an equality in the second inequality is achieved for the two-sided exponential distribution ν.

Proof Let μ be the distribution of ξ. As before, we denote by $F(x) = \mathbb{P}\{\xi \leq x\}$, $x \in \mathbb{R}$, the distribution function of ξ, and by $F^{-1} : (0,1) \to (a,b)$ its inverse, where (a,b) is the supporting interval for μ. By the definition (2.14) of the I_μ-function,

$$F^{-1}(p) - F^{-1}(q) = \int_q^p \frac{\mathrm{d}t}{I_\mu(t)}, \quad p, q \in (0,1),$$

while, by the concavity of I_μ,

$$I_\mu(p) \geq 2\, I_\mu\Big(\frac{1}{2}\Big)\, \min\{p, 1-p\} = 2\, f(m)\, I_\nu(p), \quad 0 < p < 1,$$

where ν has a two-sided exponential distribution with density $\frac{1}{2}\,\mathrm{e}^{-|x|}$ ($x \in \mathbb{R}$). Since the function F^{-1} has distribution μ under the Lebesgue measure on (0,1), we get

$$\begin{aligned}\mathrm{Var}(\xi) &= \frac{1}{2}\int_0^1\int_0^1 (F^{-1}(p) - F^{-1}(q))^2 \, \mathrm{d}p \, \mathrm{d}q \\ &= \frac{1}{2}\int_0^1\int_0^1 \left(\int_q^p \frac{\mathrm{d}t}{I_\mu(t)}\right)^2 \mathrm{d}p \, \mathrm{d}q \leq \frac{1}{2}\int_0^1\int_0^1 \left(\int_q^p \frac{\mathrm{d}t}{2f(m)\, I_\nu(t)}\right)^2 \mathrm{d}p \, \mathrm{d}q.\end{aligned}$$

For the same reason, if a random variable η has distribution ν, we have

$$\mathrm{Var}(\eta) = \frac{1}{2}\int_0^1\int_0^1 \left(\int_q^p \frac{\mathrm{d}t}{I_\nu(t)}\right)^2 \mathrm{d}p \, \mathrm{d}q.$$

Hence

$$\mathrm{Var}(\xi) \leq \frac{1}{(2f(m))^2}\,\mathrm{Var}(\eta) = \frac{1}{2f(m)^2},$$

thus proving the second inequality of the proposition.

To prove the first inequality, we should estimate the function I_μ from above. With this aim, let us again note that, by the concavity of I_μ, for some $c \in \mathbb{R}$,

$$I_\mu(p) \leq l_c(p) \equiv I_\mu\Big(\frac{1}{2}\Big) + c\left(p - \frac{1}{2}\right) = f(m) + c\left(p - \frac{1}{2}\right), \quad 0 < p < 1.$$

Because I_μ is non-negative, we have a restriction $|c| \leq 2f(m)$. From this,

$$\begin{aligned}\mathrm{Var}(\xi) &= \frac{1}{2}\int_0^1\int_0^1 \left(\int_q^p \frac{\mathrm{d}t}{I_\mu(t)}\right)^2 \mathrm{d}p \, \mathrm{d}q \\ &\geq \frac{1}{2}\int_0^1\int_0^1 \left(\int_q^p \frac{\mathrm{d}t}{l_c(t)}\right)^2 \mathrm{d}p \, \mathrm{d}q \equiv u(c).\end{aligned}$$

The function $c \to 1/l_c(t)$ is positive and convex, and so is the function u. Since the latter is also symmetric around zero, we get $u(c) \geq u(0)$ for all admissible c. Therefore,

$$\begin{aligned}\mathrm{Var}(\xi) &\geq \frac{1}{2}\int_0^1\int_0^1 \left(\int_q^p \frac{\mathrm{d}t}{l_0(t)}\right)^2 \mathrm{d}p \, \mathrm{d}q \\ &= \frac{1}{2f(m)^2}\int_0^1\int_0^1 (p-q)^2 \, \mathrm{d}p \, \mathrm{d}q = \frac{1}{12\, f(m)^2},\end{aligned}$$

where the last expression describes the variance of the uniform distribution.

Proposition 2.6.2 follows. □

Here is another interesting application.

Proposition 2.6.2 *Given a random variable ξ with a log-concave distribution,*

$$\frac{1}{\mathrm{e}} \leq \mathbb{P}\{\xi \leq \mathbb{E}\xi\} \leq 1 - \frac{1}{\mathrm{e}}. \tag{2.18}$$

Both inequalities are sharp, since on the right-hand side there is equality for ξ having a standard exponential distribution. The left inequality may be viewed as a "log-concave" or limit version of the known fact that for any convex body K in $\mathbb{R}^n$ and any half-space with boundary passing through the centroid of K,

$$\mathrm{Vol}_n(K \cap H) \geq \frac{1}{\mathrm{e}} \, \mathrm{Vol}_n(K).$$

In the space of a fixed dimension n the factor $1/\mathrm{e}$ can be replaced with $(\frac{n}{n+1})^n$. This property was first observed by Grünbaum and Hammer with a proof based on the Schwarz symmetrization, cf. [104].

Proof We use the same notations as in the proof of Proposition 2.6.1. First note that, for any positive concave function $J(p)$ in $0 < p < 1$, the function $J(p)/p$ is non-increasing, hence $J(p)/(1-p)$ is non-decreasing. Using the concavity of the function $I = I_\mu$, we conclude that, for any $p_0 \in (0,1)$,

$$\frac{I(p)}{1-p} \leq \frac{I(p_0)}{1-p_0} \ \text{for } p \in (0, p_0] \quad \text{and} \quad \frac{I(p)}{1-p} \geq \frac{I(p_0)}{1-p_0} \ \text{for } p \in [p_0, 1).$$

Hence, by Fubini's theorem,

$$\begin{aligned}
\mathbb{E}\xi = \int_0^1 F^{-1}(u)\,\mathrm{d}u &= \int_0^1 \Big[F^{-1}(p_0) + \int_{p_0}^u \frac{\mathrm{d}p}{I(p)}\Big]\,\mathrm{d}u \\
&= F^{-1}(p_0) + \int_{p_0}^1 \frac{1-p}{I(p)}\,\mathrm{d}p - \int_0^{p_0} \frac{p}{I(p)}\,\mathrm{d}p \\
&\leq F^{-1}(p_0) + \frac{1-p_0}{I(p_0)} \int_{p_0}^1 \mathrm{d}p - \frac{1-p_0}{I(p_0)} \int_0^{p_0} \frac{p}{1-p}\,\mathrm{d}p \\
&= F^{-1}(p_0) + \frac{(1-p_0)(1+\log(1-p_0))}{I(p_0)}.
\end{aligned}$$

Taking $p_0 = 1 - \frac{1}{\mathrm{e}}$, we arrive at $\mathbb{E}\xi \leq F^{-1}(1-\frac{1}{\mathrm{e}})$ which is equivalent to $F(\mathbb{E}\xi) \leq 1 - \frac{1}{\mathrm{e}}$. This is exactly the right inequality in (2.18). The left inequality follows from the right one, by applying it to the random variable $-\xi$. This proves the proposition. □

Another application deals with Khinchine-type inequalities for the moments of ξ with respect to the restricted measures.

Proposition 2.6.3 *If a random variable ξ defined on some Lebesgue probability space $(\Omega, \mathcal{F}, \mathbb{P})$ has a log-concave distribution, then for every measurable set $A \subset \Omega$,*

$$\int_A |\xi|\,\mathrm{d}\mathbb{P} \geq \frac{\mathbb{P}(A)^2}{4\sqrt{2}}\,\sqrt{\mathrm{Var}(\xi)}, \qquad \int_A \xi^2\,\mathrm{d}\mathbb{P} \geq \frac{\mathbb{P}(A)^3}{24}\,\mathrm{Var}(\xi). \tag{2.19}$$

Proof We use the previous notations, assuming that ξ has a log-concave density f. The assumption that $(\Omega, \mathcal{F}, \mathbb{P})$ is a Lebesgue space is understood in the sense of

Rokhlin, which means that $(\Omega, \mathcal{F}, \mathbb{P})$ can be transformed by virtue of a measure-preserving map to the interval $(0,1)$ with some Borel probability measure λ. Moreover, since the distribution of ξ has no atom, λ can be assumed to be the Lebesgue measure on $(0,1)$; an account on these spaces may be found in [161], [59].

Thus, let $\Omega = (0,1)$ be equipped with the Lebesgue measure $\mathbb{P}$, and let $\alpha = \mathbb{P}(A) > 0$. Putting $\xi(p) = F^{-1}(p)$, one may further assume that $0 < F(0) < 1$ (otherwise, ξ is either positive or negative, but then the inequalities in (2.19) can be made stronger by adding a constant to ξ). In this case, since the function F^{-1} is increasing and continuous, $F^{-1}(p_0) = 0$ for some $p_0 \in (0,1)$. Hence

$$|F^{-1}(p)| = \left| \int_{p_0}^{p} \frac{\mathrm{d}u}{I(u)} \right| \geq \frac{1}{\sup_u I(u)}\, |p - p_0|,$$

which yields

$$\int_A |\xi| \, \mathrm{d}\mathbb{P} = \int_A |F^{-1}(p)| \, \mathrm{d}p \geq \frac{1}{\sup_u I(u)} \int_A |p - p_0| \, \mathrm{d}p.$$

It should also be clear that within the class of all measurable sets $A \subset (0,1)$ of Lebesgue measure $\mathbb{P}(A) = \alpha$, the integral $\int_A |p - p_0| \, \mathrm{d}p$ attains its minimum at some interval (c,d) of length α containing p_0. The worst situation corresponds to the case $p_0 = \frac{1}{2}$ and $A = (\frac{1-\alpha}{2}, \frac{1+\alpha}{2})$, and then this integral is equal to $\alpha^2/4$. Hence

$$\int_A |\xi| \, \mathrm{d}\mathbb{P} \geq \frac{\alpha^2}{4 \sup_u I(u)}.$$

It remains to note that, by the concavity of the function I, we have

$$\sup_u |I(u)| \leq 2I(1/2) = 2f(m) \leq \sqrt{\frac{2}{\mathrm{Var}(\xi)}},$$

where we applied Proposition 2.6.1 in the last step.

By a similar argument,

$$\begin{aligned} \int_A \xi^2 \, \mathrm{d}\mathbb{P} &\geq \frac{1}{(\sup_u I(u))^2} \int_A (p - p_0)^2 \, \mathrm{d}p \\ &\geq \frac{1}{(\sup_u I(u))^2} \int_{\frac{1-\alpha}{2}}^{\frac{1+\alpha}{2}} \left(p - \frac{1}{2}\right)^2 \mathrm{d}p = \frac{\alpha^3}{12\,(\sup_u I(u))^2}, \end{aligned}$$

and then Proposition 2.6.1 gives the second inequality in (2.19). Note also that the first inequality implies the second one by applying the Cauchy inequality, but with a worse constant. □

Finally, let us quantify the assertion on the exponential decay of log-concave densities.

Proposition 2.6.4 *If a random variable ξ has a log-concave density f with expectation $a = \mathbb{E}\xi$ and standard deviation $\sigma = \sqrt{\mathrm{Var}(\xi)}$, then for all $x \in \mathbb{R}$,*

$$f(x) \le \frac{C}{\sigma}\,\mathrm{e}^{-|x-a|/\sigma\sqrt{12}},$$

where C is a universal constant.

Proof Let m be a median of ξ. First we show that

$$f(x) \le 2\mathrm{e}\, f(m)\,\mathrm{e}^{-f(m)\,|x-m|}, \quad x \in \mathbb{R}. \tag{2.20}$$

We may assume that $x > 0$ and $m = 0$ so that $\mathbb{P}\{\xi \ge 0\} = \int_0^\infty f(x)\,\mathrm{d}x = \frac{1}{2}$.

First consider the case $f(0)x \ge 1$. We use an upper bound

$$f(x) \le M(x) = \sup_u \mathrm{e}^{-u(x)},$$

where the supremum is taken over all convex functions $u : [0,\infty) \to (-\infty,\infty]$ such that $\mathrm{e}^{-u(0)} = f(0)$ and

$$\int_0^\infty \mathrm{e}^{-u(x)}\,\mathrm{d}x \le \frac{1}{2}. \tag{2.21}$$

By the convexity, for any such u, all these constraints will also be fulfilled for the function u_0 which is linear in $[0,x]$, is equal to ∞ on (x,∞), and which has values $u_0(0) = u(0)$, $u_0(x) = u(x)$. Hence, while computing $M(x)$, we may restrict ourselves to functions of the form $u_\alpha(y) = -\log f(0) + \alpha y$, $0 \le y \le x$ and $u_\alpha(y) = \infty$ for $y > x$. The range of α is determined by (2.21), that is, by the condition $f(0)\,\frac{1-\mathrm{e}^{-\alpha x}}{\alpha} \le \frac{1}{2}$. Since the function $\frac{1-\mathrm{e}^{-y}}{y}$ is positive and decreases on $\mathbb{R}$, we may conclude that

$$M(x) = f(0)\,\mathrm{e}^{-y} \quad \text{where } y \text{ is the solution of } \frac{1-\mathrm{e}^{-y}}{y} = \frac{1}{2f(0)x}.$$

But, by the assumption, $\frac{1}{2f(0)x} \le \frac{1}{2}$, so, $y \ge 1$, since otherwise $\frac{1-\mathrm{e}^{-y}}{y} \ge 1 - \mathrm{e}^{-1} > \frac{1}{2}$. Hence $\frac{1}{2f(0)x} = \frac{1-\mathrm{e}^{-y}}{y} \ge \frac{1}{2y}$ and thus $y \ge f(0)x$, implying

$$f(x) \le M(x) \le f(0)\,\mathrm{e}^{-f(0)x}.$$

To treat the case $f(0)x < 1$, just recall that, by the concavity of $I(t) = f(F^{-1}(t))$, we have $I(t) \le 2I(1/2) = 2f(0)$, implying $f(x) \le 2f(0) \le 2\mathrm{e}\, f(0)\,\mathrm{e}^{-f(0)x}$. In both cases, we obtain the desired estimate (2.20).

Now, starting from (2.20), we immediately get $|\mathbb{E}\xi - m| \le 4\mathrm{e}/f(m)$. Hence, applying (2.20) once more, we have

$$f(x) \le 2\mathrm{e}\cdot \mathrm{e}^{4e}\, f(m)\,\mathrm{e}^{-f(m)\,|x-a|}.$$

It remains to apply Proposition 2.6.1, connecting $f(m)$ with σ. □

2.7 Remarks

The material in Sections 2.1–2.3 is rather standard. Small ball probability bounds for independent summands like the one in Proposition 2.1.4 have been studied by many authors. Systems of pairwise independent random variables with similar deterministic properties were first constructed by Joffe [110]. The system $X_k(t,s) = \Psi(kt+s)$ as above was considered in [46].

Besides Borell's characterization of log-concave measures (in the sufficiency part of Proposition 2.4.1) and Prékopa's theorem (Corollary 2.4.3), the Prékopa–Leindler theorem is known to have numerous applications. As a simple example, one can start from a random vector $X = (X_1, \dots, X_n)$ in $\mathbb{R}_+^n$ with a log-concave density f and note that the condition (2.11) is satisfied by the functions

$$u(x) = \Big(\frac{x_1}{p_1}\Big)^{p_1} \dots \Big(\frac{x_n}{p_n}\Big)^{p_n} f(x), \quad v(y) = \Big(\frac{y_1}{q_1}\Big)^{q_1} \dots \Big(\frac{y_n}{q_n}\Big)^{q_n} f(y),$$
$$w(z) = \Big(\frac{z_1}{r_1}\Big)^{r_1} \dots \Big(\frac{z_n}{r_n}\Big)^{r_n} f(z),$$

as long as $p_k > r_k > q_k > 0$, $r_k = tp_k + sq_k$ $(t,s>0,\ t+s=1)$. The conclusion (2.12) of Proposition 2.4.2 is then equivalent to the statement that the function

$$g(p_1, \dots, p_n) = \mathbb{E}\,\Big(\frac{X_1}{p_1}\Big)^{p_1} \dots \Big(\frac{X_n}{p_n}\Big)^{p_n}$$

is log-concave on $\mathbb{R}_+^n$ (cf. [27]). In dimension $n = 1$, when X has a standard exponential distribution, we obtain in particular that the function $\Gamma(p+1)/p^p$ is log-concave in $p > 0$ (a property which was mentioned in the proof of Proposition 1.3.3).

Khinchine-type inequalities for norms as in Propositions 2.5.1 and 2.5.3 may further be strengthened to the form

$$\|X\|_p \leq C_{p,q}\, \|X\|_q, \qquad p > q > -1,$$

cf. Guédon [102]. In the class of one-dimensional symmetric log-concave distributions, best constants in Khinchine-type inequalities were found by Karlin, Proschan and Barlow [119] in their study of total positivity. Namely, they proved that

$$(\mathbb{E}\,|\xi|^p)^{1/p} \leq \Gamma(p+1)^{1/p}\, \mathbb{E}\,|\xi|, \quad p \geq 1,$$

as long as the random variable ξ has a symmetric log-concave distribution on the real line. Here, equality is attained for the two-sided exponential distribution.

This distribution plays an extremal role in some similar relations involving the values of the density f of ξ at zero. In particular, for the same class, K. Ball [10] proves that

$$f(0)^p\, \Gamma(q+1)^{p+1}\, (\mathbb{E}\,|\xi|^p)^{q+1} \leq f(0)^q\, \Gamma(p+1)^{q+1}\, (\mathbb{E}\,|\xi|^q)^{p+1}$$

whenever $0 \leq q < p < \infty$.

Relations between L^p-norms of ξ with 3 different values of p have been studied in the context of reverse Lyapunov inequalities, cf. e.g. Borell [62].

The description of log-concave probability distributions on the line as in Proposition 2.4.4 was emphasized in [18]. Let μ be a non-atomic probability measure with distribution function F which is continuous and strictly increasing on the supporting interval. In [18] it was shown that μ is log-concave if and only if, for any $h > 0$, the function $R_h(p) = F(F^{-1}(p) + h)$ is concave on $(0, 1)$. The material of Section 2.6 is based on the papers [20] and [25].

Khinchine-type inequalities for polynomials over convex bodies were studied by Bourgain [67]. The more general Proposition 2.5.4 was proved in [22], [23].

In this chapter, we did not discuss in detail the behavior of the variance functional

$$\sigma_4^2(X) = \frac{1}{n}\,\mathrm{Var}(|X|^2),$$

when the random vector X has a log-concave distribution on $\mathbb{R}^n$. The so-called thin shell (or variance) conjecture asserts that $\sigma_4(X)$ is bounded by an absolute constant in the entire class of all isotropic log-concave probability measures on $\mathbb{R}^n$, and regardless of the dimension n. In connection with the central limit theorem this conjecture was raised in [4] and [51]. Nevertheless, some positive results are known, and, for example, Proposition 6.6.2 from Chapter 6 will confirm the thin-shell conjecture for coordinatewise symmetric log-concave distributions (Klartag's theorem). In the general isotropic log-concave case, the first non-trivial bounds were obtained by Klartag [121], [122]. Later, Guédon and E. Milman [101] derived the estimate $\sigma_4(X) \leq cn^{1/3}$. Recently, Lee and Vempala [133], [134] developed Eldan's localization technique ([84]) and showed that $\sigma_4(X) \leq cn^{1/4}$. Chen [77] considerably improved this method deriving the estimate $\sigma_4(X) \leq cn^{\varepsilon_n}$ with a certain sequence $\varepsilon_n \to 0$. The next sharpening

$$\sigma_4(X) \leq c\,(\log n)^\alpha$$

with $\alpha = 4$ is due to Klartag and Lehec [124]. The preprint [108] contains a further sharpening with $\alpha = 2.2226$, which is the best known result so far.

As for lower bounds, it was shown in [51] that $\sigma_4^2(X) \geq \frac{4}{n+4}$ when the random vector X is uniformly distributed in a symmetric isotropic convex body (an equality is attained for the Euclidean ball).

Let us also mention two results about large and small ball probabilities for isotropic log-concave random vectors X in $\mathbb{R}^n$. It was shown by Paouris that, for some absolute constants $r_0 > 0$ and $c > 0$,

$$\mathbb{P}\{|X|^2 \leq rn\} \leq r^{c\sqrt{n}}, \quad 0 \leq r \leq r_0,$$

cf. [152], [69]. Whether or not one can sharpen this bound to

$$\mathbb{P}\{|X|^2 \leq r_0 n\} \leq \mathrm{e}^{-cn}$$

is equivalent to the so-called slicing problem (as was noticed by Bourgain, Klartag and V. Milman). The latter may equivalently be formulated as the property that the density f of X satisfies $\sup_x f(x) \geq c^n$ for some absolute constant $c > 0$ (still assuming that X has an isotropic log-concave distribution). As was shown by Eldan and Klartag [83], this would be indeed true if the thin shell conjecture were true. That the slicing conjecture is true for coordinatewise symmetric log-concave distributions is a well-known simple fact, cf. e.g. [58].

On the other hand, if X is isotropic and is uniformly distributed over a convex body $K \subset \mathbb{R}^n$, then we have a large deviation bound

$$\mathbb{P}\{|X| \geq cr\sqrt{n}\} \leq e^{-r\sqrt{n}}, \quad r \geq 1,$$

where $c > 0$ is an absolute constant (Paouris [151]). Thus, an essential part of mass in K is contained in the Euclidean ball of radius $c\sqrt{n}$. For coordinatewise symmetric convex bodies such bounds were previously studied in [58].

Chapter 3
Characteristic Functions

The moment functionals we discussed before may be explicitly expressed in terms of characteristic functions of linear functionals of a given random vector. However, information on various bounds on characteristic functions and their deviations from the characteristic function of another law on the real line will be needed for a different purpose – to study the Kolmogorov and Lévy distances between the corresponding distribution functions. In this chapter, we describe general tools in the form of smoothing and Berry–Esseen-type inequalities, which allow one to perform the transition from the results about closeness or smallness of Fourier–Stieltjes transforms to corresponding results about the associated functions of bounded variation.

3.1 Smoothing

For the Kolmogorov distance we use the standard notation

$$\rho(F,G) = \sup_x |F(x) - G(x)| = \|F - G\|,$$

where F and G may be arbitrary functions of bounded variation on the real line, and where $\|\cdot\|$ refers to the L^∞-norm. Denote their Fourier–Stieltjes transforms by

$$f(t) = \int_{-\infty}^{\infty} e^{itx}\, dF(x), \quad g(t) = \int_{-\infty}^{\infty} e^{itx}\, dG(x) \quad (t \in \mathbb{R}).$$

Proposition 3.1.1 *Let F be a non-decreasing bounded function, and let G be a function of bounded variation such that $F(-\infty) = G(-\infty) = 0$. For any $T > 0$,*

$$\rho(F,G) \le \int_{-T}^{T} \Big|\frac{f(t) - g(t)}{t}\Big|\, dt + T \sup_{x \in \mathbb{R}} \int_0^{\frac{2\pi}{T}} |G(x+u) - G(x)|\, du. \tag{3.1}$$

S. Bobkov et al., *Concentration and Gaussian Approximation for Randomized Sums*, Probability Theory and Stochastic Modelling 104,
https://doi.org/10.1007/978-3-031-31149-9_3

Proof Put $A = F - G$ and $\varepsilon = 1/T$. We use smoothing of the function A with the distribution function $H_\varepsilon(x) = H(x/\varepsilon)$. That is, consider the convolution $A_\varepsilon = A * H_\varepsilon$, defined by

$$A_\varepsilon(x) = \int_{-\infty}^{\infty} A(x-y)\,\mathrm{d}H_\varepsilon(y),$$

by choosing the canonical distribution function with density

$$p(x) = \frac{\mathrm{d}H(x)}{\mathrm{d}x} = \frac{1}{2\pi}\Big(\frac{\sin(x/2)}{x/2}\Big)^2 = \frac{1-\cos x}{\pi x^2}, \qquad x \in \mathbb{R}.$$

It has the triangle characteristic function $h(t) = (1-|t|)^+$, so that H_ε has the characteristic function $h_\varepsilon(t) = (1-\varepsilon|t|)^+$ supported on the interval $[-T, T]$.

We may assume that the first integral in (3.1) is convergent. Note that A_ε has the Fourier–Stieltjes transform $a_\varepsilon(t) = (f(t) - g(t))\,h_\varepsilon(t)$, which is supported on $[-T, T]$. Hence, the function A_ε is continuous and admits the representation according to the Fourier inversion formula for functions of bounded variation. That is, for all $x, y \in \mathbb{R}$, we have

$$A_\varepsilon(x) - A_\varepsilon(y) = \frac{1}{2\pi}\int_{-T}^{T} \frac{\mathrm{e}^{-itx} - \mathrm{e}^{-ity}}{-it}\,a_\varepsilon(t)\,\mathrm{d}t.$$

Here, since the function $a_\varepsilon(t)/t$ is integrable, necessarily $\int_{-T}^{T} \frac{\mathrm{e}^{-ity}}{it}\,a_\varepsilon(t)\,\mathrm{d}t \to 0$ as $y \to -\infty$ (by the Riemann–Lebesgue lemma). Since also

$$\lim_{y\to-\infty} A_\varepsilon(y) = \lim_{y\to-\infty} A(y) = F(-\infty) - G(-\infty) = 0,$$

we obtain that

$$A_\varepsilon(x) = \frac{1}{2\pi}\int_{-T}^{T} \frac{\mathrm{e}^{-itx}}{-it}\,a_\varepsilon(t)\,\mathrm{d}t, \quad x \in \mathbb{R}. \tag{3.2}$$

This equality implies that

$$|A_\varepsilon(x)| \le \frac{1}{2\pi}\int_{-T}^{T} \Big|\frac{a_\varepsilon(t)}{t}\Big|\,\mathrm{d}t \le \frac{1}{2\pi}\int_{-T}^{T} \Big|\frac{f(t)-g(t)}{t}\Big|\,\mathrm{d}t,$$

and since x is arbitrary,

$$\|A_\varepsilon\| \le \frac{1}{2\pi}\int_{-T}^{T} \Big|\frac{f(t)-g(t)}{t}\Big|\,\mathrm{d}t. \tag{3.3}$$

The next step is to properly bound $\|A_\varepsilon\|$ from below. Write

$$A_\varepsilon(x) = \int_{-\infty}^{\infty} A(x-\varepsilon y)\,\mathrm{d}H(y) = I_0(x) + I_1(x)$$

with

$$I_0(x) = \int_{|y| \le r/2} A(x - \varepsilon y)\,\mathrm{d}H(y), \quad I_1(x) = \int_{|y| > r/2} A(x - \varepsilon y)\,\mathrm{d}H(y),$$

where the parameter $r > 0$ will be chosen later on. Putting

$$\gamma = 1 - H\Big(\Big[-\frac{r}{2}, \frac{r}{2}\Big]\Big) = \int_{|x| \ge r/2} p(x)\,\mathrm{d}x,$$

where H is treated as measure, we have $|I_1(x)| \le \gamma \|A\|$ and thus $\|I_1\| \le \gamma \|A\|$.

To estimate the integral $I_0(x)$, introduce the quantity

$$\delta(h) = \sup_x \int_0^h |G(x+u) - G(x)|\,\mathrm{d}u, \qquad h > 0,$$

which appears on the right-hand side of (3.1). The monotonicity of F ensures that

$$\begin{aligned} I_0(x) &\ge \int_{|y| \le r/2} (F(x - \varepsilon r/2) - G(x - \varepsilon y))\,\mathrm{d}H(y) \\ &= (1-\gamma)\,(F(x - \varepsilon r/2) - G(x - \varepsilon r/2)) \\ &\quad + \int_{|y| \le r/2} (G(x - \varepsilon r/2) - G(x - \varepsilon y))\,p(y)\,\mathrm{d}y. \end{aligned}$$

Since $p(y) \le \frac{1}{2\pi}$, the last integral does not exceed in absolute value

$$\frac{1}{2\pi} \int_{|y| \le r/2} |G(x - \varepsilon r/2) - G(x - \varepsilon y)|\,\mathrm{d}y =$$

$$\frac{1}{2\pi\varepsilon} \int_0^{\varepsilon r} |G(x - \varepsilon r/2) - G(x - \varepsilon r/2 + u)|\,\mathrm{d}u \;\le\; \frac{1}{2\pi\varepsilon}\,\delta(\varepsilon r).$$

Hence

$$I_0(x) \ge (1-\gamma)\,A(x - \varepsilon r/2) - \frac{1}{2\pi\varepsilon}\,\delta(\varepsilon r).$$

By a similar argument, using the monotonicity of the function F, we also have

$$I_0(x) \le (1-\gamma)\,A(x + \varepsilon r/2) + \frac{1}{2\pi\varepsilon}\,\delta(\varepsilon r),$$

so that the two bounds yield

$$|I_0(x)| \ge (1-\gamma)\,\max\{A(x - \varepsilon r/2), -A(x + \varepsilon r/2)\} - \frac{1}{2\pi\varepsilon}\,\delta(\varepsilon r).$$

Taking here the supremum over all x on the left and then on the right, we arrive at

$$\|I_0\| \ge (1-\gamma)\,\|A\| - \frac{1}{2\pi\varepsilon}\,\delta(\varepsilon r).$$

But, by the triangle inequality,

$$|A_\varepsilon(x)| \geq |I_0(x)| - |I_1(x)| \geq |I_0(x)| - \gamma \|A\|,$$

so, $\|A_\varepsilon\| \geq \|I_0\| - \gamma \|A\|$ and thus

$$\|A_\varepsilon\| \geq (1 - 2\gamma)\, \|A\| - \frac{1}{2\pi\varepsilon}\, \delta(\varepsilon r).$$

This is a smoothing inequality: In the case $\gamma < \frac{1}{2}$, it allows one to estimate the Kolmogorov distance $\|A\| = \rho(F, G)$ between the original distributions in terms of the Kolmogorov distance $\|A_\varepsilon\| = \rho(F * H_\varepsilon, G * H_\varepsilon)$ between the smoothed distributions (at the expense of a potentially small error). Namely, returning to (3.3), the above inequality implies

$$(1 - 2\gamma)\, \|A\| - \frac{1}{2\pi\varepsilon}\, \delta(\varepsilon r) \leq \frac{1}{2\pi} \int_{-T}^{T} \left|\frac{f(t) - g(t)}{t}\right| \mathrm{d}t,$$

or equivalently

$$\|A\| \leq \frac{1}{2\pi\,(1 - 2\gamma)} \int_{-T}^{T} \left|\frac{f(t) - g(t)}{t}\right| \mathrm{d}t + \frac{1}{2\pi\,(1 - 2\gamma)}\, \frac{\delta(\varepsilon r)}{\varepsilon}. \tag{3.4}$$

For example, let us choose $r = 2\pi$, for which

$$\begin{aligned} \gamma &= 1 - H([-\pi, \pi]) \\ &= \int_{|x|>\pi} p(x)\, \mathrm{d}x = \frac{2}{\pi} \int_{\pi/2}^{\infty} \frac{\sin^2 x}{x^2}\, \mathrm{d}x < \frac{2}{\pi} \int_{\pi/2}^{\infty} \frac{1}{x^2}\, \mathrm{d}x = \frac{4}{\pi^2}. \end{aligned}$$

In this case,

$$\frac{1}{2\pi\,(1 - 2\gamma)} < \frac{\pi}{2\,(\pi^2 - 8)} = 0.8401\ldots$$

Hence, (3.4) yields the desired inequality (3.1). Proposition 3.1.1 is proved. □

3.2 Berry–Esseen-type Inequalities

If G is differentiable and has a bounded derivative on the real line, then (3.1) immediately yields a more familiar inequality due to Esseen (also called the Berry–Esseen inequality). Namely, from Proposition 3.1.1 we immediately get:

Proposition 3.2.1 *Let F be a non-decreasing bounded function, and let G be a differentiable function of bounded variation such that $F(-\infty) = G(-\infty) = 0$. If $|G'(x)| \le A$ for all x, then for any $T > 0$,*

$$\rho(F,G) \le \int_{-T}^{T} \left|\frac{f(t) - g(t)}{t}\right| \mathrm{d}t + \frac{2\pi^2 A}{T}. \tag{3.5}$$

Here, as before,

$$f(t) = \int_{-\infty}^{\infty} \mathrm{e}^{itx}\,\mathrm{d}F(x), \quad g(t) = \int_{-\infty}^{\infty} \mathrm{e}^{itx}\,\mathrm{d}G(x) \qquad (t \in \mathbb{R})$$

denote the Fourier–Stieltjes transforms of F and G, respectively. Indeed, under the assumptions on G, for any $x \in \mathbb{R}$,

$$T \int_0^{\frac{2\pi}{T}} |G(x+u) - G(x)|\,\mathrm{d}u \le AT \int_0^{\frac{2\pi}{T}} u\,\mathrm{d}u = \frac{2\pi^2 A}{T},$$

so that the last term in (3.1) is bounded by the last term in (3.5).

When G is a distribution function (not necessarily differentiable), one may also relate the second integral in (3.1) to the concentration function defined by

$$Q_G(h) = \sup_x \,(G(x+h) - G(x-)), \quad h \ge 0.$$

Since it is non-decreasing in h, (3.1) yields

$$\rho(F,G) \le \int_{-T}^{T} \left|\frac{f(t) - g(t)}{t}\right| \mathrm{d}t + 2\pi\, Q_G(2\pi/T), \tag{3.6}$$

which is more general than (3.5), since it may involve discontinuous functions G. The last term in (3.6) may be further bounded in terms of the characteristic function g of G by virtue of the following general relation of independent interest.

Lemma 3.2.2 *Given a distribution function G with characteristic function g, for any $\varepsilon > 0$,*

$$Q_G(\varepsilon) \le 2\left(\frac{96}{95}\right)^2 \varepsilon \int_0^{1/\varepsilon} |g(t)|\,\mathrm{d}t.$$

We also have

$$Q_G(2\pi\varepsilon) \le \frac{\pi^2}{2}\,\varepsilon \int_0^{1/\varepsilon} |g(t)|\,\mathrm{d}t.$$

Indeed, applying the last inequality with $\varepsilon = 1/T$ in (3.6), we arrive at:

Proposition 3.2.3 *If F and G are distribution functions, then for any $T > 0$,*

$$\rho(F,G) \leq \int_{-T}^{T} \left|\frac{f(t) - g(t)}{t}\right| \mathrm{d}t + \frac{\pi^3}{T} \int_0^T |g(t)| \, \mathrm{d}t.$$

Proof (of Lemma 3.2.2) Let us start with an obvious general identity

$$\int_{-\infty}^{\infty} \hat{u}(x) \, \mathrm{d}G(x) = \int_{-\infty}^{\infty} u(t)g(t) \, \mathrm{d}t, \tag{3.7}$$

which holds true for any integrable function u on the real line with Fourier transform

$$\hat{u}(x) = \int_{-\infty}^{\infty} \mathrm{e}^{itx} \, u(t) \, \mathrm{d}t.$$

Recall that the function $p(x) = \big(\frac{\sin(x/2)}{x/2}\big)^2$ appears as the Fourier transform of $q(t) = (1 - |t|)^+$. One may therefore apply (3.7) to $u(t) = q(\varepsilon t)$, in which case $\hat{u}(x) = \frac{1}{\varepsilon} \, p(\frac{x}{\varepsilon})$. Since both sides of (3.7) are real numbers, we then get

$$\int_{-\infty}^{\infty} u(t)g(t) \, \mathrm{d}t = \int_{-\infty}^{\infty} (1 - \varepsilon|t|)^+ g(t) \, \mathrm{d}t \leq \int_{-1/\varepsilon}^{1/\varepsilon} |g(t)| \, \mathrm{d}t.$$

On the other hand, for any $0 < \kappa < \pi$, the left integral in (3.7) is greater than or equal to

$$\begin{aligned} \frac{1}{\varepsilon} \int_{|x| \leq 2\kappa\varepsilon} \Big(\frac{\sin(x/2\varepsilon)}{x/2\varepsilon}\Big)^2 \mathrm{d}G(x) &\geq \frac{1}{\varepsilon} \Big(\frac{\sin \kappa}{\kappa}\Big)^2 \int_{|x| \leq 2\kappa\varepsilon} \mathrm{d}G(x) \\ &= \frac{1}{\varepsilon} \Big(\frac{\sin \kappa}{\kappa}\Big)^2 G([-2\kappa\varepsilon, 2\kappa\varepsilon]), \end{aligned}$$

where in the last step G is treated as a measure, and where we used the property that $\frac{\sin x}{x}$ is decreasing in $0 \leq x \leq \pi$. Hence

$$G([-2\kappa\varepsilon, 2\kappa\varepsilon]) \leq \Big(\frac{\kappa}{\sin \kappa}\Big)^2 \varepsilon \int_{-1/\varepsilon}^{1/\varepsilon} |g(t)| \, \mathrm{d}t.$$

Applying it to the shifted distribution functions $G(x + a)$ with arbitrary $a \in \mathbb{R}$, the above inequality remains valid for any interval of length $4\kappa\varepsilon$, so that

$$Q_G(4\kappa\varepsilon) \leq 2 \Big(\frac{\kappa}{\sin \kappa}\Big)^2 \varepsilon \int_0^{1/\varepsilon} |g(t)| \, \mathrm{d}t, \qquad 0 < \kappa < \pi, \ \ \varepsilon > 0.$$

Choosing here $\kappa = \frac{1}{4}$ and using $\frac{1}{(4\,\sin(1/4))^2} < \frac{96}{95}$, we obtain the first inequality of the lemma, and the choice $\kappa = \frac{\pi}{2}$ leads to the second one. □

3.3 Lévy Distance and Zolotarev's Inequality

Another important metric which metrizes the weak topology in the space of probability measures on the line (in contrast with the Kolmogorov metric) is given by the Lévy distance

$$L(F,G) = \inf\big\{h \geq 0 : \ G(x-h) - h \leq F(x) \leq G(x+h) + h \ \text{ for all } x \in \mathbb{R}\big\},$$

where F and G are arbitrary distribution functions. By continuity of F and G from the right, this infimum is always attained (so, it may be replaced with the minimum). In a more geometric language, L describes the size of the side of the largest square enclosed between the graphs of the two distribution functions (on the plane $\mathbb{R}^2$).

In general, $0 \leq L \leq 1$, and there is an elementary relation $L(F,G) \leq \rho(F,G)$ connecting it with the Kolmogorov distance. Conversely, if G has a bounded derivative satisfying $|G'(x)| \leq A$ for all $x \in \mathbb{R}$, we have an opposite bound

$$\rho(F,G) \leq (1+A)\,L(F,G).$$

Therefore, when the constant A is known to be not large, the problems of the estimation of the Kolmogorov and Lévy distances are in essence equivalent. Otherwise, the Berry–Esseen inequality (3.5) is not that satisfactory. As it turns out, when bounding the weaker distance $L(F,G)$ in terms of the associated characteristic functions

$$f(t) = \int_{-\infty}^{\infty} e^{itx}\,dF(x), \quad g(t) = \int_{-\infty}^{\infty} e^{itx}\,dG(x) \quad (t \in \mathbb{R}),$$

any "smoothness"-type condition imposed on G is irrelevant. This may be seen from the following statement due to Zolotarev (which we formulate and prove in a slightly different form).

Proposition 3.3.1 *Given distribution functions F and G, for any $T > 0$,*

$$L(F,G) \leq \frac{1}{\pi}\int_{-T}^{T} \left|\frac{f(t)-g(t)}{t}\right| dt + \frac{4\log(1+T)}{T}. \tag{3.8}$$

Proof For $x \in \mathbb{R}$, $h \geq 0$, it is sufficient to estimate the quantity

$$\mathcal{L}(x,h) = \max\{F(x-h) - G(x+h),\ G(x-h) - F(x+h)\},$$

in view of the implication

$$\sup_x \mathcal{L}(x,h) \leq b + 2h \implies L(F,G) \leq b + 2h \qquad (b \geq 0). \tag{3.9}$$

With this aim, one may use smoothing with the help of the normal distribution functions

$$\Phi_\varepsilon(x) = \frac{1}{\varepsilon\sqrt{2\pi}}\int_{-\infty}^{x} e^{-y^2/2\varepsilon^2}\,dy \quad (x \in \mathbb{R})$$

with a parameter $\varepsilon > 0$ which will be chosen as a function of T. Similarly to the proof of Proposition 3.1.1, for the convolutions $F_\varepsilon = F * \Phi_\varepsilon$ and $G_\varepsilon = G * \Phi_\varepsilon$, consider the deviations

$$I(x) = F_\varepsilon(x) - G_\varepsilon(x) = \int_{-\infty}^{\infty} (F(x - \varepsilon y) - G(x - \varepsilon y))\, d\Phi(y),$$

where $\Phi = \Phi_1$ is the standard normal distribution function. Assuming that the integral in (3.8) is finite, from the Fourier inversion formula (3.2) applied with $A_\varepsilon = F_\varepsilon - G_\varepsilon$, we have

$$|I(x)| \leq \frac{1}{2\pi} \int_{-\infty}^{\infty} \left|\frac{f(t) - g(t)}{t}\right| e^{-\varepsilon^2 t^2/2}\, dt. \tag{3.10}$$

Now, let us bound the integral $I(x)$ from below by splitting the integration into two parts. Namely, given $l > 0$, write $I(x) = I_0(x) + I_1(x)$ with

$$I_0(x) = \int_{|y| \leq l} (F(x - \varepsilon y) - G(x - \varepsilon y))\, d\Phi(y),$$
$$I_1(x) = \int_{|y| \geq l} (F(x - \varepsilon y) - G(x - \varepsilon y))\, d\Phi(y).$$

Clearly,

$$|I_1(x)| \leq 2\,(1 - \Phi(l)) \equiv \gamma. \tag{3.11}$$

On the other hand, using the monotonicity of both F and G,

$$I_0(x) \geq (F(x - \varepsilon l) - G(x + \varepsilon l))\,(1 - \gamma),$$
$$I_0(x) \leq (F(x + \varepsilon l) - G(x - \varepsilon l))\,(1 - \gamma),$$

which imply that

$$|I_0(x)| \geq (1 - \gamma)\, \mathcal{L}(x, \varepsilon l).$$

Since $|I_0(x)| \leq |I(x)| + |I_1(x)|$, (3.10)–(3.11) yield

$$\sup_x \mathcal{L}(x, \varepsilon l) \leq \frac{1}{2\pi\,(1 - \gamma)}\, J(\varepsilon) + \frac{\gamma}{1 - \gamma}, \tag{3.12}$$

where

$$J(\varepsilon) = \int_{-\infty}^{\infty} \left|\frac{f(t) - g(t)}{t}\right| e^{-\varepsilon^2 t^2/2}\, dt.$$

In order to estimate this integral, one may split the integration to the regions $|t| < T$ and $|t| \geq T$ and make use of the elementary inequality

$$\int_T^{\infty} \frac{1}{t}\, e^{-\varepsilon^2 t^2/2}\, dt < \frac{1}{\varepsilon^2 T^2}\, e^{-\varepsilon^2 T^2/2}.$$

Since $|f(t) - g(t)| \leq 2$, this gives

$$J(\varepsilon) \leq \int_{-T}^{T} \left|\frac{f(t) - g(t)}{t}\right| \mathrm{d}t + \frac{4}{\varepsilon^2 T^2}\, \mathrm{e}^{-\varepsilon^2 T^2/2},$$

so that, by (3.12),

$$\sup_x \mathcal{L}(x, \varepsilon l) \leq \frac{1}{2\pi(1-\gamma)} \int_{-T}^{T} \left|\frac{f(t) - g(t)}{t}\right| \mathrm{d}t + \frac{2}{\pi(1-\gamma)\, \varepsilon^2 T^2}\, \mathrm{e}^{-\varepsilon^2 T^2/2} + \frac{\gamma}{1-\gamma}.$$

Thus, according to the remark (3.9) with $h = \varepsilon l$, if we show that

$$\frac{2}{\pi(1-\gamma)\, \varepsilon^2 T^2}\, \mathrm{e}^{-\varepsilon^2 T^2/2} + \frac{\gamma}{1-\gamma} \leq 2\varepsilon l, \tag{3.13}$$

then we will get

$$L(F, G) \leq \frac{1}{2\pi(1-\gamma)} \int_{-T}^{T} \left|\frac{f(t) - g(t)}{t}\right| \mathrm{d}t + 2\varepsilon l. \tag{3.14}$$

Put

$$\varepsilon = \frac{1}{T}\sqrt{2\log(1+T)}, \quad l = \sqrt{2\log(1+T)}.$$

With this choice, using another well-known estimate $1 - \Phi(x) \leq \frac{1}{2}\,\mathrm{e}^{-x^2/2}$ $(x \geq 0)$, we have

$$\gamma = 2\,(1 - \Phi(l)) \leq \frac{1}{1+T}, \qquad \frac{\gamma}{1-\gamma} \leq \frac{1}{T}.$$

If $T \geq \mathrm{e} - 1$, then $l \geq \sqrt{2}$, $\gamma \leq 2\,(1 - \Phi(\sqrt{2})) < 0.16$, $\frac{1}{\pi(1-\gamma)} < 0.4$, hence the left-hand side of (3.13) does not exceed

$$\begin{aligned}\frac{1}{\pi(1-\gamma)\,(1+T)\log(1+T)} + \frac{1}{T} &< \frac{0.4}{(1+T)\log(1+T)} + \frac{\mathrm{e}/(\mathrm{e}-1)}{1+T} \\ &< \frac{2\log(1+T)}{1+T} = \varepsilon l.\end{aligned}$$

Therefore, the condition (3.13) is fulfilled, and we obtain (3.14), that is,

$$L(F, G) \leq \frac{1}{2\pi(1-\gamma)} \int_{-T}^{T} \left|\frac{f(t) - g(t)}{t}\right| \mathrm{d}t + \frac{4\log(1+T)}{T}.$$

This inequality also holds for $T < \mathrm{e} - 1$, since $L \leq 1$, while the last term is larger than $\frac{4}{\mathrm{e}-1} > 1$. As a result, we obtain (3.8) for all $T > 0$. Proposition 3.3.1 is proved.□

3.4 Lower Bounds for the Kolmogorov Distance

In order to obtain lower bounds for the Kolmogorov distance, characteristic functions may be used as well. In particular, we have:

Proposition 3.4.1 *Let F and G be functions of bounded variation such that $F(-\infty) = G(-\infty)$ and $F(\infty) = G(\infty)$, with Fourier–Stieltjes transforms f and g, respectively. Then*

$$\rho(F,G) \geq \frac{1}{2\sqrt{2\pi}} \left| \int_{-\infty}^{\infty} (f(t) - g(t))\, \mathrm{e}^{-t^2/2}\, \mathrm{d}t \right|. \tag{3.15}$$

Moreover, for any $T > 0$,

$$\rho(F,G) \geq \frac{1}{3T} \left| \int_{0}^{T} (f(t) - g(t)) \left(1 - \frac{t}{T}\right) \mathrm{d}t \right|. \tag{3.16}$$

Proof These bounds are consequences of a more general estimate

$$\left| \int_{-\infty}^{\infty} (f(t) - g(t))\, w(t)\, \mathrm{d}t \right| \leq \sup_x |F(x) - G(x)| \int_{-\infty}^{\infty} |\mathcal{F}(tw(t))|\, \mathrm{d}x, \tag{3.17}$$

which holds true whenever the function $(1 + |t|)w(t)$ is integrable. Here

$$(\mathcal{F}v(t))(x) = \int_{-\infty}^{\infty} \mathrm{e}^{itx}\, v(t)\, \mathrm{d}t, \quad x \in \mathbb{R},$$

denotes the Fourier transform of v.

To derive (3.17), let $A : \mathbb{R} \to \mathbb{R}$ be a function of bounded variation such that $A(-\infty) = A(\infty) = 0$, with Fourier–Stieltjes transform $a(t)$. For points $T > 0$ of continuity of A, define the truncated Fourier–Stieltjes transform

$$a_T(t) = \int_{-T}^{T} \mathrm{e}^{itx}\, \mathrm{d}A(x) = \varepsilon_T(t) - it \int_{-T}^{T} \mathrm{e}^{itx} A(x)\, \mathrm{d}x, \quad t \in \mathbb{R},$$

where we integrated by parts and put $\varepsilon_T(t) = \mathrm{e}^{itT} A(T) - \mathrm{e}^{-itT} A(-T)$. By Fubini's theorem,

$$\begin{aligned} \int_{-T}^{T} A(x)\, (\mathcal{F}tw(t))(x)\, \mathrm{d}x &= \int_{-T}^{T} A(x) \left[\int_{-\infty}^{\infty} \mathrm{e}^{itx}\, tw(t)\, \mathrm{d}t \right] \mathrm{d}x \\ &= \int_{-\infty}^{\infty} tw(t) \left[\int_{-T}^{T} \mathrm{e}^{itx} A(x)\, \mathrm{d}x \right] \mathrm{d}t \\ &= i \int_{-\infty}^{\infty} (\varepsilon_T(t) - a_T(t))\, w(t)\, \mathrm{d}t. \end{aligned}$$

From this it follows immediately that

$$\left| \int_{-\infty}^{\infty} a_T(t)\, w(t)\, \mathrm{d}t \right| \leq \|A\| \int_{-\infty}^{\infty} |\mathcal{F}(tw(t))|\, \mathrm{d}x + \int_{-\infty}^{\infty} |\varepsilon_T(t)|\, |w(t)|\, \mathrm{d}t, \tag{3.18}$$

where $\|A\| = \sup_x |A(x)|$. By the definition, $\varepsilon_T(t) \to 0$ as $T \to \infty$ uniformly over all $t \in \mathbb{R}$. In addition, $a_T(t) \to a(t)$ and, using the total variation norm of A,

$$|a_T(t)| \leq \|A\|_{\mathrm{TV}} \quad \text{for all } t \in \mathbb{R}.$$

Hence, letting $T \to \infty$ along the points of continuity of A and applying the Lebesgue dominated convergence theorem, from (3.18) we obtain that

$$\Big|\int_{-\infty}^{\infty} a(t)\, w(t)\, \mathrm{d}t\Big| \leq \|A\| \int_{-\infty}^{\infty} |\mathcal{F}(tw(t))|\, \mathrm{d}x.$$

The latter becomes the desired bound (3.17) by putting $A = F - G$ and $a = f - g$.

For example, the particular choice $w(t) = \mathrm{e}^{-t^2/2}$ in (3.17) leads us to the first lower bound (3.15) of the proposition.

To deduce the second bound, we choose the function

$$w_T(t) = (1 - t/T)\, 1_{(0,T)}(t).$$

First let $T = 1$ and write $w = w_1$. Then, for all $x \neq 0$, for $v(t) = tw(t)$, we have

$$\mathcal{F}v(x) = \int_0^1 \mathrm{e}^{itx}\, t(1-t)\, \mathrm{d}t = \frac{-\mathrm{e}^{ix} - 1}{x^2} + \frac{2(\mathrm{e}^{ix} - 1)}{ix^3} = \frac{q(x)}{x^3},$$

where $q(x) = -x\mathrm{e}^{ix} - 2i\, \mathrm{e}^{ix} - x + 2i$. Clearly, $|q(x)| \leq 2|x| + 4$. On the other hand,

$$|\mathcal{F}v(x)| \leq \int_0^1 t(1-t)\, \mathrm{d}t = \frac{1}{6},$$

and therefore

$$\begin{aligned}\int_{-\infty}^{\infty} |\mathcal{F}v(x)|\, \mathrm{d}x &= 2\int_0^4 |\mathcal{F}v(x)|\, \mathrm{d}x + 2\int_4^{\infty} |\mathcal{F}v(x)|\, \mathrm{d}x \\ &\leq \frac{4}{3} + 2\int_4^{\infty} \frac{2x+4}{x^3}\, \mathrm{d}x = \frac{31}{12} < 3.\end{aligned}$$

In the general case of $T > 0$, for $v_T(t) = tw_T(t) = tw(t/T)$, we have $\mathcal{F}v_T(x) = T^2 \cdot (\mathcal{F}v)(Tx)$, so that

$$\int_{-\infty}^{\infty} |\mathcal{F}v_T(x)|\, \mathrm{d}x = T\int_{-\infty}^{\infty} |\mathcal{F}v(x)|\, \mathrm{d}x < 3T.$$

Thus Proposition 3.4.1 is proved. □

Let also note that the choice $w(t) = (1 - |t|/T)^+$ in (3.17) would lead us to a similar bound

$$\rho(F,G) \geq \frac{1}{3.5}\Big|\int_{-T}^{T} (f(t) - g(t))\, \Big(1 - \frac{|t|}{T}\Big)\mathrm{d}t\Big|.$$

3.5 Remarks

In a slightly different form, with an arbitrary factor $b > \frac{1}{2\pi}$ in front of the first integral and with corresponding modifications on the right, the inequality (3.5) is due to Esseen [86], who also required the unnecessary assumption $F(\infty) = G(\infty)$. See also Petrov [155]. The more general relation (3.1), with implicit universal factors, is due to Fainleib [88], who additionally assumed that both F and G are distribution functions.

There are other related forms of Proposition 3.2.3 such as

$$\rho(F,G) \leq \frac{1}{2\pi} \int_{-T}^{T} \left|\frac{f(t) - g(t)}{t}\right| \mathrm{d}t + \frac{1}{T} \int_{-T}^{T} (|f(t)| + |g(t)|) \, \mathrm{d}t,$$

where $T > 0$, and f and g are characteristic functions associated to given distribution functions F and G. This bound immediately follows from the representation

$$F(x) = \frac{1}{2} + \frac{i}{2\pi} \lim_{\varepsilon \downarrow 0} \int_{\varepsilon < |t| < T} \mathrm{e}^{-itx} f(t) \, \frac{\mathrm{d}t}{t} + R$$

with a remainder term satisfying

$$|R| \leq \frac{1}{T} \int_{-T}^{T} |f(t)| \, \mathrm{d}t.$$

This representation was obtained in [12] as a consequence of smoothing inequalities due to Prawitz [159].

Zolotarev's inequality was proved in [177] by a different smoothing argument; here we followed the exposition of [32].

The lower bound (3.15) was obtained in [106], cf. also [135] for a correction. The second lower bound (3.16) was derived in [32].

Chapter 4
Sums of Independent Random Variables

In this chapter we collect basic definitions and results on the normal approximation of distributions of independent random variables (in Kolmogorov distance), and also discuss possible improved rates of approximation when replacing the normal law by corresponding Edgeworth corrections. The first section deals with moment based quantities for single random variables

4.1 Cumulants

If a random variable X has a finite absolute moment $\beta_p = \mathbb{E}\,|X|^p$ for an integer $p \geq 1$, it has finite moments of any order up to p, and, by Hölder's inequality, $\beta_r \leq \beta_p^{r/p}$ for $1 \leq r \leq p$. Since the characteristic function

$$f(t) = \mathbb{E}\,\mathrm{e}^{itX}, \quad t \in \mathbb{R},$$

is non-vanishing in some interval $|t| \leq t_0$, the principal value of the logarithm $\log f(t)$ has continuous derivatives up to order p, and one may introduce the cumulants

$$\gamma_r = \gamma_r(X) = \frac{\mathrm{d}^r}{i^r\,\mathrm{d}t^r} \log f(t)\big|_{t=0}, \quad r = 1, 2, \dots, p.$$

Moreover, the function $\log f$ is represented in the form of the power series near the origin

$$\log f(t) = \sum_{r=1}^{p-1} \gamma_r \frac{(it)^r}{r!} + \varepsilon(t). \tag{4.1}$$

Here, the remainder term may be controlled in terms of the absolute moments of X.

S. Bobkov et al., *Concentration and Gaussian Approximation for Randomized Sums*, Probability Theory and Stochastic Modelling 104,
https://doi.org/10.1007/978-3-031-31149-9_4

Proposition 4.1.1 *If* $|f(t)| \geq 1/\lambda$ *in the interval* $|t| \leq t_0$ *for some* $\lambda \geq 1$, *then in this interval the representation* (4.1) *holds true with*

$$|\varepsilon(t)| \leq \frac{(1+\lambda)^p - 1}{p} \beta_p |t|^p. \tag{4.2}$$

In particular, if $p \geq 2$, *and* $\mathbb{E}X = 0$, $\mathbb{E}X^2 = \sigma^2$ $(\sigma > 0)$, *then* $|f(t)| \geq \frac{1}{2}$ *whenever* $\sigma|t| \leq 1$, *and in this interval*

$$|\varepsilon(t)| \leq \frac{3^p}{p} \beta_p |t|^p. \tag{4.3}$$

The argument is based on the generalized chain rule: If a complex-valued function $y = y(t)$ is defined and has p derivatives in some open interval of the real line, and $z = z(y)$ is analytic in the region containing all values of y, then

$$\frac{d^p}{dt^p} z(y(t)) = p! \sum \frac{d^m z(y)}{dy^m}\Big|_{y=y(t)} \prod_{r=1}^{p} \frac{1}{k_r!} \bigg(\frac{1}{r!} \frac{d^r y(t)}{dt^r}\bigg)^{k_r}, \tag{4.4}$$

where $m = k_1 + \cdots + k_p$ and where the summation runs over all tuples $(k_1, \dots, k_p)$ of non-negative integers such that $k_1 + 2k_2 + \cdots + pk_p = p$. This identity has a number of interesting particular cases such as the following ones.

Lemma 4.1.2 *With summation as before, for any* $\lambda \in \mathbb{R}$ *and any integer* $p \geq 1$,

$$\sum (m-1)! \prod_{r=1}^{p} \frac{1}{k_r!} \lambda^{k_r} = \frac{(1+\lambda)^p - 1}{p},$$

$$\sum \prod_{r=1}^{p} \frac{1}{k_r!} \Big(\frac{\lambda^r}{r}\Big)^{k_r} = \lambda^p.$$

Proof Applying (4.4) with $z(y) = -\log(1-y)$, this identity becomes

$$-\frac{d^p}{dt^p} \log(1-y(t)) = p! \sum \frac{(m-1)!}{(1-y(t))^m} \prod_{r=1}^{p} \frac{1}{k_r!} \Big(\frac{1}{r!} \frac{d^r y(t)}{dt^r}\Big)^{k_r}. \tag{4.5}$$

Let us choose here

$$y(t) = \frac{\lambda t}{1-t} = -\lambda + \lambda(1-t)^{-1}$$

with sufficiently small values of t, so that $|y(t)| < 1$. In this case,

$$\frac{d^r y(t)}{dt^r} = r!\, \lambda (1-t)^{-(r+1)},$$

while

$$-\frac{\mathrm{d}^p}{\mathrm{d}t^p}\log(1-y(t)) = \frac{\mathrm{d}^p}{\mathrm{d}t^p}\Big[\log(1-t) - \log(1-(1+\lambda)t)\Big]$$
$$= -\frac{(p-1)!}{(1-t)^p} + (1+\lambda)^p\,\frac{(p-1)!}{(1-(1+\lambda)t)^p}.$$

Therefore, (4.5) yields

$$(p-1)!\Big[\frac{(1+\lambda)^p}{(1-(1+\lambda)t)^p} - \frac{1}{(1-t)^p}\Big] = p!\sum\frac{(m-1)!\,\lambda^m}{(1-y(t))^m\,(1-t)^{p+m}}\prod_{r=1}^p\frac{1}{k_r!}.$$

Putting $t = 0$, we obtain the first identity of the lemma.

Finally, let us apply (4.4) with $z(y) = \mathrm{e}^y$, when this identity becomes

$$\frac{\mathrm{d}^p}{\mathrm{d}t^p}\,\mathrm{e}^{y(t)} = p!\,\mathrm{e}^{y(t)}\sum\prod_{r=1}^p\frac{1}{k_r!}\Big(\frac{1}{r!}\,\frac{\mathrm{d}^r y(t)}{\mathrm{d}t^r}\Big)^{k_r}. \tag{4.6}$$

It remains to choose here $y(t) = -\log(1-\lambda t)$, so that

$$\frac{\mathrm{d}^r y(t)}{\mathrm{d}t^r} = \lambda^r\,(r-1)!\,(1-\lambda t)^{-r},$$

while

$$\frac{\mathrm{d}^p}{\mathrm{d}t^p}\,\mathrm{e}^{y(t)} = \frac{\mathrm{d}^p}{\mathrm{d}t^p}\,(1-\lambda t)^{-1} = p!\,\lambda^p\,(1-\lambda t)^{-(p+1)}.$$

Hence, at the point $t = 0$ the equality (4.6) yields the second identity of the lemma.□

Proof (of Proposition 4.1.1) Given $|t| \le t_0$, it is sufficient to estimate the p-th derivative of $\log f(s)$ in absolute value uniformly in the interval $|s| \le |t|$. According to the chain rule (4.4) with $z(y) = \log y$ and $y = f(s)$,

$$(\log f(s))^{(p)} = p!\sum\frac{(-1)^{m-1}\,(m-1)!}{f(s)^m}\prod_{r=1}^p\frac{1}{k_r!}\Big(\frac{1}{r!}\,f^{(r)}(s)\Big)^{k_r}, \tag{4.7}$$

where $m = k_1 + \dots + k_s$ and the summation runs over all non-negative integers $k_1, \dots, k_p$ such that $k_1 + 2k_2 + \dots + pk_p = p$. Using $|f^{(r)}(s)| \le \beta_r \le \beta_p^{r/p}$ together with $|f(s)| \ge \frac{1}{\lambda}$, Lemma 4.1.2 gives

$$\big|(\log f(s))^{(p)}\big| \le p!\,\beta_p\sum(m-1)!\prod_{r=1}^p\frac{1}{k_r!}\Big(\frac{\lambda}{r!}\Big)^{k_r} \le p!\,\beta_p\,\frac{(1+\lambda)^p-1}{p}.$$

Hence, by the integral Taylor formula,

$$|\varepsilon(t)| \le \sup_{|s|\le|t|}\big|(\log f(s))^{(p)}\big|\,\frac{|t|^p}{p!} \le \beta_p\,\frac{(1+\lambda)^p-1}{p}\,|t|^p,$$

as required.

In the second assertion, we have $f'(0) = 0$, $|f''(t)| \le \sigma^2$. Hence, by the Taylor formula,

$$|1 - f(t)| \le \sup_{|z| \le |t|} |f''(s)| \, \frac{t^2}{2} \le \frac{\sigma^2 t^2}{2} \le \frac{1}{2}$$

for $\sigma|t| \le 1$, so that $|f(t)| \ge \frac{1}{2}$. Proposition 4.1.1 is proved. □

Each cumulant $\gamma_p = \gamma_p(X)$ of a random variable X is determined by the first moments $\alpha_r = \mathbb{E}X^r$, $r = 1, \dots, p$. More precisely, the chain rule formula (4.7) at $s = 0$ gives the identity

$$\gamma_p = p! \sum (-1)^{m-1} \, (m-1)! \prod_{r=1}^{p} \frac{1}{k_r!} \Big(\frac{\alpha_r}{r!}\Big)^{k_r} \tag{4.8}$$

with summation as in (4.4). For example, $\gamma_1 = \alpha_1$, $\gamma_2 = \alpha_2 - \alpha_1^2$. Moreover, if $\alpha_1 = \mathbb{E}X = 0$, $\sigma^2 = \mathbb{E}X^2$, then

$$\gamma_1 = \alpha_1, \quad \gamma_2 = \alpha_2 = \sigma^2, \quad \gamma_3 = \alpha_3, \quad \gamma_4 = \alpha_4 - 3\alpha_2^2 = \beta_4 - 3\sigma^4.$$

The reverse statement is also true. Namely, applying the chain rule formula (4.6) with $y(t) = \log f(t)$ at $t = 0$, we arrive at

$$\alpha_p = p! \sum \prod_{r=1}^{p} \frac{1}{k_r!} \Big(\frac{\gamma_r}{r!}\Big)^{k_r}. \tag{4.9}$$

Using (4.1)–(4.2), the cumulants can be bounded in terms of absolute moments. Indeed, applying (4.2) at $t = 0$ and letting $t_0 \to 0$, $\lambda \to 1$, we get

$$|\gamma_p| = p! \, \lim_{t \to 0} \frac{|\varepsilon(t)|}{|t|^p} \le (2^p - 1)\,(p-1)!\,\beta_p.$$

If X has mean zero, the constant on the right-hand side can be improved, as the following statement shows. We follow a simple argument due to Bikjalis [17].

Proposition 4.1.3 *If $\beta_p < \infty$ for some integer $p \ge 1$, and $\mathbb{E}X = 0$, then*

$$|\gamma_p| \le (p-1)!\,\beta_p.$$

Proof Since $\gamma_1 = 0$ and $\gamma_2 = \alpha_2 = \beta_2$ when $\mathbb{E}X = 0$, the desired bound is immediate for $p \le 2$. So, let $p \ge 3$. Differentiating the identity $f'(t) = f(t)\,(\log f(t))'$ near zero $p - 1$ times in accordance with the binomial formula, one gets

$$\frac{d^p}{dt^p} f(t) = \sum_{r=0}^{p-1} C_{p-1}^r \, \frac{d^{p-1-r}}{dt^{p-1-r}} f(t) \, \frac{d^{r+1}}{dt^{r+1}} \log f(t),$$

where we use the notation $C_n^k = \frac{n!}{k!(n-k)!}$ for the binomial coefficients. Equivalently,

$$\frac{d^p}{dt^p} \log f(t) = \frac{1}{f(t)} \frac{d^p}{dt^p} f(t) - \frac{1}{f(t)} \sum_{r=0}^{p-2} C_{p-1}^r \frac{d^{p-1-r}}{dt^{p-1-r}} f(t) \frac{d^{r+1}}{dt^{r+1}} \log f(t).$$

At $t = 0$, using $\alpha_1 = 0$, this identity becomes

$$\gamma_p = \alpha_p - \sum_{r=0}^{p-3} C_{p-1}^r \alpha_{p-1-r} \gamma_{r+1}.$$

One can now proceed by induction on p. Since $|\alpha_{p-1-r}| \leq \beta_p^{(p-1-r)/p}$ and $\gamma_{r+1} \leq r!\beta_p^{(r+1)/p}$ (the induction hypothesis), we obtain that

$$|\gamma_p| \leq \beta_p + \beta_p \sum_{r=0}^{p-3} C_{p-1}^r r! = (p-1)!\beta_p \Big[\frac{1}{(p-1)!} + \sum_{r=0}^{p-3} \frac{1}{(p-1-r)!}\Big].$$

The expression in the square brackets $\frac{1}{(p-1)!} + (\frac{1}{2!} + \dots + \frac{1}{(p-1)!})$ is equal to 1 for $p = 3$ and is smaller than $\frac{1}{6} + (\mathrm{e} - 2) < 1$ for $p \geq 4$, thus proving Proposition 4.1.3.□

4.2 Lyapunov Coefficients

Let $X_1, \dots, X_n$ be independent random variables on a probability space $(\Omega, \mathcal{F}, \mathbb{P})$ with mean zero and variances $b_k^2 = \mathrm{Var}(X_k)$ ($b_k \geq 0$) such that $\sum_{k=1}^n b_k^2 = 1$. Many important properties of the distributions of the sums

$$S_n = X_1 + \dots + X_n$$

rely upon the behavior of moment functionals such as

$$\alpha_p = \sum_{k=1}^n \mathbb{E} X_k^p, \quad \beta_{p,k} = \mathbb{E}\,|X_k|^p, \quad L_p = \sum_{k=1}^n \beta_{p,k} \quad (p \geq 2 \text{ integer}).$$

The quantity L_p is called the Lyapunov ratio or the Lyapunov coefficient of order p. It is finite if and only if all the X_k's have finite absolute moments of order p, in which case α_p is well-defined and satisfies $|\alpha_p| \leq L_p$.

Let us mention a few general properties of these quantities.

Proposition 4.2.1 *The function $p \rightarrow L_p^{\frac{1}{p-2}}$ is non-decreasing, that is,*

$$L_q \leq L_p^{\frac{q-2}{p-2}}, \quad 2 \leq q \leq p.$$

In addition,

$$L_{p-q} L_q \leq L_{p-2}, \quad 2 \leq q \leq p-2.$$

Proof Let F_k denote the distribution of X_k. Since $L_2 = 1$, the equality $\mathrm{d}\mu(x) = \sum_{k=1}^n x^2 \, \mathrm{d}F_k(x)$ defines a (Borel) probability measure on the real line. Moreover,

$$L_p = \sum_{k=1}^n \int_{-\infty}^{\infty} |x|^p \, \mathrm{d}F_k(x) = \int_{-\infty}^{\infty} |x|^{p-2} \, \mathrm{d}\mu(x) = \mathbb{E}\,|\xi|^{p-2},$$

where ξ is a random variable distributed according to μ. Hence $L_p^{\frac{1}{p-2}} = \big(\mathbb{E}\,|\xi|^{p-2}\big)^{\frac{1}{p-2}}$ is non-decreasing on the axis $p > 2$ (not necessarily for integer values).

In the second assertion, we may assume that $2 < q < p - 2$ (real), so that $p > 4$. Let η be an independent copy of ξ. Using Young's inequality $xy \leq \frac{1}{u} x^u + \frac{1}{v} y^v$ ($x, y \geq 0$, $u, v \geq 1$ such that $\frac{1}{u} + \frac{1}{v} = 1$), we get

$$\begin{aligned} L_{p-q} L_q &= \mathbb{E}\,|\xi|^{p-q-2}\,|\eta|^{q-2} \\ &\leq \mathbb{E}\left(\frac{p-q-2}{p-4}\,|\xi|^{p-4} + \frac{q-2}{p-4}\,|\eta|^{p-4}\right) = L_{p-2}. \end{aligned}$$

The proposition is proved. □

Proposition 4.2.2 *We have* $L_p \geq n^{-\frac{p-2}{p}}$. *In particular,* $L_4 \geq n^{-1}$, $L_5 \geq n^{-3/2}$.

Proof Let $p > 2$. By Hölder's inequality with exponents $u = \frac{p}{p-2}$ and $v = \frac{p}{2}$,

$$1 = \sum_{k=1}^n b_k^2 \leq n^{1/u} \Big(\sum_{k=1}^n b_k^p\Big)^{1/v} \leq n^{1/u} \Big(\sum_{k=1}^n \beta_{p,k}\Big)^{1/v} = n^{1/u}\, L_p^{1/v}.$$

Hence $L_p \geq n^{-v/u} = n^{-(p-2)/p}$, thus proving the claim. □

On the other hand, if L_p is small, then the variances b_k^2 of X_k have to be small as well, uniformly over all k.

Proposition 4.2.3 *For any* $p > 2$, *we have* $\max_{k \leq n} b_k \leq L_p^{1/p}$.

Indeed,

$$\max_k b_k \leq \Big(\sum_{k=1}^n b_k^p\Big)^{1/p} \leq \Big(\sum_{k=1}^n \beta_{p,k}\Big)^{1/p} = L_p^{1/p}.$$

Once the Lyapunov coefficient L_p is finite, one may introduce the cumulants

$$\gamma_{p,k} = \gamma_p(X_k) = \frac{\mathrm{d}^p}{i^p\, \mathrm{d}t^p} \log v_k(t)\big|_{t=0}, \quad k = 1, \dots, n,$$

where $v_k = \mathbb{E}\, e^{itX_k}$ denote the characteristic functions of X_k. Since the characteristic function of S_n is described as the product

$$f_n(t) = \mathbb{E}\, e^{itS_n} = v_1(t) \dots v_n(t),$$

the p-th cumulant of S_n is given by

$$\gamma_p = \gamma_p(S_n) = \frac{\mathrm{d}^p}{i^p\, \mathrm{d}t^p} \log f_n(t)\big|_{t=0} = \sum_{k=1}^n \gamma_{p,k}.$$

The first values are $\gamma_1 = 0$, $\gamma_2 = 1$.

Applying the inequality of Proposition 4.1.3, we immediately obtain a similar relation between the Lyapunov coefficients and the cumulants of the sums.

Proposition 4.2.4 *For all $p \geq 2$, the cumulants $\gamma_p = \gamma_p(S_n)$ satisfy*

$$|\gamma_p| \leq (p-1)!\, L_p.$$

4.3 Rosenthal-type Inequalities

The Lyapunov coefficients

$$L_p = \sum_{k=1}^n \mathbb{E}\,|X_k|^p$$

may also be used to bound absolute moments of the sums $S_n = X_1 + \cdots + X_n$. In particular, we have the following observation due to Rosenthal [162]. As before, the random variables $X_1, \dots, X_n$ are assumed to be independent with mean zero.

Proposition 4.3.1 *If $\mathbb{E}\,|S_n|^2 = 1$, then for any real number $p \geq 2$,*

$$\mathbb{E}\,|S_n|^p \leq C_p \max\{L_p, 1\} \tag{4.10}$$

for some constant C_p depending on p only.

For even integers $p \geq 4$, we have the following simple argument. Applying the expression (4.9) to S_n and putting $r^* = \max(r, 2)$, Proposition 4.2.4 gives

$$\begin{aligned} \mathbb{E}\,|S_n|^p = \alpha_p(S_n) &= p! \sum \prod_{r=1}^p \frac{1}{k_r!} \Big(\frac{\gamma_r(S_n)}{r!}\Big)^{k_r} \\ &\leq p! \sum \prod_{r=1}^p \frac{1}{k_r!} \Big(\frac{L_{r^*}}{r}\Big)^{k_r}, \end{aligned} \tag{4.11}$$

where the summation is performed over all non-negative integers $k_1, \dots, k_p$ such that $k_1 + 2k_2 + \cdots + pk_p = p$. By Proposition 4.2.1,

$$L_r \leq L_p^{\frac{r-2}{p-2}} \leq (\max(L_p, 1))^{\frac{r}{p}}, \quad 2 \leq r \leq p.$$

Hence, using the second identity of Lemma 4.1.2, the second sum in (4.11) does not exceed

$$\sum \prod_{r=1}^p \frac{1}{k_r!} \Big(\frac{(\max(L_p^{1/p}, 1))^r}{r}\Big)^{k_r} = \max(L_p, 1).$$

Thus,(4.10) holds true with $C_p \le p!$ for $p = 2, 4, 6, 8, \dots$

Let us also describe an alternative induction argument restricting ourselves to integer values $p \ge 3$ (although the argument works for real $p > 2$ as well). We consider the following Rosenthal-type inequality

$$\mathbb{E}\,|S_n|^p \le \sum_{m=2}^{p} A_m(p)\, L_m\, L_2^{\frac{p-m}{2}}, \tag{4.12}$$

where the non-negative coefficients $A_m(p)$ will be determined by the induction. In fact, it makes sense to choose $A_p(p) = 1$ and $A_{p-1}(p) = p$. Then (4.12) holds automatically for $n = 1$. Note that this inequality is homogeneous with respect to X_k, so, the condition $L_2 = \mathbb{E}\,|S_n|^2 = 1$ is irrelevant.

To carry out the induction step, assume that $n \ge 2$ and $L_2 = 1$. By the integral Taylor formula, for all $x, h \in \mathbb{R}$, if p is odd, we have

$$|x+h|^p = \sum_{l=0}^{p-1} C_p^l \operatorname{sign}(x)\, x^{p-l} h^l + p h^p \int_0^1 \operatorname{sign}(x+th)\,(1-t)^{p-1}\,dt.$$

A similar representation without the sign function also holds when p is even. We apply it with independent summands $x = S_{n,k} = S_n - X_k$ and $h = X_k$, $1 \le k \le n$. Taking expectations of both sides and noting that the term corresponding to $l = 1$ is vanishing (due to $\mathbb{E}X_k = 0$), we get

$$\mathbb{E}\,|S_n|^p \le \mathbb{E}\,|S_{n,k}|^p + \sum_{2 \le l \le p} C_p^l\, \mathbb{E}\,|S_{n,k}|^{p-l}\, \mathbb{E}\,|X_k|^l.$$

Here, for $l = p-1$, one may use $\mathbb{E}\,|S_{n,k}| \le (\mathbb{E}\,|S_{n,k}|^2)^{1/2} \le 1$, while for $2 \le l \le p-2$, from the induction hypothesis (4.12) it follows that

$$\mathbb{E}\,|S_{n,k}|^{p-l} \le \sum_{q=2}^{p-l} A_q(p-l)\, L_q.$$

Plugging this result into the previous bound and summing over all $k \le n$, we arrive at

$$\begin{aligned} n\,\mathbb{E}\,|S_n|^p &\le \sum_{k=1}^{n} \mathbb{E}\,|S_{n,k}|^p \\ &\quad + \sum_{2 \le l \le p-2}\ \sum_{2 \le q \le p-l} C_p^l\, A_q(p-l)\, L_q L_l + p\, L_{p-1} + L_p. \end{aligned}$$

To further simplify, recall that $L_q L_l \le L_{q+l-2}$ (Proposition 4.2.1), which turns the right-hand side into a linear function with respect to the Lyapunov coefficients:

$$n\,\mathbb{E}\,|S_n|^p \le \sum_{k=1}^{n} \mathbb{E}\,|S_{n,k}|^p + \sum_{2\le m\le p-2} B_m(p)\,L_m + p\,L_{p-1} + L_p, \tag{4.13}$$

where $B_m(p) = \sum_{2\le l\le m} C_p^l\, A_{m-l+2}(p-l)$. To bound the first sum in (4.13), again by the induction hypothesis (4.12), we have

$$\mathbb{E}\,|S_{n,k}|^p \le \sum_{m=2}^{p} A_m(p) \sum_{j\neq k} \mathbb{E}\,|X_j|^m,$$

so that

$$\sum_{k=1}^{n} \mathbb{E}\,|S_{n,k}|^p \le (n-1) \sum_{m=2}^{p} A_m(p)\,L_m.$$

Plugging this into (4.13), we obtain

$$\begin{aligned} n\,\mathbb{E}\,|S_n|^p &\le n \sum_{m=2}^{p} A_m(p)\,L_m + \sum_{2\le m\le p-2} (B_m(p) - A_m(p))\,L_m \\ &\quad + (p - A_{p-1}(p))\,L_{p-1} + (1 - A_p(p))\,L_p. \end{aligned}$$

As a result, we arrive at the desired relation (4.12) for n summands, as long as $A_p(p) = 1$, $A_{p-1}(p) = p$, and if the remaining coefficients satisfy $B_m(p) \le A_m(p)$, that is,

$$A_m(p) \ge \sum_{2\le l\le m} C_p^l\, A_{m-l+2}(p-l), \quad 2 \le m \le p-2. \tag{4.14}$$

The latter is not a condition for $p = 3$. For larger values of p, one may replace the inequality symbol here with the equality symbol and use it as a recursive formula

$$A_m(p) = \sum_{2\le l\le m} C_p^l\, A_{m-l+2}(p-l), \quad 2 \le m \le p-2,$$

with respect to p. If $p = 4$, this formula is reduced to $A_2(4) = C_4^2 A_2(2) = 6$, and if $p = 5$, it is reduced to

$$A_2(5) = C_5^2 A_2(3) = 30 \quad \text{and} \quad A_3(5) = C_5^2 A_3(3) + C_5^3 A_2(2) = 20.$$

Hence, assuming that $L_2 = 1$, we get in (4.12) the Rosenthal-type inequalities

$$\begin{aligned} \mathbb{E}\,|S_n|^3 &\le L_3 + 3 \;\le\; 4\max\{L_3, 1\}, \\ \mathbb{E}\,|S_n|^4 &\le L_4 + 4L_3 + 6 \;\le\; 11\max\{L_4, 1\}, \\ \mathbb{E}\,|S_n|^5 &\le L_5 + 5L_4 + 20L_3 + 30 \;\le\; 56\max\{L_5, 1\}. \end{aligned}$$

To handle larger values, one may take $A_m(p) = p!$ $(2 \le m \le p-2)$ in (4.14) or even better $A_m(p) = (p-1)!$ (for large p). The latter choice leads to

$$\mathbb{E}\,|S_n|^p \le L_p + pL_{p-1} + (p-1)! \sum_{2 \le m \le p-2} L_m < p!\, \max\{L_p, 1\}.$$

4.4 Normal Approximation

Now, we turn to the problem of normal approximation for distribution functions $F_n(x) = \mathbb{P}\{S_n \le x\}$ of the sums

$$S_n = X_1 + \cdots + X_n.$$

We assume that the random variables $X_1, \dots, X_n$ are independent and have mean zero and variances $b_k^2 = \mathrm{Var}(X_k)$ such that $\sum_{k=1}^n b_k^2 = 1$. Thus, $\mathbb{E}S_n = 0$, $\mathrm{Var}(S_n) = 1$.

The central limit theorem aims to provide natural conditions such that the distribution of S_n can be approximated in some sense by the standard Gaussian measure on the real line, that is, with density and distribution function

$$\varphi(x) = \frac{1}{\sqrt{2\pi}}\, \mathrm{e}^{-x^2/2}, \quad \Phi(x) = \int_{-\infty}^{x} \varphi(y)\,\mathrm{d}y \quad (x \in \mathbb{R}).$$

In terms of the Kolmogorov distance, the normal approximation may be quantified by means of the Lyapunov coefficient $L_3 = \sum_{k=1}^n \mathbb{E}\,|X_k|^3$. Namely, we have the following well-known variant of the central limit theorem.

Proposition 4.4.1 *For some absolute constant $c > 0$,*

$$\rho(F_n, \Phi) \le cL_3. \tag{4.15}$$

This assertion is often referred to as the Berry–Esseen theorem, especially in the i.i.d. case $X_k = \xi_k/\sqrt{n}$ with $\mathbb{E}\,\xi_k = 0$, $\mathbb{E}\,\xi_k^2 = 1$, $\beta_3 = \mathbb{E}\,|\xi_k|^3$, when (4.15) becomes

$$\rho(F_n, \Phi) \le \frac{c\beta_3}{\sqrt{n}}.$$

Here, the best constant is unknown, but it is known that one may take $c = 0.56$ ([166]).

As a preliminary step towards the proof of Proposition 4.4.1, first let us see that the characteristic function $f_n(t) = \mathbb{E}\,\mathrm{e}^{itS_n}$ has a Gaussian decay on a large interval, whose length is controlled by L_3 (of course, this makes sense when L_3 is small).

Lemma 4.4.2 *In the interval $|t| \le \frac{1}{L_3}$, we have $|f_n(t)| \le \mathrm{e}^{-t^2/6}$.*

Proof Let v_k denote the characteristic function of X_k, and let X_k' be an independent copy of X_k. The random variable $Y_k = X_k - X_k'$ has a symmetric distribution with second moment $2b_k^2$ and a non-negative characteristic function $|v_k(t)|^2$. If L_3 is finite, so are the absolute moments $\beta_{3,k} = \mathbb{E}\,|X_k|^3$, and we have

$$\begin{aligned}\mathbb{E}\,|Y_k|^3 &= \mathbb{E}\,(X_k - X_k')^2\,|X_k - X_k'|\\ &\le \mathbb{E}\,(X_k^2 - 2X_kX_k' + X_k'^2)\,(|X_k| + |X_k'|)\\ &= 2\,\mathbb{E}\,|X_k|^3 + 2\,\mathbb{E}X_k^2\,\mathbb{E}\,|X_k| \;\le\; 4\beta_{3,k}.\end{aligned}$$

Hence, expanding the function $|v_k|^2$ around zero according to Taylor's formula up to the 3-rd term, we get

$$|v_k(t)|^2 \le 1 - b_k^2 t^2 + \frac{4}{3!}\,\beta_{3,k}|t|^3 \le \exp\Big\{-b_k^2 t^2 + \frac{2}{3}\,\beta_{3,k}|t|^3\Big\}.$$

Multiplying these inequalities and assuming that $|t| \le \frac{1}{L_3}$, we arrive at

$$|f_n(t)|^2 \le \exp\Big\{-t^2 + \frac{2}{3}\,L_3|t|^3\Big\} \le e^{-t^2/3},$$

proving the lemma. □

On a smaller interval, the statement of Lemma 4.4.2 may be sharpened by comparing $f_n(t)$ with the characteristic function of the standard Gaussian measure.

Lemma 4.4.3 *In the interval* $|t| \le L_3^{-1/3}$, *we have*

$$|f_n(t) - e^{-t^2/2}| \le cL_3\,|t|^3\,e^{-t^2/2} \tag{4.16}$$

for some absolute constant $c > 0$.

Proof Applying Proposition 4.1.1 with $p = 3$ and using $\mathbb{E}X_k = 0$, we see that the characteristic functions $v_k(t) = \mathbb{E}\,e^{itX_k}$ admit the representation

$$v_k(t) = \exp\Big\{-b_k^2\,\frac{t^2}{2} + \varepsilon_k(t)\Big\}, \quad b_k|t| \le 1,$$

with remainder terms such that $|\varepsilon_k(t)| \le 9\,\beta_{3,k}\,|t|^p$, cf. (4.3). Multiplying these equalities, we obtain a similar representation

$$f_n(t) = e^{-\frac{1}{2}t^2 + \varepsilon(t)}. \tag{4.17}$$

Moreover, according to Proposition 4.2.3, if $L_3^{1/3}\,|t| \le 1$, then $b_k|t| \le 1$ for all $k \le n$, and therefore (4.17) holds true in this interval with a remainder term satisfying

$$|\varepsilon(t)| \le \sum_{k=1}^n |\varepsilon_k(t)| \le 9L_3\,|t|^3 \le 9.$$

In particular, $|e^{\varepsilon(t)} - 1| \le c\,|\varepsilon(t)| \le 9c\,L_3\,|t|^3$ for some absolute constant $c > 0$. Hence, (4.17) implies the desired relation (4.16), thus proving the lemma. □

With a certain loss in the bound, both lemmas can be united in one relation

$$|f_n(t) - e^{-t^2/2}| \le cL_3 \min\{1, |t|^3\} e^{-t^2/8}, \quad |t| \le \frac{1}{L_3}.$$

Proof (of Proposition 4.4.1) Since $\rho(F_n, \Phi) \le 1$, we may assume that $L_3 \le 1$ in (4.15). Putting $T_0 = L_3^{-1/3}$ and $T = L_3^{-1}$, the Berry–Esseen inequality (3.5) of Proposition 3.2.1 applied with $F = F_n$ and $G = \Phi$ yields

$$\rho(F_n, \Phi) \le 2\int_0^{T_0} \frac{|f_n(t) - e^{-t^2/2}|}{t}\, dt$$
$$+ 2\int_{T_0}^{T} \frac{|f_n(t)| + e^{-t^2/2}}{t}\, dt + \frac{2\pi^2}{\sqrt{2\pi}} \frac{1}{T}.$$

Here, the first integral is bounded by $c_1 L_3$, by Lemma 4.4.3, while the second integral is bounded according to Lemma 4.4.2 by

$$2\int_{T_0}^{\infty} \frac{e^{-t^2/6}}{t}\, dt \le c_2 e^{-T_0^2/6} \le \frac{c_3}{T_0^3} = c_3 L_3.$$

Here, c_j are positive absolute constants. Proposition 4.4.1 is proved. □

4.5 Expansions for the Product of Characteristic Functions

One may wonder whether or not it is possible to sharpen the rate of approximation in Proposition 4.4.1 when replacing the standard normal distribution function by a certain correction. In general this is not possible unless one uses rather non-universal approximations (for example, when all random variables X_k are discrete and are identically distributed), but a certain sharpening is indeed possible in a weaker sense, assuming that the Lyapunov coefficient L_p is finite (and actually small) for a fixed integer $p \ge 4$. We now focus on the refinement of the inequality (4.16) of Lemma 4.4.3, which will be used when applying the Berry–Esseen inequality (3.5).

Thus, keeping the same notations and assumptions as in the previous section, let us return to the characteristic function $f_n(t) = \mathbb{E}\, e^{itS_n}$ of $S_n = X_1 + \cdots + X_n$. As we know from Proposition 4.1.1, in terms of the cumulants $\gamma_{r,k}$ of X_k, the characteristic functions $v_k(t) = \mathbb{E}\, e^{itX_k}$ admit the representation

$$v_k(t) = \exp\left\{\sum_{r=2}^{p-1} \gamma_{r,k} \frac{(it)^r}{r!} + \varepsilon_k(t)\right\}, \quad b_k|t| \le 1,$$

with remainder terms satisfying $|\varepsilon_k(t)| \le \frac{3^p}{p} \beta_{p,k} |t|^p$. Here, we used the assumption $\gamma_{1,k} = \mathbb{E}X_k = 0$. Multiplying these equalities, we obtain a similar representation

$$f_n(t) = \mathrm{e}^{-t^2/2} \exp\left\{\sum_{r=3}^{p-1} \gamma_r \frac{(it)^r}{r!} + \varepsilon(t)\right\} \tag{4.18}$$

involving the cumulants γ_r of S_n (recall that $\gamma_2 = \mathbb{E}S_n^2 = 1$). In view of Proposition 4.2.3, the interval where this equality holds is also controlled in terms of L_p. Namely, if $L_p^{1/p}|t| \le 1$, then $b_k|t| \le 1$ for all $k \le n$, and therefore

$$|\varepsilon(t)| \le \sum_{k=1}^{n} |\varepsilon_k(t)| \le \frac{3^p}{p}\, L_p|t|^p \le \frac{3^p}{p}.$$

In particular, in this interval, $|\mathrm{e}^{\varepsilon(t)} - 1| \le c_p\, |\varepsilon(t)| \le c'_p L_p|t|^p$ for some constants depending on p only. As a result, (4.18) is simplified to

$$f_n(t) = \mathrm{e}^{-t^2/2}\, \mathrm{e}^{Q_{p-1}(it)}\, (1 + \varepsilon(t)), \quad |\varepsilon(t)| \le c_p L_p|t|^p, \tag{4.19}$$

where

$$Q_{p-1}(z) = \sum_{r=3}^{p-1} \frac{\gamma_r}{r!}\, z^r = \frac{\gamma_3}{3!}\, z^3 + \cdots + \frac{\gamma_{p-1}}{(p-1)!}\, z^{p-1}, \quad z \in \mathbb{C}.$$

That is, we arrive at:

Lemma 4.5.1 *The representation* (4.19) *holds in the interval* $L_p^{1/p}\,|t| \le 1$ $(p \ge 3)$ *with a constant* $c_p > 0$ *depending on p only.*

Note that, when $p = 3$, the sum in (4.19) is vanishing, and we return to the inequality (4.16) of Lemma 4.4.3.

In the next step, assuming that $p \ge 4$, it is natural to simplify the expression $\mathrm{e}^{Q_{p-1}(it)}\, (1 + \varepsilon(t))$ to the form $1 + R_{p-1}(it) + \varepsilon(t)$ with a certain polynomial R_{p-1} and with a new remainder term, which would be still as small as the Lyapunov coefficient L_p. This may indeed be possible on a smaller interval. Namely, write

$$\exp\{Q_{p-1}(it)\} = \sum_{k=0}^{\infty} a_k\, (it)^k$$

with coefficients

$$a_k = \sum_{3k_1+\cdots+(p-1)k_{p-3}=k} \frac{1}{k_1!\cdots k_{p-3}!} \left(\frac{\gamma_3}{3!}\right)^{k_1} \cdots \left(\frac{\gamma_{p-1}}{(p-1)!}\right)^{k_{p-3}}.$$

Clearly, all these series are absolutely convergent for all t. Part of the last infinite series represents the desired polynomial R_{p-1}.

Recall that by γ_r we denote the cumulants of S_n of orders r, which are well defined for all $r = 1, \dots, p-1$ as long as the Lyapunov coefficient L_{p-1} is finite.

Definition 4.5.2 If L_{p-1} is finite ($p \geq 4$), put

$$R_{p-1}(it) = \sum \frac{1}{k_1! \dots k_{p-3}!} \left(\frac{\gamma_3}{3!}\right)^{k_1} \cdots \left(\frac{\gamma_{p-1}}{(p-1)!}\right)^{k_{p-3}} (it)^{3k_1+\cdots+(p-1)k_{p-3}},$$

where the summation runs over all collections of non-negative integers $(k_1, \dots, k_{p-3})$ that are not all zero and such that $d \equiv k_1 + 2k_2 + \cdots + (p-3)k_{p-3} \leq p-3$.

Here the constraint $d \leq p-3$ has the aim to involve only the terms that might not be small in comparison with L_p. Indeed, as we know from Propositions 4.2.4 and 4.2.1,

$$|\gamma_l| \leq (l-1)!\, L_l \leq (l-1)!\, L_{p-1}^{\frac{l-2}{p-3}}, \quad 3 \leq l \leq p-1, \tag{4.20}$$

which gives

$$\left|\left(\frac{\gamma_3}{3!}\right)^{k_1} \cdots \left(\frac{\gamma_{p-1}}{(p-1)!}\right)^{k_{p-3}}\right| \leq \frac{1}{3^{k_1} \cdots (p-1)^{k_{p-3}}}\, L_{p-1}^{\frac{d}{p-3}}. \tag{4.21}$$

Here, $L_{p-1}^{\frac{d}{p-3}} \leq L_p^{\frac{d}{p-2}}$. So, the left product in (4.21) is at least as small as L_p in the case $d \geq p-2$, when L_p is small. Of course, this should be justified when comparing $e^{Q_{p-1}(it)} - 1$ and $R_{p-1}(it)$ on a proper interval of the t-axis.

The index $p-1$ in R_{p-1} indicates that all cumulants up to γ_{p-1} participate in the constructions of these polynomials. In their definition, it is raised to the power

$$k = 3k_1 + \cdots + (p-1)k_{p-3} = d + 2(k_1 + k_2 + \cdots + k_{p-3}),$$

which may vary from 3 to $3(p-3)$, with maximum $3(p-3)$ attainable when $k_1 = p-3$ and all other $k_r = 0$. Anyway, $\deg(R_{p-1}) \leq 3(p-3)$.

These observations imply a simple bound on the growth of R_{p-1}, which will be needed in the sequel. First, using $|t|^k \leq \max\left\{|t|^3, |t|^{3(p-3)}\right\}$, we have, by (4.21),

$$|R_{p-1}(it)| \leq \max\left\{|t|^3, |t|^{3(p-3)}\right\} \sum \frac{1}{k_1! \cdots k_{p-3}!}\, \frac{1}{3^{k_1} \cdots (p-1)^{k_{p-3}}}\, L_{p-1}^{\frac{d}{p-3}}.$$

Applying the elementary bound

$$\sum \frac{1}{k_1! \cdots k_{p-3}!}\, \frac{1}{3^{k_1} \cdots (p-1)^{k_{p-3}}} < e^{1/3} \cdots e^{1/(p-1)} < p-1, \tag{4.22}$$

we arrive at:

Proposition 4.5.3 *For all $t \in \mathbb{R}$,*

$$|R_{p-1}(it)| \leq (p-1) \max\left\{|t|^3, |t|^{3(p-3)}\right\} \max\left\{L_{p-1}, 1\right\}.$$

4.6 Higher Order Approximations of Characteristic Functions

Using the polynomials R_{p-1} from Definition 4.5.2, the approximation of the characteristic functions f_n given in (4.19) may be simplified in terms of the functions

$$g_{p-1}(t) = \mathrm{e}^{-t^2/2}\,(1 + R_{p-1}(it)). \tag{4.23}$$

Proposition 4.6.1 *If $L_p < \infty$ $(p \geq 4)$, in the interval $|t|\max\big\{L_p^{\frac{1}{p-2}}, L_p^{\frac{1}{3(p-2)}}\big\} \leq 1$, we have for some constant $c_p > 0$ depending on p*

$$|f_n(t) - g_{p-1}(t)| \leq c_p L_p \max\big\{|t|^p, |t|^{3(p-2)}\big\}\,\mathrm{e}^{-t^2/2}. \tag{4.24}$$

This inequality is also true for $p = 3$, when it becomes

$$|f_n(t) - \mathrm{e}^{-t^2/2}| \leq cL_3\,|t|^3\,\mathrm{e}^{-t^2/2},$$

that is, the bound (4.16) of Lemma 4.4.3 (recall that $R_2 = 0$, so that g_2 is the standard normal characteristic function). But, if L_p is smaller than L_3 for $p \geq 4$ (which is typical), (4.24) is better than (4.16). This may be seen in the i.i.d. case where $X_k = \frac{1}{\sqrt{n}}\,\xi_k$ with $\mathbb{E}\,\xi_k = 0$, $\mathbb{E}\,\xi_k^2 = 1$, $\beta_p = \mathbb{E}\,|\xi_k|^p < \infty$. Then (4.24) becomes

$$|f_n(t) - g_{p-1}(t)| \leq c_p\beta_p\,n^{-\frac{p-2}{2}}\,\max\big\{|t|^p, |t|^{3(p-2)}\big\}\,\mathrm{e}^{-t^2/2},$$

which holds on the interval $|t| = O(n^{\frac{1}{3(p-2)}})$. That is why the function $g_{p-1}(t)$ is often called the corrected normal "characteristic" function of order $p-1$.

Let us mention that an inequality similar to (4.24) is also true for the first $p-1$ derivatives of f_n. Namely, in the same interval as in Proposition 4.6.1,

$$\Big|\frac{\mathrm{d}^r}{\mathrm{d}t^r}\,f_n(t) - \frac{\mathrm{d}^r}{\mathrm{d}t^r}\,g_{p-1}(t)\Big| \leq c_p L_p \max\big\{|t|^{p-r}, |t|^{3(p-2)+r}\big\}\,\mathrm{e}^{-t^2/2} \tag{4.25}$$

for any $r = 1, \dots, p-1$.

Proof (of Proposition 4.6.1) In order to apply (4.19), we need to relate R_{p-1} to the cumulant polynomials Q_{p-1}. Recall that

$$|\gamma_l| \leq (l-1)!\,L_l \leq (l-1)!\,L_p^{\frac{l-2}{p-2}}, \quad 3 \leq l \leq p-1,$$

like in (4.20). Hence, for any complex number z in the disc $L_p|z|^{p-2} \leq 1$,

$$\begin{aligned}|Q_m(z)| &\leq \sum_{l=3}^{p-1}\frac{1}{l}\,L_p^{\frac{l-2}{p-2}}\,|z|^l \\ &= L_p^{\frac{1}{p-2}}\,|z|^3\sum_{l=3}^{p-1}\frac{1}{l}\Big(L_p^{\frac{1}{p-2}}\,|z|\Big)^{l-3} \leq \log(p-1)\,L_p^{\frac{1}{p-2}}\,|z|^3.\end{aligned} \tag{4.26}$$

where we used $\sum_{l=3}^{p-1} \frac{1}{l} < \log(p-1)$. Moreover, if additionally $L_p^{\frac{1}{p-2}} |z|^3 \leq 1$, we get a uniform bound $|Q_{p-1}(z)| \leq \log(p-1)$.

Now, consider the function

$$\Psi(w) = \mathrm{e}^w - \sum_{k=0}^{p-3} \frac{1}{k!} w^k = w^{p-2} \sum_{k=p-2}^{\infty} \frac{w^{k-(p-2)}}{k!}$$

of the complex variable w. If $|w| \leq \log(p-1)$, then

$$|\Psi(w)| \leq |w|^{p-2} \sum_{k=p-2}^{\infty} \frac{1}{k!} \left(\log(p-1)\right)^{k-(p-2)} \leq |w|^{p-2}.$$

Indeed, the last sum, call it $S(p)$, does not exceed

$$\frac{1}{(\log(p-1))^{p-2}} \left(\mathrm{e}^{\log(p-1)} - 1 - \log(p-1)\right) = \frac{p-2-\log(p-1)}{(\log(p-1))^{p-2}} \equiv T(p).$$

Clearly, $T(4) < 1$ and $T(5) < 1$. For $p \geq 6$, we have

$$\log T(p) < \log(p-2) - (p-2) \log\log(p-1) < 0.$$

Here, since the function $\frac{\log x}{x}$ is decreasing for $x \geq \mathrm{e}$, we only need to verify the last inequality for $p = 6$, when it is also true. Thus, $S(p) < 1$ for all $p \geq 4$.

The inequality $\Psi(w) \leq |w|^{p-2}$ can be used with $w = Q_{p-1}(z)$, $z = it$. Since $|Q_{p-1}(it)| \leq \log(p-1)$, applying the non-uniform estimate (4.26), we get

$$|\Psi(Q_{p-1}(it))| \leq |Q_{p-1}(it)|^{p-2} \leq \log^{p-2}(p-1)\, L_p |t|^{3(p-2)},$$

that is,

$$\left| \mathrm{e}^{Q_p(it)} - \sum_{k=0}^{p-3} \frac{1}{k!} Q_{p-1}(it)^k \right| \leq \log^{p-2}(p-1)\, L_p |t|^{3(p-2)}. \tag{4.27}$$

As a result, we are only concerned with the remainder term

$$\begin{aligned} r(it) &= \sum_{k=1}^{p-3} \frac{1}{k!} Q_{p-1}(it)^k - R_{p-1}(it) \\ &= \sum_{k=1}^{p-3} \frac{1}{k!} \left(\sum_{l=3}^{p-1} \gamma_l \frac{(it)^l}{l!} \right)^k - R_{p-1}(it). \end{aligned}$$

Using the polynomial formula, let us write down the double sum (with $z = it$) as

$$\sum_{k=1}^{p-3} \sum_{k_1+\cdots+k_{p-3}=k} \frac{1}{k_1! \cdots k_{p-3}!} \left(\frac{\gamma_3}{3!}\right)^{k_1} \cdots \left(\frac{\gamma_{p-1}}{(p-1)!}\right)^{k_{p-3}} z^{3k_1+\cdots+(p-1)k_{p-3}}.$$

This expression defines $R_{p-1}(it)$ with the difference that Definition 4.5.2 contains the constraint $k_1 + 2k_2 + \cdots + (p-3)k_{p-3} \le p-3$, while here we have a weaker constraint $k_1 + k_2 + \cdots + k_{p-3} \le p-3$. Hence, all terms appearing in $R_{p-1}(it)$ are present in the above double sum, so,

$$r(it) = \sum \frac{1}{k_1! \cdots k_{p-3}!} \left(\frac{\gamma_3}{3!}\right)^{k_1} \cdots \left(\frac{\gamma_{p-1}}{(p-1)!}\right)^{k_{p-3}} (it)^{3k_1+\cdots+(p-1)k_{p-3}}$$

with summation subject to

$$k_1 + k_2 + \cdots + k_{p-3} \le p-3, \quad k_1 + 2k_2 + \cdots + (p-3)k_{p-3} \ge p-2.$$

Necessarily, all $k_j \le p-3$ and at least one $k_j \ge 1$. Using $|\gamma_l| \le (l-1)!\, L_p^{\frac{l-2}{p-2}}$, we get

$$|r(z)| \le \sum \prod_{l=3}^{p-1} \frac{1}{k_{l-2}!} \left(\frac{1}{l} L_p^{\frac{l-2}{p-2}}\right)^{k_{l-2}} |z|^N = \sum L_p^M |z|^N \prod_{l=3}^{p-1} \frac{1}{k_{l-2}!} \left(\frac{1}{l}\right)^{k_{l-2}},$$

where

$$M = \frac{1}{p-2}\,(k_1 + 2k_2 + \cdots + (p-3)k_{p-3}),$$

$$\begin{aligned} N &= 3k_1 + \cdots + (p-1)k_{p-3} \\ &= (k_1 + 2k_2 + \cdots + (p-3)k_{p-3}) + 2k, \quad k = k_1 + k_2 + \cdots + k_{p-3}. \end{aligned}$$

Note that $(p-2)M = N - 2k$. If $L_p^{\frac{1}{p-2}} |z| \le 1$, using $1 \le k \le p-2$, we have

$$L_p^{M-1} |z|^N \le |z|^{N-(p-2)(M-1)} = |z|^{(p-2)+2k} \le \max\left\{|z|^p, |z|^{3(p-2)}\right\}.$$

Hence

$$|r(z)| \le L_p \max\left\{|z|^p, |z|^{3(p-2)}\right\} \sum \prod_{l=3}^{p-1} \frac{1}{k_{l-2}!} \left(\frac{1}{l}\right)^{k_{l-2}}.$$

The latter sum is dominated by $\exp\{\sum_{l=3}^{p-1} \frac{1}{l}\} \le p-1$, so

$$|r(z)| \le (p-1)\, L_p \max\left\{|z|^p, |z|^{3(p-2)}\right\}.$$

In view of (4.27), this gives the representation

$$e^{Q_{p-1}(it)} = 1 + R_{p-1}(it) + \delta(t)$$

with an error term satisfying

$$|\delta(t)| \le c_p\, L_p \max\left\{|t|^p, |t|^{3(p-2)}\right\} \quad \text{for } |t| \max\left\{L_p^{\frac{1}{p-2}}, L_p^{\frac{1}{3(p-2)}}\right\} \le 1,$$

where we denote by c_p a constant depending on p only.

Note that $|\delta(t)| \le c_p$ in the indicated interval. Since $|Q_{p-1}(it)| \le \log(p-1)$, we also have $|R_{p-1}(it)| \le c_p + p$. Returning to (4.19), we may write

$$\begin{aligned} f_n(t) &= \mathrm{e}^{-t^2/2}\Big(1 + R_{p-1}(it) + \delta(t)\Big)\,(1+\varepsilon(t)) \\ &= \mathrm{e}^{-t^2/2}\Big(1 + R_{p-1}(it)\Big) + \mathrm{e}^{-t^2/2}\Big((1 + R_{p-1}(it))\,\varepsilon(t) + (1+\varepsilon(t))\,\delta(t)\Big). \end{aligned}$$

Here, by Lemma 4.5.1, $|\varepsilon(t)| \le c_p L_p |t|^p$ under the condition $L_p^{\frac{1}{p}}|t| \le 1$. This is indeed fulfilled, since $L_p^{\frac{1}{p}} \le \max\{L_p^{\frac{1}{p-2}}, L_p^{\frac{1}{3(p-2)}}\}$. Moreover, $L_p|t|^p \le 1$, so that $|\varepsilon(t)| \le c_p$. Since both $\varepsilon(t)$ and $\delta(t)$ have been properly bounded, Proposition 4.6.1 is proved. □

4.7 Edgeworth Corrections

Again, let $X_1, \dots, X_n$ be independent random variables with mean zero and variances $b_k^2 = \mathrm{Var}(X_k)$ such that $\sum_{k=1}^n b_k^2 = 1$. If the Lyapunov coefficient L_p is small for an integer $p \ge 4$, Proposition 4.6.1 tells us that the characteristic function $f_n(t)$ of

$$S_n = X_1 + \dots + X_n$$

is close to the function $g_{p-1}(t)$, defined in (4.23), on a relatively long interval. Therefore, it is reasonable to believe that in some sense the distribution of S_n is close to the signed (Lebesgue–Stieltjes) measure μ_{p-1} on the real line, whose Fourier–Stieltjes transform is exactly g_{p-1}, that is, with

$$\int_{-\infty}^{\infty} \mathrm{e}^{itx}\,\mathrm{d}\mu_{p-1}(x) = g_{p-1}(t), \quad t \in \mathbb{R}. \tag{4.28}$$

In order to describe these measures explicitly, let us recall the Chebyshev–Hermite polynomials

$$H_k(x) = (-1)^k\,(\mathrm{e}^{-x^2/2})^{(k)}\,\mathrm{e}^{x^2/2}, \qquad k = 0, 1, 2, \dots \ (x \in \mathbb{R}).$$

Equivalently, $\varphi^{(k)}(x) = (-1)^k H_k(x)\varphi(x)$ in terms of the standard normal density φ. Each H_k is a polynomial of degree k with leading coefficient 1. For example,

$$\begin{aligned} &H_0(x) = 1, \quad H_2(x) = x^2 - 1, \quad H_4(x) = x^4 - 6x^2 + 3, \\ &H_1(x) = x, \quad H_3(x) = x^3 - 3x, \quad H_5(x) = x^5 - 10x^3 + 15x. \end{aligned}$$

These polynomials are orthogonal on the real line with weight $\varphi(x)$ and form a complete orthogonal system in the Hilbert space $L^2(\mathbb{R}, \varphi(x)\,\mathrm{d}x)$. By repeated integration by parts (with $t \ne 0$)

$$e^{-t^2/2} = \int_{-\infty}^{\infty} e^{itx} \varphi(x)\,dx = \frac{1}{(-it)^k} \int_{-\infty}^{\infty} e^{itx} \varphi^{(k)}(x)\,dx,$$

we have the identity

$$\int_{-\infty}^{\infty} e^{itx} H_k(x)\varphi(x)\,dx = (it)^k\, e^{-t^2/2}.$$

Using the inverse Fourier transform, one may therefore write

$$H_k(x)\,\varphi(x) = \frac{1}{2\pi} \int_{-\infty}^{\infty} e^{-itx}\,(it)^k\, e^{-t^2/2}\,dt,$$

which may be taken as another definition of H_k. Returning to Definition 4.5.2, we therefore obtain:

Proposition 4.7.1 *Let $L_{p-1} < \infty$ for an integer $p \geq 4$. The measure μ_{p-1} with Fourier–Stieltjes transform g_{p-1} has density*

$$\varphi_{p-1}(x) = \varphi(x) + \varphi(x) \sum \frac{1}{k_1! \cdots k_{p-3}!} \left(\frac{\gamma_3}{3!}\right)^{k_1} \cdots \left(\frac{\gamma_{p-1}}{(p-1)!}\right)^{k_{p-3}} H_k(x),$$

where $k = 3k_1 + \cdots + (p-1)k_{p-3}$ and where the summation runs over all non-negative integers $k_1, \dots, k_{p-3}$ not all zero and such that $k_1 + 2k_2 + \cdots + (p-3)k_{p-3} \leq p-3$.

Recall that the cumulants γ_r of S_n are well defined for all $1 \leq r \leq p-1$. If $p = 3$, the above sum is empty, and we put $\varphi_2 = \varphi$.

Since $(H_{k-1}(x)\varphi(x))' = -H_k(x)\varphi(x)$, the corresponding "distribution" function

$$\Phi_{p-1}(x) = \mu_{p-1}((-\infty, x]) = \int_{-\infty}^{x} \varphi_{p-1}(y)\,dy, \qquad x \in \mathbb{R},$$

is described as

$$\Phi_{p-1}(x) = \Phi(x) - \varphi(x) \sum \frac{1}{k_1! \cdots k_{p-3}!} \left(\frac{\gamma_3}{3!}\right)^{k_1} \cdots \left(\frac{\gamma_{p-1}}{(p-1)!}\right)^{k_{p-3}} H_{k-1}(x)$$

with summation as before.

Definition 4.7.2 The signed measure μ_{p-1} is called the Edgeworth approximation of order $p-1$ for the distribution of S_n (or an Edgeworth correction of the normal law). Thus,

$$\varphi_{p-1}(x) = \varphi(x) + \varphi(x) U_{p-1}(x),$$
$$\Phi_{p-1}(x) = \Phi(x) - \varphi(x) V_{p-1}(x),$$

where U_{p-1} and V_{p-1} are polynomials of degree at most $3(p-3)$ and $3(p-3)-1$, respectively.

For the first values, we have

$$\begin{aligned}
V_3(x) &= \frac{\gamma_3}{3!}\,H_2(x),\\
V_4(x) &= \frac{\gamma_3}{3!}\,H_2(x) + \frac{\gamma_4}{4!}\,H_3(x) + \frac{\gamma_3^2}{2!\,3!^2}\,H_5(x),\\
V_5(x) &= \frac{\gamma_3}{3!}\,H_2(x) + \frac{\gamma_4}{4!}\,H_3(x) + \frac{\gamma_5}{5!}\,H_4(x)\\
&\quad + \frac{\gamma_3^2}{2!\,3!^2}\,H_5(x) + \frac{\gamma_3\gamma_4}{3!\,4!}\,H_6(x) + \frac{\gamma_3^3}{3!^4}\,H_8(x).
\end{aligned}$$

Let us briefly describe a few basic properties of these measures.

Proposition 4.7.3 *The moments of S_n and μ_{p-1} coincide up to order $p-1$, or equivalently*

$$f_n^{(r)}(0) = g_{p-1}^{(r)}(0), \quad r = 0, 1, \dots, p-1.$$

In particular,

$$\mu_{p-1}(\mathbb{R}) = \int_{-\infty}^{\infty} \varphi_{p-1}(x)\,\mathrm{d}x = 1.$$

The latter immediately follows from the Fourier transform formula (4.28) applied at $t = 0$. The more general assertion may be obtained as a consequence of (4.25), which gives $|f_n^{(r)}(t) - g_{p-1}^{(r)}(t)| = O(|t|^{p-r})$ as $t \to 0$.

Proposition 4.7.4 *If $L_{p-1} < \infty$ for an integer $p \geq 4$, then for some positive constant c_p depending on p,*

$$\int_{-\infty}^{\infty} (1+|x|)^p\,|\mu_{p-1}(\mathrm{d}x) - \mu(\mathrm{d}x)| \leq c_p \max\{L_{p-1}, 1\},$$

where μ is the standard Gaussian measure. In addition,

$$\sup_x |\varphi_{p-1}(x) - \varphi(x)| \leq c_p \max\{L_{p-1}, 1\}.$$

Proof Denote by I the integral appearing in the first inequality. It describes the weighted total variation distance between the measures μ_{p-1} and μ with weight $(1+|x|)^p$ and is explicitly given as

$$I = \int_{-\infty}^{\infty} (1+|x|^p)\,|U_{p-1}(x)|\,\varphi(x)\,\mathrm{d}x.$$

In Proposition 4.7.1, the tuples $(k_1, \dots, k_{p-3})$ from the sum satisfy $1 \leq d \leq p-3$, where $d = k_1 + 2k_2 + \cdots + (p-3)k_{p-3}$. Hence, by (4.21), I is bounded by

$$\max\{L_{p-1}, 1\} \sum \frac{1}{k_1! \dots k_{p-3}!}\,\frac{1}{3^{k_1} \dots (p-1)^{k_{p-3}}} \int_{-\infty}^{\infty} (1+|x|^p)\,|H_k(x)|\,\varphi(x)\,\mathrm{d}x,$$

where $k = 3k_1 + \cdots + (p-1)k_{p-3}$ may vary from 3 to $3(p-3)$. Let Z be a standard normal random variable. Using $\mathbb{E}\, H_k(Z)^2 = k!$ we get, by the Cauchy inequality,

$$\int_{-\infty}^{\infty} (1+|x|)^p \, |H_k(x)| \, \varphi(x) \, dx = \mathbb{E}\,(1+|Z|)^p \, |H_k(Z)| \\ \le \sqrt{\mathbb{E}\,(1+|Z|)^{2p}} \sqrt{k!} \le c_p.$$

Hence I does not exceed

$$c_p \max\{L_{p-1}, 1\} \sum \frac{1}{k_1! \cdots k_{p-3}!} \, \frac{1}{3^{k_1} \cdots (p-1)^{k_{p-3}}}.$$

The latter sum is smaller than $p-1$, cf. (4.22), and we obtain the first inequality.

The second bound follows from Proposition 4.5.3. Since

$$\varphi_{p-1}(x) - \varphi(x) = \int_{-\infty}^{\infty} e^{itx} \, (g_{p-1}(t) - e^{-t^2/2}) \, dt = \int_{-\infty}^{\infty} e^{itx} \, R_{p-1}(it) \, e^{-t^2/2} \, dt,$$

it gives

$$|\varphi_{p-1}(x) - \varphi(x)| \le (p-1) \max\big\{L_{p-1}, 1\big\} \int_{-\infty}^{\infty} \max\big\{|t|^3, |t|^{3(p-3)}\big\} \, e^{-t^2/2} \, dt.$$

Proposition 4.7.4 is proved. □

4.8 Rates of Approximation

We are prepared to derive a Berry–Esseen-type bound quantifying the approximation of the distribution function $F_n(x) = \mathbb{P}\{S_n \le x\}$ by the corrected normal "distribution" function Φ_{p-1}. Throughout, we denote by c_p a positive constant depending on p only, which may vary from place to place. Continuing the setting of the previous section, we have:

Proposition 4.8.1 *If $L_p < \infty$ $(p \ge 3)$, then for any $\delta \ge 0$,*

$$c_p \, \rho(F_n, \Phi_{p-1}) \le L_p + \delta + 1_{\{\delta \le L_3 \le 1\}} \int_{1/L_3}^{1/\delta} \frac{|f_n(t)|}{t} \, dt. \tag{4.29}$$

As before, we denote by $f_n(t)$ the characteristic function of S_n.

If $p = 3$ and $\delta = L_3$, we return in (4.29) to the Berry–Esseen bound (4.15) in Proposition 4.4.1 (so, this value of δ is optimal). If $p \ge 4$, choosing $\delta = L_p$, (4.29) simplifies to

$$c_p \, \rho(F_n, \Phi_{p-1}) \le L_p + 1_{\{L_p \le L_3 \le 1\}} \int_{1/L_3}^{1/L_p} \frac{|f_n(t)|}{t} \, dt.$$

However, in further applications a slightly different choice of δ will be more useful.

As a preliminary step, here we extend the inequality (4.24), although with a worse Gaussian decay on the right-hand side, to the larger interval in full analogy with Lemma 4.4.2. The next inequality is of independent interest.

Proposition 4.8.2 *If $L_p < \infty$ $(p \geq 3)$, then in the interval $|t| \leq 1/L_3$, we have*

$$|f_n(t) - g_{p-1}(t)| \leq c_p L_p \min\left\{1, |t|^p\right\} \mathrm{e}^{-t^2/8}. \tag{4.30}$$

In the proof, we employ Lemma 4.4.2 in order to bound the characteristic function $f_n(t)$ outside the interval of Proposition 4.6.1, together with a similar observation about the decay of the corrected normal characteristic function.

Lemma 4.8.3 *In the region $|t| \max\{L_p^{\frac{1}{p-2}}, L_p^{\frac{1}{3(p-2)}}\} \geq 1$, we have*

$$|g_{p-1}(t)| \leq c_p L_p \mathrm{e}^{-t^2/8}.$$

Proof We use (4.21), implying that

$$|R_{p-1}(it)| \leq \sum \frac{1}{k_1! \cdots k_{p-3}!} \frac{1}{3^{k_1} \cdots (p-1)^{k_{p-3}}} L_p^{\frac{d}{p-2}} |t|^k,$$

where the summation runs over all collections of non-negative integers $(k_1, \dots, k_{p-3})$ that are not all zero and such that

$$k = 3k_1 + \cdots + (p-1)k_{p-3}, \quad d = k_1 + 2k_2 + \cdots + (p-3)k_{p-3} \leq p-3.$$

All tuples that are involved satisfy $1 \leq k \leq 3d \leq 3(p-3)$.

If $L_p \geq 1$, then $L_p^{\frac{d}{p-2}} \leq L_p$, and using $|t|^k \mathrm{e}^{-3t^2/8} \leq c_p$, we get $L_p^{\frac{d}{p-2}} |t|^k \mathrm{e}^{-t^2/2} \leq c_p L_p \mathrm{e}^{-t^2/8}$, and the inequality of the lemma follows.

If $L_p \leq 1$, it will be sufficient to bound the products $L_p^{\frac{d}{p-2}-1} |t|^k \mathrm{e}^{-3t^2/8}$ uniformly for all admissible tuples. Put $x = L_p^{-\frac{1}{p-2}}$. Using the hypothesis $|t| \geq x^{1/3}$, let us rewrite every such product and then estimate it as follows:

$$\begin{aligned} L_p^{\frac{d}{p-2}-1} |t|^k \mathrm{e}^{-3t^2/8} &= x^{(p-2)-d} \mathrm{e}^{-t^2/4} \cdot |t|^k \mathrm{e}^{-t^2/8} \\ &\leq x^{(p-2)-d} \mathrm{e}^{-\frac{1}{4} x^{2/3}} \cdot |t|^k \mathrm{e}^{-t^2/8} \\ &= (4y)^{\frac{3}{2}((p-2)-d)} \mathrm{e}^{-y} \cdot (8u)^{k/2} \mathrm{e}^{-u}, \end{aligned}$$

where we changed the variables $x = (4y)^{3/2}$, $t = (8u)^{1/2}$. Both functions in the variables $y \geq 0$ and $u \geq 0$ are bounded by constants depending p, only. Thus,

$$L_p^{\frac{d}{p-2}} |t|^k \mathrm{e}^{-t^2/2} \leq c_p L_p \mathrm{e}^{-t^2/8},$$

proving the lemma. □

Proof (of Proposition 4.8.2) We distinguish between several cases.

Case 1. Moderate interval: $|t|\ \max\{L_p^{\frac{1}{p-2}}, L_p^{\frac{1}{3(p-2)}}\} \leq 1$. In this case, the inequality (4.30) follows from (4.24).

Case 2. Large region: $|t|\ \max\{L_p^{\frac{1}{p-2}}, L_p^{\frac{1}{3(p-2)}}\} \geq 1$ with $1 \leq |t| \leq \frac{1}{L_3}$. In this case, we bound both $f_n(t)$ and $g_{p-1}(t)$ in absolute value by appropriate quantities. First, by Lemma 4.4.2, $|f_n(t)| \leq \mathrm{e}^{-t^2/6}$, which is valid for $|t| \leq \frac{1}{L_3}$. Let us now consider an estimate of the form

$$\mathrm{e}^{-t^2/24} \leq C_p L_p.$$

If $L_p \geq 1$, this holds with $C_p = 1$. If $L_p \leq 1$, then necessarily $|t| \geq t_0 \equiv L_p^{-\frac{1}{3(p-2)}}$, so, one may take

$$C_p = \frac{1}{L_p}\,\exp\Big\{-\frac{1}{24}\,L_p^{-\frac{2}{3(p-2)}}\Big\} = t_0^{3(p-2)}\,\mathrm{e}^{-\frac{1}{24}t_0^2},$$

which is bounded by a p-dependent constant c_p. As a result, we arrive at the desired upper bound

$$|f_n(t)| \leq \mathrm{e}^{-t^2/24}\cdot \mathrm{e}^{-t^2/8} \leq c_p L_p\, \mathrm{e}^{-t^2/8}.$$

A similar bound also holds for the approximating function $g_{p-1}(t)$, by Lemma 4.8.3 whenever $|t| \geq 1$, implying (4.30).

Case 3. Consider the region: $|t|\ \max\{L_p^{\frac{1}{p-2}}, L_p^{\frac{1}{3(p-2)}}\} \geq 1$ with $|t| \leq \min\{1, \frac{1}{L_3}\}$. Necessarily $L_p \geq 1$. Hence, by Proposition 4.7.4,

$$\int_{-\infty}^{\infty} |x|^p\ |\mathrm{d}\mu_{p-1}(x)| \leq c_p L_p,$$

implying that, for all $t \in \mathbb{R}$, we have a similar bound on the p-th derivative of the corrected normal characteristic function, namely

$$|g_{p-1}^{(p)}(t)| = \Big|\int_{-\infty}^{\infty} x^p\, \mathrm{e}^{itx}\, \mathrm{d}\mu_{p-1}(x)\Big| \leq c_p L_p.$$

Similarly, by the Rosenthal inequality, $\big|f_n^{(p)}(t)\big| \leq \mathbb{E}\,|S_n|^p \leq c_p L_p$. Since also $f_n^{(r)}(0) = g_{p-1}^{(r)}(0)$ for $r = 1, \dots, p-1$ (Proposition 4.7.3), it follows from the Taylor integral formula that

$$|f_n(t) - g_{p-1}(t)| \leq 2c_p L_p\, \frac{|t|^p}{p!}.$$

Proposition 4.8.2 is proved. □

Proof (of Proposition 4.8.1) As was mentioned, we may assume that $p \geq 4$.

If $L_p \geq 1$, there is nothing to prove. Indeed, applying Proposition 4.7.4 (the first inequality for the total variation norm without weight), we conclude that

$$\begin{aligned}\rho(F_n, \Phi_{p-1}) &\le \rho(F_n, \Phi) + \|\mu_{p-1} - \mu\|_{\mathrm{TV}} \\ &\le 1 + c_p \max\{L_{p-1}, 1\} \le 1 + c_p L_p^{\frac{p-3}{p-2}} \le (1 + c_p) L_p.\end{aligned}$$

Thus, we may assume that $L_p \le 1$, in which case necessarily $L_3 \le L_p^{\frac{1}{p-2}} \le 1$, by Proposition 4.2.1. First let $\delta \le L_3$. We are in position to apply Proposition 3.2.1 with $F = F_n$ and $G = \Phi_{p-1}$. Choosing $T = 1/\delta$, the inequality (3.5) yields

$$\rho(F_n, \Phi_{p-1}) \le 2 \int_0^{1/\delta} \frac{|f_n(t) - g_{p-1}(t)|}{t}\, dt + 2\pi^2 A\delta, \tag{4.31}$$

where $A = \sup_x |\Phi'_{p-1}(x)| = \sup_x |\varphi_{p-1}(x)|$ is bounded by a p-dependent constant, by Proposition 4.7.4. The bound (4.30) may be used to estimate the integral over the interval $0 \le t \le 1/L_3$, which gives

$$\int_0^{1/L_3} \frac{|f_n(t) - g_{p-1}(t)|}{t}\, dt \le c_p L_p \int_0^\infty \min\{t^{p-1}, 1\}\, e^{-t^2/8}\, dt = c'_p L_p.$$

Next, by Proposition 4.5.3, and using $L_{p-1} \le L_p^{\frac{p-3}{p-2}} \le 1$, we have

$$|g_{p-1}(t)| \le e^{-t^2/2} + (p-1) \max\left\{|t|^3, |t|^{3(p-3)}\right\} e^{-t^2/2} \le c_p\, e^{-t^2/4},$$

which implies

$$\int_{1/L_3}^\infty \frac{|g_{p-1}(t)|}{t}\, dt \le 2c_p\, e^{-1/(4L_3^2)} \le c'_p L_3^{p-2} \le c'_p L_p$$

(where we used Proposition 4.2.1 in the last step). The two integral bounds give

$$\int_0^{1/\delta} \frac{|f_n(t) - g_{p-1}(t)|}{t}\, dt \le c_p L_p + \int_{1/L_3}^{1/\delta} \frac{|f_n(t)|}{t}\, dt.$$

Hence, the right-hand side of (4.31) is bounded by the right-hand side of (4.29).

In the case $L_3 < \delta$, the integral term in (4.29) is not present. Choosing $T = 1/L_3$ in Proposition 3.2.1, the bound (4.30) similarly yields

$$\begin{aligned}\rho(F_n, \Phi_{p-1}) &\le 2 \int_0^{1/L_3} \frac{|f_n(t) - g_{p-1}(t)|}{t}\, dt + 2\pi^2 A L_3 \\ &\le c_p\,(L_p + L_3) \le c_p\,(L_p + \delta).\end{aligned}$$

This corresponds to (4.29) without the integral on the right-hand side.

Proposition 4.8.1 is proved. □

4.9 Remarks

Assuming that the random variable X has mean zero and standard deviation σ, the constant in the inequality (4.3) can be improved for a smaller interval $\sigma|t| \le \frac{1}{5}$. As was shown in [33],

$$\left|(\log f(t))^{(p)}\right| \le (p-1)!\,\beta_p \quad (p \ge 3),$$

and thus $|\varepsilon(p)| \le \frac{1}{p}\beta_p$. This extends the Bikjalis inequality $|\gamma_p| \le (p-1)!\,\beta_p$ from Proposition 4.1.3. The factorial growth of the constant is optimal, up to an exponentially growing factor as p tends to infinity, which was emphasized by Bulinskii [72] in his study of upper bounds in a more general scheme of random vectors and associated mixed cumulants. To illustrate possible lower bounds, he considered the symmetric Bernoulli distribution assigning the mass $\frac{1}{2}$ to the points ± 1. In this case, the characteristic function is $f(t) = \cos t$, and one may use the Taylor expansion

$$\log f(t) = \log\cos t = -\sum_{s=1}^{\infty} \frac{2^{2s}\,(2^{2s}-1)}{(2s)!}\,B_s\,\frac{t^{2s}}{2s}, \qquad |t| < \frac{\pi}{2},$$

involving Bernoulli numbers $B_p = \frac{2\,(2p)!}{(2\pi)^{2p}}\,d_{2p}$, where $d_{2p} = \sum_{n=1}^{\infty} n^{-2p}$. Thus, for even integer values of p,

$$|\gamma_p| = \frac{2^p\,(2^p-1)}{p}\,B_{p/2} = \frac{2\,(2^p-1)}{\pi^p}\,(p-1)!\,d_p \sim 2\left(\frac{2}{\pi}\right)^p (p-1)!$$

To compare with the upper bound, note that in this Bernoulli case, $\beta_p = 1$ for all p.

The study of the best value C_p in Rosenthal's inequality (4.10) has a long history, and here we only mention some selected results. Define C_p^* to be an optimal constant in (4.10), when additionally assuming that the distributions of X_k are symmetric around the origin. By Jensen's inequality, there is a simple general relation

$$C_p^* \le C_p \le 2^{p-1} C_p^*,$$

which reduces in essence the study of Rosenthal-type inequalities to the symmetric case. Johnson, Schechtman and Zinn [109] have derived the two-sided bounds

$$\frac{p}{\sqrt{2}\,\mathrm{e}\,\log(\max(p,\mathrm{e}))} \le (C_p^*)^{1/p} \le \frac{7.35\,p}{\log(\max(p,\mathrm{e}))}.$$

Hence, asymptotically $C_p^{1/p}$ is of order $p/\log p$ for growing values of p. They have also obtained an upper bound with a better numerical factor, $(C_p^*)^{1/p} \le p/\sqrt{\log\max(p,\mathrm{e})}$, which implies a simple bound $C_p \le (2p)^p$, $p > 2$.

As for the best constant in the symmetric case, it was shown by Ibragimov and Sharakhmetov [107] that $C_p^* = \mathbb{E}\,|\xi-\eta|^p$ for $p > 4$, where ξ and η are independent Poisson random variables with parameter $\lambda = \frac{1}{2}$ (cf. also Pinelis [156] for a similar description without the symmetry assumption). In particular, $(C_p^*)^{1/p} \sim \frac{p}{\mathrm{e}\log p}$ as p tends to infinity. This result easily yields $C_p^* \le p!$ for $p = 3, 4, 5, \dots$

Standard sources about Edgeworth expansions for distributions of the sums of independent random variables are the books by Petrov [154], [155], and by Bhattacharya and Ranga Rao [16]. The survey [33] extends several results in this direction to non-integer values of the parameter p. In particular, the inequalities (4.24) and (4.30) are derived there for $p = m + \alpha$ with integers $m \ge 2$ and $0 < \alpha \le 1$, with g_{p-1} replaced by g_m, and for the derivatives of order $r = 0, 1, \dots, m$.

Part II
Selected Topics on Concentration

Chapter 5
Standard Analytic Conditions

Many interesting classes of probability distributions on $\mathbb{R}^n$ may be described by means of various analytic, that is, integro-differential inequalities, usually called Sobolev-type inequalities. Keeping the setting of an abstract metric space, in this chapter we introduce the most frequently used ones known as Poincaré-type and (stronger) Cheeger-type inequalities. The second type appears as a particular case of more general Sobolev-type inequalities which can be used as an equivalent form for isoperimetric inequalities. At the end of the chapter, we briefly describe basic examples of probability distributions on Euclidean space satisfying these inequalities.

5.1 Moduli of Gradients in the Continuous Setting

Analytic conditions imposed on a given Borel probability measure μ on $\mathbb{R}^n$ are usually expressed in the form of Sobolev-type inequalities which connect the distribution of an arbitrary smooth function f under μ with the distribution of its gradient ∇f, more often – with the distribution of its Euclidean length, or the modulus of the gradient

$$|\nabla f(x)| = \Big(\sum_{k=1}^{n} |\partial_k f(x)|^2\Big)^{1/2}, \tag{5.1}$$

where $\partial_k f(x) = \frac{\partial f(x)}{\partial x_k}$ are partial derivatives of f at the point x. For various reasons, it is convenient to extend such inequalities to larger classes of functions, such as the class of all locally Lipschitz f. Moreover, to keep a higher level of generality and thus get more freedom in applications, it is also useful to have results that may be derived from Sobolev-type inequalities in the setting of abstract "continuous" metric spaces rather than on $\mathbb{R}^n$, only. Therefore, let us start with an arbitrary metric space $(\mathbb{M}, d)$ without isolated points – a property which serves well enough the idea of "continuity" of the space.

A function f on $\mathbb{M}$ is said to be Lipschitz on a given nonempty subset A of $\mathbb{M}$ if, for some constant $L \geq 0$, we have

S. Bobkov et al., *Concentration and Gaussian Approximation for Randomized Sums*, Probability Theory and Stochastic Modelling 104,
https://doi.org/10.1007/978-3-031-31149-9_5

$$|f(x) - f(y)| \le L d(x, y) \quad \text{for all } x, y \in A.$$

The optimal value of L is called the Lipschitz semi-norm of f on A, and when $A = \mathbb{M}$, it will be denoted $\|f\|_{\mathrm{Lip}}$. The function f will be called locally Lipschitz if it has a finite Lipschitz semi-norm on every ball in $\mathbb{M}$. When $\mathbb{M}$ is locally compact, and every ball is compact in $\mathbb{M}$, the latter is equivalent to the property that any point in $\mathbb{M}$ has a neighborhood where f has a finite Lipschitz semi-norm.

For locally Lipschitz functions f on $\mathbb{M}$, the modulus of the gradient may be understood in the generalized sense as the finite function

$$|\nabla f|(x) = |\nabla f(x)| = \limsup_{y \to x} \frac{|f(x) - f(y)|}{d(x, y)}, \qquad x \in \mathbb{M}. \tag{5.2}$$

One may also involve the operators

$$\Delta_\varepsilon f(x) = \sup_{d(x,y)<\varepsilon} |f(y) - f(x)|, \qquad \varepsilon > 0,$$

which represent lower semi-continuous (hence Borel measurable) functions, as long as f is continuous (as the supremum of a family of lower semi-continuous functions $x \to |f(y) - f(x)|\, 1_{\{d(x,y)<\varepsilon\}}$). It then follows from (5.2) that, for all $x \in \mathbb{M}$,

$$|\nabla f(x)| = \limsup_{\varepsilon \to 0} \frac{\Delta_\varepsilon f(x)}{\varepsilon} = \limsup_{n \to \infty} \Big[n \Delta_{1/n} f(x) \Big],$$

implying that the function $|\nabla f|$ is Borel measurable.

The definition (5.2) extends by the same formula to complex-valued functions $f = u + iv$, $u = \mathrm{Re}(f)$, $v = \mathrm{Im}(f)$, and then we have the obvious relations

$$|\nabla u| \le |\nabla f|, \quad |\nabla v| \le |\nabla f|, \quad |\nabla f|^2 \le |\nabla u|^2 + |\nabla v|^2.$$

When f is defined on an open set $\mathbb{M}$ in $\mathbb{R}^n$ (with the Euclidean distance d) and is differentiable at the point $x \in \mathbb{M}$, this definition leads to the modulus of the gradient in the usual sense (5.1), and in the complex-valued case we have $|\nabla f(x)|^2 = |\nabla u(x)|^2 + |\nabla v(x)|^2$. But, if f is not differentiable at x, it is safer to understand $|\nabla f(x)|$ according to (5.2).

Note that when f is locally Lipschitz on the open set $\mathbb{M} \subset \mathbb{R}^n$, it must be differentiable at almost all points with respect to the Lebesgue measure (by a well-known theorem of Rademacher). In dimension one, any locally Lipschitz function f on $\mathbb{M} = (a, b)$ is absolutely continuous (locally). Conversely, if f is (locally) absolutely continuous on that interval, then (5.2) defines the modulus of the Radon–Nikodym derivative of f. Another important particular space which will be used intensively is the unit sphere $\mathbb{M} = \mathbb{S}^{n-1}$, where (5.2) defines the length of the spherical gradient of f at x.

In some problems/Sobolev-type inequalities, it makes sense to slightly modify the notion of the generalized modulus of gradient. For example, on product spaces $\mathbb{M} = \mathbb{M}_1 \times \cdots \times \mathbb{M}_n$ with Euclidean-type metrics

$$d(x, y) = \Big(\sum_{k=1}^{n} d_k(x_k, y_k)^2 \Big)^{1/2},$$

where $x = (x_1, \dots, x_n)$, $y = (y_1, \dots, y_n)$, $x_k, y_k \in \mathbb{M}_k$, and where d_k is a metric in $\mathbb{M}_k$, a natural alternative is given by

$$|\widetilde{\nabla} f(x)| = \Big(\sum_{k=1}^{n} |\nabla_{x_k} f(x)|^2 \Big)^{1/2}. \tag{5.3}$$

Here $|\nabla_{x_k} f(x)|$ is defined on $(\mathbb{M}_k.d_k)$ according to (5.2) for the function $x_k \to f(x)$ with fixed x_i, $i \neq k$. If f is locally Lipschitz on $\mathbb{M} = \mathbb{R}^n$, we have again $|\tilde{\nabla} f(x)| = |\nabla f(x)|$ for all points $x \in \mathbb{R}^n$ where f is differentiable (hence, for almost all x).

The definition (5.2) saves many relations from calculus or, at least, it transforms them into corresponding inequalities. For example, for any family of functions $(f_t)_{t \in T}$ on $\mathbb{M}$,

$$\Big|\nabla \sup_{t \in T} f_t(x)\Big| \leq \sup_{t \in T} |\nabla f_t(x)|.$$

The chain rule formula turns into

$$|\nabla T(f)| \leq |T'(f)|\, |\nabla f(x)|,$$

where $|T'|$ may also be understood according to (5.2) on the real line $\mathbb{R}$. It will be called the chain rule inequality.

An application of the operator $f \to |\nabla f|$ to the function $|\nabla f|$ leads to the second order modulus of the gradient

$$|\nabla^2 f(x)| = |\nabla\, |\nabla f(x)|\,| = \limsup_{y \to x} \frac{|\,|\nabla f(x)| - |\nabla f(y)|\,|}{d(x, y)}. \tag{5.4}$$

Note that the Lipschitz property $\|f\|_{\mathrm{Lip}} \leq 1$ implies that $|\nabla f(x)| \leq 1$ for all $x \in \mathbb{M}$, and the converse is also often true, including the case where $\mathbb{M}$ is an open convex subset of $\mathbb{R}^n$ (with the induced Euclidean distance). Hence, in this case, the requirement $|\nabla^2 f(x)| \leq 1$ for every $x \in \mathbb{M}$ means that the function $|\nabla f|$ is Lipschitz. If $|\nabla f|$ is locally Lipschitz, then f is of course locally Lipschitz as well.

If f is C^2-smooth on an open subset $\mathbb{M}$ of $\mathbb{R}^n$, we denote by

$$f''(x) = \left(\frac{\partial^2 f(x)}{\partial x_i \partial x_j} \right)_{i,j=1}^{n}$$

the $n \times n$ matrix of second order partial derivatives of f at x (the Hessian of f). In this case, the function $|\nabla f|$ will be locally Lipschitz, and as is easy to check,

$$|\nabla^2 f(x)| = |\nabla f(x)|^{-1} |f''(x) \nabla f(x)|$$

for all points at which $|\nabla f(x)| > 0$. Here the right-hand side should be understood as the operator norm $\|f''(x)\|$ in the case $|\nabla f(x)| = 0$. In particular, in all cases

$$|\nabla^2 f(x)| \le \|f''(x)\| \quad \text{for all } x \in \mathbb{M}.$$

For example, for the quadratic function $f(x) = \frac{1}{2}\sum_{k=1}^n \lambda_k x_k^2$, $x = (x_1, \dots, x_n)$,

$$|\nabla^2 f(x)| = \frac{\sqrt{\sum_{k=1}^n \lambda_k^4 x_k^2}}{\sqrt{\sum_{k=1}^n \lambda_k^2 x_k^2}} \le \max_k |\lambda_k|.$$

5.2 Perimeter and Co-area Inequality

Let μ be a Borel probability measure on the metric space (M, d). Integrals involving moduli of gradients are closely related to geometric characteristics such as the perimeter of sets (via the co-area formula). Given a non-empty set A in $\mathbb{M}$ and $\varepsilon > 0$, denote by

$$A_\varepsilon = \{x \in \mathbb{M} : d(x, y) < \varepsilon \text{ for some } y \in A\}$$

an ε-neighborhood of A. Being a union of open balls in $\mathbb{M}$, it is an open set as well. The (outer) μ-perimeter of A is defined by

$$\mu^+(A) = \liminf_{\varepsilon \to 0} \frac{\mu(A_\varepsilon) - \mu(A)}{\varepsilon}.$$

In the abstract metric probability space setting the co-area formula takes the form of the following inequality.

Proposition 5.2.1 *For any Lipschitz function f on $\mathbb{M}$,*

$$\int |\nabla f| \, d\mu \ge \int_0^\infty \mu^+\{|f| > t\} \, dt. \tag{5.5}$$

Proof Since $|\nabla |f|| \le |\nabla f|$, we may assume that $f \ge 0$. First, let f be bounded and satisfy $|f(x) - f(y)| \le L d(x, y)$ for all $x, y \in \mathbb{M}$ with some constant L. Put $A(t) = \{x \in \mathbb{M} : f(x) > t\}$, $t \in \mathbb{R}$, and

$$f_\varepsilon(x) = \sup\{f(y) : d(x, y) < \varepsilon\}, \quad \varepsilon > 0.$$

Clearly, $\{x \in \mathbb{M} : f_\varepsilon(x) > t\} = A(t)_\varepsilon$, so, these sets are open, which implies that the function f_ε is lower semi-continuous. We can also write

$$\int f_\varepsilon \, d\mu = \int_0^\infty \mu(A(t)_\varepsilon) \, dt,$$

and with a similar representation for f we get

$$\int \frac{f_\varepsilon(x) - f(x)}{\varepsilon} \, d\mu(x) = \int_0^\infty \frac{\mu(A(t)_\varepsilon) - \mu(A(t))}{\varepsilon} \, dt.$$

By the Lipschitz condition, $0 \le f_\varepsilon(x) - f(x) \le L\varepsilon$ for all $x \in \mathbb{M}$, so, the left integrand is bounded by L. Therefore, using the Lebesgue dominated convergence theorem and Fatou's lemma, and noting that

$$\limsup_{\varepsilon\to 0} \frac{f_\varepsilon(x) - f(x)}{\varepsilon} = \limsup_{y\to x} \frac{f(y) - f(x)}{\varepsilon} \le |\nabla f(x)|,$$

we have

$$\begin{aligned}\int |\nabla f| \, \mathrm{d}\mu &\ge \int \limsup_{\varepsilon\to 0} \frac{f_\varepsilon - f}{\varepsilon} \, \mathrm{d}\mu \\ &\ge \limsup_{\varepsilon\to 0} \int \frac{f_\varepsilon - f}{\varepsilon} \, \mathrm{d}\mu \ge \liminf_{\varepsilon\to 0} \int \frac{f_\varepsilon - f}{\varepsilon} \, \mathrm{d}\mu \\ &= \liminf_{\varepsilon\to 0} \int_0^\infty \frac{\mu(A(t)_\varepsilon) - \mu(A(t))}{\varepsilon} \, \mathrm{d}t \\ &\ge \int_0^\infty \liminf_{\varepsilon\to 0} \frac{\mu(A(t)_\varepsilon) - \mu(A(t))}{\varepsilon} \, \mathrm{d}t = \int_0^\infty \mu^+(A(t)) \, \mathrm{d}t.\end{aligned}$$

Hence, we obtain (5.5).

To remove the boundedness assumption, one may use a truncation argument. Define $f_r(x) = \max\{f(x), r\}$, where r tends to infinity along the values such that $\mu\{|f| = r\} = 0$. Then f has a Lipschitz constant at most L, with $|\nabla f_r| = |\nabla f|$ on the set $\{f < r\}$ and $|\nabla f_r| = 0$ on $\{f > r\}$. Hence, by the previous step applied to f_r,

$$\int_{\{f<r\}} |\nabla f| \, \mathrm{d}\mu \ge \int_0^r \mu^+\{f > t\} \, \mathrm{d}t.$$

It remains to send $r \to \infty$. Proposition 5.2.1 is proved. □

An opposite inequality to (5.5), which is usually stated in the setting of Riemannian manifolds in the class of C^∞-smooth functions, requires more assumptions. Keeping the general setting, let us give a weaker assertion.

Proposition 5.2.2 *For any non-empty Borel set A in $\mathbb{M}$, there exists a sequence of Lipschitz functions $f_n : \mathbb{M} \to [0, 1]$ such that $f_n \to 1_{\mathrm{clos}(A)}$ pointwise and*

$$\limsup_{n\to\infty} \int |\nabla f_n| \, \mathrm{d}\mu \le \mu^+(A).$$

Proof Here $\mathrm{clos}(A)$ denotes the closure of A. If $\mu(\mathrm{clos}(A)) > \mu(A)$, then, by the definition of the perimeter, $\mu^+(A) = \infty$, so there is nothing to prove.

Suppose that $\mu(\mathrm{clos}(A)) = \mu(A)$. Since $\mathrm{clos}(A_\varepsilon) \subset A_{\varepsilon'}$ for $\varepsilon < \varepsilon'$, one has

$$\mu^+(A) = \liminf_{\varepsilon\to 0} \frac{\mu(\mathrm{clos}(A_\varepsilon)) - \mu(A)}{\varepsilon}.$$

Hence there exists a sequence $\varepsilon_n \downarrow 0$ such that

$$\frac{\mu(\mathrm{clos}(A_{\varepsilon_n})) - \mu(A)}{\varepsilon_n} \to \mu^+(A) \quad \text{as } n \to \infty.$$

Now, take a sequence $c_n \in (0,1)$, $c_n \to 0$, and let

$$f_n(x) = 1 - \min\left\{1, \frac{d(A_{c_n\varepsilon_n}, x)}{(1-c_n)\,\varepsilon_n}\right\},$$

where $d(B,x) = \inf\{d(y,x) : y \in B\}$ denotes the distance from a set B in $\mathbb{M}$ to a point $x \in \mathbb{M}$. Any such function $d(B,x)$ has Lipschitz semi-norm at most 1, hence

$$|\nabla f_n(x)| \leq \|f_n\|_{\mathrm{Lip}} \leq \frac{1}{(1-c_n)\,\varepsilon_n}, \qquad x \in \mathbb{M}.$$

Necessarily $d(A_{c_n\varepsilon_n}, x) \geq (1-c_n)\,\varepsilon_n$ for $x \notin A_{\varepsilon_n}$. Indeed, in the other case, we have $d(y,x) < (1-c_n)\,\varepsilon_n$ for some $y \in A_{c_n\varepsilon_n}$ which means that $d(a,y) < c_n\varepsilon_n$ for some $a \in A$. Hence, by the triangle inequality, $d(a,x) \leq d(a,y) + d(y,x) < \varepsilon_n$, implying $x \in A_{\varepsilon_n}$. Thus, $f_n(x) = 0$ for $x \notin A_{\varepsilon_n}$ and therefore $|\nabla f_n(x)| = 0$ on the open set $\mathbb{M} \setminus \mathrm{clos}(A_{\varepsilon_n})$. On the other hand, $f_n(x) = 1$ on the open set $A_{c_n\varepsilon_n}$, where $|\nabla f_n(x)| = 0$ as well. As a result, $1_{A_{c_n\varepsilon_n}} \leq f_n \leq 1_{A_{\varepsilon_n}}$ so that

$$\lim_{n\to\infty} f_n(x) = \lim_{\varepsilon\to 0} 1_{A_\varepsilon}(x) = 1_{\mathrm{clos}(A)}(x)$$

for all $x \in \mathbb{M}$. On the other hand,

$$\int |\nabla f_n|\,d\mu = \int_{\mathrm{clos}(A_{\varepsilon_n})\setminus A_{c_n\varepsilon_n}} |\nabla f_n|\,d\mu \leq \frac{\mu(\mathrm{clos}(A_{\varepsilon_n})) - \mu(A)}{(1-c_n)\,\varepsilon_n} \to \mu^+(A).$$

Proposition 5.2.2 is proved. □

5.3 Poincaré-type Inequalities

Analytic conditions imposed on a measure are often stated in the form of Poincaré-type inequalities. Let us start with a standard variant of such integro-differential relations, assuming that $(\mathbb{M}, d)$ is an arbitrary metric space (without isolated points).

Definition 5.3.1 A Borel probability measure μ on $\mathbb{M}$ is said to satisfy a Poincaré-type inequality with constant $\lambda_1 > 0$ if, for any bounded, Lipschitz function f on $\mathbb{M}$,

$$\lambda_1 \mathrm{Var}_\mu(f) \leq \int |\nabla f|^2\,d\mu. \tag{5.6}$$

The best value $\lambda_1 = \lambda_1(\mu) = \lambda_1(\mathbb{M}, d, \mu)$ in (5.6) is called the Poincaré constant. If X is a random element in $\mathbb{M}$ with distribution μ, (5.6) may be rewritten in the form

$$\lambda_1 \operatorname{Var}(f(X)) \le \mathbb{E}\,|\nabla f(X)|^2,$$

and then we also write $\lambda_1 = \lambda_1(X)$.

Before giving any examples and comments about properties of such measures, first let us make sure that (5.6) may be extended to a larger class of functions, or in the case of Euclidean space – may be reduced to the class of smooth functions in terms of the usual gradient.

Proposition 5.3.2 *If the inequality* (5.6) *holds true for any Lipschitz function f supported on some ball in $\mathbb{M}$, then it may be extended to the class of all locally Lipschitz functions f on $\mathbb{M}$.*

More precisely, if f is locally Lipschitz, while $|\nabla f|$ belongs to $L^2(\mu)$, then f is in $L^2(\mu)$ as well, and the inequality (5.6) is true with the same constant λ_1.

Recall that the property of being locally Lipschitz is understood as having finite Lipschitz semi-norms on every ball in $\mathbb{M}$ (which is consistent with the traditional understanding, when $\mathbb{M}$ is locally compact, with balls being compact). Also, if f is Lipschitz and is supported on some ball in $\mathbb{M}$, then it has to be bounded on $\mathbb{M}$.

Note that, in the current setting of an abstract metric space, the inequality (5.6) may be extended to the class of all locally Lipschitz complex-valued functions $f = u + iv$ on $\mathbb{M}$ in a weaker form as the relation

$$\lambda_1 \int |f|^2 \, d\mu \le 2 \int |\nabla f|^2 \, d\mu,$$

with the assumption $\int f \, d\mu = 0$ (by applying (5.6) to u and v).

Proof (of Proposition 5.3.2) Assuming that f is Lipschitz on every ball and is bounded so that $|f(x)| \le A$, $x \in \mathbb{M}$, for some constant A, consider the functions

$$f_r(x) = f(x)\Big(1 - \frac{1}{r}\, d(x, x_0)\Big)_+$$

with a fixed point $x_0 \in \mathbb{M}$. They are supported on the closed balls $B(x_0, r)$ with center at x_0 and radius $r > 0$, and moreover, $|\nabla f_r(x)| \le (|\nabla f(x)| + \frac{1}{r}\, A)\, 1_{B(x_0,r)}$. Hence, an application of (5.6) to f_r leads to the same inequality for f as $r \to \infty$. Indeed, since $f_r(x) \to f(x)$ for all $x \in \mathbb{M}$, we may apply Fatou's lemma, which gives

$$\begin{aligned}
\lambda_1 \operatorname{Var}_\mu(f) &= \frac{\lambda_1}{2} \iint (f(x) - f(y))^2 \, d\mu(x)\, d\mu(y) \\
&\le \frac{\lambda_1}{2} \liminf_{r\to\infty} \iint (f_r(x) - f_r(y))^2 \, d\mu(x)\, d\mu(y) \le \liminf_{r\to\infty} \int |\nabla f_r|^2 \, d\mu.
\end{aligned}$$

On the other hand, assuming that $|\nabla f| \in L^2(\mu)$, the last integral does not exceed

$$\int \Big(|\nabla f|^2 + \frac{2A}{r}\, |\nabla f| + \frac{A^2}{r^2}\Big)\, d\mu \to \int |\nabla f|^2 \, d\mu \quad (r \to \infty).$$

To remove the boundedness assumption, assume that f is locally Lipschitz and define

$$f_r(x) = \max\Big\{-r, \min\{f, r\}\Big\} = \begin{cases} f(x), & \text{if } |f(x)| < r, \\ r, & \text{if } f(x) \geq r, \\ -r, & \text{if } f(x) \leq -r, \end{cases}$$

where r will tend to infinity along the values such that $\mu\{|f| = r\} = 0$. In that case, f_r is locally Lipschitz for any $r > 0$, while

$$|\nabla f_r(x)| = |\nabla f(x)|\, 1_{U_r}(x)$$

for μ-almost all x, where $U_r = \{x : |f(x)| < r\}$ (which is an open set). Hence, an application of (5.6) to f_r gives

$$\begin{aligned} \frac{\lambda_1}{2} \int_{U_r} \int_{U_r} |f(x) - f(y)|^2 \, d\mu(x) \, d\mu(y) &\leq \frac{\lambda_1}{2} \int_{\mathbb{M}} \int_{\mathbb{M}} |f_r(x) - f_r(y)|^2 \, d\mu(x) \, d\mu(y) \\ &= \lambda_1 \mathrm{Var}_\mu(f_r) \leq \int_{\mathbb{M}} |\nabla f_r|^2 \, d\mu = \int_{U_r} |\nabla f|^2 \, d\mu. \end{aligned}$$

Sending $r \to \infty$ completes the proof of the proposition. □

5.4 The Euclidean Setting

In the case of the Euclidean space $\mathbb{M} = \mathbb{R}^n$, under certain regularity assumptions imposed on the density of the measure μ, the number λ_1 is described as the smallest positive eigenvalue of the Sturm–Liouville operator associated to the density. Hence, λ_1 is also called the spectral gap.

The functional $\lambda_1 = \lambda_1(X)$, where the random vector X is distributed according to μ, is translation invariant, homogeneous of order -2, and is invariant under all orthogonal transformations of the space, i.e.,

$$\lambda_1(a + bT(X)) = b^{-2}\lambda_1(X) \quad (a \in \mathbb{R}^n,\ b \neq 0,\ T \in O(n)).$$

Many important probability distributions on $\mathbb{R}^n$ are known to satisfy Poincaré-type inequalities, although the problem of bounding λ_1 from below is often non-trivial. Note that an application of (5.6) to linear functionals leads to the simple upper bound

$$\lambda_1(X) \leq \inf_{|\theta|=1} \frac{1}{\mathrm{Var}(\langle X, \theta \rangle)}.$$

The aim of this section is to complement Proposition 5.3.2 with the following.

Proposition 5.4.1 *Suppose that a Borel probability measure μ is absolutely continuous on $\mathbb{R}^n$. If the inequality* (5.6) *is fulfilled in the class of all C^∞-smooth, compactly supported functions $f : \mathbb{R}^n \to \mathbb{R}$, then it continues to hold for all locally Lipschitz f on $\mathbb{R}^n$.*

Proof By Proposition 5.3.2, we may assume that f is compactly supported and has a finite Lipschitz semi-norm $L = \|f\|_{\mathrm{Lip}}$. By Rademacher's theorem, f is differentiable at almost all points $x \in \mathbb{R}^n$, and for every such point, the generalized modulus of the gradient $|\nabla f(x)|$ represents the Euclidean length of the usual gradient $\nabla f(x)$. The same is true of the modified modulus of the gradient

$$|\widetilde{\nabla} f(x)| = \limsup_{y_1, y_2 \to x,\ y_1 \neq y_2} \frac{|f(y_1) - f(y_2)|}{|y_1 - y_2|}. \tag{5.7}$$

By the definition, if $|\widetilde{\nabla} f(x)| < a$, then, for some $\varepsilon > 0$,

$$\sup \left\{ \frac{|f(y_1) - f(y_2)|}{|y_1 - y_2|} : |y_1 - x| < \varepsilon,\ |y_2 - x| < \varepsilon,\ y_1 \neq y_2 \right\} < a.$$

Hence $|\widetilde{\nabla} f(x')| < a$ whenever $|x' - x| < \varepsilon$. This implies $\limsup_{x' \to x} |\widetilde{\nabla} f(x')| \leq |\widetilde{\nabla} f(x)|$, which means that the function $|\widetilde{\nabla} f|$ is upper semi-continuous.

Next, we employ a smoothing argument: Given a non-negative, C^∞-smooth function w on $\mathbb{R}^n$ supported on the Euclidean ball $B(0, r)$ and such that $\int w \, \mathrm{d}x = 1$, consider the convolutions

$$\begin{aligned} f_\delta(x) &= \int f(x - \delta z) w(z) \, \mathrm{d}z \\ &= \delta^{-n} \int w\Big(\frac{1}{\delta} x - z\Big) f(z) \, \mathrm{d}z \quad (x \in \mathbb{R}^n,\ \delta > 0). \end{aligned}$$

Every such function f_δ is compactly supported (since f has a compact support) and is C^∞-smooth, so that (5.6) is fulfilled:

$$\lambda_1 \mathrm{Var}_\mu(f_\delta) \leq \int |\nabla f_\delta|^2 \, \mathrm{d}\mu. \tag{5.8}$$

By the Lipschitz assumption, $|f_\delta(x) - f(x)| \leq C\delta$ for some constant C independent of x, so, the variances $\mathrm{Var}_\mu(f_\delta)$ tend to $\mathrm{Var}_\mu(f)$ as $\delta \to 0$.

To handle the right integrals in (5.8), recall the operators

$$\Delta_\varepsilon f(x) = \sup_{|y - x| < \varepsilon} |f(y) - f(x)| \quad (x \in \mathbb{R}^n,\ \varepsilon > 0).$$

As was stressed, $\Delta_\varepsilon f$ is lower semi-continuous. Since $|f(y - \delta z) - f(x - \delta z)| \leq \Delta_\varepsilon f(x - \delta z)$ for $|y - x| \leq \varepsilon$, we get

$$|f_\delta(y) - f_\delta(x)| \leq \int |f(y - \delta z) - f(x - \delta z)| \, w(z) \, \mathrm{d}z \leq \int \Delta_\varepsilon f(x - \delta z) \, w(z) \, \mathrm{d}z$$

and therefore

$$\frac{1}{\varepsilon}\,\Delta_\varepsilon f_\delta(x) \,\le\, \frac{1}{\varepsilon}\int \Delta_\varepsilon f(x-\delta z)\,w(z)\,dz.$$

Using $\frac{1}{\varepsilon}\,\Delta_\varepsilon f \le L$ and applying the dominated convergence theorem, we obtain that

$$\begin{aligned}|\nabla f_\delta(x)| &= \limsup_{\varepsilon\to 0}\frac{\Delta_\varepsilon f_\delta(x)}{\varepsilon} \le \limsup_{\varepsilon\to 0}\Big[\frac{1}{\varepsilon}\int \Delta_\varepsilon f(x-\delta z)\,w(z)\,dz\Big]\\ &\le \int \limsup_{\varepsilon\to 0}\Big[\frac{1}{\varepsilon}\,\Delta_\varepsilon f(x-\delta z)\Big]\,w(z)\,dz\\ &= \int |\nabla f(x-\delta z)|\,w(z)\,dz = \int |\widetilde{\nabla} f(x-\delta z)|\,w(z)\,dz.\end{aligned}$$

The last two integrands coincide for almost all z, by the Lipschitz property of f.

One may now use the upper semi-continuity of $|\widetilde{\nabla} f|$ and apply the Lebesgue dominated convergence theorem once more. This gives

$$\begin{aligned}\limsup_{\delta\to 0^+}|\nabla f_\delta(x)| &\le \limsup_{\delta\to 0^+}\int |\widetilde{\nabla} f(x-\delta z)|\,w(z)\,dz\\ &\le \int \limsup_{\delta\to 0^+}|\widetilde{\nabla} f(x-\delta z)|\,w(z)\,dz \,=\, |\widetilde{\nabla} f(x)|.\end{aligned}$$

As a result, from (5.8) we derive the desired inequality

$$\lambda_1 \mathrm{Var}_\mu(f) \le \int |\widetilde{\nabla} f|^2\,d\mu = \int |\nabla f|^2\,d\mu.$$

Proposition 5.4.1 is proved. □

Without the assumption that μ is absolutely continuous, we have thus derived a slightly weaker variant of (5.6), namely,

$$\lambda_1 \mathrm{Var}_\mu(f) \le \int |\widetilde{\nabla} f|^2\,d\mu \tag{5.9}$$

with a modified modulus of gradient defined in (5.7).

For various applications, one would like to apply (5.6) to functions of the form $T(f)$ with f being Lipschitz. Starting with the relation (5.9), note first that, according to the definition (5.7), we still have a chain rule inequality

$$|\widetilde{\nabla} T(f)| \le |(\widetilde{\nabla} T)(f)|\,|\widetilde{\nabla} f|,$$

where $|\widetilde{\nabla} T|$ is defined in (5.7) on the real line. Since also $|\widetilde{\nabla} f(x)| \le \|f\|_{\mathrm{Lip}}$ for all $x \in \mathbb{R}^n$, we arrive at the following distributional inequality on the basis of (5.9).

Proposition 5.4.2 *Suppose that a Borel probability measure μ on $\mathbb{R}^n$ admits a Poincaré-type inequality* (5.6) *in the class of all C^∞-smooth, compactly supported functions on $\mathbb{R}^n$. Then, for any function f on $\mathbb{R}^n$ with $\|f\|_{\mathrm{Lip}} \leq 1$ and any locally Lipschitz function T on $\mathbb{R}$,*

$$\lambda_1 \mathrm{Var}_\mu(T(f)) \leq \int |(\widetilde{\nabla} T)(f)|^2 \, \mathrm{d}\mu. \tag{5.10}$$

Here, $|\widetilde{\nabla} T| = |T'|$ as long as T has a derivative T'.

Remark 5.4.3 Under the assumption of Proposition 5.4.1, the inequality (5.6) may be extended to the class of all complex-valued, locally Lipschitz functions $f = u + iv$ on $\mathbb{R}^n$. Indeed, f is differentiable a.e., so that $|\nabla f(x)|^2 = |\nabla u(x)|^2 + |\nabla v(x)|^2$ for almost all points $x \in \mathbb{R}^n$. If $\mathbb{M}$ is an open subset of $\mathbb{R}^n$ with the Euclidean metric, and μ is absolutely continuous, such a generalization should be modified as follows: If the inequality (5.6) is fulfilled in the class of all C^∞-smooth, Lipschitz functions $f : \mathbb{M} \to \mathbb{R}$, then it continues to hold for all locally Lipschitz $f : \mathbb{M} \to \mathbb{C}$.

The same remark applies to many other spaces including the unit sphere $\mathbb{M} = \mathbb{S}^{n-1}$, which we equip with the normalized spherical Lebesgue measure $\mu = \mathfrak{s}_{n-1}$.

5.5 Isoperimetry and Cheeger-type Inequalities

Let us return to the setting of an abstract metric space $(\mathbb{M}, d)$ with a Borel probability measure μ. In some problems it is desirable to use an L^1-version of the Poincaré-type inequality such as

$$h \int |f - m| \, \mathrm{d}\mu \leq \int |\nabla f| \, \mathrm{d}\mu. \tag{5.11}$$

It is required to hold with some constant $h > 0$ for all bounded, Lipschitz functions f on $\mathbb{M}$ (and then it will hold for all locally Lipschitz f like in Proposition 5.3.2). Here $m = m(f)$ denotes a median of f under μ, that is, a real number m such that

$$\mu\{f \leq m\} \geq \frac{1}{2}, \quad \mu\{f \geq m\} \geq \frac{1}{2}.$$

Note that the first integral in (5.11) does not depend on the choice of the median.

The inequality (5.11) is related to problems of an isoperimetric flavor. As in Section 5.2, denote by A_ε an open ε-neighborhood of a set A in $\mathbb{M}$. The isoperimetric problem in $(\mathbb{M}, d, \mu)$ amounts finding lower bounds for the quantity $\mu(A_\varepsilon)$ uniformly over all sets $A \subset \mathbb{M}$ of measure $\mu(A) \geq p$ for fixed $0 < p < 1$ and $\varepsilon > 0$. As another variant, this problem is stated in terms of the μ-perimeter as the relation

$$\mu^+(A) \geq I(\mu(A))$$

in the class of all Borel subsets A in $\mathbb{M}$. An optimal function $I = I_\mu$ in such a relation is called an isoperimetric profile or an isoperimetric function of μ.

Definition 5.5.1 The quantity

$$h(\mu) = \inf \frac{\mu^+(A)}{\min\{\mu(A), 1 - \mu(A)\}},$$

where the infimum runs over all Borel sets A in $\mathbb{M}$ of measure $0 < \mu(A) < 1$, is called the Cheeger isoperimetric constant for the metric probability space $(\mathbb{M}, d, \mu)$.

That is, $h = h(\mu)$ is an optimal value in the inequality $I_\mu(p) \geq h \min\{p, 1 - p\}$ with arbitrary $0 < p < 1$.

Proposition 5.5.2 *The Cheeger isoperimetric constant is an optimal non-negative value h for which the L^1-Poincaré-type inequality* (5.11) *holds true.*

Proof In one direction, in order to derive an isoperimetric-type inequality

$$\mu^+(A) \geq h \min\{\mu(A), 1 - \mu(A)\} \tag{5.12}$$

on the basis of (5.11), we may assume that $\mu(\mathrm{clos}(A)) = \mu(A)$ (since otherwise $\mu^+(A) = \infty$). We apply Proposition 5.2.2 and take a sequence of Lipschitz functions $f_n : \mathbb{M} \to [0, 1]$ such that $f_n \to 1_{\mathrm{clos}(A)}$ pointwise and

$$\limsup_{n\to\infty} \int |\nabla f_n| \, d\mu \leq \mu^+(A).$$

By (5.11), the integral on the left is greater than or equal to $h \int |f_n - m_n| \, d\mu$, where $m_n \in [0, 1]$ are medians for f_n. If $p = \mu(A) < 1/2$, then necessarily $m_n \to 0$. Indeed, suppose that $m_n \geq m > 0$ for infinitely many values of n. By the definition of medians, the sets $A_n = \{f_n \geq m_n\}$ have μ-measures at least $\frac{1}{2}$. Hence, $\mu(B) \geq \frac{1}{2}$ for the set $B = \bigcap_{n\geq 1} \bigcup_{k\geq n} A_k$. But, for any point $x \in B$, we have $f_n(x) \geq m_n$ for infinitely many values of n and hence $1_{\mathrm{clos}(A)}(x) \geq m$, that is, $x \in \mathrm{clos}(A)$. This means that $B \subset \mathrm{clos}(A)$, which would imply that $\mu(A) \geq \frac{1}{2}$. Thus, by the Lebesgue dominated convergence theorem,

$$\lim_{n\to\infty} \int |f_n - m_n| \, d\mu = \lim_{n\to\infty} \int f_n \, d\mu = \mu(\mathrm{clos}(A)) = \mu(A),$$

thus proving (5.12). Similarly, if $p > 1/2$, then necessarily $m_n \to 1$, so that

$$\lim_{n\to\infty} \int |f_n - m_n| \, d\mu = \lim_{n\to\infty} \int (1 - f_n) \, d\mu = 1 - \mu(\mathrm{clos}(A)) = 1 - \mu(A),$$

which also proves (5.12). Finally, in the case $p = 1/2$, take a convergent subsequence $m_{n'} \to m \in [0, 1]$ and note that any $m \in [0, 1]$ represents a median for the indicator function $1_{\mathrm{clos}(A)}$ under μ. Hence,

$$\lim_{n\to\infty} \int |f_{n'} - m_{n'}| \, d\mu = \lim_{n\to\infty} \int |f_n - m| \, d\mu = \mu(\mathrm{clos}(A)) = \mu(A) = \frac{1}{2},$$

and we obtain (5.12) again.

For the opposite direction, we may assume that $m = 0$ (in view of the translation invariance of (5.11)). First let $|\nabla f| = 0$ on the set $\{f = 0\}$ and define

$$f^+ = \max\{f, 0\}, \quad f^- = \max\{-f, 0\},$$

so that $f = f^+ - f^-$ and $|f| = f^+ + f^-$. Note that $|\nabla f^+| = |\nabla f|$ and $|\nabla f^-| = |\nabla f|$ on the open sets $\{f > 0\}$ and $\{f < 0\}$, respectively. Hence, applying (5.12) and the co-area inequality of Proposition 5.2.1 to the function f^+, we get

$$\int_{\{f>0\}} |\nabla f| \, \mathrm{d}\mu \geq \int_0^\infty \mu^+\{f > t\} \, \mathrm{d}t \geq h \int_0^\infty \mu\{f > t\} \, \mathrm{d}t.$$

Similarly for f^- we have

$$\int_{\{f<0\}} |\nabla f| \, \mathrm{d}\mu \geq \int_0^\infty \mu^+\{f < -t\} \, \mathrm{d}t \geq h \int_0^\infty \mu\{f < -t\} \, \mathrm{d}t.$$

Adding these inequalities we obtain that

$$\int |\nabla f| \, \mathrm{d}\mu \geq h \int_0^\infty \mu\{|f| > t\} \, \mathrm{d}t = h \int |f| \, \mathrm{d}\mu.$$

which is the required inequality (5.11).

In order to remove the assumption that $|\nabla f| = 0$ on the set $\{f = 0\}$, one may apply the previous step to the functions $T_\varepsilon(f)$ with $T_\varepsilon(x) = \sqrt{\varepsilon^2 + x^2} - \varepsilon$ ($\varepsilon > 0$). Since in general $|\nabla T_\varepsilon(f)| \leq T_\varepsilon'(|f|)\, |\nabla f|$, the modulus of the gradient $|\nabla T_\varepsilon(f)|$ is vanishing when $f = 0$, hence when $T_\varepsilon(f) = 0$. Since also $m(T_\varepsilon(f)) = 0$, we conclude that

$$\int T_\varepsilon'(|f|)\, |\nabla f| \, \mathrm{d}\mu \geq h \int T_\varepsilon(|f|) \, \mathrm{d}\mu.$$

Here, $T_\varepsilon' \leq 1$, and letting $\varepsilon \to 0$, in the limit we arrive at (5.11) for f.

Thus Proposition 5.5.2 is proved. □

5.6 Rothaus Functionals

One may ask whether or not one may replace the median m in (5.11) with more tractable quantities such as the expectation. The answer is affirmative, and moreover, one may consider more general homogeneous functionals including

$$\mathcal{L}(f) = \sup_{(g_1, g_2) \in \mathcal{G}} \left[\int f^+ g_1 \, \mathrm{d}\mu + \int f^- g_2 \, \mathrm{d}\mu \right], \tag{5.13}$$

where the supremum runs over some non-empty family $\mathcal{G}$ of pairs of μ-integrable functions (g_1, g_2) on $\mathbb{M}$. The value $\mathcal{L}(f)$, finite or not, is well defined when $f^+ g_1$ and

$f^- g_2$ are μ-integrable for any couple $(g_1, g_2) \in \mathcal{G}$. In connection with isoperimetric inequalities, these functionals were introduced by Rothaus [163].

If $g_2 = -g_1$, the functional becomes

$$\mathcal{L}(f) = \sup_{g \in G} \int fg \, d\mu$$

for some family G, which includes all L^p-norms. Another important example is given by the entropy functional $\mathcal{L}(f) = \mathrm{Ent}_\mu(|f|) = \mathbb{E}_\mu \, |f| \log |f| - \mathbb{E}_\mu \, |f| \, \log \mathbb{E}_\mu \, |f|$.

Proposition 5.6.1 *The inequality*

$$\int |\nabla f| \, d\mu \geq \mathcal{L}(f) \tag{5.14}$$

holds true for all bounded Lipschitz functions f on $\mathbb{M}$ (and then for all locally Lipschitz f for which $\mathcal{L}(f)$ is well defined) if and only if for all Borel sets A in $\mathbb{M}$,

$$\mu^+(A) \geq \max\{\mathcal{L}(1_A), \mathcal{L}(-1_A)\}. \tag{5.15}$$

Proof The implication (5.14) $\Rightarrow$ (5.15) is based on the approximation of indicator functions by Lipschitz functions; it is very similar to the analogous step in the proof of Proposition 5.5.2.

To argue in the other direction, let us return to the co-area inequality (5.5) and apply it to the functions f^+ and f^-. It then yields a similar relation

$$\int |\nabla f| \, d\mu \geq \int_0^\infty \mu^+\{f > t\} \, dt + \int_{-\infty}^0 \mu^+\{f < t\} \, dt \tag{5.16}$$

(if the property $|\nabla f| = 0$ on the set $\{f = 0\}$ is not fulfilled, one may use the transforms T_ε as in the proof of Proposition 5.4.2).

To derive (5.14) from (5.15), take a couple (g_1, g_2) participating in (5.13). Then, using (5.15)–(5.16) and introducing the sets $A_t = \{f > t\}$, $B_t = \{f < t\}$, we have

$$\begin{aligned}
\int |\nabla f| \, d\mu &\geq \int_0^\infty \mathcal{L}(1_{A_t}) \, dt + \int_{-\infty}^0 \mathcal{L}(-1_{B_t}) \, dt \\
&\geq \int_0^\infty \int 1_{A_t} g_1 \, d\mu \, dt + \int_{-\infty}^0 \int 1_{B_t} g_2 \, \mu \, dt \\
&= \int 1_{A_0} f g_1 \, d\mu + \int 1_{B_0} f g_2 \, d\mu \; = \; \int (f^+ g_1 + f^- g_2) \, d\mu.
\end{aligned}$$

Note that we have made use of Fubini's theorem, which was possible due to the integrability of g_1 and g_2. It remains to take the supremum over all admissible (g_1, g_2), and then we arrive at (5.14). The extension of this inequality to locally Lipschitz f may be justified similarly to the proof of Proposition 5.3.2. □

In the particular case

$$\mathcal{L}(f) = \widetilde{h}\,\|f - \mathbb{E}_\mu f\|_1 = \widetilde{h}\,\sup\Big\{\int fg\,\mathrm{d}\mu : \int g\,\mathrm{d}\mu = 0,\ |g| \le 1\Big\}$$

with parameter $\widetilde{h} \ge 0$, (5.14) turns into

$$\widetilde{h}\int |f|\,\mathrm{d}\mu \le \int |\nabla f|\,\mathrm{d}\mu. \tag{5.17}$$

Corollary 5.6.2 *The best (largest) value $\widetilde{h} = \widetilde{h}(\mu)$ in the inequality* (5.17)*, holding for any μ-integrable, locally Lipschitz function f on $\mathbb{M}$ with μ-mean zero, represents an optimal one in the isoperimetric-type inequality*

$$\mu^+(A) \ge 2\widetilde{h}\,\mu(A)(1 - \mu(A)), \quad A \subset \mathbb{M}\ (\mathit{Borel}) \tag{5.18}$$

and is connected with the Cheeger isoperimetric constant $h = h(\mu)$ via $\frac{1}{2}\,h \le \widetilde{h} \le h$. Moreover, $\widetilde{h} = h$, if $I_\mu(p)/p(1-p)$ is minimized at $p = 1/2$.

Proof Indeed, in this case $\mathcal{L}(1_A) = \mathcal{L}(-1_A) = 2\,\mu(A)(1 - \mu(A))$. Hence, by Proposition 5.6.1, the relations (5.17) and (5.18) are equivalent for any fixed value $\widetilde{h} \ge 0$. In terms of the isoperimetric profile, one may rewrite (5.18) as the property $I_\mu(p) \ge 2\widetilde{h}\,p(1-p)$, where the optimal value is described as

$$\widetilde{h} = \inf_{0<p<1} \frac{I_\mu(p)}{2p(1-p)}.$$

Since $\frac{1}{2}\min\{p, 1-p\} \le p(1-p) \le \min\{p, 1-p\}$, it follows that $\frac{1}{2}\,h \le \widetilde{h} \le h$. Moreover, if the above ratio is minimized at $p = 1/2$, we get

$$\widetilde{h} = 2I_\mu\Big(\frac{1}{2}\Big) \ge \inf_{0<p<1} \frac{I_\mu(p)}{\min\{p, 1-p\}} = h,$$

thus proving the last assertion of Corollary 5.6.2. □

Finally, let us now connect the L^1 and the usual L^2-Poincaré-type inequalities.

Proposition 5.6.3 *The spectral gap $\lambda_1 = \lambda_1(\mu)$ and the Cheeger isoperimetric constant $h = h(\mu)$ are connected by the relation $\lambda_1 \ge \frac{1}{4}\,h^2$.*

Proof According to Proposition 5.5.2, we need to derive from (5.11) the Poincaré-type inequality (5.6) with $\lambda_1 = \frac{1}{4}\,h^2$. Since both inequalities are translation invariant, we may assume that $m(f) = 0$. Moreover, first suppose that a Lipschitz function f is bounded and non-negative. Then $m(f^2) = 0$, so, by (5.11) applied to f^2, and using Cauchy's inequality, we have

$$h\int f^2\,\mathrm{d}\mu \le \int |\nabla f^2|\,\mathrm{d}\mu \le 2\int f\,|\nabla f|\,\mathrm{d}\mu$$
$$\le 2\Big(\int f^2\,\mathrm{d}\mu\Big)^{1/2}\Big(\int_{\{f>0\}} |\nabla f|^2\,\mathrm{d}\mu\Big)^{1/2}.$$

Hence

$$\int_{\{f>0\}} |\nabla f|^2\,\mathrm{d}\mu \ge \frac{h^2}{4}\int f^2\,\mathrm{d}\mu.$$

If f is not necessarily non-negative, one may apply the previous step to the functions f^+ and f^-. Since they have median at the origin, we have

$$\int_{\{f>0\}} |\nabla f|^2\,\mathrm{d}\mu \ge \frac{h^2}{4}\int_{\{f>0\}} f^2\,\mathrm{d}\mu, \quad \int_{\{f<0\}} |\nabla f|^2\,\mathrm{d}\mu \ge \frac{h^2}{4}\int_{\{f<0\}} f^2\,\mathrm{d}\mu,$$

and adding these inequalities, we get

$$\int |\nabla f|^2\,\mathrm{d}\mu \ge \frac{h^2}{4}\int f^2\,\mathrm{d}\mu \ge \frac{h^2}{4}\,\mathrm{Var}_\mu(f).$$

□

5.7 Standard Examples and Conditions

Let us describe several examples of probability measures on the Euclidean space $\mathbb{R}^n$ satisfying Poincaré- and Cheeger-type inequalities (when the space is endowed with the canonical Euclidean distance d). We start with dimension one. As usual, $\mathbb{E}_\mu$ is used to denote the integral with respect to μ.

5.7.1 Exponential distributions. For the (one-dimensional) exponential distribution μ with parameter $\lambda > 0$, i.e., with density $\lambda \mathrm{e}^{-\lambda x}$ $(x > 0)$, one has

$$h(\mu) = \lambda, \quad \lambda_1(\mu) = \frac{1}{4}\,\lambda^2.$$

This example shows that the Cheeger inequality of Proposition 5.6.3 is optimal.

To prove this, assume without loss of generality that $\lambda = 1$. Given a smooth function f on the real line with bounded derivative, put $g(x) = f(0) + \int_0^x |f'(y)|\,\mathrm{d}y$. Integrating by parts, we have

$$\mathbb{E}_\mu|f - m(f)| \le \mathbb{E}_\mu|f - f(0)| \le \mathbb{E}_\mu(g - g(0))$$
$$= \lambda\int_0^\infty g'(x)\,\mathrm{e}^{-\lambda x}\,\mathrm{d}x = \mathbb{E}_\mu\,|f'|.$$

Hence, the inequality (5.11) holds true with $h = 1$, which can be seen to be optimal (by testing it on exponential functions). Thus, $h(\mu) = \lambda$, and, by Proposition 5.6.3, $\lambda_1(\mu) \ge \frac{1}{4}\,\lambda^2$. For the opposite inequality, one may test the inequality

$$\mathbb{E}_\mu f^2 - (\mathbb{E}_\mu f)^2 \leq \frac{1}{\lambda_1}\, \mathbb{E}_\mu f'^2$$

on the functions $f_t(x) = e^{tx}$ with $t < \frac{\lambda}{2}$. Inserting $f = f_t$ with $t \to \frac{\lambda}{2}$ yields $\lambda_1 \leq \frac{\lambda^2}{4}$.

5.7.2 Two-sided exponential distributions. A similar argument leads to the same relations $h(\mu) = \lambda$ and $\lambda_1(\mu) = \frac{1}{4}\lambda^2$ for the symmetric probability measures with densities $\frac{1}{2}\lambda e^{-\lambda|x|}$ $(x \in \mathbb{R})$ with parameter $\lambda > 0$.

5.7.3 Tail-type sufficient conditions. Following Borovkov and Utev [66], suppose that the probability measure μ on $\mathbb{R}$ may be decomposed as the mixture of two probability measures, $\mu = \alpha\mu_0 + (1-\alpha)\mu_1$, $0 < \alpha \leq 1$, where μ_0 has density p satisfying

$$\int_x^\infty (y - x_0)\, p(y)\, dy \leq cp(x) \quad \text{for } x > x_0,$$

$$-\int_{-\infty}^x (y - x_0)\, p(y)\, dy \leq cp(x) \quad \text{for } x < x_0$$

for some fixed $x_0 \in \mathbb{R}$ and $c > 0$. Then $\lambda_1(\mu) \geq \alpha/c$.

We leave this assertion as an exercise. Note that, when μ_0 has mean x_0, we have $\int_{-\infty}^{\infty} (y - x_0)\, p(y)\, dy = 0$, and the two tail conditions can be united as one inequality

$$\int_x^\infty (y - x_0)\, p(y)\, dy \leq cp(x) \quad \text{for all } x \in \mathbb{R}.$$

5.7.4 Gaussian measures. If μ is a probability measure with density

$$p(x) = \frac{1}{\sigma\sqrt{2\pi}}\, e^{-(x-a)^2/2\sigma^2}, \qquad x \in \mathbb{R} \ (a \in \mathbb{R},\ \sigma > 0),$$

the last tail condition is fulfilled with $\alpha = 1$, $\mu_0 = \mu$, $c = \sigma^2$. Hence, by the previous example and also taking into account the general upper bound $\lambda_1(X) \leq 1/\mathrm{Var}(X)$, we conclude that $\lambda_1(\mu) = 1/\sigma^2$.

5.7.5 Log-concave measures. Recall that a probability measure μ on $\mathbb{R}$ is log-concave (or logarithmically concave), if it has a log-concave density p, i.e., such that $\log p$ is concave. Necessarily, p decays at infinity exponentially fast, so μ has a finite exponential moment. One can show that if σ^2 is the variance of μ, then

$$\frac{1}{3\sigma^2} \leq h^2(\mu) \leq \frac{2}{\sigma^2}, \qquad \frac{1}{12\sigma^2} \leq \lambda_1(\mu) \leq \frac{1}{\sigma^2},$$

where the constants in the bounds for the Cheeger constant are optimal (cf. [20]).

5.7.6 Necessary and sufficient conditions. Any probability measure μ on $\mathbb{R}$ may be decomposed in a unique way as the sum $\mu = \mu_0 + \mu_1$, where μ_0 is an absolutely continuous component with respect to the Lebesgue measure, with some density p, and where μ_1 is an orthogonal component in the sense of Measure Theory. For the property $\lambda_1 > 0$ (hence for $h > 0$ as well) it is necessary that $p(x) > 0$ almost everywhere on the support interval $(a, b) \subset \mathbb{R}$ of μ_0. More precisely, the Cheeger isoperimetric constant admits a simple description

$$h = \operatorname{ess\,inf}_{a<x<b} \frac{p(x)}{\min\{F(x), 1 - F(x)\}},$$

where $F(x) = \mu(-\infty, x]$ is the distribution function associated to μ (cf. [50]).

As for the spectral gap, introduce the quantities

$$A = \sup_{a<x<m} \left[F(x) \int_x^m \frac{1}{p(y)}\, dy \right], \quad B = \sup_{m<x<b} \left[(1 - F(x)) \int_m^x \frac{1}{p(y)}\, dy \right],$$

where m is a median of μ. Then, λ_1 is positive if and only if both A and B are finite, in which case

$$\frac{1}{2} \max\{A, B\} \le \frac{1}{\lambda_1} \le 4 \max\{A, B\}.$$

These relations follow from a similar description of Hardy-type inequalities obtained by Kac and Krein [113], cf. also [148], [47].

5.7.7 Product measures. If $\mu = \mu_1 \otimes \cdots \otimes \mu_n$ is the product of n probability measures on some measurable spaces $\mathbb{M}_k$, then, for any measurable function f on the product space $\mathbb{M} = \mathbb{M}_1 \times \cdots \times \mathbb{M}_n$,

$$\mathrm{Var}_\mu(f) \le \int \left[\sum_{k=1}^n \mathrm{Var}_{x_k}(f) \right] d\mu(x_1, \dots, x_n). \tag{5.19}$$

Here, $\mathrm{Var}_{x_k}(f)$ denotes the variance of $y \to f(x_1, \dots, x_{k-1}, y, x_{k+1}, \dots, x_n)$ under μ_k, which depends on the coordinates x_i, $i \neq k$. This assertion is called the subadditivity property of the variance functional, while an application of (5.19) is referred to as the tensorization of the variance.

Suppose that each $\mathbb{M}_k$ is a separable metric space with metric d_k. The product of the Borel σ-algebras in $\mathbb{M}_k$ coincides with the Borel σ-algebra in $\mathbb{M}$ generated, for example, by the Euclidean-type metric

$$d(x, y) = \Big(\sum_{k=1}^n d_k(x_k, y_k)^2 \Big)^{1/2}, \quad x = (x_1, \dots, x_n),\ y = (y_1, \dots, y_n),\ x_k, y_k \in \mathbb{M}_k.$$

If μ_k is a Borel measure on $\mathbb{M}_k$ satisfying a Poincaré-type inequality with constant $\lambda_1(\mu_k)$, then, for any locally Lipschitz function f on $\mathbb{M}$, we therefore obtain that

$$\lambda\, \mathbb{E}_\mu \mathrm{Var}_{x_k}(f) \le \mathbb{E}_\mu \int_{\mathbb{M}_k} |\nabla_{x_k} f(x)|^2\, d\mu_k(x_k) = \int_{\mathbb{M}} |\nabla_{x_k} f(x)|^2\, d\mu(x)$$

with $\lambda = \min_k \lambda_1(\mu_k)$. Here $|\nabla_{x_k} f(x)|$ denotes the generalized modulus of gradient with respect to the variable $x_k \in \mathbb{M}_k$. Summing these inequalities over $k \leq n$, we obtain from (5.19) a Poincaré-type inequality on $\mathbb{M}$,

$$\lambda \operatorname{Var}_\mu(f) \leq \int |\widetilde{\nabla} f|^2 \, d\mu,$$

for a modified modulus of gradient defined in (5.3), that is, with

$$|\widetilde{\nabla} f(x)|^2 = |\nabla_{x_1} f(x)|^2 + \cdots + |\nabla_{x_k} f(x)|^2.$$

Actually, testing this inequality on functions $f = f(x_k)$ depending only on the k-th variable, we get necessarily that $\lambda \leq \lambda_1(\mu_k)$.

Recall also that, by Proposition 5.4.1, when each $\mathbb{M}_k$ is a Euclidean space $\mathbb{R}^{n_k}$, $|\widetilde{\nabla} f(x)|$ can be replaced with $|\nabla f(x)|$, as long as the marginal distributions μ_k are absolutely continuous. Therefore,

$$\lambda_1(\mu) = \min_k \lambda_1(\mu_k).$$

This relation may be used to construct more examples, starting from one-dimensional measures. According to item 5.7.1, for the n-dimensional exponential distribution μ with density $p(x) = e^{-(x_1+\cdots+x_n)}\, 1_{\{x_1,\dots,x_n>0\}}$ we thus have $\lambda_1(\mu) = \frac{1}{4}$.

As for the Cheeger isoperimetric constant, one can derive a similar relation in the form of the two-sided bounds (cf. [50]), namely

$$\frac{1}{\sqrt{6}} \min_k h(\mu_k) \leq h(\mu) \leq \min_k h(\mu_k).$$

5.7.8 Lipschitz transforms. If a probability measure ν on $\mathbb{R}^N$ satisfies a Poincaré-type inequality, one can create more examples by a push forward of ν into a probability measure $\mu = \nu T^{-1}$ on $\mathbb{R}^n$ with the help of an arbitrary map $T : \mathbb{M} \to \mathbb{R}^n$ having a finite Lipschitz semi-norm

$$\|T\|_{\rm Lip} = \sup_{x,y\in\mathbb{M},\ x\neq y} \frac{|T(x) - T(y)|}{|x - y|}$$

on a subset $\mathbb{M}$ of $\mathbb{R}^N$ of ν-measure 1. Indeed, given a locally Lipschitz function $f : \mathbb{R}^n \to \mathbb{R}$, the chain rule inequality $|\nabla f(T)|(x) \leq \|T\|_{\rm Lip}\, |\nabla f(T(x))|$ implies that

$$\begin{aligned}\operatorname{Var}_\mu(f) &= \operatorname{Var}_\nu(f(T)) \\ &\leq \frac{1}{\lambda_1(\nu)}\, \|T\|_{\rm Lip}^2 \int |\nabla f(T)|^2 \, d\nu = \frac{1}{\lambda_1(\nu)}\, \|T\|_{\rm Lip}^2 \int |\nabla f|^2 \, d\mu.\end{aligned}$$

This gives a lower bound

$$\lambda_1(\mu) \geq \frac{1}{\|T\|_{\rm Lip}^2}\, \lambda_1(\nu).$$

By a similar argument, based on Proposition 5.4.2, we also have

$$h(\mu) \geq \frac{1}{\|T\|_{\mathrm{Lip}}}\, h(\nu).$$

5.7.9 Convolutions. Let $X_1, \dots, X_k$ be independent random vectors in $\mathbb{R}^n$ satisfying a Poincaré-type inequality, and let $X = a_1 X_1 + \dots + a_k X_k$, $a_i \in \mathbb{R}$. Since $\lambda_1(b_i X_i) = b_i^{-2} \lambda_1(X_i) = 1$ for $b_i = \sqrt{\lambda_1(X_i)}$, the distribution of the random vector

$$Y = (b_1 X_1, \dots, b_k X_k) \ \text{ in } \mathbb{R}^{nk}$$

satisfies a Poincaré-type inequality with constant 1 (according to item 5.7.7). The linear map $T : \mathbb{R}^{nk} \to \mathbb{R}^n$ defined by $T(y_1, \dots, y_k) = \sum_{i=1}^k a_i y_i / b_i$ ($y_i \in \mathbb{R}^n$) has a Lipschitz semi-norm given by $\|T\|_{\mathrm{Lip}}^2 = \sum_{i=1}^k (a_i/b_i)^2$. Since $X = T(Y)$, by item 5.7.8, we get that

$$\lambda_1^{-1}(X) \leq \sum_{i=1}^k a_i^2\, \lambda_1^{-1}(X_i).$$

5.7.10 Rotationally invariant measures. Suppose that a random vector $X = (X_1, \dots, X_n)$ in $\mathbb{R}^n$, $n \geq 2$, is distributed according to an absolutely continuous probability measure μ with density of the form $p(|x|)$, where $p = \mathrm{e}^{-V}$ for some convex function $V : (0, \infty) \to (-\infty, \infty]$. Then the Poincaré constant of μ can be effectively bounded in terms of $\mathbb{E}X_1^2$. As shown in [27], we have

$$\mathbb{E}X_1^2 \leq \lambda_1^{-1}(X) \leq 13\, \mathbb{E}X_1^2.$$

This class includes all rotationally invariant log-concave probability measures on $\mathbb{R}^n$. In this case, the functions V are convex and non-decreasing, and the constant 13 may be improved to $\frac{n}{n-1}$, cf. [61].

5.8 Canonical Gaussian Measures

One important example should be discussed separately and in more detail. In the sequel, we denote by $\gamma_n = \gamma_1 \otimes \dots \otimes \gamma_1$ the canonical Gaussian measure on $\mathbb{R}^n$, i.e., with density

$$\frac{d\gamma_n(x)}{dx} = \frac{1}{(2\pi)^{\frac{n}{2}}}\, \mathrm{e}^{-|x|^2/2}, \quad x \in \mathbb{R}^n.$$

Being a product measure, it is rotationally invariant. So, a number of the previously mentioned general observations are applicable to γ_n. In particular, according to items 5.7.4 and 5.7.7, we have $\lambda_1(\gamma_n) = 1$, which is a short way to formulate the Gaussian Poincaré-type inequality:

Proposition 5.8.1 *For any locally Lipschitz function f on $\mathbb{R}^n$,*

$$\mathrm{Var}_{\gamma_n}(f) \le \int |\nabla f|^2 \, \mathrm{d}\gamma_n. \tag{5.20}$$

Modulo an absolute factor, one may strengthen this result in terms of the L^1-Poincaré-type inequality.

Proposition 5.8.2 *For any locally Lipschitz function f on $\mathbb{R}^n$ with γ_n-mean zero,*

$$\int |f| \, \mathrm{d}\gamma_n \le \sqrt{\frac{\pi}{2}} \int |\nabla f| \, \mathrm{d}\gamma_n, \tag{5.21}$$

where the constant on the right is the best possible.

The proof of (5.21) can be based on Corollary 5.6.2. Indeed, a solution to the isoperimetric problem for the Gauss space $(\mathbb{R}^n, d, \gamma_n)$ was given by Sudakov and Tsirelson [170] and Borell [65], proving that in the class of all Borel measurable sets A in $\mathbb{R}^n$ with a fixed value $\gamma_n(A) = p$, the measure $\gamma_n(A_\varepsilon)$ is minimized for the half-space $H(a) = \{x \in \mathbb{R}^n : x_1 \le a\}$. Equivalently,

$$\gamma_n(A) = \gamma_n(H(a)) \;\Rightarrow\; \gamma_n^+(A) \ge \gamma_n^+(H(a)).$$

Note that $\gamma_n(H(a)) = p$ for $a = \Phi^{-1}(p)$, where Φ^{-1} is the inverse of the distribution function of the standard normal law (in dimension one),

$$\Phi(x) = \gamma_1((-\infty, x]) = \int_{-\infty}^{x} \varphi(y) \, \mathrm{d}y, \quad x \in \mathbb{R},$$

with density $\varphi(y) = \frac{1}{\sqrt{2\pi}} \, \mathrm{e}^{-y^2/2}$. The associated isoperimetric profile is thus given by

$$I_{\gamma_n}(p) = \gamma_n^+(H(a)) = \varphi(\Phi^{-1}(p)), \quad 0 < p < 1.$$

This function is concave on $(0, 1)$ and is symmetric about $p = \frac{1}{2}$. So, the ratio $I_{\gamma_n}(p)/\min\{p, 1-p\}$ is minimized at this point, and thus

$$h(\gamma_n) = 2I_{\gamma_n}\Big(\frac{1}{2}\Big) = \sqrt{2/\pi}.$$

One can check that $I_{\gamma_n}(p)/p(1-p)$ is minimized for $p = \frac{1}{2}$ as well. Therefore, applying Corollary 5.6.2, we obtain (5.21), which thus also holds under the assumption that $m(f) = 0$.

There is also a simple alternative approach to Proposition 5.8.2 proposed by Pisier [157], which actually yields a larger family of Sobolev-type inequalities. For $x, y \in \mathbb{R}^n$, consider the curve $x(t) = x\cos(t) + y\sin(t)$, $0 \le t \le \frac{\pi}{2}$, so that $x(0) = x$ and $x(\frac{\pi}{2}) = y$. Assuming that f is C^1-smooth on $\mathbb{R}^n$, we then have

$$f(y) - f(x) = \int_0^{\frac{\pi}{2}} \frac{\mathrm{d}}{\mathrm{d}t} f(x(t)) \, \mathrm{d}t = \int_0^{\frac{\pi}{2}} \langle \nabla f(x(t)), x'(t) \rangle \, \mathrm{d}t.$$

If $\Psi : \mathbb{R} \to [0, \infty)$ is a convex function, by Jensen's inequality, we get

$$\Psi(f(y) - f(x)) \le \frac{2}{\pi} \int_0^{\frac{\pi}{2}} \Psi\Big(\frac{\pi}{2} \langle \nabla f(x(t)), x'(t)\rangle\Big)\,dt.$$

Integrating this inequality over $\gamma_n \otimes \gamma_n(dx, dy)$ and using the property that, for each t, the map $(x, y) \to (x(t)), x'(t))$ pushes forward $\gamma_n \otimes \gamma_n$ to itself, we arrive at

$$\iint \Psi(f(y) - f(x))\,d\gamma_n(x)\,d\gamma_n(dy) \le \iint \Psi\Big(\frac{\pi}{2} \langle \nabla f(x), y\rangle\Big)\,d\gamma_n(x)\,d\gamma_n(dy).$$

Moreover, if f has γ_n-mean zero, another application of Jensen's inequality to the above integral on the left leads to

$$\int \Psi(f)\,d\gamma_n \le \iint \Psi\Big(\frac{\pi}{2} \langle \nabla f(x), y\rangle\Big)\,d\gamma_n(x)\,d\gamma_n(dy). \tag{5.22}$$

For example, choosing here $\Psi(r) = |r|$ yields (5.21). Other choices of Ψ in (5.22) are also of interest.

5.9 Remarks

Much of the material in this chapter may be found in [49] and [50], where, in particular, the definition (5.2) of the generalized modulus of the gradient was proposed for abstract metric spaces (the so-called continuous setting does not actually pose an essential constraint). Proposition 5.5.2 concerning L^1 Poincaré-type inequalities, as well as many other connections between isoperimetric and Sobolev-type inequalities, goes back to the work of Mazya [138], cf. also [139]. The well-known relation $\lambda_1 \ge h^2/4$ of Proposition 5.6.3 is due to Cheeger [76] (in the setting of Riemannian manifolds). The class of Sobolev-type inequalities (5.14) and their isoperimetric characterization (5.15) were considered by Rothaus [163].

For different approaches to the isoperimetric inequality for the standard Gaussian measure, see also [82], [9], [19], [24]. This theorem can be shown to imply the Poincaré-type inequality (5.20).

Chapter 6
Poincaré-type Inequalities

In connection with the isoperimetric problem in metric probability spaces, Poincaré-type inequalities have been already discussed in Chapter 5, where we also described basic examples. Such inequalities also serve as the most convenient route to concentration results. In this chapter, first we derive deviation bounds in terms of the spectral gap λ_1, and then analyze the relationship between λ_1 and moment and correlation-type functionals. The second part of the chapter deals with Poincaré-type inequalities with weights. The powerful Brascamp–Lieb inequality is proved and applied to the problem of bounding the variance functional σ_4 for the class of coordinatewise symmetric log-concave probability distributions.

6.1 Exponential Integrability

In the presence of Poincaré-type inequalities, Lipschitz functions are known to have a finite exponential moment. This property may be quantified in the following assertion. Let μ be a Borel probability measure on the metric space $(\mathbb{M}, d)$ with a positive Poincaré constant $\lambda_1 = \lambda_1(\mu)$, so that, for any locally Lipschitz function f on $\mathbb{M}$,

$$\lambda_1 \mathrm{Var}_\mu(f) \leq \int |\nabla f|^2 \, \mathrm{d}\mu. \tag{6.1}$$

Proposition 6.1.1 *Any Lipschitz function f on $\mathbb{M}$ has finite absolute moments of all orders. Moreover, if $\|f\|_{\mathrm{Lip}} \leq 1$, then for any $r \geq 0$,*

$$\mu\{f - \mathbb{E}_\mu f \geq r\} \leq 3\,\mathrm{e}^{-2\sqrt{\lambda_1}\, r}. \tag{6.2}$$

As a preliminary step, first we derive an estimate for exponential moments of Lipschitz functions.

S. Bobkov et al., *Concentration and Gaussian Approximation for Randomized Sums*, Probability Theory and Stochastic Modelling 104,
https://doi.org/10.1007/978-3-031-31149-9_6

Lemma 6.1.2 *For any bounded f on $\mathbb{M}$ with $\|f\|_{\mathrm{Lip}} \leq 1$ and μ-mean zero,*

$$\int e^{\sqrt{\lambda_1} f}\, d\mu \leq K_0 = \prod_{n=0}^{\infty} \frac{1}{(1-4^{-n-1})^{2^n}} = 1.72\ldots$$

Proof The function

$$u(t) = \int e^{2t\sqrt{\lambda_1} f}\, d\mu, \quad t \in \mathbb{R},$$

is finite and smooth on the real line. An application of (6.1) to the functions $e^{t\sqrt{\lambda_1} f}$ together with the chain rule inequality leads to

$$\lambda_1 \left[\int e^{2t\sqrt{\lambda_1} f}\, d\mu - \left(\int e^{t\sqrt{\lambda_1} f}\, d\mu \right)^2 \right] \leq \lambda_1 t^2 \int e^{2t\sqrt{\lambda_1} f}\, |\nabla f|^2\, d\mu.$$

Since $|\nabla f| \leq 1$, we obtain a functional inequality $u(t) \leq \frac{1}{1-t^2}\, u(t/2)^2$, $|t| < 1$. Applying it to the points $t, \frac{1}{2}t, \dots, \frac{1}{2^{N-1}}\, t$ with integers $N \geq 1$, we then get

$$u(t) \leq u(t/2^N)^{2^N} \prod_{n=0}^{N-1} \frac{1}{(1-t^2/4^n)^{2^n}}.$$

Since $\mathbb{E}_\mu f = 0$, $u(\varepsilon) = 1 + O(\varepsilon^2)$ as $\varepsilon \to 0$, so, in the limit as $N \to \infty$ we arrive at

$$u(t) \leq \prod_{n=0}^{\infty} \frac{1}{(1-t^2/4^n)^{2^n}}.$$

It remains to apply it with $t = \frac{1}{2}$. □

Proof (of Proposition 6.1.1) Without loss of generality, we may assume that the function f is bounded and has μ-mean zero. Given $r \in \mathbb{R}$, let us apply the Poincaré-type inequality to the functions

$$f_r(x) = \exp\left\{ \sqrt{\lambda_1}\, \min\{f(x), r\} \right\},$$

when it becomes

$$\lambda_1\, \mathbb{E}_\mu f_r^2 \leq \lambda_1\, (\mathbb{E}_\mu f_r)^2 + \mathbb{E}_\mu\, |\nabla f_r|^2.$$

By Lemma 6.1.2 applied to the Lipschitz function $\min\{f, r\} - \mathbb{E}_\mu \min\{f, r\}$, we have

$$\mathbb{E}_\mu f_r \leq K_0\, e^{\sqrt{\lambda_1}\, \mathbb{E}_\mu \min\{f,r\}} \leq K_0\, e^{\sqrt{\lambda_1}\, \mathbb{E}_\mu f} = K_0.$$

Note that f_r is constant on the open set $U_r = \{f(x) > r\}$, so that $|\nabla f_r| = 0$ on it. In addition, since $|\nabla f| \leq 1$ on the whole space (the Lipschitz assumption), by the chain rule inequality,

$$|\nabla f_r| \leq \sqrt{\lambda_1} f_r \leq \sqrt{\lambda_1}\, e^{\sqrt{\lambda_1}\, f}.$$

Hence

$$\lambda_1 \, \mathbb{E}_\mu f_r^2 \le \lambda_1 K_0^2 + \lambda_1 \, \mathbb{E}_\mu \, e^{2\sqrt{\lambda_1} f} \, 1_{\{f \le r\}}.$$

But, the left expectation may be represented as $\mathbb{E}_\mu \, e^{2\sqrt{\lambda_1} f} \, 1_{\{f \le r\}} + e^{2\sqrt{\lambda_1} r} \, \mu(U_r)$, so that the above inequality is simplified to

$$\mu\{f > r\} \le K_0^2 \, e^{-2\sqrt{\lambda_1} r}.$$

Here, the right-hand side continuously depends on r, and we arrive at (6.2) (since $K_0^2 < 3$). Combined with a similar bound for $-f$, (6.1) implies the finiteness of the exponential moments $\mathbb{E}_\mu e^{2t\sqrt{\lambda_1}\,|f|}$ for all $0 < t < 1$. The proposition is proved. □

Remarks. 1) The right-hand side in (6.2) is optimal up to a factor, which may be seen in the example of the two-sided exponential measures $d\mu(x) = \frac{\lambda}{2} \, e^{-\lambda |x|} \, dx$ on $\mathbb{R}$ as in item 5.7.2. In this case, $\lambda_1 = \frac{1}{4} \lambda^2$, and taking $f(x) = x$, we have

$$\mu\{f \ge r\} = \frac{1}{2} \, e^{-\lambda r} = \frac{1}{2} \, e^{-2\sqrt{\lambda_1} r}$$

for all $r \ge 0$. This shows that the two-sided exponential measures play an almost extremal role in (6.2).

2) Combining (6.2) with a similar inequality for $-f$, we get

$$\mu\{|f - \mathbb{E}_\mu f| \ge r\} \le 6 \, e^{-2\sqrt{\lambda_1} r} \quad (r \ge 0),$$

which provides an exponential bound

$$\int e^{2t\sqrt{\lambda_1}\,|f|} \, d\mu \le \frac{6}{1-t}, \qquad 0 < t < 1,$$

assuming that f has μ-mean zero. This implies that for some absolute constant $c > 0$, the Orlicz norm of f generated by the Young function $\psi_1(t) = e^{|t|} - 1$ satisfies

$$\|f\|_{\psi_1} \equiv \inf \Big\{ t > 0 : \int \psi_1(f/t) \, d\mu \le 1 \Big\} \le \frac{c}{\sqrt{\lambda_1}}.$$

That is, the collection of all μ-mean zero functions on $\mathbb{M}$ with $\|f\|_{\mathrm{Lip}} \le 1$ represents a bounded (convex) subset in the Orlicz space $L^{\psi_1}(\mu)$ with diameter $\le \frac{2c}{\sqrt{\lambda_1}}$.

3) As we can see from the proof, the Lipschitz assumption of Proposition 6.1.1 and Lemma 6.1.2 may be replaced by the potentially weaker condition that f is a continuous function on $\mathbb{M}$ satisfying $|\nabla f| \le 1$ μ-a.e.

4) In the case of the Euclidean space $\mathbb{M} = \mathbb{R}^n$, for the validity of (6.2) it is sufficient that the Poincaré-type inequality (6.1) holds true in a smaller class of functions f compared to the one of Proposition 5.4.1. The argument employs Proposition 5.4.2; let us restate Proposition 6.1.1 for product measures.

Proposition 6.1.3 *The assertion of Proposition 6.1.1 remains valid for any product measure $\mu = \mu_1 \otimes \cdots \otimes \mu_p$ on $\mathbb{R}^n$, $n = n_1 + \cdots + n_p$, whose marginals μ_k are Borel probability measures on $\mathbb{R}^{n_k}$ satisfying* (6.1) *in the class of all C^∞-smooth, compactly supported functions.*

Proof Applying the subadditivity property of the variance functional (cf. item 5.7.7) and using the hypothesis on the μ_k, we have that, for any C^∞-smooth, compactly supported function f on $\mathbb{R}^n$,

$$\begin{aligned}\lambda_1 \mathrm{Var}_\mu(f) &\leq \lambda_1 \int_{\mathbb{R}^n} \Big[\sum_{k=1}^p \mathrm{Var}_{x_k}(f)\Big]\, d\mu(x_1,\dots,x_p) \\ &\leq \lambda_1 \int_{\mathbb{R}^n} \Big[\sum_{k=1}^p \int_{\mathbb{R}^{n_k}} |\nabla_{x_k} f(x_1,\dots,x_p)|^2\, d\mu_k(x_k)\Big]\, d\mu(x_1,\dots,x_p) \\ &= \lambda_1 \int_{\mathbb{R}^n} |\nabla f|^2\, d\mu.\end{aligned}$$

Here, $\nabla_{x_k} f(x)$ denotes the usual gradient for the function $x_k \to f(x)$ with fixed x_i, $i \neq k$. In other words, the measure μ satisfies (6.1) in the class of all C^∞-smooth, compactly supported functions on $\mathbb{R}^n$. According to Proposition 5.4.2, we then arrive at (5.10) for any function f on $\mathbb{R}^n$ with $\|f\|_{\mathrm{Lip}} \leq 1$. Hence, all arguments used in the proof of Proposition 6.1.1 may be repeated to obtain the desired inequality (6.2). □

6.2 Growth of L^p-norms

If the function f is not necessarily Lipschitz (for example, a polynomial), the exponential bound in (6.2) is no longer true, and a more general variant of Proposition 6.1.1 is needed. This would allow us to control probabilities of large deviations. With this aim, proper bounds on the L^p-norms of f in terms of the L^p-norms of the modulus of the gradient may be useful. As before, we assume that a Borel probability measure μ on the metric space $(\mathbb{M}, d)$ satisfies the Poincaré-type inequality (6.1) in the class of all locally Lipschitz functions f on $\mathbb{M}$.

Proposition 6.2.1 *Given a locally Lipschitz function f on $\mathbb{M}$, let the integral $\int |\nabla f|^p\, d\mu$ be finite for $p \geq 2$. Then f is integrable, and if it has μ-mean zero,*

$$\int |f|^p\, d\mu \leq \Big(\frac{p}{\sqrt{2\lambda_1}}\Big)^p \int |\nabla f|^p\, d\mu. \tag{6.3}$$

Proof By the subadditivity property of the variance (cf. item 5.7.7), the Poincaré-type inequality for the probability measure μ on $\mathbb{M}$ extends to $\mathbb{M} \times \mathbb{M}$ in the form

$$\lambda_1 \mathrm{Var}_{\mu\otimes\mu}(u) \leq \iint |\widetilde{\nabla} u(x,y)|^2\, d\mu(x) d\mu(y) \tag{6.4}$$

with respect to the product measure $\mu^2 = \mu \otimes \mu$ and for the modulus of gradient defined by

$$|\widetilde{\nabla} u(x,y)|^2 = |\nabla_x u(x,y)|^2 + |\nabla_y u(x,y)|^2.$$

The function u should be locally Lipschitz in each coordinate and measurable with respect to the product of the Borel σ-algebra in $\mathbb{M}$ by itself.

Assuming that $p > 2$, let us apply the inequality (6.3) to

$$u(x,y) = |f(x) - f(y)|^{\frac{p}{2}} \operatorname{sign}(f(x) - f(y)), \quad x, y \in \mathbb{M}.$$

By the chain rule inequality in $\mathbb{M}$,

$$\begin{aligned} |\nabla_x u(x,y)| &\le \frac{p}{2}\, |f(x) - f(y)|^{\frac{p}{2}-1}\, |\nabla f(x)|, \\ |\nabla_y u(x,y)| &\le \frac{p}{2}\, |f(x) - f(y)|^{\frac{p}{2}-1}\, |\nabla f(y)|, \end{aligned}$$

so that

$$|\widetilde{\nabla} u(x,y)| \le \frac{p}{2}\, |f(x) - f(y)|^{\frac{p}{2}-1} \sqrt{|\nabla f(x)|^2 + |\nabla f(y)|^2}.$$

Since u has a symmetric distribution under μ^2, applying (6.3) together with Hölder's inequality, we conclude that

$$\begin{aligned} \lambda_1 &\iint |f(x) - f(y)|^p \, d\mu^2(x,y) \\ &\le \frac{p^2}{4} \iint |f(x) - f(y)|^{p-2} \left(|\nabla f(x)|^2 + |\nabla f(y)|^2\right) d\mu^2(x,y) \\ &\le \frac{p^2}{4} \left(\iint |f(x) - f(y)|^p \, d\mu^2\right)^{\frac{p-2}{p}} \left(\iint \left(|\nabla f(x)|^2 + |\nabla f(y)|^2\right)^{\frac{p}{2}} d\mu^2\right)^{\frac{2}{p}}. \end{aligned}$$

By Jensen's inequality, the last integral does not exceed

$$2^{\frac{p}{2}-1} \iint \left(|\nabla f(x)|^p + |\nabla f(y)|^p\right) d\mu^2(x,y) = 2^{\frac{p}{2}} \int |\nabla f|^p \, d\mu.$$

Hence

$$\lambda_1 \left(\iint |f(x) - f(y)|^p \, d\mu^2(x,y)\right)^{\frac{2}{p}} \le \frac{p^2}{2} \left(\int |\nabla f|^p \, d\mu\right)^{\frac{2}{p}}.$$

Equivalently,

$$\iint |f(x) - f(y)|^p \, d\mu^2(x,y) \le \left(\frac{p}{\sqrt{2\lambda_1}}\right)^p \int |\nabla f|^p \, d\mu.$$

If the right integral is finite, then so is the left one, thus necessarily f is integrable. Moreover, if f has μ-mean zero, the left integral is greater than or equal to $\int |f|^p \, d\mu$ (by Jensen's inequality). Proposition 6.2.1 is proved. □

6.3 Moment Functionals. Small Ball Probabilities

Let us now connect the Poincaré constant with the functionals which we considered in Chapter 1. Thus, let X be a random vector in $\mathbb{R}^n$ satisfying

$$\lambda_1 \mathrm{Var}(f(X)) \leq \mathbb{E}\,|\nabla f(X)|^2. \tag{6.5}$$

It will be sufficient to require that this inequality holds true in the class of all C^∞-smooth functions f on $\mathbb{R}^n$ with $\lambda_1 = \lambda_1(X) > 0$. First, it makes sense to emphasize that λ_1 is bounded from above under natural moment assumptions.

Proposition 6.3.1 *a) If X isotropic in $\mathbb{R}^n$ $(n \geq 2)$, then $\lambda_1 \leq 1$.*
b) If X has mean zero and $\mathbb{E}\,|X|^2 = n$, then $\lambda_1 \leq 1$.

The second assertion generalizes the first one for mean zero random vectors.

Note that $a)$ is no longer true for $n = 1$; for example, the constant random variable $X = 1$ is isotropic and has the Poincaré constant $\lambda_1 = \infty$.

Proof Applying (6.5) to the linear functions $f(x) = \langle x, \theta\rangle$, $\theta \in \mathbb{S}^{n-1}$, we get

$$\lambda_1 \mathrm{Var}(f(X)) = \lambda_1 \left[1 - \langle v, \theta\rangle^2\right] \leq 1,$$

where $v = \mathbb{E}X$. Choosing θ to be orthogonal to v, the conclusion in $a)$ follows.

To prove $b)$, let $X = (X_1, \dots, X_n)$. We may assume that the eigenvectors of the covariance operator of X form a canonical basis in $\mathbb{R}^n$, so that $\mathbb{E}X_iX_j = a_i\delta_{ij}$, $1 \leq i, j \leq n$, with eigenvalues a_i. Hence $\mathbb{E}\,|X|^2 = a_1 + \dots + a_n = n$, which implies that $\max_{i \leq n} a_i \geq 1$. On the other hand, by (6.5) with $f(x) = x_i$, we have $\lambda_1 a_i \leq 1$ for all $i \leq 1$. As a result, $\lambda_1 \leq 1$. □

Proposition 6.3.2 *If X has mean zero, then $\mathbb{E}\,|X|^2 \leq \frac{n}{\lambda_1}$ and $M_2(X) \leq \frac{1}{\sqrt{\lambda_1}}$. Moreover,*

$$M_p(X) \leq \frac{p}{\sqrt{\lambda_1}}, \quad p \geq 1.$$

Proof Again, by (6.5) applied with linear functions f, we have $\lambda_1 \mathbb{E}\,\langle X, \theta\rangle^2 \leq 1$, which means that $M_2(X) \leq 1/\sqrt{\lambda_1}$. By Proposition 1.2.2, $\mathbb{E}\,|X|^2 \leq \frac{n}{\lambda_1}$.

An application of the inequality (6.3) gives a bound on the L^p-norms,

$$\mathbb{E}\,|\langle X, \theta\rangle|^p \leq \left(\frac{p}{\sqrt{2\lambda_1}}\right)^p$$

for $p \geq 2$. Increasing the constant a bit, one may also cover the region $1 \leq p < 2$ as well, by estimating the L^p-norm by the L^2-norm. Proposition 6.3.2 follows. □

Proposition 6.3.3 *a) If X has mean zero, then*

$$\sigma_4^2(X) \le \Lambda(X) \le \frac{4}{\lambda_1^2}.$$

b) If X isotropic, then

$$\sigma_4^2(X) \le \Lambda(X) \le \frac{4}{\lambda_1}.$$

Proof Applying (6.5) to the quadratic function $f(x) = \sum_{i,j=1}^n a_{ij}\, x_i x_j$ with coefficients such that $a_{ij} = a_{ji}$, we get, by Cauchy's inequality,

$$\mathrm{Var}\bigg(\sum_{i,j=1}^n a_{ij} X_i X_j\bigg) \le \frac{4}{\lambda_1} \sum_{i=1}^n \mathbb{E}\bigg(\sum_{j=1}^n a_{ij} X_j\bigg)^2 \le \frac{4}{\lambda_1} \sum_{i=1}^n \sum_{j=1}^n a_{ij}^2\, \mathbb{E}X_j^2.$$

Since $\mathbb{E}X_j^2 \le 1/\lambda_1$, the right-hand side does not exceed $\frac{4}{\lambda_1^2}$ subject to $\sum_{i,j=1}^n a_{ij}^2 \le 1$. Hence $\Lambda(X) \le 4/\lambda_1^2$. In the isotropic case (without the mean zero assumption), we have $\mathbb{E}X_j^2 = 1$, so that the latter bound may be improved to $\Lambda(X) \le 4/\lambda_1$. □

We now turn to small ball probabilities.

Proposition 6.3.4 *If* $\mathbb{E}\,|X|^2 = n$, *then*

$$\mathbb{P}\Big\{|X|^2 \le \frac{1}{4}\, n\Big\} \le 3\, e^{-\frac{1}{2}\sqrt{\lambda_1 n}}.$$

Proof One may appeal to the bound (6.2) via Proposition 6.1.1, which for $f(x) = -|x|$ yields

$$\mathbb{P}\{|X| - \mathbb{E}\,|X| \le -r\} \le 3\, e^{-2\sqrt{\lambda_1} r}, \quad r \ge 0. \tag{6.6}$$

Here, one can bound $\mathbb{E}\,|X|$ from below by virtue of the Poincaré-type inequality (5.6) for this particular function f. It gives

$$n - (\mathbb{E}\,|X|)^2 = \mathrm{Var}(|X|) \le \frac{1}{\lambda_1},$$

which in the case $\lambda_1 n \ge \frac{4}{3}$ implies $\mathbb{E}\,|X| \ge \sqrt{n - \frac{1}{\lambda_1}} \ge \frac{1}{2}\sqrt{n}$. Hence, applying (6.6) with $r = \mathbb{E}|X| - \frac{1}{2}\sqrt{n}$, we get

$$\mathbb{P}\Big\{|X| \le \frac{1}{2}\sqrt{n}\Big\} \le 3\, e^{-2\sqrt{\lambda_1}\, r}.$$

Here $r \ge \sqrt{n - \frac{1}{\lambda_1}} - \frac{1}{2}\sqrt{n} \ge \frac{1}{4}\sqrt{n}$ whenever $\lambda_1 n \ge \frac{16}{7}$, in which case the bound simplifies to

$$\mathbb{P}\Big\{|X| \le \frac{1}{2}\sqrt{n}\Big\} \le 3\, e^{-\frac{1}{2}\sqrt{\lambda_1 n}}.$$

But, it remains to hold when $\lambda_1 n < 16/7$ as well, since then the right-hand side is greater than $3\, \exp\{-2/\sqrt{7}\} > 1$. Proposition 6.3.4 proved. □

Note that the bound in this proposition is not as sharp, with respect to n, as in the case of independent components (cf. Proposition 2.1.4). Hence, it is natural to ask if it is possible to sharpen the bound when λ_1 is bounded away from zero. Nevertheless, one has to keep in mind the following "degenerate" situation covered by Proposition 6.3.4. Let ξ be a random variable with $\mathbb{E}\xi^2 = 1$ and a positive λ_1. Then the random vector $X = (\xi, \dots, \xi)$ satisfies $\mathbb{E}\,|X|^2 = n$ and $\lambda_1(X) = \frac{\lambda_1}{n}$. In this case, $\mathbb{P}\{|X|^2 \leq \frac{1}{4}\,n\} = \mathbb{P}\{\xi^2 \leq \frac{1}{4}\}$ may not necessarily be small.

Let us give another version of this statement for convolutions, namely, for sums

$$\Sigma_p = \sum_{k=1}^{p} (X^{(k)} - Y^{(k)}),$$

where $X^{(k)}, Y^{(k)}$ $(1 \leq k \leq p)$ are independent copies of X.

Proposition 6.3.5 *If X has mean zero and $\mathbb{E}\,|X|^2 = n$, then*

$$\mathbb{P}\Big\{|\Sigma_p|^2 \leq \frac{np}{2}\Big\} \leq 3\, \mathrm{e}^{-\frac{1}{3}\sqrt{\lambda_1 n}}.$$

Proof We now apply Proposition 6.1.1 to the product measure $\mu^{\otimes 2p}$ on $(\mathbb{R}^n)^{2p} = \mathbb{R}^{2pn}$, where μ is the distribution of X. The function

$$f(x_1, \dots, x_p, y_1, \dots, y_p) = -\Big|\sum_{k=1}^{p} (x_k - y_k)\Big|, \quad x_k, y_k \in \mathbb{R}^n,$$

has Lipschitz semi-norm $\sqrt{2p}$ with respect to the Euclidean distance on $\mathbb{R}^{2pn}$, and therefore it admits an exponential inequality

$$\mu^{\otimes 2p}\{f - m \geq r\} \leq 3\mathrm{e}^{-2\sqrt{\lambda_1}\, r/\sqrt{2p}} \quad (r > 0),$$

where m is the $\mu^{\otimes 2p}$-mean of f. Equivalently,

$$\mathbb{P}\{|\Sigma_p| - \mathbb{E}\,|\Sigma_p| \leq -r\} \leq 3\mathrm{e}^{-2\sqrt{\lambda_1}\, r/\sqrt{2p}}. \tag{6.7}$$

But, by the Poincaré-type inequality on the product space applied to the function f, and using $\mathbb{E}\,|\Sigma_p|^2 = 2pn$, we have

$$2pn - (\mathbb{E}\,|\Sigma_p|)^2 \leq \frac{2p}{\lambda_1} \leq pn,$$

where the last inequality holds true when $\lambda_1 n \geq 2$. In this case, $\mathbb{E}\,|\Sigma_p| \geq \sqrt{pn}$, and applying (6.7) with $r = (1 - \frac{1}{\sqrt{2}})\sqrt{pn}$, we obtain

$$\mathbb{P}\Big\{|\Sigma_p| \le \frac{1}{\sqrt{2}}\sqrt{np}\Big\} \le 3\,\mathrm{e}^{-(\sqrt{2}-1)\sqrt{\lambda_1 n}} < 3\,\mathrm{e}^{-\frac{1}{3}\sqrt{\lambda_1 n}}.$$

In the case $\lambda_1 n \le 2$, this inequality is fulfilled automatically. □

6.4 Weighted Poincaré-type Inequalities

Since Poincaré-type inequalities imply exponential integrability of Lipschitz functions, they may not be used in the study of probability distributions whose densities have a slow (for example, polynomial) decay at infinity on $\mathbb{R}^n$. Therefore, it is natural to consider more general Sobolev-type inequalities, in particular, such as

$$\mathrm{Var}_\mu(f) \le \int |\nabla f(x)|^2\, w(x)^2\, \mathrm{d}\mu(x), \tag{6.8}$$

where w is a non-negative Borel measurable function on a given metric probability space $(\mathbb{M}, d, \mu)$. These relations, where f may be an arbitrary locally Lipschitz function on $\mathbb{M}$, are called weighted Poincaré-type inequalities with weights w. On $\mathbb{R}^n$, they can further be generalized to the form

$$\mathrm{Var}_\mu(f) \le \int |T(x)\nabla f(x)|^2\, \mathrm{d}\mu(x), \tag{6.9}$$

involving a Borel measurable map $T : \mathbb{M} \to \mathbb{R}^n$ on a subset $\mathbb{M}$ of $\mathbb{R}^n$ of μ-measure 1. Such a map may be properly associated to μ.

Examples

6.4.1 Cauchy measures. The generalized Cauchy (or Student) distribution $\mu_{\beta,n}$ on $\mathbb{R}^n$ has density

$$\frac{\mathrm{d}\mu_{\beta,n}(x)}{\mathrm{d}x} = \frac{1}{Z_{\beta,n}}\,(1+|x|^2)^{-\beta}, \quad x \in \mathbb{R}^n,$$

where $Z_{\beta,n}$ is a normalizing factor. If $\beta > n$, this measure can be shown to satisfy the weighted Poincaré-type inequality

$$\mathrm{Var}_{\mu_{\beta,n}}(f) \le \frac{C_\beta}{2(\beta-1)} \int |\nabla f(x)|^2\,(1+|x|^2)\, \mathrm{d}\mu_{\beta,n}(x) \tag{6.10}$$

with $C_\beta = \big(\sqrt{1+\frac{2}{\beta-1}} + \sqrt{\frac{2}{\beta-1}}\,\big)^2$ (cf. [55], [56]). Note that, after the linear change of the variable $y = \sqrt{2\beta}\,x$, the rescaled densities of $\mu_{\beta,n}$ converge as $\beta \to \infty$ to the density of the standard normal law. Since also $C_\beta \to 1$, (6.10) returns us in the limit to the Poincaré-type inequality for the standard Gaussian measure γ_n on $\mathbb{R}^n$,

$$\mathrm{Var}_{\gamma_n}(f) \le \int |\nabla f(x)|^2 \, \mathrm{d}\gamma_n(x). \tag{6.11}$$

6.4.2 Exponential measures. If Z_1, Z_2 are independent random variables with a standard normal distribution, the random variable $X = (Z_1^2 + Z_2^2)/2$ has the standard exponential distribution μ with density $\mathrm{e}^{-x}, x > 0$. Starting from (6.11) in dimension $n = 2$, we therefore obtain that, for any C^1-smooth function f on $\mathbb{R}$,

$$\begin{aligned} \mathrm{Var}(f(X)) &= \mathrm{Var}\Big(f\Big(\frac{Z_1^2 + Z_2^2}{2}\Big)\Big) \\ &\le \mathbb{E}\, Z_1^2 f'\Big(\frac{Z_1^2 + Z_2^2}{2}\Big)^2 + \mathbb{E}\, Z_2^2 f'\Big(\frac{Z_1^2 + Z_2^2}{2}\Big)^2 = 2\,\mathbb{E} X^2 f'(X)^2. \end{aligned}$$

In other words, μ satisfies the weighted Poincaré-type inequality

$$\mathrm{Var}_\mu(f) \le 2 \int_{\mathbb{R}} f'(x)^2 x^2 \, \mathrm{d}\mu(x),$$

thus with a similar weight as in the case of the Cauchy measures in dimension one. After tensorization, we obtain that any product measure $\mu_n = \mu^{\otimes n}$ satisfies, for any smooth function f on $\mathbb{R}^n$,

$$\mathrm{Var}_{\mu_n}(f) \le 2 \int_{\mathbb{R}^n} \sum_{k=1}^n x_k^2 \, (\partial_k f(x))^2 \, \mathrm{d}\mu_n(x).$$

6.4.3 Gaussian measures. With a proper choice of the weight function w, (6.8) may sharpen the usual Poincaré-type inequalities. For example, for the standard Gaussian measure $\mu = \gamma_1$ on the real line, one can show (cf. [55]) that

$$\mathrm{Var}_{\gamma_1}(f) \le 6 \int_{\mathbb{R}} f'(x)^2 \, \frac{1}{1 + x^2} \, \mathrm{d}\gamma_1(x).$$

6.4.4 Beta distributions. The beta distribution $\mu_{\alpha,\beta}$ with parameters $\alpha, \beta > 0$ is supported on the unit interval $(0, 1)$ and has density

$$\frac{\mathrm{d}\mu_{\alpha,\beta}(x)}{\mathrm{d}x} = \frac{1}{B(\alpha, \beta)} \, x^{\alpha-1} (1 - x)^{\beta-1}, \quad 0 < x < 1,$$

where $B(\alpha, \beta) = \frac{\Gamma(\alpha)\,\Gamma(\beta)}{\Gamma(\alpha+\beta)}$. It admits the weighted Poincaré-type inequality

$$\mathrm{Var}_{\mu_{\alpha,\beta}}(f) \le \frac{1}{\alpha + \beta} \int_0^1 f'(x)^2 x(1 - x) \, \mathrm{d}\mu_{\alpha,\beta}(x).$$

Here an equality is attained for all linear functions $f(x) = a + bx$ (cf. [57]).

In analogy with the limiting Gaussian model, the weighted Poincaré-type inequalities (6.8) fully describe the measure concentration and large deviations bounds for the class of Lipschitz functions f. In particular, we have the following assertion on the growth of L^p-norms $\|f\|_p = (\int |f|^p \, d\mu)^{1/p}$ in terms of the L^p-norms $\|w\|_p$. We restrict it to the Euclidean space $\mathbb{M} = \mathbb{R}^n$.

Proposition 6.4.5 *Assume that w has a finite p-th moment under μ $(p \ge 2)$. If the weighted Poincaré-type inequality* (6.8) *holds, then any Lipschitz function f on $\mathbb{R}^n$ with $\|f\|_{\mathrm{Lip}} \le 1$ has a finite p-th moment. Moreover, if f has μ-mean zero,*

$$\|f\|_p \le \frac{p}{\sqrt{2}} \, \|w\|_p.$$

The argument is very much similar to the one used in the proof of Proposition 6.2.1, so we omit the proof (cf. [55]).

6.5 The Brascamp–Lieb Inequality

As an important example of (6.9), let us derive here the Brascamp–Lieb inequality for the class of log-concave probability measures. Any such (absolutely continuous) measure μ is supported on an open convex set $\mathbb{M}$ in $\mathbb{R}^n$, where it has density

$$\frac{d\mu(x)}{dx} = e^{-V(x)}$$

for some convex function $V : \mathbb{M} \to \mathbb{R}$.

Proposition 6.5.1 *Suppose that V is C^2-smooth and the matrix $V''(x)$ is positive definite at every point $x \in \mathbb{M}$, with inverse matrix $V''(x)^{-1}$. Then, for any C^1-smooth function f on $\mathbb{M}$ with gradient ∇f,*

$$\mathrm{Var}_\mu(f) \le \int \langle V''^{-1} \nabla f, \nabla f \rangle \, d\mu. \tag{6.12}$$

Note that (6.12) corresponds to (6.9) with $T(x) = V''(x)^{-1/2}$.

This result contains in particular the Poincaré-type inequality for the standard Gaussian measure $\mu = \gamma_n$. More generally, if $V''(x) \ge c\, \mathrm{I}_n$ in the matrix sense on $\mathbb{M}$ for some $c > 0$, then we get $\lambda_1(\mu) \ge c$. In fact, this conclusion can be sharpened in terms of logarithmic Sobolev constants.

Proof Let us describe an approach relying on the Prékopa–Leindler theorem (cf. Proposition 2.4.2): If $t, s > 0$, $t + s = 1$, and $u, v, w : \mathbb{R}^n \to [0, \infty)$ are Borel measurable functions satisfying $w(tx + sy) \ge u(x)^t \, v(y)^s$ for all $x, y \in \mathbb{R}^n$, then

$$\int w(z)\, dz \ge \Big(\int u(x)\, dx\Big)^t \Big(\int v(y)\, dy\Big)^s. \tag{6.13}$$

Given a small parameter $\varepsilon > 0$, we apply it with $t = s = \frac{1}{2}$ to the functions

$$u(x) = e^{\varepsilon f(x) - V(x)}\, 1_K(x), \quad v(y) = e^{-V(y)}\, 1_K(y),$$
$$w(z) = e^{f_\varepsilon(z) - V(z)}\, 1_K(z),$$

where K is a fixed convex body in $\mathbb{M}$ (which will approximate $\mathbb{M}$ from below). In this case, the conclusion (6.13) becomes

$$\Big(\int e^{f_\varepsilon}\, d\mu_K\Big)^2 \geq \int e^{\varepsilon f}\, d\mu_K \tag{6.14}$$

in terms of the normalized restriction of μ to K for the function

$$f_\varepsilon(z) = \sup_{z = \frac{x+y}{2},\ x,y \in K} \Big[\frac{\varepsilon}{2}\, f(x) - \Big(\frac{1}{2}\, V(x) + \frac{1}{2}\, V(y) - V(z)\Big)\Big], \quad z \in K.$$

There is an equality in (6.14) at $\varepsilon = 0$, and we claim that comparing the coefficients in the Taylor expansion of both sides up to the quadratic term ε^2 yields the desired inequality (6.12) for μ_K.

With this aim, choose $\varepsilon_0 > 0$ such that an ε_0-neighborhood K' of K is contained in $\mathbb{M}$ together with its closure (in particular K' is a convex body). Put $C_1 = \sup_{K'} |f|$, $C_2 = \sup_{K'} |\nabla f|$, and choose a constant $c > 0$ such that $V'' \geq c\,\mathrm{Id}$ on K' (where Id is the identity $n \times n$ matrix). On the set K we consider the function

$$\tilde{f}_\varepsilon(z) = \sup_{z \pm h \in K'} \Big[\frac{\varepsilon}{2}\, f(z+h) - \Big(\frac{1}{2}\, V(z+h) + \frac{1}{2}\, V(z-h) - V(z)\Big)\Big], \tag{6.15}$$

so that $\tilde{f}_\varepsilon(z) \geq f_\varepsilon(z)$. The choice $h = 0$ under the sup-sign implies $\tilde{f}_\varepsilon(z) \geq \frac{\varepsilon}{2}\, f(z)$. On the other hand,

$$\frac{1}{2}\, V(z+h) + \frac{1}{2}\, V(z-h) - V(z) \geq \frac{c}{2}\, |h|^2,$$

so that

$$\tilde{f}_\varepsilon(z) \leq \frac{\varepsilon}{2}\, C_1 - \frac{c}{2}\, |h|^2.$$

Hence, for $|h| > r_\varepsilon = \sqrt{2\varepsilon C_1/c}$, the expression in the square brackets in (6.15) is smaller than $-\frac{\varepsilon}{2}\, C_1$ and therefore smaller than $\frac{\varepsilon}{2} f(z)$. Hence, one may restrict the supremum in (6.15) to the vectors with $|h| < 2r_\varepsilon$, i.e., to the ball $B(0, 2r_\varepsilon)$. Moreover, this supremum may not be attained on the boundary of the ball. Note that $B(z, 2r_\varepsilon)$ is contained in K' for ε small enough, namely, when $2r_\varepsilon < \varepsilon_0$. Consequently, in this case the supremum is attained at some $h = h_\varepsilon$ inside $B(0, 2r_\varepsilon)$, which thus tends to zero as ε tends to zero. Since the derivative of the expression in the square brackets in (6.15) over h should be vanishing at the point of maximum, we also get

$$\frac{\varepsilon}{2}\, \nabla f(z + h_\varepsilon) = \frac{1}{2}\, \nabla V(z + h_\varepsilon) - \frac{1}{2}\, \nabla V(z - h_\varepsilon) = \frac{1}{2} \int_{-1}^{1} V''(z + t h_\varepsilon)\, h_\varepsilon\, dt.$$

In particular,

$$\frac{\varepsilon}{2}\langle\nabla f(z+h_\varepsilon),h_\varepsilon\rangle=\frac{1}{2}\int_{-1}^{1}\langle V''(z+th_\varepsilon)h_\varepsilon,h_\varepsilon\rangle\,\mathrm{d}t\geq c|h_\varepsilon|^2,$$

which gives $|h_\varepsilon|\leq\frac{\varepsilon}{2c}C_2$. In addition, since

$$\frac{1}{2}\nabla V(z+h_\varepsilon)-\frac{1}{2}\nabla V(z-h_\varepsilon)=V''(z)h_\varepsilon+o(|h_\varepsilon|),$$

we find an asymptotic expression

$$h_\varepsilon=\frac{\varepsilon}{2}V''(z)^{-1}\nabla f(z)+o(\varepsilon),$$

which is uniform over all $z\in K$. Therefore, by Taylor's expansion, we get from (6.15) that uniformly on K

$$\begin{aligned}\tilde{f}_\varepsilon(z)&=\frac{\varepsilon}{2}f(z+h_\varepsilon)-\Big(\frac{1}{2}V(z+h_\varepsilon)+\frac{1}{2}V(z-h_\varepsilon)-V(z)\Big)\\&=\frac{\varepsilon}{2}f(z)+\frac{\varepsilon}{2}\langle\nabla f(z),h_\varepsilon\rangle-\frac{1}{2}\langle V''(z)\,h_\varepsilon,h_\varepsilon\rangle+o(\varepsilon^2)\\&=\frac{\varepsilon}{2}f(z)+\frac{\varepsilon^2}{8}\big\langle V''(z)^{-1}\nabla f(z),\nabla f(z)\big\rangle+o(\varepsilon^2).\end{aligned}$$

This implies that

$$\int\mathrm{e}^{\tilde{f}_\varepsilon}\,\mathrm{d}\mu_K=1+\frac{\varepsilon}{2}\int f\,\mathrm{d}\mu_K+\frac{\varepsilon^2}{8}\int\big\langle V''^{-1}\nabla f,\nabla f\big\rangle\,\mathrm{d}\mu_K+\frac{\varepsilon^2}{8}\int f^2\,\mathrm{d}\mu_K+o(\varepsilon^2)$$

and

$$\begin{aligned}\Big(\int\mathrm{e}^{\tilde{f}_\varepsilon}\,\mathrm{d}\mu_K\Big)^2&=1+\varepsilon\int f\,\mathrm{d}\mu_K+\frac{\varepsilon^2}{4}\int\big\langle V''^{-1}\nabla f,\nabla f\big\rangle\,\mathrm{d}\mu_K\\&\quad+\frac{\varepsilon^2}{4}\int f^2\,\mathrm{d}\mu_K+\frac{\varepsilon^2}{4}\Big(\int f\,\mathrm{d}\mu_K\Big)^2+o(\varepsilon^2).\end{aligned}$$

Also,

$$\int\mathrm{e}^{\varepsilon f}\,\mathrm{d}\mu_K=1+\varepsilon\int f\,\mathrm{d}\mu_K+\frac{\varepsilon^2}{2}\int f^2\,\mathrm{d}\mu_K+o(\varepsilon^2).$$

But, by (6.14),

$$\Big(\int\mathrm{e}^{\tilde{f}_\varepsilon}\,\mathrm{d}\mu_K\Big)^2\geq\int\mathrm{e}^{\varepsilon f}\,\mathrm{d}\mu_K,$$

and sending $\varepsilon\to0$, we obtain that

$$\int\big\langle V''^{-1}\nabla f,\nabla f\big\rangle\,\mathrm{d}\mu_K\geq\int f^2\,\mathrm{d}\mu_K-\Big(\int f\,\mathrm{d}\mu_K\Big)^2.$$

It remains to apply this inequality to a sequence $K = K_l \uparrow \mathbb{M}$.

Proposition 6.5.1 is proved. □

6.6 Coordinatewise Symmetric Log-concave Distributions

Let us now describe one interesting application of the Brascamp–Lieb inequality, due to Klartag, extending the weighted Poincaré-type inequality for the exponential measures (which was mentioned in item 6.4.2).

Proposition 6.6.1 *Suppose that a probability measure μ is supported on $\mathbb{R}^n_+$, where it has a log-concave coordinatewise non-increasing density. Then, for any smooth function f on $\mathbb{R}^n$,*

$$\mathrm{Var}_\mu(f) \leq 4 \int \sum_{k=1}^n x_k^2\,(\partial_k f(x))^2\,\mathrm{d}\mu(x). \tag{6.16}$$

Proof The proof is based on a transport argument combined with the inequality (6.12). Without loss of generality, let the density p of μ be positive and C^2-smooth on the positive octant $\mathbb{R}^n_+ = (0,\infty)^n$. By the assumption, it is log-concave, that is,

$$p(tx+sy) \geq p(x)^t p(y)^s \quad \text{for all } \ x, y \in \mathbb{R}^n_+, \ t, s > 0, \ t+s = 1,$$

and $p(y) \leq p(x)$, whenever $0 < y_k \leq x_k$, $y = (y_1,\dots,y_n)$, $x = (x_1,\dots,x_n)$.

Consider the map

$$T(x) = \big(\sqrt{x_1},\dots,\sqrt{x_n}\big), \quad x = (x_1,\dots,x_n) \in \mathbb{R}^n_+.$$

It transforms μ to some probability measure $\nu = \mu T^{-1}$. To find its density explicitly, note that, for any bounded Borel measurable function v on $\mathbb{R}^n_+$,

$$\begin{aligned}\int v\,\mathrm{d}\nu = \int v(T)\,\mathrm{d}\mu &= \int v\big(\sqrt{x_1},\dots,\sqrt{x_n}\big)\,p(x)\,\mathrm{d}x \\ &= 2^n \int v(y)\,y_1\dots y_n\,p(y_1^2,\dots,y_n^2)\,\mathrm{d}y_1\dots\mathrm{d}y_n.\end{aligned}$$

This means that ν has density

$$q(y) = 2^n\,y_1\dots y_n\,p(y_1^2,\dots,y_n^2), \quad y = (y_1,\dots,y_n) \in \mathbb{R}^n_+.$$

It is also log-concave, and moreover, the function $\log p(y_1^2,\dots,y_n^2)$ is concave on the positive octant R^n_+. Indeed, given two points $y = (y_1,\dots,y_n)$, $z = (z_1,\dots,z_n) \in \mathbb{R}^n_+$, using the inequality

$$(ty_k + sz_k)^2 \leq ty_k^2 + sz_k^2 \quad (t, s > 0,\ t+s = 1)$$

together with the monotonicity and log-concavity of p, we obtain that

$$p\big((ty_1 + sz_1)^2, \dots, (ty_n + sz_n)^2\big) \ge p\big(ty_1^2 + sz_1^2, \dots, ty_n^2 + sz_n^2\big)$$
$$\ge p(y_1^2, \dots, y_n^2)^t \, p(z_1^2, \dots, z_n^2)^s.$$

Thus, $q(y) = \mathrm{e}^{-V(y)}$ with the convex function

$$V(y) = -n \log 2 - \sum_{k=1}^n \log y_k - \log p(y_1^2, \dots, y_n^2).$$

The measure μ is obtained from ν via the inverse map $T^{-1}(y) = \big(y_1^2, \dots, y_n^2\big)$, $y = (y_1, \dots, y_n) \in \mathbb{R}_+^n$. Hence, for any bounded Borel measurable function f on $\mathbb{R}^n$,

$$\int f \, \mathrm{d}\mu = \int f(T^{-1}) \, \mathrm{d}\nu, \quad \mathrm{Var}_\mu(f) = \mathrm{Var}_\nu(f(T^{-1})).$$

Since the function $-\log p(y_1^2, \dots, y_n^2)$ is convex, the second derivative $V''(y)$ of V is bounded from below by the diagonal matrix $W(y)$ with entries $W(y)_{kk} = 1/y_k^2$ on the diagonal. The inverse matrix $V''^{-1}(y)$ is therefore bounded from above by $W^{-1}(y)$, which is also diagonal and has entries $W^{-1}(y)_{kk} = y_k^2$. Hence, the Brascamp–Lieb inequality implies

$$\mathrm{Var}_\nu(u) \le \int \sum_{k=1}^n y_k^2 \, (\partial_k u(y))^2 \, \mathrm{d}\nu(y).$$

We need to apply it to the composition $u(y) = f(T^{-1}(y)) = f\big(y_1^2, \dots, y_n^2\big)$ with a smooth f, in which case $\partial_k u(y) = 2y_k \, \partial_k f(T^{-1}(y))$. This leads to the bound

$$\mathrm{Var}_\nu(f(T^{-1})) \le 4 \int \sum_{k=1}^n y_k^4 \, \Big(\partial_k f(T^{-1}(y))\Big)^2 \, \mathrm{d}\nu(y)$$
$$= 4 \int \sum_{k=1}^n x_k^2 \, (\partial_k f(x))^2 \, \mathrm{d}\mu(x).$$

As a result, we obtain the required bound (6.16). □

In particular, we have the following assertion.

Proposition 6.6.2 *Assume that the random vector X in $\mathbb{R}^n$ is isotropic and has a coordinatewise symmetric, log-concave distribution. Then for some absolute constant C,*

$$\sigma_4^2(X) \le \Lambda(X) \le C.$$

Proof The distribution μ of the random vector $(|X_1|, \dots, |X_n|)$ has a log-concave, coordinatewise non-increasing density. Applying Proposition 6.6.1, we get

$$\mathrm{Var}(f(X)) \le 4\,\mathbb{E}\sum_{k=1}^{n} X_k^2\,(\partial_i f(X_k))^2$$

for any C^1-smooth function f on $\mathbb{R}^n$ such that $f(\pm x_1, \dots, \pm x_n) = f(x_1, \dots, x_n)$. Here, the choice $f(x) = \theta_1 x_1^2 + \dots + \theta_n x_n^2$ with $\theta_1^2 + \dots + \theta_n^2 = 1$ leads to

$$\mathrm{Var}(f(X)) \le 16 \sum_{k=1}^{n} \theta_k^2\,\mathbb{E}X_k^4 \le 16 \max_{k\le n} \mathbb{E}X_k^4.$$

One can now involve Proposition 2.3.3, which gives

$$\Lambda(X) \le 2\max_k\, \mathbb{E}X_k^4 + \sup_{\theta\in S^{n-1}} \mathrm{Var}(\theta_1 X_1^2 + \dots + \theta_n X_n^2) \le 18 \max_i\, \mathbb{E}X_k^4.$$

It remains to recall that the L^p-norms of random variables with log-concave distributions are equivalent to each other (cf. Section 2.5). In particular, for isotropic log-concave X_k's, we have $\mathbb{E}X_k^4 \le c\,(\mathbb{E}X_k^2)^2 = c$, thus proving Proposition 6.6.2. □

6.7 Remarks

In the presence of the Poincaré-type inequality with constant λ_1, the validity of the exponential bounds such as

$$\int e^{t_0\sqrt{\lambda_1}\,|f|}\,d\mu \le C_0$$

(for some universal $t_0 > 0$ and C_0) was first established by Gromov and Milman (in the setting of Riemannian manifolds, [100]). Later on, Borovkov and Utev [66], considered the setting of the real line. Lemma 6.1.2 is due to Aida and Strook [2]. The current formulation and proof of the inequality (6.1) with optimal constant in the exponent is based on the paper [21].

Weighted Poincaré-type inequalities of the form (6.8)–(6.9) have been considered in several papers, cf. [55], [56]. Proposition 6.5.1 was obtained by Brascamp and Lieb [68]; the proof reproduced here was proposed in [53]; its refinements and extensions to larger classes of probability distributions may be found in [55]. Propositions 6.6.1 and 6.6.2 are due to Klartag [123].

All log-concave probability distributions on $\mathbb{R}^n$ are known to satisfy Poincaré-type and, moreover – Cheeger-type inequalities (regardless of whether or not these measures are coordinatewise symmetric). To mention a few results, first let us stress that the two problems are equivalent, in the sense that the Cheeger inequality $\lambda_1^2 \ge \frac{1}{4}\,h^2$ can be reversed to $\lambda_1^2 \le ch^2$ up to an absolute constant c in the entire class of log-concave probability measures on $\mathbb{R}^n$. This result, due to Ledoux [132], was refined by E. Milman [146]: The deviation inequalities that are obtained for mean zero Lipschitz functions in the presence of the Poincaré-type inequalities are also sufficient to control the Cheeger constants.

However, the problem of bounding the Cheeger constant $h = h(\mu)$ and the spectral gap $\lambda_1 = \lambda_1(\mu)$ from below is often non-trivial. Given a random vector X in $\mathbb{R}^n$ with uniform distribution μ over a convex body K, Payne and Weinberger [153] derived an estimate in terms of the diameter of the body,

$$\lambda_1^{-1}(X) \le \frac{1}{\pi^2}\,\mathrm{diam}^2(K).$$

In fact, here the constant is optimal and is achieved in dimension $n = 1$. Kannan, Lovász and Simonovits [116] sharpened this estimate in terms of the distribution of $|X|$,

$$\lambda_1^{-1}(X) \le c\,\mathbb{E}\,|X|^2$$

up to an absolute constant c. Later on, this result was extended in [19] to arbitrary log-concave distributions μ (cf. also [29] for an extension of the Payne–Weinberger inequality). In [30] the above inequality was further sharpened in terms of the variance:

$$\lambda_1^{-1}(X) \le c\sqrt{\mathrm{Var}(|X|^2)} = c\sigma_4(X)\sqrt{n}.$$

If $\sigma_4(X)$ is bounded, this gives an upper bound of order $\sqrt{n}$. See also [42] for related results reflecting the strength of log-concavity of densities.

While the thin shell conjecture asserts that $\sigma_4(X)$ is indeed bounded by an absolute constant in the entire class of all isotropic log-concave probability measures on $\mathbb{R}^n$, a stronger conjecture due to Kannan, Lóvasz and Simonovits [116] indicates that $\lambda_1(\mu) \ge c$ for some absolute constant $c > 0$ in the same class of probability measures. We already mentioned at the end of Chapter 2 the logarithmic estimate $\sigma_4(X) \le c\,(\log n)^\alpha$ with $\alpha = 4$ due to Klartag and Lehec [124] and its improvement by Jambulapati, Lee and Vempala [108] with $\alpha = 2.2226$. Using a result due to Eldan [84], these estimates imply that

$$\lambda_1^{-1}(X) \le c\,(\log n)^{2\alpha+1}.$$

In particular, $\lambda_1^{-1} \le c(\log n)^2$ assuming that the thin shell conjecture were true. Whether or not λ_1^{-1} is bounded by a universal constant is unknown already for coordinatewise symmetric isotropic log-concave distributions; cf. the handbook by Alonso-Gutiérrez and Bastero [5] for an introduction to this research area.

Let us also mention that the thin shell conjecture would imply the hyperplane conjecture in the slicing problem raised by Bourgain in the 1980s (cf. Eldan and Klartag [83]).

Chapter 7
Logarithmic Sobolev Inequalities

Logarithmic Sobolev inequalities strengthen Poincaré-type inequalities, which allows one to derive sharper deviation inequalities for various classes of functions, not necessarily under the Lipschitz hypothesis. To introduce this class of analytic inequalities, first we briefly mention basic properties of the involved entropy functional and then describe several important examples of measures satisfying logarithmic Sobolev inequalities. The remaining part of the chapter deals with various bounds that are valid in the presence of logarithmic Sobolev inequalities.

7.1 The Entropy Functional and Relative Entropy

One natural (potential) sharpening of Poincaré-type inequalities may be achieved, when replacing the variance by the entropy functional

$$\mathrm{Ent}(f) = \mathbb{E}\, f \log f - \mathbb{E} f \log \mathbb{E} f,$$

where $\mathbb{E} f = \int f \, \mathrm{d}\mu$ is the expectation of f. This quantity is well-defined in the setting of an abstract probability space $(\mathbb{M}, \mu)$ in the class of all non-negative measurable functions f on $\mathbb{M}$ (the assumption that $\mathbb{E} f$ is finite may be removed, which leads to an infinite value of $\mathrm{Ent}(f)$). As in the case of expectation and variance, to emphasize the presence or dependence of the entropy functional on the measure μ, we also write $\mathrm{Ent}_\mu(f)$ for $\mathrm{Ent}(f)$, similarly to $\mathbb{E}_\mu f$ and $\mathrm{Var}_\mu(f)$.

The entropy functional is closely related to the relative entropy. Namely, if X and Y are random elements in $\mathbb{M}$ with distributions μ and ν such that $\mathrm{d}\nu = f \mathrm{d}\mu$ (i.e., ν has density f with respect to μ and thus $\mathbb{E} f = 1$), then the value

$$D(Y||X) = D(\nu||\mu) = \mathrm{Ent}(f)$$

represents the Kullback–Leibler distance from ν to μ. It is also called the informational divergence or the relative entropy of Y with respect to X.

S. Bobkov et al., *Concentration and Gaussian Approximation for Randomized Sums*, Probability Theory and Stochastic Modelling 104,
https://doi.org/10.1007/978-3-031-31149-9_7

In this section we collect basic properties of the entropy functional.

1) $\mathrm{Ent}(f) \geq 0$ for all $f \geq 0$.
2) $\mathrm{Ent}(f) = 0$ if and only if $f = c$ a.e. for some constant $c \geq 0$.
3) $\mathrm{Ent}(cf) = c\,\mathrm{Ent}(f)$ for all $f \geq 0$ and all constants $c \geq 0$.
4) $\mathrm{Ent}(f)$ is finite, if and only if $\mathbb{E}\, f \log(1+f)$ is finite. Moreover, if $\mathbb{E}f = 1$,

$$\mathrm{Ent}(f) \leq \mathbb{E}\, f \log(1+f) \leq \mathrm{Ent}(f) + 1.$$

Indeed, in this case $\mathrm{Ent}(f) = \mathbb{E}\, f \log f$. But the function $u(t) = t\log(1+t) - t\log t$ is non-negative and concave in $t \geq 0$, with $u(t) \to 1$ as $t \to \infty$.
5) Relationship with L^p-norms $\|f\|_p = (\mathbb{E}\, f^p)^{1/p}$: If $\mathbb{E}\, f^p < \infty$ for some $p > 1$, then

$$\mathrm{Ent}(f) = \lim_{p \to 1} \frac{\|f\|_p - \|f\|_1}{p-1} = \frac{\mathrm{d}}{\mathrm{d}p}\, \|f\|_p \Big|_{p=1}.$$

More generally, if $\|f\|_{p+\varepsilon} < \infty$ for $p > 0$ and some $\varepsilon > 0$, then

$$\mathrm{Ent}(f^p) = p^2\, \|f\|_p^{p-1}\, \frac{\mathrm{d}}{\mathrm{d}p}\, \|f\|_p,$$

where the derivative is evaluated at the point p.
6) Variational formula:

$$\mathrm{Ent}(f) = \sup\big\{\mathbb{E}\, fg : \mathbb{E}\, e^g \leq 1\big\}.$$

Here the supremum is taken over all measurable functions $g : \mathbb{M} \to [-\infty, \infty)$ with $\mathbb{E}\, e^g \leq 1$. It is attained at $g = \log(f/\mathbb{E}f)$, provided that $\mathbb{E}f > 0$, $\mathbb{E}\, f \log(1+f) < \infty$.
7) As an immediate consequence of the variational formula, we conclude that the functional $f \to \mathrm{Ent}(f)$ is convex (as a supremum of linear functionals).
8) To remove the requirement $\mathbb{E}\, e^g \leq 1$ in the variational formula, the latter may equivalently be formulated as the inequality

$$\mathbb{E}\, fg \leq \mathrm{Ent}(f) + \mathbb{E}\, f \log \mathbb{E}\, e^g.$$

This makes sense and holds true whenever $\mathrm{Ent}(f)$ and $\mathbb{E}\, e^g$ are finite.
9) Another equivalent form of the variational formula is the entropic representation for the Laplace transform: If $\mathbb{E}\, e^g < \infty$, then

$$\log \mathbb{E}\, e^g = \sup_{\mathbb{E}f=1} \big[\,\mathbb{E}\, fg - \mathrm{Ent}(f)\big],$$

where the supremum runs over all measurable $f \geq 0$ on $\mathbb{M}$ such that $\mathbb{E}f = 1$.
10) One useful application of the variational formula is the following characterization: Given $\sigma > 0$, a measurable function g on $\mathbb{M}$ with $\mathbb{E}g = 0$ satisfies

$$\mathbb{E}\, e^{tg} \leq e^{\sigma^2 t^2/2} \quad \text{for all } t \in \mathbb{R}$$

if and only if, for any (bounded) measurable function $f \geq 0$ on $\mathbb{M}$,

$$(\mathbb{E}\, fg)^2 \leq 2\sigma^2\, \mathbb{E} f\, \mathrm{Ent}_\mu(f).$$

In particular, for the indicator function $f = 1_A$ of a measurable set $A \subset \mathbb{M}$, we get

$$\Big|\int_A g\, \mathrm{d}\mu\Big| \leq \sqrt{2\sigma^2\, \mu(A) \log \frac{1}{\mu(A)}}.$$

11) There is another representation of the entropy (cf. [105], [129], [130]):

$$\mathrm{Ent}(f) \,=\, \inf_{c>0}\, \Big[\mathbb{E}\, f \log f - (\log c + 1)\, \mathbb{E} f + c\Big].$$

Here, the infimum is attained at $c = \mathbb{E} f$. This equality shows that the functional $\mu \to \mathrm{Ent}_\mu(f)$ is concave on the space of probability measures on $\mathbb{M}$.

12) Rothaus' inequality [163]: For any $f \in L^2(\mu)$ and all $c \in \mathbb{R}$,

$$\mathrm{Ent}_\mu((f+c)^2) \leq \mathrm{Ent}_\mu(f^2) + 2\, \mathbb{E}_\mu f^2.$$

13) Relationship with variance: $\mathrm{Ent}(f^2) \geq \mathrm{Var}(f)$ for any $f \in L^2(\mu)$. Moreover, if $\mathbb{E}\, f^2 \log(1 + f^2)$ is finite, then

$$\lim_{|a| \to \infty} \mathrm{Ent}\big((f+a)^2\big) = 2\, \mathrm{Var}(f).$$

14) Relationship with the Orlicz norm [44]: For any f from the Orlicz space $L^\psi(\mu)$ generated by the Young function $\psi(t) = t^2 \log(1 + t^2)$, we have

$$c_1\, \|f - \mathbb{E} f\|_\psi^2 \leq \sup_{a \in \mathbb{R}}\, \mathrm{Ent}\big((f+a)^2\big) \leq c_2\, \|f - \mathbb{E} f\|_\psi^2,$$

where c_1, c_2 are positive absolute constants. Let us recall that

$$\|g\|_\psi = \inf\{\lambda > 0 : \mathbb{E}\, \psi(g/\lambda) \leq 1\}.$$

15) The following subadditivity property of the entropy is crucial in the theory of logarithmic Sobolev inequalities and the study of the concentration of measure phenomenon. Let $(\mathbb{M}, \mu) = (\mathbb{M}_1, \mu_1) \otimes \cdots \otimes (\mathbb{M}_n, \mu_n)$ be a product probability space. Then, for any measurable $f \geq 0$ on $\mathbb{M}$,

$$\mathrm{Ent}(f) \leq \mathbb{E} \sum_{k=1}^n \mathrm{Ent}_{x_k}(f).$$

Here $\mathrm{Ent}_{x_k}(f)$ denotes the entropy of the function

$$y \to f(x_1, \dots, x_{k-1}, y, x_{k+1}, x_k)$$

on $\mathbb{M}_k$ with respect to the measure μ_k with fixed remaining coordinates x_j, $j \neq k$ (cf. [131]). This property is related to the subadditivity of the classical Shannon entropy for random vectors. A similar property holds for other functionals, e.g., for the variance $\mathrm{Var}(f)$ in place of $\mathrm{Ent}(f)$, cf. item 5.7.7.

16) Finally, let us note that, if $f = u + iv$ is complex-valued, then

$$\mathrm{Ent}(|f|^2) \leq \mathrm{Ent}(u^2) + \mathrm{Ent}(v^2).$$

This immediately follows from the convexity and homogeneity of the entropy functional (Properties 3 and 7).

7.2 Definitions and Examples

Let us now return to the setting of the metric space $(\mathbb{M}, d)$.

Definition 7.2.1 A Borel probability measure μ on $\mathbb{M}$ is said to satisfy a logarithmic Sobolev inequality with constant $\rho > 0$ if, for any bounded, Lipschitz function f on $\mathbb{M}$,

$$\rho\, \mathrm{Ent}_\mu(f^2) \leq 2 \int |\nabla f|^2\, d\mu. \tag{7.1}$$

Like the Poincaré-type inequality, the relation (7.1) may be extended to all locally Lipschitz functions f (with arguments used in the proof of Proposition 5.3.2). Moreover, with the same constant ρ it continues to hold for complex-valued functions (cf. Corollary 7.5.5 below). When $\mathbb{M}$ is an open subset of $\mathbb{R}^n$ and μ is absolutely continuous with respect to the Lebesgue measure, it is sufficient to require that (7.1) be fulfilled in the smaller class of all C^∞-smooth f (like in Proposition 5.4.1).

The best value $\rho = \rho(\mu)$ in (7.1) is called the logarithmic Sobolev constant. When X is a random vector in $\mathbb{R}^n$ with distribution μ, (7.1) may be rewritten in the form

$$\rho\, \mathrm{Ent}(f(X)^2) \leq 2\, \mathbb{E}\, |\nabla f(X)|^2.$$

As a functional of random vectors, it is translation invariant, homogeneous of order -2, and is invariant under all orthogonal transformations of the space, i.e.,

$$\rho(a + bT(X)) = b^{-2} \rho(X) \quad (a \in \mathbb{R},\ b \neq 0,\ T \in O(n)),$$

in full analogy with Poincaré constants.

The inequality (7.1) does not lose generality if we restrict it to positive functions (this follows from the chain rule inequality $|\nabla |f|\,| \leq |\nabla f|$). Applying (7.1) to $e^{f/2}$ in place of f, we arrive at the equivalent (so-called modified or exponential) form

$$\rho\, \mathrm{Ent}_\mu(e^f) \leq \frac{1}{2} \int |\nabla f|^2 e^f\, d\mu, \tag{7.2}$$

which holds in the class of all locally Lipschitz functions f on $\mathbb{M}$. Another popular equivalent form is obtained from (7.1) when replacing f^2 with f. It then becomes

$$\rho \operatorname{Ent}_\mu(f) \le \frac{1}{2} \int \frac{|\nabla f|^2}{f}\, d\mu,$$

which holds for all positive locally Lipschitz f on $\mathbb{M}$. In view of the two-sided bounds for the entropy functional described in property 14 of Section 7.1, the logarithmic Sobolev inequality (7.1) is also equivalent, up to an absolute factor $c > 0$, to

$$c\rho \, \|f - \mathbb{E}_\mu f\|_\psi^2 \le \int |\nabla f|^2 \, d\mu.$$

The latter may be viewed as a Poincaré-type inequality for the Orlicz norm in the space $L^\psi(\mathbb{M}, \mu)$ generated by the Young function $\psi(t) = t^2 \log(1 + t^2)$.

Since $\|\cdot\|_\psi$ is stronger than the L^2-norm, (7.1) is stronger than the usual Poincaré-type inequality. In fact, applying (7.1) to $f + a$ with $a \to \infty$ and using the property 13 of the entropy functional (Section 7.1), we arrive in the limit at

$$\rho \operatorname{Var}_\mu(f) \le \int |\nabla f|^2 \, d\mu.$$

Hence, we have a general relation $\rho(\mu) \le \lambda_1(\mu)$.

Let us now turn to some examples and conditions under which logarithmic Sobolev inequalities are fulfilled.

7.2.2 Necessary and sufficient conditions on the line. Given a Borel probability measure μ on $\mathbb{R}$, write a canonical decomposition $\mu = \mu_0 + \mu_1$, where μ_0 is an absolutely continuous component with density p, and μ_1 is an orthogonal component. For the property $\rho(\mu) > 0$, it is necessary that $p(y) > 0$ almost everywhere on the support interval $(a, b) \subset \mathbb{R}$ of μ_0. More precisely, introduce the quantities

$$A = \sup_{a<x<m} \left[\mu(-\infty, x) \log \frac{1}{\mu(-\infty, x)} \int_x^m \frac{1}{p(y)}\, dy \right],$$
$$B = \sup_{m<x<b} \left[\mu(x, \infty) \log \frac{1}{\mu(x, \infty)} \int_m^x \frac{1}{p(y)}\, dy \right],$$

where m is a median of μ. Then (cf. [44]), $\rho(\mu)$ is positive if and only if both A and B are finite, in which case, up to some numerical values $c_2 > c_1 > 0$,

$$c_1 \max\{A, B\} \le \frac{1}{\rho(\mu)} \le c_2 \max\{A, B\}.$$

7.2.3 Product measures. If the probability measures μ_i on $\mathbb{R}^{n_i}$ have positive logarithmic Sobolev constants $\rho(\mu_i)$, $i = 1, \dots, k$, then the product measure $\mu = \mu_1 \otimes \cdots \otimes \mu_k$ is supported on the Euclidean space $\mathbb{R}^n$ of dimension $n = n_1 + \cdots + n_k$ and also satisfies the logarithmic Sobolev inequality. Moreover, the best value in (7.1) is given by

$$\rho(\mu) = \min_i \rho(\mu_i),$$

in full analogy with Poincaré constants. This follows from the subadditivity of the entropy functional (property 15, Section 7.1), using the same arguments as in item 5.7.7.

7.2.4 Lipschitz transforms and convolutions. Like in items 5.7.8–5.7.9, if a random vector $X = T(Y)$ in $\mathbb{R}^n$ represents a Lipschitz transform of a random vector in $\mathbb{R}^N$, then

$$\rho^{-1}(X) \le \|T\|_{\mathrm{Lip}}^2 \, \rho^{-1}(Y).$$

Using item 7.2.3, this implies a corresponding bound on the logarithmic Sobolev constant for convolutions. Namely, if $X = a_1 X_1 + \dots + a_k X_k$ ($a_i \in \mathbb{R}$) is a weighted sum of independent random vectors in $\mathbb{R}^n$, then

$$\rho^{-1}(X) \le \sum_{i=1}^k a_i^2 \, \rho^{-1}(X_i).$$

7.2.5 Gaussian measures. The standard Gaussian measure γ_1 on $\mathbb{R}$ satisfies the logarithmic Sobolev inequality with constant $\rho = 1$, which is optimal. Hence, the same is true for the standard n-dimensional Gaussian measure $\gamma_n = \gamma_1 \otimes \dots \otimes \gamma_1$ on $\mathbb{R}^n$. This result can also be derived from the isoperimetric inequality for γ_n.

7.2.6 Uniform distributions on balls. Let μ_n denote the normalized Lebesgue measure (a uniform distribution) on the unit ball $B_n = \{x \in \mathbb{R}^n : |x| \le 1\}$. One can push forward γ_n into μ_n with the help of a rotationally invariant map $T(x) = u(|x|)\,\frac{x}{|x|}$ on $\mathbb{M} = \mathbb{R}^n \setminus \{0\}$. Here, the function $u : (0,\infty) \to (0,1)$ is unique, if we require that it is non-decreasing. In that case, the property $\mu_n = \gamma_n T^{-1}$ is equivalent to

$$\gamma_n\{x \in \mathbb{R} : u(|x|) < r\} = \mu_n(rB_n) = r^n, \quad 0 < r < 1.$$

It is easy to check that u is smooth, concave, and

$$\|T\|_{\mathrm{Lip}} = \|u\|_{\mathrm{Lip}} = u'(0^+) = \frac{\omega_n^{1/n}}{\sqrt{2\pi}},$$

where $\omega_n = \mathrm{vol}_n(B_n) = \pi^{n/2}/\Gamma(\frac{n}{2}+1)$ is the volume of the unit ball. This gives an asymptotically correct lower bound $\rho(\mu_n) \ge cn$ for some absolute constant $c > 0$. We refer to [53] for technical details.

One important class of Lipschitz images of γ_n is described in the following assertion, known as the Bakry–Emery criterion.

Proposition 7.2.7 *Suppose that a probability measure μ on $\mathbb{R}^n$ is absolutely continuous and has a log-concave density with respect to γ_n. Then $\rho(\mu) \ge 1$.*

The log-concavity of the density $p = \mathrm{d}\mu/\mathrm{d}\gamma_n$ means that $p(tx+sy) \ge p(x)^t p(y)^s$ for all $x, y \in \mathbb{R}^n$ and $t, s > 0$ such that $t + s = 1$. In this case, the property that $\mu = \gamma_n T^{-1}$ for some Lipschitz T with $\|T\|_{\mathrm{Lip}} \le 1$ was established by Caffarelli [74].

Proposition 7.2.7 may equivalently be stated as follows. Assume that μ is supported on an open convex set $\mathbb{M}$ in $\mathbb{R}^n$, where it has a C^2-smooth positive density $p = \mathrm{e}^{-V}$ with respect to the Lebesgue measure, such that the second derivative of V satisfies $V''(x) \geq c\,\mathbb{I}_n$ for all $x \in \mathbb{M}$ with some constant $c > 0$ (in the matrix sense, where $\mathbb{I}_n$ is the unit matrix). Then $\rho(\mu) \geq c$.

Note that uniform distributions on balls are not covered by Proposition 7.2.7.

7.2.8 Uniform distribution on the sphere. Let $\mathfrak{s}_{n-1}$ denote the normalized Lebesgue measure (a uniform distribution) on the unit sphere $\mathbb{S}^{n-1} = \{x \in \mathbb{R}^n : |x| = 1\}, n \geq 2$. This case is very similar to the uniform distribution on the unit ball. And indeed, a deep result due to Mueller and Weissler [149] indicates that

$$\rho(\mathfrak{s}_{n-1}) = \lambda_1(\mathfrak{s}_{n-1}) = n - 1.$$

In this statement it does not matter whether we take for $\mathbb{M}$ the whole Euclidean space $\mathbb{R}^n$ with its Euclidean metric, or just the sphere $\mathbb{S}^{n-1}$ with its geodesic metric (which is infinitesimally equivalent to the Euclidean one). Note that $\mathfrak{s}_{n-1}$ cannot be obtained as a Lipschitz image of the Gaussian measure.

7.3 Exponential Bounds

Since logarithmic Sobolev inequalities are stronger in comparison with Poincaré-type inequalities, it is natural to expect that (7.1) should yield stronger results in comparison with those obtained on the basis of Poincaré-type inequalities. This may be seen already on the examples of deviation inequalities and various exponential bounds which can be derived from logarithmic Sobolev inequalities like the following ones. Keeping the setting of a metric space $(\mathbb{M}, d)$, we assume that a Borel probability measure μ on $\mathbb{M}$ has a positive logarithmic Sobolev constant $\rho = \rho(\mu)$.

Proposition 7.3.1 *Given a bounded, locally Lipschitz function f on $\mathbb{M}$ with μ-mean zero, for any $\alpha > 1/(2\rho)$,*

$$\int \mathrm{e}^f \, \mathrm{d}\mu \leq \bigg(\int \mathrm{e}^{\alpha\,|\nabla f|^2} \, \mathrm{d}\mu\bigg)^{1/(2\alpha\rho - 1)}. \tag{7.3}$$

In addition, for any $0 < \lambda \leq 1$,

$$\int \mathrm{e}^f \, \mathrm{d}\mu \leq \bigg(\int \mathrm{e}^{\lambda f + (1-\lambda)\,|\nabla f|^2/(2\rho)} \, \mathrm{d}\mu\bigg)^{1/\lambda}. \tag{7.4}$$

There is no assumption about the mean in (7.4). For $\lambda = 1$, (7.4) becomes an equality, and comparing the derivatives of both sides at this point, we return to (7.1). Hence, the bound (7.4) represents an equivalent form of the logarithmic Sobolev inequality with constant ρ. The form (7.4) is convenient to derive the following "Gaussian" bound for the class of Lipschitz functions.

Corollary 7.3.2 *Given a locally Lipschitz function f on $\mathbb{M}$ such that $|\nabla f| \leq 1$ on the support of μ, for any $0 < t < \rho/2$,*

$$\int e^{tf^2}\,d\mu \leq \exp\Big\{\frac{t}{1-2t/\rho}\int f^2\,d\mu\Big\}. \tag{7.5}$$

As for the bound (7.3), choosing there $\alpha = 1/\rho$ or letting $p \to \infty$ in the case $|\nabla f| \leq 1$, we obtain the following popular exponential bounds, where the boundedness assumption is easily removed (likewise in (7.5)).

Corollary 7.3.3 *Given a locally Lipschitz function f on $\mathbb{M}$ with μ-mean zero,*

$$\int e^{tf}\,d\mu \leq \int e^{t^2|\nabla f|^2/\rho}\,d\mu \tag{7.6}$$

for all $t \in \mathbb{R}$. Moreover, in the case $\|f\|_{\mathrm{Lip}} \leq 1$,

$$\int e^{tf}\,d\mu \leq e^{t^2/2\rho}. \tag{7.7}$$

More precisely, the finiteness of the right integral in (7.6) guarantees that the function f is μ-integrable, and the left integral is finite both for f and $-f$.

Proof (of Proposition 7.3.1) We may assume that $\beta = \log\int e^{\alpha|\nabla f|^2}\,d\mu$ is finite. By the variational formula for the entropy functional (property 8 in Section 7.1),

$$\int e^f g\,d\mu \leq \mathrm{Ent}_\mu(e^f), \tag{7.8}$$

whenever $\int e^g\,d\mu = 1$. The latter is fulfilled for the function $g = \alpha\,|\nabla f|^2 - \beta$, so,

$$\alpha\int |\nabla f|^2 e^f\,d\mu - \beta\int e^f d\mu \leq \mathrm{Ent}_\mu(e^f).$$

According to the exponential form (7.2), the above left integral is bounded from below by $2\rho\,\mathrm{Ent}_\mu(e^f)$, hence

$$2\alpha\rho\,\mathrm{Ent}_\mu(e^f) - \beta\int e^f d\mu \leq \mathrm{Ent}_\mu(e^f),$$

which in the case $2\alpha\rho > 1$ is equivalent to

$$\mathrm{Ent}_\mu(e^f) \leq \frac{\beta}{2\alpha\rho - 1}\int e^f\,d\mu.$$

Let us now apply this inequality to the functions tf, so as to get similar relations

$$\mathrm{Ent}_\mu(e^{tf}) \leq \frac{\beta(t^2)}{2\alpha\rho - 1}\int e^{tf}\,d\mu, \tag{7.9}$$

where $\beta(t) = \log \int e^{\alpha t\,|\nabla f|^2}\,d\mu$ is finite for $0 \le t \le 1$. The equality

$$\int e^{tf}\,d\mu = e^{tu(t)}$$

defines a continuous function $u(t)$ on $[0,\infty)$ with $u(0) = 0$ (since f has μ-mean zero), and which is smooth on $(0,\infty)$. As is easy to check, $\mathrm{Ent}_\mu(e^{tf}) = t^2 u'(t) e^{tu(t)}$. Hence, (7.9) takes the form of the differential inequality

$$u'(t) \le \frac{1}{2\alpha\rho - 1}\,\frac{\beta(t^2)}{t^2}, \qquad 0 < t \le 1.$$

But the function $\beta(t)$ is convex on $[0,1]$ with $\beta(0) = 0$, so that $\beta(t)/t$ does not decrease. Since $\beta(1) = \beta$, it follows that $u'(t) \le \frac{1}{2\alpha\rho-1}\,\beta$, implying $u(1) \le \frac{1}{2\alpha\rho-1}\,\beta$, which is the same as (7.3).

To derive (7.4), we may assume that

$$\beta = \log \int e^{\lambda f + (1-\lambda)|\nabla f|^2/(2\rho)}\,d\mu$$

is finite. Choosing $g = \lambda f + \frac{1-\lambda}{2\rho}\,|\nabla f|^2 - \beta$ in (7.8) and applying (7.2), we get

$$\begin{aligned}\mathrm{Ent}_\mu(e^f) &\ge \int (\lambda f - \beta)\,e^f\,d\mu + \frac{1}{2\rho}\,(1-\lambda)\int |\nabla f|^2\,e^f\,d\mu\\ &\ge \int (\lambda f - \beta)\,e^f\,d\mu + (1-\lambda)\,\mathrm{Ent}_\mu(e^f).\end{aligned}$$

The latter is equivalent to $\lambda \log \int e^f\,d\mu \le \beta$, which is exactly (7.4). □

Proof (of Corollary 7.3.2) We may assume that f is bounded (by using the truncation argument). Let us apply the inequality (7.4) to $g = s\rho f^2/2$ in place of f with $0 < s < 1$. Since $|\nabla g| \le s\rho\,|f|\,|\nabla f| \le s\rho\,|f|$ μ-almost everywhere, it gives

$$\int e^{s\rho f^2/2}\,d\mu \le \Big(\int \exp\Big\{\frac{\rho f^2}{2}\,s\,(\lambda + (1-\lambda)s)\Big\}\,d\mu\Big)^{1/\lambda}, \qquad 0 < \lambda \le 1.$$

Given $s < p < 1$, we choose $\lambda = \frac{p-s}{1-s}$ so that $\lambda + (1-\lambda)s = p$, in which case the above inequality turns into

$$\int e^{s\rho f^2/2}\,d\mu \le \Big(\int e^{ps\rho f^2/2}\,d\mu\Big)^{\frac{1-s}{p-s}}.$$

It becomes an equality for $p = 1$, and comparing the derivatives of the logarithm of both sides at $p = 1$, we obtain that

$$0 \ge (1-s)\int \frac{s\rho f^2}{2}\,e^{\frac{s\rho f^2}{2}}\,d\mu - \int e^{\frac{s\rho f^2}{2}}\,d\mu\,\log \int e^{\frac{s\rho f^2}{2}}\,d\mu.$$

Equivalently, in terms of $u(s) = \log \int e^{\frac{s\rho f^2}{2}}\, d\mu$, the latter may be stated as a differential inequality $(1-s)s\,u'(s) \leq u(s)$, which in turn expresses the fact that

$$v(s) = \exp\Big\{\frac{1-s}{s}\,u(s)\Big\} = \Big(\int e^{\frac{s\rho f^2}{2}}\, d\mu\Big)^{\frac{1-s}{s}}$$

does not increase in $0 < s < 1$. Comparing $v(s)$ with $v(0+)$ and replacing $t = \rho s/2$, we arrive at (7.5) thus proving Corollary 7.3.2. □

7.4 Bounds Involving Relative Entropy

Some of the exponential bounds considered in the previous section have nice reformulations in terms of the relative entropy (Kullback–Leibler distance)

$$D(\nu||\mu) = \mathrm{Ent}_\mu(g) = \int g \log g\, d\mu.$$

Here, ν is an arbitrary probability measure on $\mathbb{M}$ which is absolutely continuous with respect to μ with density $g = d\nu/d\mu$.

Let us return to Corollary 7.3.2. Using a variational description of the entropy functional (cf. Section 7.1), the bound (7.5) with a fixed function $f \in L^2(\mu)$ is equivalent to the property that the inequality

$$\int \Big(tf^2 - \frac{t}{1-2t/\rho}\int f^2\, d\mu\Big)\, g\, d\mu \leq \mathrm{Ent}_\mu(g), \quad 0 \leq t < \frac{\rho}{2}, \tag{7.10}$$

holds true for any (say, bounded measurable) function $g \geq 0$ on $\mathbb{M}$. This inequality may be optimized over all admissible t. Note that any function of the form

$$u(t) = at - \frac{bt}{1-2t/\rho} \quad (a, b > 0)$$

is concave and satisfies $u(0) = 0$, $u(\rho/2-) = -\infty$. If $a \leq b$, it is non-increasing, so $u(t) \leq 0$ for all t. In the other case $a > b$, there is a unique point of maximum $t_0 = \frac{\rho}{2}\,(1 - \frac{\sqrt{b}}{\sqrt{a}})$, at which $u(t_0) = \frac{\rho}{2}\,(\sqrt{a} - \sqrt{b})^2$. That is,

$$\max_{0 \leq t < \frac{\rho}{2}} u(t) = \begin{cases} 0, & \text{if } a \leq b, \\ \frac{\rho}{2}\,(\sqrt{a} - \sqrt{b})^2, & \text{if } a \geq b. \end{cases}$$

Taking $a = \int f^2 g\, d\mu$, $b = \int f^2\, d\mu$, (7.10) is therefore equivalent to

$$\Big(\int f^2 g\, d\mu\Big)^{1/2} - \Big(\int f^2\, d\mu\Big)^{1/2} \leq \sqrt{\frac{2}{\rho}\,\mathrm{Ent}_\mu(g)}.$$

At this step, once this bound is obtained for any bounded measurable $g \geq 0$, it is extended to all measurable functions $g \geq 0$.

We are in position to give another (equivalent) version of Corollary 7.3.2 for the class of all Lipschitz functions f on $\mathbb{M}$ with $\|f\|_{\mathrm{Lip}} \leq 1$.

Corollary 7.4.1 *Suppose that the metric probability space (M, d, μ) has a positive logarithmic Sobolev constant $\rho = \rho(\mu)$. Then, for any Borel probability measure ν on $\mathbb{M}$ which is absolutely continuous with respect to μ,*

$$\sup_{\|f\|_{\mathrm{Lip}} \leq 1} \Big[\|f\|_{L^2(\nu)} - \|f\|_{L^2(\mu)}\Big] \leq \sqrt{\frac{2}{\rho}\, D(\nu||\mu)}. \tag{7.11}$$

If we apply (7.11) to $f_a = f + a$ with $a \to \pm\infty$, we get in the limit another (a priori weaker) relation

$$\sup_{\|f\|_{\mathrm{Lip}} \leq 1} \Big|\int f\,\mathrm{d}\nu - \int f\,\mathrm{d}\mu\Big| \leq \sqrt{\frac{2}{\rho}\, D(\nu||\mu)}. \tag{7.12}$$

Using the variational formula for the entropy and similar arguments as in the proof of (7.11), one may recognize (7.12) as an equivalent variant of the inequality

$$\int e^{tf - t^2/2\rho}\,\mathrm{d}\mu \leq 1, \quad t \in \mathbb{R}.$$

But this is exactly the exponential bound (7.7) of Corollary 7.3.3. Therefore, with these arguments, (7.7) can be obtained as a consequence of (7.5).

7.5 Orlicz Norms and Growth of L^p-norms

Let us now comment on the relationship of the exponential bounds (7.5)–(7.7) with Poincaré-type inequalities for the Orlicz norms

$$\|f\|_\psi = \inf\{\lambda > 0 : \mathbb{E}_\mu\, \psi(f/\lambda) \leq 1\},$$

which we consider here for the Young functions $\psi_1(t) = e^{|t|} - 1$ and $\psi_2(t) = e^{t^2} - 1$. These specific Orlicz-norms are connected with the L^p-norms $\|f\|_p = (\int |f|^p\,\mathrm{d}\mu)^{1/p}$ through the well-known inequalities

$$c_0 \sup_{p\geq 1} \frac{\|f\|_p}{p} \leq \|f\|_{\psi_1} \leq c_1 \sup_{p\geq 1} \frac{\|f\|_p}{p},$$

$$c_0 \sup_{p\geq 1} \frac{\|f\|_p}{\sqrt{p}} \leq \|f\|_{\psi_2} \leq c_1 \sup_{p\geq 1} \frac{\|f\|_p}{\sqrt{p}},$$

holding true for some absolute constants $c_1 > c_0 > 0$. Thus, $\|f\|_{\psi_\alpha}$ is of order 1 ($\alpha = 1, 2$) if and only if the norms $\|f\|_p$ grow at rate at most p and $\sqrt{p}$, respectively, as functions of the growing parameter p.

As in the previous section, suppose that a Borel probability measure μ on the metric space $\mathbb{M}$ has a positive logarithmic Sobolev constant $\rho = \rho(\mu)$. Assume that f has μ-mean zero and $\|\nabla f\|_{\psi_2} \equiv \|\,|\nabla f|\,\|_{\psi_2} = 1$, so that $\mathbb{E}_\mu\, e^{|\nabla f|^2} = 2$. Putting $a = \mu\{f \leq 0\}$, $b = \mu\{f > 0\}$, and applying Cauchy's inequality together with (7.6), we get, for $t > 0$,

$$\begin{aligned}\mathbb{E}_\mu\, e^{t|f|} &= \mathbb{E}_\mu\, e^{-tf}\, 1_{\{f \leq 0\}} + \mathbb{E}_\mu\, e^{tf}\, 1_{\{f>0\}} \\ &\leq \sqrt{a}\,\Big(\mathbb{E}_\mu\, e^{-2tf}\Big)^{1/2} + \sqrt{b}\,\Big(\mathbb{E}_\mu\, e^{2tf}\Big)^{1/2} \leq \big(\sqrt{a}+\sqrt{b}\big)\Big(\mathbb{E}_\mu\, e^{4t^2\,|\nabla f|^2/\rho}\Big)^{1/2}.\end{aligned}$$

Here $\sqrt{a} + \sqrt{b} \leq \sqrt{2}$, while the last expectation is equal to 2 for $t = \frac{1}{2}\sqrt{\rho}$. Therefore, in this case $\mathbb{E}_\mu\, e^{t|f|} \leq 2$, meaning that $\|f\|_{\psi_1} \leq \frac{1}{t} = \frac{2}{\sqrt{\rho}}$. By the homogeneity of the Orlicz norm, the latter relation may be stated in a more general form.

Corollary 7.5.1 *For any locally Lipschitz function f on $\mathbb{M}$ with μ-mean zero,*

$$\|f\|_{\psi_1} \leq \frac{2}{\sqrt{\rho}}\, \|\nabla f\|_{\psi_2}.$$

If $|\nabla f| \leq 1$, this bound can be sharpened by replacing the ψ_1-norm of f on the left with the ψ_2-norm, which can be achieved by (7.5). Indeed, the expression on the right-hand side of (7.5) is equal to 2 for the value $t = \big(\frac{2}{\rho} + \frac{1}{\log 2}\, \|f\|_2^2\big)^{-1}$, in which case $\|f\|_{\psi_2}^2 \leq 1/t$. Therefore, we get:

Corollary 7.5.2 *For any function f on $\mathbb{M}$ with finite Lipschitz semi-norm,*

$$\|f\|_{\psi_2}^2 \leq \frac{2}{\rho}\, \|f\|_{\mathrm{Lip}}^2 + \frac{1}{\log 2}\, \|f\|_2^2. \tag{7.13}$$

Note that, due to the homogeneity of this inequality with respect to f, one may assume that $\|f\|_{\mathrm{Lip}} \leq 1$. In this case, if f has μ-mean zero, then, by the Poincaré-type inequality, $\|f\|_2^2 \leq \frac{1}{\lambda_1} \leq \frac{1}{\rho}$, and (7.13) yields

$$\|f\|_{\psi_2} \leq \frac{2}{\sqrt{\rho}}.$$

Thus, the collection of all mean zero, Lipschitz functions on $\mathbb{M}$ represents a bounded (convex) subset in the Orlicz space $L^{\psi_2}(\mu)$ with diameter $\leq \frac{4}{\sqrt{\rho}}$. In fact, a more precise statement can be made on the basis of (7.7), which implies (by applying Markov's inequality) the following sub-Gaussian large deviation bounds improving exponential bounds on tails under Poincaré-type inequalities.

Corollary 7.5.3 *For any function f on $\mathbb{M}$ with $\|f\|_{\mathrm{Lip}} \le 1$ and μ-mean zero, for any $r \ge 0$,*

$$\mu\{f \ge r\} \le \mathrm{e}^{-\rho r^2/2}, \quad \mu\{|f| \ge r\} \le 2\,\mathrm{e}^{-\rho r^2/2}.$$

The advantage of (7.13) is however the fact that this bound does not require f to have mean zero.

As we know from Proposition 6.2.1, the Poincaré-type inequality gives rise to the whole family of Poincaré-type inequalities for L^p-norms with $p > 2$. Namely, according to (6.3), if f has μ-mean zero, then

$$\|f\|_p \le \frac{p}{\sqrt{2\lambda_1}}\,\|\nabla f\|_p. \tag{7.14}$$

The same is true in the presence of logarithmic Sobolev inequalities, and actually the behavior of the p-dependent constant may be properly sharpened.

Proposition 7.5.4 *Given a locally Lipschitz function $f : \mathbb{M} \to \mathbb{C}$, for any $p > 2$,*

$$\|f\|_p^2 - \|f\|_2^2 \le \frac{p-2}{\rho}\,\|\nabla f\|_p^2 \tag{7.15}$$

as long as the right-hand side is finite. In particular, if f is real-valued and has μ-mean zero, then

$$\|f\|_p \le \frac{\sqrt{p-1}}{\sqrt{\rho}}\,\|\nabla f\|_p. \tag{7.16}$$

Note that the inequality (7.16) sharpens (7.14) with respect to p. An advantage of the form (7.15) is that, after dividing by $p-2$ and letting $p \to 2$, it returns us to the original logarithmic Sobolev inequality (7.1), even for a larger class of functions. That is, we obtain:

Corollary 7.5.5 *If* (7.1) *is fulfilled in the class of all bounded Lipschitz functions $f : \mathbb{M} \to \mathbb{R}$, then for any locally Lipschitz function $f : \mathbb{M} \to \mathbb{C}$,*

$$\rho\,\mathrm{Ent}_\mu(|f|^2) \le 2\int |\nabla f|^2\,\mathrm{d}\mu.$$

Proof (of Proposition 7.5.4) One may assume that f is bounded. Let us recall the identity in property 5 (Section 7.1), which implies

$$\frac{\mathrm{d}}{\mathrm{d}p}\,\|f\|_p^2 = \frac{2}{p^2}\,\|f\|_p^{2-p}\,\mathrm{Ent}_\mu(|f|^p), \quad p > 2.$$

To estimate the entropy term, we may apply the logarithmic Sobolev inequality to the function $u = |f|^{p/2}$. By the chain rule inequality, $|\nabla u|^2 \le \frac{p^2}{4}\,|f|^{p-2}\,|\nabla f|^2$, and applying Hölder's inequality, we get

$$\mathbb{E}_\mu\,|\nabla u|^2 \le \frac{p^2}{4}\,\big(\mathbb{E}_\mu\,|f|^p\big)^{\frac{p-2}{p}}\,\big(\mathbb{E}_\mu\,|\nabla f|^p\big)^{\frac{2}{p}} = \frac{p^2}{4}\,\|f\|_p^{p-2}\,\|\nabla f\|_p^2.$$

Hence, by (7.1),

$$\mathrm{Ent}_\mu(|f|^p) = \mathrm{Ent}_\mu(u^2) \le \frac{2}{\rho}\,\mathbb{E}_\mu\,|\nabla u|^2 \le \frac{p^2}{2\rho}\,\|f\|_p^{p-2}\,\|\nabla f\|_p^2,$$

so that $\frac{d}{dp}\,\|f\|_p^2 \le \frac{1}{\rho}\,\|\nabla f\|_p^2$. It remains to integrate this differential inequality from 2 to p.

The second bound (7.16) follows from the first one by involving the Poincaré-type inequality which gives $\|f\|_2^2 \le \frac{1}{\rho}\,\|\nabla f\|_2^2 \le \frac{1}{\rho}\,\|\nabla f\|_p^2$. As an alternative approach, one may also apply (7.15) to the functions $f + c$ and turn to the limit as $c \to \infty$. □

Remark 7.5.6 If f is complex-valued, the above bound on the L^2-norm takes the slightly weaker form $\|f\|_2^2 \le \frac{2}{\rho}\,\|\nabla f\|_p^2$. Hence, (7.16) holds true with an additional factor on the right-hand side,

$$\|f\|_p \le \frac{\sqrt{2(p-1)}}{\sqrt{\rho}}\,\|\nabla f\|_p.$$

Here, the constant 2 may be removed, so that we still have (7.16) for the class of all μ-mean zero, locally Lipschitz $f : \mathbb{M} \to \mathbb{C}$, as long as the Poincaré-type inequality $\|f\|_2^2 \le \frac{1}{\lambda_1}\,\|\nabla f\|_2^2$ is fulfilled in this class. In particular, this is true when $\mathbb{M}$ is an open subset of $\mathbb{R}^n$ and μ is absolutely continuous with respect to the Lebesgue measure, as well as for other spaces such as the unit sphere $\mathbb{M} = \mathbb{S}^{n-1}$ equipped with the normalized Lebesgue measure $\mu = \mathfrak{s}_{n-1}$, cf. Remark 5.4.3.

7.6 Bounds Involving Second Order Derivatives

The moment and exponential bounds which we considered so far can further be refined by involving the second order modulus of the gradient. In particular, assuming that the metric probability space $(\mathbb{M}, d, \mu)$ satisfies a logarithmic Sobolev inequality with positive constant $\rho = \rho(\mu)$, we have the following:

Proposition 7.6.1 *For any locally Lipschitz function f on $\mathbb{M}$ with μ-mean zero, such that $|\nabla f|$ is locally Lipschitz and $|\nabla^2 f| \le 1$ on the support of μ, we have*

$$\int e^{\rho f/2}\,d\mu \le \exp\Big\{\frac{\rho}{2}\int |\nabla f|^2\,d\mu\Big\}. \tag{7.17}$$

Proof The argument is based on the combination of the exponential bounds (7.5) and (7.6). Assuming that the last integral in (7.17) is finite, the function f must belong to $L^2(\mu)$, so that the μ-mean of f is finite as well. Now, choosing $t = \rho/2$ in (7.6), we have

$$\int e^{\rho f/2}\,d\mu \le \int e^{\rho\,|\nabla f|^2/4}\,d\mu.$$

Here the function $u = |\nabla f|$ is assumed to be locally Lipschitz on $\mathbb{M}$ and to satisfy $|\nabla u| \leq 1$ μ-almost everywhere. Hence, one may apply Corollary 7.3.2, and, by (7.5) with $t = \rho/4$,

$$\int e^{\rho\,|\nabla f|^2/4}\,d\mu = \int e^{tu^2}\,d\mu$$
$$\leq \exp\Big\{\frac{t}{1-2t/\rho}\int u^2\,d\mu\Big\} = \exp\Big\{\frac{\rho}{2}\int |\nabla f|^2\,d\mu\Big\}.$$

Proposition 7.6.1 is proved. □

Let us state the inequality (7.17) for Euclidean spaces, by making use of the estimate $|\nabla^2 f(x)| \leq \|f''(x)\|$ in terms of the operator norm of the second derivative of f (cf. Section 5.1).

Corollary 7.6.2 *Let a probability measure μ on $\mathbb{R}^n$ have a positive logarithmic Sobolev constant ρ, and let a function f be C^2-smooth in an open neighborhood of the support of μ. If it has μ-mean zero, and $\|f''\| \leq 1$ on the support of μ, then*

$$\int e^{\rho f/2}\,d\mu \leq \exp\Big\{\frac{\rho}{2}\int |\nabla f|^2\,d\mu\Big\}. \tag{7.18}$$

One may further estimate the last integral in (7.18) by involving the Poincaré-type inequality (5.6), where necessarily $\lambda_1 \geq \rho$. To describe the general situation, suppose that the measure μ has a C^1-smooth positive density p on $\mathbb{R}^n$, vanishing at infinity and such that $\partial_i p(x)/p(x)$ belong to $L^2(\mu)$ (that is, μ has finite Fisher information, where ∂_i means differentiation with respect to the i-th coordinate x_i). In this case, if f is bounded and C^2-smooth, one may integrate by parts to get

$$\int \partial_i f(x)\,d\mu(x) = \int \partial_i f(x)\,p(x)\,dx$$
$$= -\int f(x)\,\partial_i p(x)\,dx = -\int f(x)\,\frac{\partial_i p(x)}{p(x)}\,d\mu(x).$$

If the last integrals are vanishing for all $i \leq n$, an application of (5.6) to the partial derivatives $\partial_i f$ yields

$$\int (\partial_i f(x))^2\,d\mu(x) \leq \frac{1}{\lambda_1}\int \sum_{j=1}^n (\partial_{ij} f(x))^2\,d\mu(x).$$

Summing over all $i \leq n$ and using $\lambda_1 \geq \rho$, we get

$$\int |\nabla f|^2\,d\mu \leq \frac{1}{\rho}\int \sum_{i,j=1}^n (\partial_{ij} f(x))^2\,d\mu(x) = \frac{1}{\rho}\int \|f''\|_{\rm HS}^2\,d\mu,$$

where $\|f''(x)\|_{\rm HS}$ stands for the Hilbert–Schmidt norm of the matrix $f''(x)$. Denoting by b the last integral, we therefore obtain from (7.18) that

$$\int e^{\rho|f|/2}\,d\mu \le \int e^{\rho f/2}\,d\mu + \int e^{-\rho f/2}\,d\mu \le 2e^{b/2},$$

and thus for any $\lambda \ge 1$,

$$\int e^{\rho|f|/2\lambda}\,d\mu \le \left(\int e^{\rho f/2}\,d\mu\right)^{1/\lambda} \le \left(2e^{b/2}\right)^{1/\lambda}.$$

Here, the last expression is equal to 2 for $\lambda = 1 + \frac{b}{\log 4} \le 1 + b$. Hence, with these assumptions on the density p of μ we arrive at:

Corollary 7.6.3 *Assume that a C^2-smooth, bounded function f on $\mathbb{R}^n$ is orthogonal in $L^2(\mu)$ to the functions $\frac{\partial_i p}{p}$ $(i \le n)$. If f has μ-mean zero, and $\|f''(x)\| \le 1$ for all $x \in \mathbb{R}^n$, then*

$$\int e^{\rho|f|/2(1+b)}\,d\mu \le 2, \quad b = \int \|f''\|_{\mathrm{HS}}^2\,d\mu.$$

In other words,

$$\|f\|_{\psi_1} \le \frac{2}{\rho}\left(1 + \int \|f''\|_{\mathrm{HS}}^2\,d\mu\right).$$

7.7 Remarks

Logarithmic Sobolev inequalities were introduced in the mid 1970s by Gross [103], who established a sharp log-Sobolev inequality for the standard Gaussian measure. Ledoux wrote two surveys [129] and [130] which discuss many interesting aspects of the subject studied by various authors before the end of the 90s. In particular, the famous sufficient condition on $\mathbb{R}^n$ described in Proposition 7.2.7 was proposed by Bakry and Emery [7] in the 80s; it is applicable to the standard Gaussian measure and thus extends Gross' result.

In the class of log-concave probability measures μ on $\mathbb{R}^n$, the logarithmic Sobolev constant $\rho(\mu)$ is positive if and only if the Euclidean norm $|x|$ has a finite ψ_2-norm over μ, i.e., $\int e^{\varepsilon\,|x|^2}\,d\mu(x) < \infty$ for some $\varepsilon > 0$, cf. [20]. As shown by Wang [175], the latter condition remains sufficient for a larger class of densities $\frac{d\mu(x)}{dx} = e^{-V(x)}$ on $\mathbb{R}^n$ such that $V''(x) \ge c\,\mathbb{I}_n$ for some $c \in \mathbb{R}$ in the sense of positive matrices (where V'' denotes the Hessian of V and $\mathbb{I}_n$ is the identity matrix). An explicit description of all probability measures on the real line that satisfy logarithmic Sobolev inequalities was given in [44].

An interesting line of investigation has been conducted around the following natural question (in analogy with Poincaré-type inequalities): What can one say about the probability measure, once we know that it satisfies a logarithmic Sobolev inequality? The Gaussian bound (7.5) is due to Aida, Masuda and Shigekawa [1]. Apparently, it was the first rigorous justification of the property that Lipschitz functions must have sub-Gaussian tails in the presence of logarithmic Sobolev inequalities (although the assertion was stressed by Herbst already in the mid 1970s). The alternative exponential bound (7.7) was proposed by Ledoux [128], cf. also [129]. Here, we followed the exposition of [44] with the unifying approach described in Proposition 7.3.1.

Although logarithmic Sobolev inequalities are stronger than Poincaré-type inequalities, one may derive an exponential form (7.2) in the class of Lipschitz f by means of λ_1 only, cf. [52]. In this way one may explore a concentration phenomenon for product measures discovered by Talagrand [171], [172]. In this connection, let us mention that the notion of concentration was introduced and emphasized using the example of the uniform distribution on the Euclidean sphere in the early 1970s by V. Milman [142], [143]. For an account of various concentration results, we refer the interested reader to the books by Milman and Schechtman [145], Ledoux [131], and Artstein-Avidan, Giannopoulos, and Milman [6].

Chapter 8
Supremum and Infimum Convolutions

Logarithmic Sobolev inequalities are closely related to another important class of inequalities in terms of the so-called supremum- and infimum-convolutions, whose advantage is that they do not require smoothness or even continuity of the functions. It is therefore not surprising that supremum- and infimum-convolution inequalities find a wide range of applications. We restrict the exposition of these operators and associated inequalities to the Euclidean space $\mathbb{M} = \mathbb{R}^n$.

8.1 Regularity and Analytic Properties

Let us start with a definition and basic properties of the supremum- and infimum-convolution inequalities.

Definition 8.1.1 Given a function f on $\mathbb{R}^n$, for $t > 0$ and $x \in \mathbb{R}^n$, put

$$P_t f(x) = \sup_{y \in \mathbb{R}^n} \Big[f(y) - \frac{1}{2t}\, |x - y|^2 \Big],$$

$$Q_t f(x) = \inf_{y \in \mathbb{R}^n} \Big[f(y) + \frac{1}{2t}\, |x - y|^2 \Big].$$

The functions $P_t f$ and $Q_t f$ are respectively called the supremum- and the infimum-convolution of f (with the normalized square of the Euclidean norm in $\mathbb{R}^n$).

In view of the elementary identity

$$\inf_{x \in \mathbb{R}^n} \Big[\frac{1}{t_1}\, |x_1 - x|^2 + \frac{1}{t_2}\, |x - x_2|^2 \Big] = \frac{1}{t_1 + t_2}\, |x_1 - x_2|^2 \quad (x_1, x_2 \in \mathbb{R}^n),$$

both families of these operators represent semi-groups:

$$P_{t_1+t_2} f = P_{t_1} P_{t_2} f \quad \text{and} \quad Q_{t_1+t_2} f = Q_{t_1} Q_{t_2} f \quad \text{for all } t_1, t_2 > 0.$$

S. Bobkov et al., *Concentration and Gaussian Approximation for Randomized Sums*, Probability Theory and Stochastic Modelling 104,
https://doi.org/10.1007/978-3-031-31149-9_8

These semi-groups are generated by certain operators (generators), which can easily be evaluated: For smooth f, we have (to be made more precise later on),

$$\frac{d}{dt} P_t f \,\big|_{t=0} = \frac{1}{2}\,|\nabla f|^2, \qquad \frac{d}{dt} Q_t f \,\big|_{t=0} = -\frac{1}{2}\,|\nabla f|^2. \tag{8.1}$$

The operators P_t and Q_t are related by the identity $P_t(-f) = -Q_t f$, and we also have $g \geq P_t f \Longleftrightarrow f \leq Q_t g$ which implies

$$P_t Q_t f \leq f \leq Q_t P_t f.$$

By the definition, $P_t f$ and $Q_t f$ represent the supremum and respectively the infimum of the continuous functions $y \to f(y) \pm \frac{1}{2t}\,|x - y|^2$. Hence, the function $P_t f$ is always lower semi-continuous, while $Q_t f$ is upper semi-continuous (hence, Borel measurable). One can show that, if f_+ and f_- are respectively the smallest upper semi-continuous function majorizing f and the largest lower semi-continuous function majorized by f, i.e.,

$$f_+(x) = \limsup_{y \to x} f(y) \quad \text{and} \quad f_-(x) = \liminf_{y \to x} f(y),$$

then $P_t f_+ = P_t f$ and $Q_t f_- = Q_t f$.

In some sense, an application of these operators improves the regularity properties. If f is Lipschitz, then both $P_t f$ and $Q_t f$ are Lipschitz as well, and moreover,

$$\|P_t f\|_{\mathrm{Lip}} \leq \|f\|_{\mathrm{Lip}}, \qquad \|Q_t f\|_{\mathrm{Lip}} \leq \|f\|_{\mathrm{Lip}}.$$

But, starting, for example, from the indicator (thus discontinuous) function $f(x) = 1_{(-\infty,0]}(x)$, $x \in \mathbb{R}$, we obtain again a Lipschitz function, namely,

$$P_t f(x) = \begin{cases} 1, & x \leq 0, \\ 1 - \frac{1}{2t}\,x^2, & 0 \leq x \leq \sqrt{2t}, \\ 0, & x \geq \sqrt{2t}. \end{cases}$$

Let us describe in more detail basic analytic properties of the supremum- and infimum convolution operators. We write $\|f\|_\infty = \sup_x |f(x)|$.

Proposition 8.1.2 *Let f be a bounded function on $\mathbb{R}^n$.*

a) For any $t > 0$, the functions $P_t f, Q_t f$ are bounded and Lipschitz. Moreover,

$$\|P_t f\|_\infty \leq \|f\|_\infty, \qquad \|Q_t f\|_\infty \leq \|f\|_\infty,$$

$$\|P_t f\|_{\mathrm{Lip}}^2 \leq \frac{4}{t}\,\|f\|_\infty, \qquad \|Q_t f\|_{\mathrm{Lip}}^2 \leq \frac{4}{t}\,\|f\|_\infty.$$

b) For any $x \in \mathbb{R}^n$, the functions $t \to P_t f(x)$ and $t \to Q_t f(x)$ are Lipschitz on the half-axes (t_0, ∞), $t_0 > 0$, where they have Lipschitz semi-norms at most $\frac{2}{t_0}\,\|f\|_\infty$.

c) *Hence, for any* $x \in \mathbb{R}^n$, *the functions* $u(t) = tP_t f(x)$ *and* $u(t) = tQ_t f(x)$ *are Lipschitz on* $(0, \infty)$ *and satisfy* $\|u\|_{\mathrm{Lip}} \le 3\,\|f\|_\infty$.

Combining *a*) and *b*), one may also conclude that the functions $(x,t) \to P_t f(x)$ and $(x,t) \to Q_t f(x)$ are Lipschitz on the half-spaces $\mathbb{R}^n \times (t,\infty)$ of $\mathbb{R}^{n+1}$.

Proof It is sufficient to restrict ourselves to the operators P_t. Put $C = \|f\|_\infty$.

a) The uniform bounds are obvious; they imply that the supremum in Definition 8.1.1 may be restricted to the closed ball $B(x,r)$ of radius $r = 2\sqrt{Ct}$ with center at x. Indeed, if $|h| > r$, then $f(x+h) - \frac{1}{2t}\,|h|^2 < C - \frac{1}{2t}\,r^2 = -C$. Thus,

$$P_t f(x) = \sup_{|h|\le r} \Big[f(x+h) - \frac{1}{2t}\,|h|^2\Big], \quad x \in \mathbb{R}^n, \ \ r = 2\sqrt{t\,\|f\|_\infty}. \tag{8.2}$$

Given $u, h \in \mathbb{R}^n$ with $|h| \le r$ and putting $\bar h = h + u$, we have $|\bar h| \le r + |u|$ and

$$\begin{aligned} f(x+h+u) - \frac{1}{2t}\,|h|^2 &= f(x+\bar h) - \frac{1}{2t}\,|\bar h - u|^2 \\ &= f(x+\bar h) - \frac{1}{2t}\,|\bar h|^2 + \frac{1}{t}\,\langle \bar h, u\rangle - \frac{1}{2t}\,|u|^2 \\ &\le f(x+\bar h) - \frac{1}{2t}\,|\bar h|^2 + \frac{(r+|u|)\,|u|}{t} \\ &\le P_t f(x) + \frac{(r+|u|)\,|u|}{t}. \end{aligned}$$

Taking the supremum on the left over all h such that $|h| \le r$ and applying (8.2) with $x+u$ in place of x, we get $P_t f(x+u) \le P_t f(x) + \frac{(r+|u|)\,|u|}{t}$. In the same way, with $\bar h = h - u$,

$$\begin{aligned} P_t f(x+u) &\ge f(x+\bar h+u) - \frac{1}{2t}\,|\bar h|^2 \ = \ f(x+h) - \frac{1}{2t}\,|h-u|^2 \\ &\ge f(x+h) - \frac{1}{2t}\,|h|^2 - \frac{(r+|u|)|u|}{t}. \end{aligned}$$

Taking on the right the supremum over all h, we get $P_t f(x+u) \ge P_t f(x) - \frac{(r+|u|)\,|u|}{t}$. The two estimates give, for any $x, u \in \mathbb{R}^n$,

$$|P_t f(x+u) - P_t f(x)| \le \frac{(r+|u|)|u|}{t}.$$

It remains to let $|u| \to 0$, which yields $|\nabla P_t f(x)| \le r/t$, that is, $\|P_t f\|_{\mathrm{Lip}} \le r/t$.

b) Given a Lipschitz function g on $\mathbb{R}^n$, one has $g(x+h) \le g(x) + \|g\|_{\mathrm{Lip}}\,|h|$, so,

$$P_\varepsilon g(x) \le \sup_{h\in\mathbb{R}^n} \Big[g(x) + \|g\|_{\mathrm{Lip}}\,|h| - \frac{1}{2\varepsilon}\,|h|^2\Big] = g(x) + \frac{\varepsilon}{2}\,\|g\|_{\mathrm{Lip}}^2.$$

Applying this bound to $g = P_t f$ and $g = P_{t-\varepsilon} f$ $(0 < \varepsilon < t)$, and using the previous step, we get

$$P_{t+\varepsilon}f(x) = P_\varepsilon P_t f(x) \le P_t f(x) + \frac{2\,\|f\|_\infty}{t}\,\varepsilon,$$
$$P_t f(x) = P_\varepsilon P_{t-\varepsilon} f(x) \le P_{t-\varepsilon} f(x) + \frac{2\,\|f\|_\infty}{t-\varepsilon}\,\varepsilon,$$

thus proving $b)$ and Proposition 8.1.2. □

With similar arguments, one can also show that the functions $(x,t) \to P_t f(x)$ and $(x,t) \to Q_t f(x)$ are Lipschitz on the whole subspace $\mathbb{R}^n \times [0,\infty)$, as long as f is Lipschitz.

We now turn to the differentiability properties. By $C^k(\mathbb{R}^n)$, we denote the class of all functions on $\mathbb{R}^n$ which have continuous partial derivatives up to order k. As before, $\|A\| = \sup\{|A\theta| : |\theta| \le 1\}$ denotes the operator norm of the matrix A.

Proposition 8.1.3 *Let $f \in C^k(\mathbb{R}^n)$, $k \ge 2$, satisfy $\|f''(x)\| \le M$ on $\mathbb{R}^n$. Then $P_t f$ and $Q_t f$ belong to $C^k(\mathbb{R}^n)$ whenever $0 < t < 1/M$. Moreover, for all $x \in \mathbb{R}^n$,*

$$\|(P_t f)''(x)\| \le \frac{M}{1-Mt}, \qquad \|(Q_t f)''(x)\| \le \frac{M}{1-Mt}.$$

The example of the quadratic functions $f(x) = \frac{1}{2}Mx^2$ $(M > 0)$ shows that the above bounds are optimal. In this case, $P_t f(x) = \frac{M}{2(1-Mt)}\,x^2$ for $0 < t < \frac{1}{M}$, $x \in \mathbb{R}$, and $(P_t f)''(x) = \frac{M}{1-Mt}$, while $P_t f(x) = \infty$ for $t \ge \frac{1}{M}$.

Proof Since

$$\begin{aligned} f(x+h) &= f(x) + \langle \nabla f(x), h\rangle + \int_0^1 \langle f''(x+sh)h, h\rangle\,(1-s)\mathrm{d}s \\ &\le f(x) + |\nabla f(x)|\,|h| + \frac{1}{2}\,M|h|^2, \end{aligned}$$

we see that $f(x+h) - \frac{1}{2t}\,|h|^2 \to -\infty$ as $|h| \to \infty$, as soon as $0 < t < 1/M$. In this case, for any $x \in \mathbb{R}^n$, the function $u_x(y) = f(y) - \frac{1}{2t}\,|x-y|^2$ attains its maximum $P_t f(x)$ at some point $y = y(x)$.

Let us now show that such a point y is unique. A necessary condition for u_x to attain its maximum at the point y is that $\nabla u_x(y) = \nabla f(y) + (x-y)/t = 0$. If y and z are two distinct points of maximum, we therefore obtain that

$$t\,(\nabla f(y) - \nabla f(z)) = y - z,$$

so that $|\nabla f(y) - \nabla f(z)| > |y-z|/M$. But

$$|\nabla f(y) - \nabla f(z)| = \Big|\int_0^1 f''(z + s(y-z))(y-z)\,\mathrm{d}s\Big| \le M|y-z|,$$

which leads to a contradiction. Thus, for any $x \in \mathbb{R}^n$, there exists a unique point $y = y(x)$, depending also on the parameter t, such that

$$P_t f(x) = f(y(x)) - \frac{1}{2t}\,|x - y(x)|^2, \tag{8.3}$$

and moreover, this point satisfies an implicit equation $\nabla u_x(y) = 0$, i.e.,

$$y - t\nabla f(y) = x. \tag{8.4}$$

Consider the map $U(y) = y - t\nabla f(y)$ on $\mathbb{R}^n$. Its components $U(y)_i = y_i - t\,\partial_i f(y)$ represent functions in $C^{k-1}(\mathbb{R}^n)$, and the derivative $U'(y) = \mathbb{I}_n - tf''(y)$ is an invertible matrix for any $y \in \mathbb{R}^n$ due to the property $\|tf''(y)\| < 1$ (where $\mathbb{I}_n$ denotes the identity $n \times n$ matrix). Since the equation $U(y) = x$ has a unique solution $y(x)$, $x \in \mathbb{R}^n$, according to the theorem on invertible maps, the map $y = y(x)$ has components which are in $C^{k-1}(\mathbb{R}^n)$ as well. Now, by the chain rule applied to (8.3),

$$\nabla P_t f(x) = \nabla f(y(x))\, y'(x) - \frac{1}{t}\,(y(x) - x)(y'(x) - \mathbb{I}_n),$$

where $y' = \nabla y$. In view of (8.4), we have $\nabla f(y(x)) = (y(x) - x)/t$, so

$$(\nabla P_t f)(x) = \frac{1}{t}\,(y(x) - x) = \nabla f(y(x)). \tag{8.5}$$

Hence, $\nabla P_t f$ has components which are in $C^{k-1}(\mathbb{R}^n)$. This proves the first claim of the proposition. One can further differentiate the identity (8.5) to obtain that

$$(P_t f)''(x) = \frac{1}{t}\,(y'(x) - \mathbb{I}_n) = f''(y(x))\, y'(x).$$

It follows that, for every $\theta \in \mathbb{R}^n$, $|\theta| \le 1$,

$$|(P_t f)''(x)\,\theta| \le \|f''(y(x))\|\;|y'(x)\theta| \le M\,\|y'(x)\|.$$

As we noted before, $|\nabla f(y_1) - \nabla f(y_2)| \le M\,|y_1 - y_2|$ for all $y_1, y_2 \in \mathbb{R}^n$. Hence,

$$|U(y_1) - U(y_2)| \ge |y_1 - y_2| - t\,|\nabla f(y_1) - \nabla f(y_2)| \ge (1 - Mt)\,|y_1 - y_2|.$$

Setting $y_1 = y(x_1)$, $y_2 = y(x_2)$ and recalling that $y = y(x)$ is the inverse to the map U, we get $|y(x_1) - y(x_2)| \le |x_1 - x_2|/(1 - Mt)$. This implies $\|y'(x)\| \le 1/(1 - Mt)$ and thus $|(P_t f)''(x)\,\theta| \le M/(1 - Mt)$ for all $x \in \mathbb{R}^n$. Hence the second claim of the proposition is proved as well. □

Remark 8.1.4 It should be clear from (8.3)–(8.4) that the solution $y(x) = y_t(x)$ depends continuously on t. Hence, by (8.5), the derivative $\nabla P_t f(x)$ represents a continuous vector-valued function in t as well.

8.2 Generators

In order to make the identities in (8.1) more precise, let us look at the behavior of $P_t f(x)$ and $Q_t f(x)$ when t is close to zero. Recall the notations

$$f_+(x) = \limsup_{y\to x} f(y) = \lim_{\varepsilon\to 0^+} \sup_{|x-y|\le\varepsilon} f(y),$$
$$f_-(x) = \liminf_{y\to x} f(y) = \lim_{\varepsilon\to 0^+} \inf_{|x-y|\le\varepsilon} f(y).$$

In particular, $f_-(x) \le f(x) \le f_+(x)$ for all x.

Proposition 8.2.1 *Given a bounded function f on $\mathbb{R}^n$, for any point $x \in \mathbb{R}^n$,*

$$\lim_{t\to 0^+} P_t f(x) = f_+(x), \qquad \lim_{t\to 0^+} Q_t f(x) = f_-(x).$$

In particular, $P_t f(x) \to f(x)$, respectively $Q_t f(x) \to f(x)$ as $t \to 0^+$, if and only if f is upper semi-continuous, resp., if and only if f is lower semi-continuous at x.

Proof First

$$P_t f(x) \ge \limsup_{y\to x} \Big[f(y) - \frac{1}{2t}\,|x-y|^2\Big] = f_+(x),$$

which implies $\liminf_{t\to 0^+} P_t f(x) \ge f_+(x)$. For the opposite direction, define $f_\varepsilon(x) = \sup_{|y-x|\le\varepsilon} f(y)$, $\varepsilon > 0$. For $|h| < \varepsilon$,

$$f(x+h) - \frac{1}{2t}\,|h|^2 \le f(x+h) \le f_\varepsilon(x),$$

while for $|h| \ge \varepsilon$, we have $f(x+h) - \frac{1}{2t}\,|h|^2 \le \|f\|_\infty - \frac{1}{2t}\,|\varepsilon|^2$. Hence,

$$P_t f(x) \le \max\Big\{f_\varepsilon(x),\ \|f\|_\infty - \frac{1}{2t}\,|\varepsilon|^2\Big\}.$$

It follows that $\limsup_{t\to 0^+} P_t f(x) \le f_\varepsilon(x)$. It remains to let $\varepsilon \to 0^+$. Similar arguments apply to $Q_t f$. This finishes the proof of Proposition 8.2.1. □

Thus, in the class of all bounded, upper semi-continuous functions on $\mathbb{R}^n$, the natural extension of $(P_t)_{t>0}$ to the point $t = 0$ leads to the identity operator (and similarly for $(Q_t)_{t>0}$ in the class of all bounded, lower semi-continuous functions). To get more information about the rate of convergence, we need to introduce the operators

$$|\nabla^+ f(x)| = \limsup_{y\to x} \frac{(f(y)-f(x))^+}{|y-x|}, \qquad |\nabla^- f(x)| = \limsup_{y\to x} \frac{(f(y)-f(x))^-}{|y-x|},$$

where we use the standard notations $a^+ = \max(a,0)$, $a^- = \max(-a,0)$ for $a \in \mathbb{R}$.

Proposition 8.2.2 *Given a bounded function f on $\mathbb{R}^n$, for every point $x \in \mathbb{R}^n$,*

$$\limsup_{t\to 0^+} \frac{P_t f(x) - f(x)}{t} = \frac{1}{2}\,|\nabla^+ f(x)|^2,$$
$$\liminf_{t\to 0^+} \frac{Q_t f(x) - f(x)}{t} = -\frac{1}{2}\,|\nabla^- f(x)|^2.$$

Moreover, if f is differentiable at x, then these lim sup *and* lim inf *are just limits that are equal to $\frac{1}{2}\,|\nabla f(x)|^2$ and $-\frac{1}{2}\,|\nabla f(x)|^2$, respectively.*

Proof It is sufficient to consider the supremum convolution. For $\varepsilon > 0$, put

$$D_\varepsilon f(x) = \frac{1}{\varepsilon} \sup_{0<|y-x|\le\varepsilon} (f(y) - f(x))^+,$$

so that $\limsup_{\varepsilon\to 0^+} D_\varepsilon f(x) = |\nabla^+ f(x)|$. Then, for any $h \neq 0$,

$$f(x+h) - f(x) - \frac{1}{2t}\,|h|^2 \le D_{|h|} f(x)\,|h| - \frac{1}{2t}\,|h|^2 \le \frac{t}{2}\,(D_{|h|} f(x))^2.$$

Hence, by (8.2), putting $r(t) = 2\sqrt{t\,\|f\|_\infty}$, we get

$$\frac{P_t f(x) - f(x)}{t} = \frac{1}{t} \sup_{|h|\le r(t)} \Big[f(x+h) - f(x) - \frac{1}{2t}\,|h|^2 \Big] \le \frac{1}{2} \sup_{|h|\le r(t)} (D_{|h|} f(x))^2,$$

and since $r(t) \to 0$ as $t \to 0$,

$$\limsup_{t\to 0^+} \frac{P_t f(x) - f(x)}{t} \le \frac{1}{2} \limsup_{t\to 0^+} \sup_{\varepsilon\le r(t)} (D_\varepsilon f(x))^2 = \frac{1}{2}\,|\nabla^+ f(x)|^2.$$

For the lower bound, we may assume that $|\nabla^+ f(x)| > 0$ and choose a sequence $h_k \to 0$, $h_k \neq 0$, such that $\frac{f(x+h_k)-f(x)}{|h_k|} \to |\nabla^+ f(x)|$ as $k \to \infty$. Putting $t_k = \frac{|h_k|}{|\nabla^+ f(x)|}$, we then have

$$\begin{aligned} \frac{P_{t_k} f(x) - f(x)}{t_k} &\ge \frac{f(x+h_k) - f(x)}{t_k} - \frac{1}{2t_k^2}\,|h_k|^2 \\ &= |\nabla^+ f(x)|\,\frac{f(x+h_k) - f(x)}{|h_k|} - \frac{1}{2}\,|\nabla^+ f(x)|^2 \to \frac{1}{2}\,|\nabla^+ f(x)|^2. \end{aligned}$$

Hence, $\limsup_{t\to 0^+} \frac{P_t f(x) - f(x)}{t} \ge \frac{1}{2}\,|\nabla^+ f(x)|^2$.

Finally, assuming that f is differentiable at x and putting $h_t = t\,\nabla f(x)$, we have

$$\lim_{t\to 0^+} \frac{f(x+h_t) - f(x)}{t} = |\nabla f(x)|^2.$$

Since

$$P_t f(x) \ge f(x+h_t) - \frac{1}{2t}\,|h_t|^2 = f(x+h_t) - \frac{t}{2}\,|\nabla f(x)|^2,$$

we conclude that

$$\liminf_{t\to 0^+} \frac{P_t f(x) - f(x)}{t} \geq \lim_{t\to 0^+} \frac{f(x+h_t) - f(x)}{t} - \frac{1}{2}\,|\nabla f(x)|^2 = \frac{1}{2}\,|\nabla f(x)|^2.$$

Thus Proposition 8.2.2 is proved. □

One may additionally refine the last assertion of this proposition within the class of functions with bounded derivatives of second order.

Proposition 8.2.3 *Let $f \in C^2(\mathbb{R}^n)$ satisfy $\|f''\| \leq M$ on $\mathbb{R}^n$. Then, whenever $0 < t < 1/M$, for all $x \in \mathbb{R}^n$,*

$$\frac{t}{2(1+Mt)}\,|\nabla f(x)|^2 \leq P_t f(x) - f(x) \leq \frac{t}{2(1-Mt)}\,|\nabla f(x)|^2,$$
$$\frac{t}{2(1+Mt)}\,|\nabla f(x)|^2 \leq f(x) - Q_t f(x) \leq \frac{t}{2(1-Mt)}\,|\nabla f(x)|^2.$$

Proof The Taylor integral formula

$$f(x+h) = f(x) + \langle \nabla f(x), h\rangle + \int_0^1 \langle f''(x+sh)h, h\rangle\,(1-s)\,ds \quad (x, h \in \mathbb{R}^n)$$

yields

$$P_t f(x) - f(x) = \sup_{h\in\mathbb{R}^n} \Big[\, \langle \nabla f(x), h\rangle + \int_0^1 \langle f''(x+sh)h, h\rangle\,(1-s)\,ds - \frac{1}{2t}\,|h|^2\Big].$$

Since $|\,\langle f''(y)h, h\rangle\,| \leq M|h|^2$, we get

$$\sup_{h\in\mathbb{R}^n} \Big[\, \langle \nabla f(x), h\rangle - \frac{M}{2}\,|h|^2 - \frac{1}{2t}\,|h|^2\Big] \leq P_t f(x) - f(x)$$
$$\leq \sup_{h\in\mathbb{R}^n} \Big[\, \langle \nabla f(x), h\rangle + \frac{M}{2}\,|h|^2 - \frac{1}{2t}\,|h|^2\Big],$$

proving the claim. □

8.3 Hamilton–Jacobi Equations

The behavior of $P_t f(x)$ and $Q_t f(x)$ with respect to the variable t can be studied with the help of the semi-group property. In general, any function $t \to P_t f(x)$ is non-decreasing, but it may have an infinite jump at some point (which is, however, impossible, when f is bounded). Similarly, $t \to Q_t f(x)$ is always non-increasing. Moreover, these functions are absolutely continuous on the intervals where they are finite. Indeed, the function

$$u(t) = \sup_{y \in \mathbb{R}^n} \left[t f(y) - \frac{1}{2} |x-y|^2 \right]$$

is convex and thus absolutely continuous, which implies that $P_t f(x) = u(t)/t$ is absolutely continuous as well (locally). In particular, we may freely speak of the derivatives of $P_t f(x)$ and $Q_t f(x)$ with respect to t, distinguishing if necessary the right (upper) and the left (lower) derivatives

$$\frac{\partial^+}{\partial t} P_t f(x) = \lim_{\varepsilon \to 0^+} \frac{P_{t+\varepsilon} f(x) - P_t f(x)}{\varepsilon},$$
$$\frac{\partial^-}{\partial t} P_t f(x) = \lim_{\varepsilon \to 0^-} \frac{P_{t+\varepsilon} f(x) - P_t f(x)}{\varepsilon},$$

and similarly for Q_t. Thus, for bounded functions f, these limits exist, finite for all $t > 0$, and satisfy

$$0 \le \frac{\partial^-}{\partial t} P_t f(x) \le \frac{\partial^+}{\partial t} P_t f(x), \quad \frac{\partial^+}{\partial t} Q_t f(x) \le \frac{\partial^-}{\partial t} Q_t f(x) \le 0.$$

Proposition 8.3.1 *Given a bounded function f on $\mathbb{R}^n$ and a point $x \in \mathbb{R}^n$, for any $t > 0$,*

$$\frac{\partial^+}{\partial t} P_t f(x) = \frac{1}{2} |\nabla^+ P_t f(x)|^2, \quad \frac{\partial^+}{\partial t} Q_t f(x) = -\frac{1}{2} |\nabla^- Q_t f(x)|^2. \tag{8.6}$$

In particular, for any $t > 0$, for almost all x,

$$\frac{\partial^+}{\partial t} P_t f(x) = \frac{1}{2} |\nabla P_t f(x)|^2, \quad \frac{\partial^+}{\partial t} Q_t f(x) = -\frac{1}{2} |\nabla Q_t f(x)|^2. \tag{8.7}$$

Indeed, by Proposition 8.2.2, applied to the function $P_t f$ (in place of f), we have

$$\begin{aligned}\lim_{\varepsilon \to 0^+} \frac{P_{t+\varepsilon} f(x) - P_t f(x)}{\varepsilon} &= \lim_{\varepsilon \to 0^+} \frac{P_\varepsilon(P_t f(x)) - P_t f(x)}{\varepsilon} \\ &= \limsup_{\varepsilon \to 0^+} \frac{P_\varepsilon(P_t f(x)) - P_t f(x)}{\varepsilon} = \frac{1}{2} |\nabla^+ P_t f(x)|^2,\end{aligned}$$

which gives the first identity in (8.6). Applying it to $-f$ and using $P_t(-f) = -Q_t f$, we arrive at the second identity.

To get (8.7), recall that, by Proposition 8.1.2, the functions $P_t f$ and Q_f are Lipschitz and therefore almost everywhere differentiable (by Rademacher's theorem). Hence, for almost all $x \in \mathbb{R}^n$, $|\nabla^+ P_t f(x)| = |\nabla P_t f(x)|$ and $|\nabla^+ Q_t f(x)| = |\nabla Q_t f(x)|$, which may also be understood as the usual moduli of gradients (a.e.)

Let us now refine these assertions in the "smooth" case.

Proposition 8.3.2 *If $f \in C^2(\mathbb{R}^n)$ satisfies $\sup_x \|f''(x)\| \le M$, then for all $x \in \mathbb{R}^n$, the functions $t \to P_t f(x)$ and $t \to Q_t f(x)$ are differentiable in the interval $0 < t < 1/M$ and have derivatives*

$$\frac{\partial}{\partial t} P_t f(x) = \frac{1}{2} |\nabla P_t f(x)|^2, \quad \frac{\partial}{\partial t} Q_t f(x) = -\frac{1}{2} |\nabla Q_t f(x)|^2. \tag{8.8}$$

Proof According to Proposition 8.1.3, for all $t \in (0, 1/M)$, the function $P_t f$ belongs to $C^2(\mathbb{R}^n)$, and its partial derivatives of the second order are uniformly bounded, namely, $\|(P_t f)''\| \le M_t = \frac{M}{1-Mt}$. Hence, by Proposition 8.2.3, for all $x \in \mathbb{R}^n$ and $0 < \varepsilon < \min(\frac{1}{M_t}, t)$,

$$P_{t+\varepsilon} f(x) = P_\varepsilon(P_t f(x)) \le P_t f(x) + \frac{\varepsilon}{2(1 - M_t \varepsilon)} |\nabla P_t f|^2.$$

Letting here $\varepsilon \to 0^+$, we obtain an upper bound $\frac{\partial^+}{\partial t} P_t f(x) \le \frac{1}{2} |\nabla P_t f(x)|^2$ for the right derivative at t. Applying Proposition 8.2.3 once more, we also get

$$P_t f(x) = P_\varepsilon(P_{t-\varepsilon} f(x)) \ge P_{t-\varepsilon} f(x) + \frac{\varepsilon}{2(1 + M_t \varepsilon)} |\nabla P_{t-\varepsilon} f|^2.$$

Here $|\nabla P_{t-\varepsilon} f| \to |\nabla P_t f|$ as $\varepsilon \to 0^+$ (cf. Remark 8.1.4). Hence, in the limit the above inequality yields a lower bound $\frac{\partial^-}{\partial t} P_t f(x) \ge \frac{1}{2} |\nabla P_t f(x)|^2$ for the left derivative. Proposition 8.3.2 is proved. □

The second equality in (8.8) represents an important particular case of the Hamilton–Jacobi equations with initial condition (although not on a finite time interval),

$$\begin{cases} u_t = -\frac{1}{2} |\nabla u|^2 & \text{in } \mathbb{R}^n \times (0, \infty), \\ u = f & \text{on } \mathbb{R}^n \times \{0\}, \end{cases}$$

where $f(x)$ is a given function on $\mathbb{R}^n$ and $u = u(x, t)$ is an unknown function on $\mathbb{R}^n \times [0, \infty)$, with the usual notations $u_t = \frac{\partial}{\partial t} u$, $\nabla u = \nabla_x u$ (the gradient with respect to the space variable). The solution to this initial value problem (under certain conditions) may therefore be given in the form of the infimum-convolution

$$u(x, t) = Q_t f(x) = \inf_{y \in \mathbb{R}^n} \Big[f(y) + \frac{1}{2t} |x - y|^2 \Big],$$

which is also called the Hopf–Lax formula. In view of the example $f(x) = Mx^2/2$, the smoothness assumption for f only is insufficient to construct a C^1-solution $u(x, t)$ such that $u_t = -\frac{1}{2t} |\nabla u|^2$ for all points (x, t). Nevertheless, one may require less – that u should be differentiable almost everywhere on $\mathbb{R}^n \times (0, \infty)$ and satisfy the Hamilton–Jacobi equation for almost all (x, t).

The next statement, which can be found in [87], pp. 125–129, describes a typical situation. Recall that, when f is Lipschitz on $\mathbb{R}^n$, both $u(x,t) = Q_t f(x)$ and $u(x,t) = P_t f(x)$ are Lipschitz on $\mathbb{R}^n \times [0,\infty)$, and therefore they are almost everywhere differentiable (by Rademacher's theorem).

Proposition 8.3.3 *If f is Lipschitz, then the equalities in* (8.8) *make sense and are fulfilled for almost all points $(x,t) \in \mathbb{R}^n \times (0,\infty)$.*

8.4 Supremum/Infimum Convolution Inequalities

The relationship between logarithmic Sobolev inequalities and the supremum- and infimum convolution operators is described in the following statement.

Proposition 8.4.1 *Suppose that a probability measure μ on $\mathbb{R}^n$ has a positive logarithmic Sobolev constant ρ. Then, for any μ-integrable function f on $\mathbb{R}^n$,*

$$\int e^f \, d\mu \le \exp\Big\{\int P_{1/\rho} f \, d\mu\Big\}, \tag{8.9}$$

$$\int e^{Q_{1/\rho} f} d\mu \le \exp\Big\{\int f \, d\mu\Big\}. \tag{8.10}$$

Since in general $P_t Q_t f \le f \le Q_t P_t f$, (8.9) and (8.10) are equivalent.

These inequalities may be recognized as generalizations of the exponential bound (7.7) for the class of Lipschitz functions. Indeed, if $\|f\|_{\mathrm{Lip}} \le L$, then, for any $t > 0$,

$$\begin{aligned} P_t f(x) &= \sup_{h \in \mathbb{R}^n} \Big[f(x+h) - \frac{1}{2t}\,|h|^2\Big] \\ &\le \sup_{h \in \mathbb{R}^n} \Big[f(x) + L|h| - \frac{1}{2t}\,|h|^2\Big] = f(x) + \frac{tL^2}{2}. \end{aligned}$$

Hence, taking $t = 1/\rho$, if f has μ-mean zero, the bound (8.9) yields $\mathbb{E}_\mu\, e^f \le e^{L^2/2\rho}$. A similar conclusion can be made on the basis of (8.10) as well.

Replacing f with $\rho t f$ and using $P_{1/\rho}(\rho t f) = \rho t P_t f$ and similarly for Q_t, the inequalities (8.9)–(8.10) take the form

$$\int e^{\rho t f} \, d\mu \le \exp\Big\{\rho t \int P_t f \, d\mu\Big\}, \quad \int e^{\rho t\, Q_t f} \, d\mu \le \exp\Big\{\rho t \int f \, d\mu\Big\} \tag{8.11}$$

(moreover, for each fixed t, (8.11) are still equivalent to (8.9)–(8.10)). Letting here $t \to 0^+$ and applying Proposition 8.2.2, both relations yield in the limit the Poincaré-type inequality (5.6) with constant ρ. Therefore, the supremum- and infimum convolution inequalities such as (8.9)–(8.11) occupy an intermediate position between Poincaré-type and logarithmic Sobolev inequalities.

For the proof of Proposition 8.4.1 we need an auxiliary statement, which will also be used later on.

Lemma 8.4.2 *Suppose that a Borel probability measure μ on $\mathbb{R}^n$ satisfies*

$$\int e^{Q_t f} d\mu \le \exp\left\{\int f\, d\mu\right\} \tag{8.12}$$

for some $t > 0$ in the class of all bounded Lipschitz functions f on $\mathbb{R}^n$. Then this inequality continues to hold for all μ-integrable functions f on $\mathbb{R}^n$.

Proof If f is bounded and μ-measurable, then $Q_\varepsilon f$ ($\varepsilon > 0$) is bounded and Lipschitz (Proposition 8.1.2). Hence, one may apply (8.12) to $Q_\varepsilon f$, which gives

$$\int e^{Q_{t+\varepsilon} f} d\mu \le \exp\left\{\int Q_\varepsilon f\, d\mu\right\} \le \exp\left\{\int f\, d\mu\right\}.$$

Since $Q_{t+\varepsilon} f \uparrow Q_t f$ as $\varepsilon \downarrow 0$, the above left integral converges to $\int e^{Q_t f} d\mu$, and we arrive at (8.12).

It remains to remove the boundedness assumption, which can be done by virtue of a simple truncation argument. First put $f_k = \min(f, k)$, $k = 1, 2\dots$ If (8.12) holds for f_k, then, since $f_k \le f$,

$$\log \int e^{Q_t f_k} d\mu \le \int f_k\, d\mu \le \int f\, d\mu.$$

Using the property $Q_t(\min(f, g)) = \min(Q_t f, Q_t g)$, we have

$$Q_t f_k = \min(Q_t f, k) \uparrow Q_t f \quad \text{as } k \to \infty,$$

and therefore $\int e^{Q_t f_k} d\mu \uparrow \int e^{Q_t f}$, by the monotone convergence theorem. As a result, we obtain (8.12) for f.

Thus, we may assume that f is bounded from above. In this case, put $f_k = \max(f, -k)$. These functions are bounded, so (8.12) holds for all f_k. Since $Q_t f \le Q_t f_k$, we get

$$\log \int e^{Q_t f}\, d\mu \le \log \int e^{Q_t f_k}\, d\mu \le \int f_k\, d\mu.$$

By Lebesgue's dominated convergence theorem, the right-hand side tends to $\int f\, d\mu$, and we again arrive at (8.12) for f. This proves Lemma 8.4.2. □

Proof (of Proposition 8.4.1) Let f be bounded on $\mathbb{R}^n$.

Step 1. First, assume that μ is absolutely continuous with respect to the Lebesgue measure on $\mathbb{R}^n$. To derive (8.10), we apply the logarithmic Sobolev inequality in the exponential form

$$\mathrm{Ent}_\mu(e^f) \le \frac{1}{2\rho} \int |\nabla f|^2\, e^f\, d\mu$$

to the Lipschitz functions $f(t, x) = \rho t\, Q_t f(x)$ on $\mathbb{R}^n$ ($t > 0$). That is, we have

$$\mathrm{Ent}_\mu(e^{\rho t\, Q_t f}) \le \frac{\rho t^2}{2} \int |\nabla Q_t f|^2\, e^{\rho t\, Q_t f}\, d\mu.$$

Since $Q_t f$ is Lipschitz and therefore differentiable a.e., we have $|\nabla^- Q_t f| = |\nabla Q_t f|$ a.e., where we recall that

$$|\nabla^- Q_t f(x)| = \limsup_{y \to x} \frac{(Q_t f(x) - Q_t f(y))^+}{|x - y|}.$$

Hence, since μ is absolutely continuous,

$$\mathrm{Ent}_\mu(\mathrm{e}^{\rho t\, Q_t f}) \leq \frac{\rho t^2}{2} \int |\nabla^- Q_t f|^2 \, \mathrm{e}^{\rho t\, Q_t f} \, \mathrm{d}\mu. \tag{8.13}$$

The functions $f(t,x)$ are Lipschitz in t and have Lipschitz semi-norms bounded by $3\rho \, \|f\|_\infty$ (thus, independent of x, cf. Proposition 8.1.2). In addition, by Proposition 8.3.1, for all x, the right derivatives $\frac{\partial^+}{\partial t} Q_t f(x)$ exist and are equal to $-\frac{1}{2} \, |\nabla^- Q_t f(x)|^2$. Hence, the right derivatives $\frac{\partial^+ f(t,x)}{\partial t}$ exist, are finite and satisfy

$$\frac{\rho t}{2} \, |\nabla^- Q_t f(x)|^2 \; = \; \rho \, Q_t f(x) - \frac{\partial^+ f(t,x)}{\partial t}.$$

Plugging this equality into (8.13), we then get

$$\mathrm{Ent}_\mu(\mathrm{e}^{\rho t\, Q_t f}) \leq \rho t \int Q_t f \, \mathrm{e}^{\rho t\, Q_t f} \, \mathrm{d}\mu - t \int \frac{\partial^+ f(t,x)}{\partial t} \, \mathrm{e}^{\rho t\, Q_t f} \, \mathrm{d}\mu,$$

that is,

$$\int \mathrm{e}^{\rho t\, Q_t f} \, \mathrm{d}\mu \, \log \int \mathrm{e}^{\rho t\, Q_t f} \, \mathrm{d}\mu \geq t \int \frac{\partial^+ f(t,x)}{\partial t} \, \mathrm{e}^{\rho t\, Q_t f} \, \mathrm{d}\mu.$$

But, by the chain rule on the real line for the operator $\partial^+/\partial t$, the above inequality is the same as

$$\int \mathrm{e}^{\rho t\, Q_t f} \, \mathrm{d}\mu \, \log \int \mathrm{e}^{\rho t\, Q_t f} \, \mathrm{d}\mu \geq t \int \frac{\partial^+}{\partial t} \, \mathrm{e}^{\rho t\, Q_t f} \, \mathrm{d}\mu. \tag{8.14}$$

Now, we need to justify the possibility of differentiation with respect to t outside the integral on the right-hand side of (8.14). Consider the family of functions

$$g_\varepsilon(t,x) = \frac{1}{\varepsilon} \Big(\mathrm{e}^{\rho (t+\varepsilon)\, Q_{t+\varepsilon} f(x)} - \mathrm{e}^{\rho t\, Q_t f(x)} \Big), \quad 0 < \varepsilon < 1, \; t > 0, \; x \in \mathbb{R}^n.$$

By Proposition 8.1.2,

$$|(t + \varepsilon) \, Q_{t+\varepsilon} f(x) - t \, Q_t f(x)| \leq 3\varepsilon \, \|f\|_\infty.$$

Therefore, $|g_\varepsilon(t,x)| \leq 3 \, \|f\|_\infty \, \mathrm{e}^{\rho(t+1)\, \|f\|_\infty}$. Thus, the g_ε are uniformly bounded in x, and by the Lebesgue dominated convergence theorem,

$$\int \frac{\partial^+}{\partial t}\, e^{\rho t\, Q_t f}\, d\mu = \int \lim_{\varepsilon \to 0^+} g_\varepsilon(t,x)\, d\mu(x)$$
$$= \lim_{\varepsilon \to 0^+} \int g_\varepsilon(t,x)\, d\mu(x) = \frac{\partial^+}{\partial t} \int e^{\rho t\, Q_t f}\, d\mu.$$

In particular, $t \to \int e^{\rho t Q_t f}\, d\mu$ is differentiable from the right at any point $t > 0$. Note that this function is Lipschitz on $(0, \infty)$, since the function $t \to tQ_t f(x)$ has a bounded Lipschitz constant independent of x. Thus, we can rewrite (8.14) as

$$\int e^{\rho t\, Q_t f} d\mu \, \log \int e^{\rho t\, Q_t f}\, d\mu \geq t\, \frac{\partial^+}{\partial t} \int e^{\rho t\, Q_t f}\, d\mu. \tag{8.15}$$

Define the function $u = u(t)$ by $e^{tu(t)} = \int e^{\rho t\, Q_t f}\, d\mu$, so that, by (8.15),

$$tu(t)\, e^{tu(t)} \geq t\, \frac{\partial^+}{\partial t}\, e^{tu(t)} = t\,\Big(u(t) + t\, \frac{\partial^+}{\partial t}\, u(t)\Big)\, e^{tu(t)}.$$

But this is equivalent to $v(t) \equiv \frac{\partial^+}{\partial t}\, u(t) \leq 0$. Since v represents a Radon–Nikodym derivative of u, we may conclude that u is non-increasing. In particular,

$$\log \int e^{Q_{1/\rho} f}\, d\mu = \frac{1}{\rho}\, u(1/\rho) \leq \frac{1}{\rho}\, u(0^+).$$

By Proposition 8.2.1, $Q_\varepsilon f(x) \to f_-(x)$ for all $x \in \mathbb{R}^n$, and since f_- and all $Q_\varepsilon f$ are bounded by $\|f\|_\infty$,

$$u(\varepsilon) = \frac{1}{\varepsilon}\, \log \int e^{\rho\varepsilon\, Q_\varepsilon f}\, d\mu \longrightarrow \rho \int f_-\, d\mu \quad \text{as } \varepsilon \to 0^+.$$

As a result,

$$\log \int e^{Q_{1/\rho} f}\, d\mu \leq \int f_-\, d\mu \leq \int f\, d\mu, \tag{8.16}$$

which gives the desired bound (8.10).

Step 2. To remove the assumption that μ is absolutely continuous, one may consider the convolution μ_ε of μ with the Gaussian measure on $\mathbb{R}^n$ with mean zero and covariance matrix $\varepsilon^2\, \mathrm{I}_n$. Then, according to item 7.2.4, $\rho_\varepsilon^{-1} \equiv \rho^{-1}(\mu_\varepsilon) \leq \rho^{-1} + \varepsilon^2$. Given a bounded function f on $\mathbb{R}^n$, we apply (8.10) to μ_ε and $Q_t f$ with $t > 0$, so that, by the previous step,

$$\int \exp\big\{Q_{t+\rho_\varepsilon^{-1}} f\big\}\, d\mu_\varepsilon \leq \exp\Big\{\int Q_t f\, d\mu_\varepsilon\Big\}. \tag{8.17}$$

Since the functions $Q_t f$ are uniformly bounded and continuous, the last integral tends to $\int Q_t f\, d\mu$ as $\varepsilon \to 0^+$. In addition, by Proposition 8.1.2,

$$|Q_{t+\rho_\varepsilon^{-1}} f - Q_{t+\rho^{-1}} f| \leq \frac{2\,(\rho_\varepsilon^{-1} - \rho^{-1})}{t}\, \|f\|_\infty \leq \frac{2\varepsilon^2}{t}\, \|f\|_\infty.$$

Hence, letting $\varepsilon \to 0^+$ in (8.17) and using the continuity of the function $Q_{t+\rho^{-1}} f$ on $\mathbb{R}^n$, we obtain in the limit the bounds

$$\int \exp\left\{Q_{t+\rho^{-1}} f\right\} d\mu \le \exp\left\{\int Q_t f \, d\mu\right\} \le \exp\left\{\int f \, d\mu\right\}.$$

Sending here $t \to 0^+$ and using once more the Lipschitz property of the function $s \to Q_s f$, we arrive in the limit again at (8.16).

In view of Lemma 8.4.2, this proves Proposition 8.4.1. □

8.5 Transport-Entropy Inequalities

The supremum/infimum convolution inequalities considered in the previous section may equivalently be stated as a certain relation between the quadratic transport distance and the relative entropy. To describe this relationship, first let us recall that the power transport distance of order $p \ge 1$ between Borel probability measures μ and ν on $\mathbb{R}^n$ is defined by

$$W_p(\mu, \nu) = \inf_{\pi} \Big(\int\!\!\int |x - y|^p \, d\lambda(x, y)\Big)^{1/p} = \inf \left(\mathbb{E}\,|X - Y|^p\right)^{1/p}. \tag{8.18}$$

Here the infimum runs over all random vectors X and Y with respective distributions μ and ν, or over all Borel probability measures λ on $\mathbb{R}^n \times \mathbb{R}^n$ having μ and ν as marginals, i.e., such that

$$\lambda(A \times \mathbb{R}^n) = \mu(A), \quad \lambda(\mathbb{R}^n \times A) = \nu(A)$$

for all Borel sets A in $\mathbb{R}^n$. It is also called the L^p–Kantorovich distance or the minimal distance of order p. A more general quantity

$$W = \inf_{\pi} \int\!\!\int c(x, y) \, d\lambda(x, y)$$

was introduced by Kantorovich in [117, 118] with the following natural interpretation: W describes the minimal price needed to pay in order to transport the "mass" μ into ν, if it costs $c(x, y)$ to move a particle at the point x to the location y (in this model $c(x, y)$ is often called the cost function).

The quantity W_p represents a metric in the space of all Borel probability measures on $\mathbb{R}^n$ with finite p-th absolute moments. It generates a topology which is closely related to the topology of weak convergence of probability distributions. The weak topology is metrized, for example, with the help of the Lévy–Prokhorov distance

$$\pi(\mu, \nu) = \inf\{\varepsilon > 0 : \mu(A_\varepsilon) \le \nu(A) + \varepsilon \text{ for all Borel } A \subset \mathbb{R}^n\}$$

(where A_ε denotes an ε-neighborhood of A), and it is connected with the power transport distances via the elementary inequality

$$\pi(\mu, \nu) \leq \big(W_p(\mu, \nu)\big)^{p/(p+1)}.$$

This shows that the convergence in W_p is stronger than the convergence in π (although the converse is also true under proper moment assumptions, cf. e.g. [174]).

On the real line, the power transport distance of order p admits an explicit description

$$W_p(\mu, \nu) = \bigg(\int_0^1 |F_\mu^{-1}(t) - F_\nu^{-1}(t)|^p \, dt\bigg)^{1/p}$$

in terms of the inverse distribution functions $F_\mu^{-1}(t) = \min\{x \in \mathbb{R} : F_\mu(x) \geq t\}$, where $F_\mu(x) = \mu(-\infty, x]$ denotes the distribution function associated to μ. When $p = 1$, it is further simplified to the L^1-distance

$$W_1(\mu, \nu) = \int_{-\infty}^{\infty} |F_\mu(x) - F_\nu(x)| \, dx.$$

However, for $p > 1$ and even for $p = 1$ in dimensions $n \geq 2$, the distance W_p has a complicated structure, and the problem of bounding it from above is not obvious. Nevertheless, there is the following representation which is dual to the definition (8.18)

$$W_p^p(\mu, \nu) = \sup_{f, g} \bigg[\int f \, d\mu - \int g \, d\nu\bigg], \tag{8.19}$$

where the supremum extends over all bounded Borel measurable functions f and g on $\mathbb{R}^n$ such that

$$f(x) - g(y) \leq |x - y|^p \quad \text{for all } x, y \in \mathbb{R}^n. \tag{8.20}$$

We refer to the review [60] for a comprehensive account and history of such dual representations, and here only mention the following. In the important case $p = 2$, the optimal function g in (8.20) coincides with the supremum convolution $P_{1/2} f$, while the optimal f is the infimum convolution $Q_{1/2}\, g$. Hence (8.19) becomes

$$\begin{aligned} W_2^2(\mu, \nu) &= \sup_f \bigg[\int f \, d\mu - \int P_{1/2} f \, d\nu\bigg] \\ &= \sup_f \bigg[\int Q_{1/2} f \, d\mu - \int f \, d\nu\bigg]. \end{aligned} \tag{8.21}$$

Now, in the presence of a logarithmic Sobolev inequality for the measure μ, let us recall the infimum convolution inequality in (8.11) with $t = 1/2$, when it turns into

$$\int e^{\frac{\rho}{2} Q_{1/2} f} \, d\mu \leq \exp\Big\{\frac{\rho}{2} \int f \, d\mu\Big\}.$$

It is valid in the class of all bounded Borel measurable f, or equivalently, for all μ-integrable f, according to Lemma 8.4.2. By the variational formula for the entropy functional, this inequality is equivalent to the relation

$$\frac{\rho}{2}\int\Big(Q_{1/2}f-\int f\,\mathrm{d}\mu\Big)\,h\,\mathrm{d}\mu\le \mathrm{Ent}_\mu(h),$$

valid for all Borel measurable functions $h\ge 0$ on $\mathbb{R}^n$. Here, by homogeneity, one may require that $\int h\,\mathrm{d}\mu=1$. That is, in terms of the probability measure $\mathrm{d}\nu=h\,\mathrm{d}\mu$ and using the relative entropy distance, we have

$$\frac{\rho}{2}\Big[\int Q_{1/2}f\,\mathrm{d}\nu-\int f\,\mathrm{d}\mu\Big]\le D(\nu||\mu).$$

But, the expression in the brackets on the left is exactly $W_2^2(\nu,\mu)$, according to the dual formula (8.21). Therefore, we arrive at the following equivalent variant of Proposition 8.4.1.

Proposition 8.5.1 *Suppose that a probability measure μ on $\mathbb{R}^n$ has a positive logarithmic Sobolev constant ρ. Then, for any probability measure ν on $\mathbb{R}^n$ which is absolutely continuous with respect to μ, we have*

$$W_2^2(\nu,\mu)\le\frac{2}{\rho}\,D(\nu||\mu). \tag{8.22}$$

Note that the finiteness of the relative entropy on the right-hand side forces ν to have a finite second moment and thus a finite quadratic Kantorovich distance to μ.

For example, the standard Gaussian measure $\mu=\gamma_n$ has the logarithmic Sobolev constant $\rho=1$, so that we get the transport-entropy inequality

$$W_2^2(\nu,\gamma_n)\le 2\,D(\nu||\gamma_n) \tag{8.23}$$

in the class of all absolutely continuous probability measures ν on $\mathbb{R}^n$.

8.6 Remarks

Supremum and infimum convolution operators were intensively studied in the field of concentration inequalities following the pioneering work by Maurey [137]. The first Sections 8.1–8.3 are devoted to the systematic analysis of these operators, so as to rigorously justify further applications. The current exposition of the associated Hamilton–Jacobi equations is based on the paper [43], where Proposition 8.4.1 was proved. Gozlan, Roberto, and Samson [99] have developed an extension of this approach to abstract (geodesic) metric spaces.

The Gaussian transport-entropy inequality (8.23) is due to Talagrand [173]. A different approach based on the application of the Prékopa–Leindler theorem was proposed in [44]. Its generalization (8.22) in the presence of logarithmic Sobolev inequalities was obtained by Otto and Villani [150] (in the setting of Riemannian manifolds). We refer to the book by Villani [174] and the surveys [98], [60] providing detailed accounts of the transport problems.

Supremum and infimum convolution operators may also be used to derive sharp Sobolev inequalities, as well as to explore concentration properties of empirical distributions; cf. [57], [48].

Part III
Analysis on the Sphere

Chapter 9
Sobolev-type Inequalities

This chapter deals with an important special space, namely

$$\mathbb{M} = \mathbb{S}^{n-1} = \big\{\theta = (\theta_1, \dots, \theta_n) \in \mathbb{R}^n : \theta_1^2 + \dots + \theta_n^2 = 1\big\}, \quad n \geq 2,$$

the sphere in $\mathbb{R}^n$ of radius 1 and with center at the origin. We equip it with the geodesic metric d and the uniform distribution $\mu = \mathfrak{s}_{n-1}$. The aim of the chapter is to describe the analysis of functions and their derivatives on the sphere up to order 2, and briefly state basic analytic inequalities.

9.1 Spherical Derivatives

According to the general approach (cf. Section 5.1 and the equation (5.2)), the generalized modulus of the gradient of a function f on the unit sphere is defined by

$$|\nabla_{\mathbb{S}} f(\theta)| = \limsup_{\theta' \to \theta} \frac{|f(\theta) - f(\theta')|}{d(\theta, \theta')} \quad (\theta, \theta' \in \mathbb{S}^{n-1}). \tag{9.1}$$

This definition is consistent with usual differentiation rules. If f is defined on $\mathbb{S}^{n-1}$ and is differentiable at a given point $\theta \in \mathbb{S}^{n-1}$, then it admits the Taylor expansion up to the linear term

$$f(\theta') = f(\theta) + \langle v, \theta' - \theta\rangle + o\big(|\theta' - \theta|\big) \quad \text{as } \theta' \to \theta, \ \ \theta' \in \mathbb{S}^{n-1}, \tag{9.2}$$

for some $v \in \mathbb{R}^n$. If v has the smallest length (the Euclidean norm) among all such vectors (obviously, it exists and is unique), then it is called the spherical (first) derivative or the gradient of f at the point θ and is denoted $\nabla_{\mathbb{S}} f(\theta)$. The length of this vector may be determined by (9.1) as well.

S. Bobkov et al., *Concentration and Gaussian Approximation for Randomized Sums*, Probability Theory and Stochastic Modelling 104,
https://doi.org/10.1007/978-3-031-31149-9_9

This notion of the derivative of f is independent of the choice of a smooth extension of f in an open neighborhood of the sphere in $\mathbb{R}^n$. If f is C^1-smooth in a neighborhood of $\mathbb{S}^{n-1}$, then (9.2) holds with the usual (Euclidean) gradient $v = \nabla f(\theta)$, and the spherical gradient may be described as follows:

Proposition 9.1.1 *The spherical first derivative of f at each point $\theta \in \mathbb{S}^{n-1}$ is given by the orthogonal projection*

$$\nabla_{\mathbb{S}} f(\theta) = P_{\theta^\perp} \nabla f(\theta) = \nabla f(\theta) - \langle \nabla f(\theta), \theta \rangle \, \theta,$$

where $P_{\theta^\perp}$ is the orthogonal projection operator from $\mathbb{R}^n$ to $\theta^\perp$ (the tangent space).

In particular, $\langle \nabla_{\mathbb{S}} f(\theta), \theta \rangle = 0$ and $|\nabla_{\mathbb{S}} f(\theta)| \leq |\nabla f(\theta)|$ for any $\theta \in \mathbb{S}^{n-1}$.

The spherical gradient of any C^1-function represents a continuous vector-valued function on $\mathbb{S}^{n-1}$.

Analogously, the second derivative of any C^2-smooth function f on the unit sphere at a given point $\theta \in \mathbb{S}^{n-1}$ may be introduced via the Taylor expansion up to the quadratic term

$$f(\theta') = f(\theta) + \langle \nabla_S f(\theta), \theta' - \theta \rangle + \frac{1}{2} \langle B(\theta' - \theta), \theta' - \theta \rangle + o\big(|\theta' - \theta|^2\big), \tag{9.3}$$

where $\theta' \to \theta$, $\theta' \in \mathbb{S}^{n-1}$, and B is some $n \times n$ matrix (with real entries).

The space $\mathbb{M}_n$ of all $n \times n$ matrices is naturally identified with the Euclidean space $\mathbb{R}^{n \times n}$ with its inner product and the Euclidean norm

$$\|B\|_{\mathrm{HS}} = \bigg(\sum_{i,j=1}^n B_{ij}^2 \bigg)^{1/2}$$

called the Hilbert–Schmidt norm of B. The collection of all B's satisfying (9.3) represents an affine subspace of $\mathbb{M}_n$. Therefore, among all of them, there exists a unique matrix which has the smallest Hilbert–Schmidt norm. It will be called the (spherical) second derivative of f at the point θ and will be denoted $f''_{\mathbb{S}}(\theta)$.

If f is C^2-smooth in an open neighborhood of $\mathbb{S}^{n-1}$, then in accordance with the usual Taylor expansion, (9.3) holds with matrix $B_0 = f''(\theta) - \langle \nabla f(\theta), \theta \rangle \, \mathbb{I}_n$, where

$$f''(\theta) = \Big(\frac{\partial^2 f(\theta)}{\partial \theta_i \, \partial \theta_j} \Big)_{i,j=1}^n$$

is the matrix of the usual second partial derivatives of f at the point $\theta \in \mathbb{S}^{n-1}$. More generally, given $A \in \mathbb{M}_n$, the matrix $B_0 - A$ satisfies (9.3) if and only if

$$\langle A(\theta' - \theta), \theta' - \theta \rangle = o\big(|\theta' - \theta|^2\big) \qquad (\theta' \to \theta, \;\; \theta' \in \mathbb{S}^{n-1}).$$

But this is equivalent to saying that $\langle Ax, x \rangle = 0$ for all $x \in \theta^\perp$. This condition defines a linear subspace L of $\mathbb{M}_n$, and the problem

$$\|B_0 - A\|_{\mathrm{HS}} \to \min \;\; \text{over all } A \in L$$

is then solved uniquely for $B = B_0 - A$, being the orthogonal projection in $\mathbb{M}_n$ of B_0 to the linear space $L^\perp$ of all matrices orthogonal to L. In fact, since B_0 is symmetric, in this minimization problem we may restrict ourselves to symmetric matrices, and by simple algebra, we arrive at the following description.

Proposition 9.1.2 *The spherical second derivative of f at each point $\theta \in \mathbb{S}^{n-1}$ is a symmetric matrix, which is given by the orthogonal projection*

$$f''_{\mathbb{S}}(\theta) = P_{L_\theta^\perp} B, \quad B = f''(\theta) - \langle \nabla f(\theta), \theta \rangle \, \mathbb{I}_n,$$

to the orthogonal complement of the linear subspace L_θ of all symmetric matrices A in $\mathbb{M}_n$ such that $Ax = 0$ for all $x \in \theta^\perp$. Equivalently,

$$f''_{\mathbb{S}}(\theta) = P_{\theta^\perp} B P_{\theta^\perp}.$$

One immediate consequence of this description is that $f''_{\mathbb{S}}(\theta)\,\theta = 0$, and hence the vectors $f''_{\mathbb{S}}(\theta)v$ are orthogonal to θ for all $\theta \in \mathbb{S}^{n-1}$ and $v \in \mathbb{R}^n$.

One should also emphasize the contraction property

$$\|f''_{\mathbb{S}}(\theta)\|_{\mathrm{HS}} \le \|f''(\theta) - \langle \nabla f(\theta), \theta \rangle \, \mathbb{I}_n\|_{\mathrm{HS}}$$

and similarly for the operator norm.

In the usual Euclidean setting, the second derivative appears as the double application of the differentiation operator, so that

$$f''(x)v = \nabla \, \langle \nabla f(x), v \rangle \quad (v \in \mathbb{R}^n).$$

We have a similar property for the spherical derivatives – however for v in the tangent space, only.

Proposition 9.1.3 *Given a C^2-smooth function f on $\mathbb{S}^{n-1}$, for all $\theta \in \mathbb{S}^{n-1}$ and $v \in \mathbb{R}^n$,*

$$f''_{\mathbb{S}}(\theta)v = \nabla_{\mathbb{S}} \, \langle \nabla_{\mathbb{S}} f(\theta), v \rangle + \langle v, \theta \rangle \, \nabla_{\mathbb{S}} f(\theta). \tag{9.4}$$

Proof We may assume that f is defined and C^2-smooth on an open subset G of $\mathbb{R}^n$ containing the unit sphere. To compute the spherical gradient for the function

$$\psi(\theta) = \langle \nabla_{\mathbb{S}} f(\theta), v \rangle = \langle \nabla f(\theta), v \rangle - \langle \nabla f(\theta), \theta \rangle \, \langle v, \theta \rangle,$$

let us extend it smoothly to all points $x \in G$ by

$$\psi(x) = \langle \nabla f(x), v \rangle - \langle \nabla f(x), x \rangle \, \langle v, x \rangle$$

and write $\nabla_{\mathbb{S}} \psi(\theta) = \nabla \psi(\theta) - \langle \nabla \psi(\theta), \theta \rangle \, \theta$. By the Euclidean differentiation rules, it follows that

$$\begin{aligned} \nabla \psi(x) &= \nabla \, \langle \nabla f(x), v \rangle - \nabla \big(\langle \nabla f(x), x \rangle \, \langle v, x \rangle \big) \\ &= f''(x)v - \langle \nabla f(x), x \rangle \, v - \nabla \big(\langle \nabla f(x), x \rangle \big) \, \langle v, x \rangle \, . \end{aligned}$$

In addition, the function $u(x) = \langle \nabla f(x), x\rangle$ has gradient $\nabla u(x) = f''(x)x + \nabla f(x)$, so

$$\nabla\psi(x) = f''(x)v - \langle \nabla f(x), x\rangle\, v - \big(f''(x)x + \nabla f(x)\big)\, \langle v, x\rangle\,.$$

Restricting this equality to the unit sphere and using the notation $P = P_{\theta^\perp}$, we get

$$\nabla\psi(\theta) = f''(\theta)Pv - \langle \nabla f(\theta), \theta\rangle\, v - \langle v, \theta\rangle\, \nabla f(\theta),$$

which also implies

$$\langle \nabla\psi(\theta), \theta\rangle\, \theta = \langle f''(\theta)Pv, \theta\rangle\, \theta - \langle \nabla f(\theta), \theta\rangle\, \langle v, \theta\rangle\, \theta - \langle v, \theta\rangle\, \langle \nabla f(\theta), \theta\rangle\, \theta.$$

It remains to subtract this equality from the previous one. First note that

$$f''(\theta)Pv - \langle f''(\theta)Pv, \theta\rangle\, \theta = Pf''(\theta)Pv,$$

which is deduced from the general formula $Pw = w - \langle w, \theta\rangle\, \theta$ with $w = f''(\theta)Pv$. Since also $v - \langle v, \theta\rangle\, \theta = Pv$, we find that

$$\nabla f(\theta) - \langle \nabla f(\theta), \theta\rangle\, \theta = P\, \nabla f(\theta) = \nabla_{\mathbb{S}} f(\theta).$$

Therefore, using the matrix B from Proposition 9.1.2, the difference is exactly

$$\begin{aligned} Pf''(\theta)Pv - \langle \nabla f(\theta), \theta\rangle\, Pv - \langle v, \theta\rangle\, \nabla_{\mathbb{S}} f(\theta) &= PBPv - \langle v, \theta\rangle\, \nabla_{\mathbb{S}} f(\theta) \\ &= f''_{\mathbb{S}}(\theta)v - \langle v, \theta\rangle\, \nabla_{\mathbb{S}} f(\theta). \end{aligned}$$

Thus,

$$\nabla_{\mathbb{S}}\, \langle \nabla_{\mathbb{S}} f(\theta), v\rangle = f''_{\mathbb{S}}(\theta)v - \langle v, \theta\rangle\, \nabla_{\mathbb{S}} f(\theta),$$

which is the desired equality (9.4). □

9.2 Second Order Modulus of Gradient

According to the general equation (5.4), and since the geodesic and Euclidean distances are infinitesimally equivalent, the second order modulus of the gradient for functions f on the unit sphere is defined by

$$|\nabla^2_{\mathbb{S}} f(\theta)| = |\nabla_{\mathbb{S}}\, |\nabla_{\mathbb{S}} f(\theta)|\,| = \limsup_{\theta' \to \theta} \frac{|\,|\nabla_{\mathbb{S}} f(\theta)| - |\nabla_{\mathbb{S}} f(\theta')|\,|}{|\theta - \theta'|}, \quad \theta \in \mathbb{S}^{n-1}.$$

For example, the Euclidean derivatives of the linear function $f(x) = \langle v, x\rangle$ are just $\nabla f(x) = v$ and $f''(x) = 0$. As for the first and second order modulus of its spherical gradient, we have

$$|\nabla_{\mathbb{S}} f(\theta)| = \sqrt{|v|^2 - \langle v, \theta\rangle^2},$$

and, by the chain rule,

$$\begin{aligned}\nabla_{\mathbb{S}}|\nabla_{\mathbb{S}}f(\theta)| &= -\frac{1}{2\sqrt{|v|^2 - \langle v, \theta\rangle^2}}\,\nabla_{\mathbb{S}}\left(\langle v, \theta\rangle^2\right) \\ &= -\frac{1}{\sqrt{|v|^2 - \langle v, \theta\rangle^2}}\,\langle v, \theta\rangle\,\nabla_{\mathbb{S}}\,\langle v, \theta\rangle \qquad (\theta \neq v).\end{aligned}$$

Hence, $|\nabla^2_{\mathbb{S}} f(\theta)| = |\,\langle v, \theta\rangle\,|$ in contrast with $|\nabla^2 f(\theta)| = 0$.

To simplify the condition $|\nabla^2_{\mathbb{S}} f| \leq 1$, one may use the following equality, which is a full analog of the formula for the second order modulus of gradient on open regions in $\mathbb{R}^n$.

Proposition 9.2.1 *Given a C^2-smooth function f on $\mathbb{S}^{n-1}$, $|\nabla_S f|$ has a finite Lipschitz semi-norm and, for all $\theta \in \mathbb{S}^{n-1}$,*

$$|\nabla^2_{\mathbb{S}} f(\theta)| = |\nabla_{\mathbb{S}} f(\theta)|^{-1}\,|f''_{\mathbb{S}}(\theta)\nabla_{\mathbb{S}} f(\theta)|,$$

where the right-hand side should be understood as $\|f''_{\mathbb{S}}(\theta)\|$ in the case $|\nabla_{\mathbb{S}} f(\theta)| = 0$. In particular, in all cases

$$|\nabla^2_{\mathbb{S}} f(\theta)| \leq \|f''_{\mathbb{S}}(\theta)\|.$$

Proof First let us show that the function $|\nabla_{\mathbb{S}} f|$ has a finite Lipschitz semi-norm. Since the first two spherical derivatives of f are continuous and therefore bounded on the unit sphere, it follows from (9.4) that

$$|\nabla_{\mathbb{S}}\,\langle\nabla_{\mathbb{S}} f(\theta), v\rangle\,| \leq C, \quad |v| = 1,$$

for some constant C (independent of θ and v). Hence, the function $\theta \longrightarrow \langle\nabla_{\mathbb{S}} f(\theta), v\rangle$ has Lipschitz semi-norm at most C, so that, for all $\theta, \theta' \in \mathbb{S}^{n-1}$,

$$|\,\langle\nabla_{\mathbb{S}} f(\theta'), v\rangle - \langle\nabla_{\mathbb{S}} f(\theta), v\rangle\,| \leq C d(\theta', \theta)$$

(where we recall that d denotes the geodesic distance). Taking here the supremum over all unit vectors v and applying the triangle inequality, we get

$$\big|\,|\nabla_{\mathbb{S}} f(\theta')| - |\nabla_{\mathbb{S}} f(\theta)|\,\big| \leq |\nabla_{\mathbb{S}} f(\theta') - \nabla_{\mathbb{S}} f(\theta)| \leq C d(\theta', \theta),$$

which is the Lipschitz property (with constant C).

Next, to derive the required identity for the second order modulus of the gradient, we fix $\theta \in \mathbb{S}^{n-1}$ and apply the identity (9.4) once more. By the definition of the spherical gradient, it yields the Taylor expansion up to the linear term,

$$\langle\nabla_{\mathbb{S}} f(\theta'), v\rangle = \langle\nabla_{\mathbb{S}} f(\theta), v\rangle + \langle V, \theta' - \theta\rangle + o(|\theta' - \theta|)$$

as $\theta' \longrightarrow \theta$, $\theta' \in \mathbb{S}^{n-1}$, where

$$V = f''_{\mathbb{S}}(\theta)v - \langle v, \theta\rangle\,\nabla_{\mathbb{S}} f(\theta).$$

Moreover, by the Taylor formula in the integral form, and since any continuous function on a compact metric space is uniformly continuous, the o-term can be bounded by a quantity which is independent of $v \in \mathbb{S}^{n-1}$. That is,

$$\sup_{v \in \mathbb{S}^{n-1}} |\langle \nabla_{\mathbb{S}} f(\theta') - \nabla_{\mathbb{S}} f(\theta), v\rangle - \langle V, \theta' - \theta\rangle| \le \varepsilon(|\theta' - \theta|)$$

for some function $\varepsilon(t)$ such that $\varepsilon(t) \to 0$ as $t \to 0$.

Now, let us rewrite the Taylor formula as

$$\langle \nabla_{\mathbb{S}} f(\theta'), v\rangle = \langle \nabla_{\mathbb{S}} f(\theta) + L, v\rangle + o(|\theta' - \theta|),$$

where

$$\langle L, v\rangle = \langle V, \theta' - \theta\rangle = \left\langle f''_{\mathbb{S}}(\theta)v - \langle v, \theta\rangle \nabla_{\mathbb{S}} f(\theta), \theta' - \theta\right\rangle,$$

that is, with

$$L = f''_{\mathbb{S}}(\theta)(\theta' - \theta) - \langle \nabla_{\mathbb{S}} f(\theta), \theta' - \theta\rangle\, \theta.$$

Taking the absolute value of both sides in the Taylor formula and turning to the supremum over all $v \in \mathbb{S}^{n-1}$, we obtain that

$$|\nabla_{\mathbb{S}} f(\theta')| = |\nabla_{\mathbb{S}} f(\theta) + L| + o(|\theta' - \theta|).$$

Next, write

$$|\nabla_{\mathbb{S}} f(\theta) + L|^2 = |\nabla_{\mathbb{S}} f(\theta)|^2 + 2\langle \nabla_{\mathbb{S}} f(\theta), L\rangle + |L|^2.$$

Since $\nabla_{\mathbb{S}} f(\theta)$ is orthogonal to the vector θ, we have

$$\langle \nabla_{\mathbb{S}} f(\theta), L\rangle = \left\langle \nabla_{\mathbb{S}} f(\theta), f''_{S}(\theta)(\theta' - \theta)\right\rangle = \langle w, \theta' - \theta\rangle,$$

where $w = f''_{\mathbb{S}}(\theta)\nabla_{\mathbb{S}} f(\theta)$. Since also $|L|^2 = O(|\theta' - \theta|^2)$, we get

$$|\nabla_{\mathbb{S}} f(\theta) + L|^2 = |\nabla_{\mathbb{S}} f(\theta)|^2 + 2\langle w, \theta' - \theta\rangle + o(|\theta' - \theta|),$$

and therefore in the case $|\nabla_{\mathbb{S}} f(\theta)| > 0$,

$$|\nabla_{\mathbb{S}} f(\theta) + L| = |\nabla_{\mathbb{S}} f(\theta)| + |\nabla_{\mathbb{S}} f(\theta)|^{-1}\langle w, \theta' - \theta\rangle + o(|\theta' - \theta|).$$

As a result, we find that

$$|\nabla_{\mathbb{S}} f(\theta')| - |\nabla_{\mathbb{S}} f(\theta)| = |\nabla_{\mathbb{S}} f(\theta)|^{-1}\langle w, \theta' - \theta\rangle + o(|\theta' - \theta|)$$

and hence

$$\begin{aligned}\limsup_{\theta' \to \theta} \frac{\big||\nabla_{\mathbb{S}} f(\theta')| - |\nabla_{\mathbb{S}} f(\theta)|\big|}{|\theta' - \theta|} &= |\nabla_{\mathbb{S}} f(\theta)|^{-1} \limsup_{\theta' \to \theta} \frac{|\langle w, \theta' - \theta\rangle|}{|\theta' - \theta|} \\ &= |\nabla_{\mathbb{S}} f(\theta)|^{-1} P_{\theta^\perp} w.\end{aligned}$$

Thus,

$$|\nabla_{\mathbb{S}}^2 f(\theta)| = |\nabla_{\mathbb{S}} f(\theta)|^{-1} P_{\theta^\perp} w.$$

But, as was noted before, the vector w is always orthogonal to θ. Therefore, $P_{\theta^\perp} w = w$, and we arrive at the required identity

$$|\nabla_{\mathbb{S}}^2 f(\theta)| = |\nabla_{\mathbb{S}} f(\theta)|^{-1}\, w \quad \big(|\nabla_{\mathbb{S}} f(\theta)| > 0\big).$$

Finally, consider the remaining case $|\nabla_{\mathbb{S}} f(\theta)| = 0$. Then $L = f_{\mathbb{S}}''(\theta)(\theta' - \theta)$, and the Taylor formula is simplified and yields

$$|\nabla_{\mathbb{S}} f(\theta')| = |L| + o(|\theta' - \theta|).$$

Again, by the very definition, and using the orthogonality of $f_S''(\theta)h$ to θ, we obtain that

$$\begin{aligned} |\nabla_{\mathbb{S}}^2 f(\theta)| &= \limsup_{\theta' \to \theta} \frac{|\nabla_{\mathbb{S}} f(\theta')|}{|\theta' - \theta|} \\ &= \limsup_{\theta' \to \theta} \frac{|f_{\mathbb{S}}''(\theta)(\theta' - \theta)|}{|\theta' - \theta|} \;=\; \limsup_{h \to 0,\ h \in \theta^\perp} \frac{|f_{\mathbb{S}}''(\theta)h|}{|h|} \;=\; \|f_{\mathbb{S}}''(\theta)\|. \end{aligned}$$

Thus, Proposition 9.2.1 is proved. □

9.3 Spherical Laplacian

We denote by $\mathrm{Tr}A = \sum_{i=1}^n a_{ii}$ the trace of an $n \times n$ matrix A with entries a_{ij}.

The Laplacian operator is defined by the equality

$$\Delta_{\mathbb{S}} f = \mathrm{Tr}\, f_{\mathbb{S}}''.$$

Acting on the class of all C^2-smooth functions f on $\mathbb{S}^{n-1}$, it can be related to the "spherical partial derivatives" $D_i f(\theta) = \langle \nabla_{\mathbb{S}} f(\theta), e_i \rangle$, where $e_1, \dots, e_n$ is the canonical basis in $\mathbb{R}^n$. Thus,

$$\nabla_{\mathbb{S}} f(\theta) = \sum_{i=1}^n D_i f(\theta)\, e_i.$$

As further partial derivatives, consider the "second order" differential operators

$$D_{ij} f = D_i(D_j f) = \big\langle \nabla_{\mathbb{S}} \big\langle \nabla_{\mathbb{S}} f, e_j \big\rangle, e_i \big\rangle, \quad i, j = 1, \dots, n.$$

Proposition 9.3.1 $\Delta_{\mathbb{S}} = \sum_{i=1}^n D_{ii}$.

In fact, any orthonormal basis in $\mathbb{R}^n$ could be used in place of the e_i's in the definition of D_{ii}, and the above statement will continue to hold. To prove the claim, we may appeal to Proposition 9.1.3, which tells us that the identity

$$\nabla_{\mathbb{S}} \langle \nabla_{\mathbb{S}} f(\theta), v \rangle = f''_{\mathbb{S}}(\theta) v - \langle v, \theta \rangle \nabla_{\mathbb{S}} f(\theta)$$

holds true for all $v \in \mathbb{R}^n$. Hence

$$D_{ii} f(\theta) = \left\langle f''_{\mathbb{S}}(\theta) e_i, e_i \right\rangle - \langle \theta, e_i \rangle \langle \nabla_{\mathbb{S}} f(\theta), e_i \rangle$$

and thus

$$\sum_{i=1}^{n} D_{ii} f(\theta) = \operatorname{Tr} f''_{\mathbb{S}}(\theta) - \langle \nabla_{\mathbb{S}} f(\theta), \theta \rangle = \operatorname{Tr} f''_{\mathbb{S}}(\theta).$$

Let us also indicate how one may evaluate the Laplacian of a given function in terms of the Euclidean derivatives.

Proposition 9.3.2 *If f is C^2-smooth in an open region G containing the unit sphere, then for any $\theta \in \mathbb{S}^{n-1}$,*

$$\Delta_{\mathbb{S}} f(\theta) = \Delta f(\theta) - (n-1) \langle \nabla f(\theta), \theta \rangle - \langle f''(\theta) \theta, \theta \rangle. \tag{9.5}$$

Proof Equation (9.5) can be obtained from the explicit formula

$$\begin{aligned} D_{ij} f(\theta) = \; & \partial_{ij} f(\theta) - \theta_j \, \partial_i f(\theta) - \delta_{ij} \langle \nabla f(\theta), \theta \rangle + 2\theta_i \theta_j \langle \nabla f(\theta), \theta \rangle \\ & - \theta_j \langle f''(\theta) \theta, e_i \rangle - \theta_i \left\langle f''(\theta) \theta, e_j \right\rangle + \theta_i \theta_j \langle f''(\theta) \theta, \theta \rangle. \end{aligned} \tag{9.6}$$

In particular,

$$\begin{aligned} D_{ii} f(\theta) = \; & \partial_{ii} f(\theta) - \langle \nabla f(\theta), \theta \rangle - \theta_i \, \partial_i f(\theta) + 2\theta_i^2 \langle \nabla f(\theta), \theta \rangle \\ & - 2\theta_i \langle f''(\theta) \theta, e_i \rangle + \theta_i^2 \langle f''(\theta) \theta, \theta \rangle. \end{aligned}$$

Summing the latter equality over all $i \le n$, we arrive at (9.5).

Note that the operators D_i and D_j are not commutative, i.e. we do not have the identity $D_{ij} f = D_{ji} f$ in the entire class C^2. Indeed, by (9.6), $D_{ij} f(\theta) = D_{ji} f(\theta)$ if and only if $\theta_i \, \partial_j f(\theta) = \theta_j \, \partial_i f(\theta)$.

To derive (9.6), fix $j \le n$ and consider the smooth function in n real variables

$$u(x) = \left\langle \nabla f(x), e_j \right\rangle - \langle \nabla f(x), x \rangle \left\langle x, e_j \right\rangle = \partial_j f(x) - x_j \sum_{k=1}^{n} x_k \, \partial_k f(x)$$

for $x = (x_1, \dots, x_n) \in G$. In particular, $u(\theta) = D_j f(\theta)$ for $\theta \in \mathbb{S}^{n-1}$ and therefore $D_{ij} f = D_i u$. Again, following the definition of D_i, we have

$$D_i u(x) = \partial_i u(x) - x_i \sum_{l=1}^{n} x_l \, \partial_l u(x).$$

Now, by direct differentiation,

$$\partial_i u(x) = \partial_{ij} f(x) - \delta_{ij} \sum_{k=1}^{n} x_k\, \partial_k f(x) - x_j\, \partial_i f(x) - x_j \sum_{k=1}^{n} x_k\, \partial_{ik} f(x),$$
$$\partial_l u(x) = \partial_{lj} f(x) - \delta_{lj} \sum_{k=1}^{n} x_k\, \partial_k f(x) - x_j\, \partial_l f(x) - x_j \sum_{k=1}^{n} x_k\, \partial_{lk} f(x),$$

so that

$$\begin{aligned} D_i u(x) &= \partial_{ij} f(x) - \delta_{ij} \sum_{k=1}^{n} x_k\, \partial_k f(x) - x_j\, \partial_i f(x) - x_j \sum_{k=1}^{n} x_k\, \partial_{ik} f(x) \\ &\quad -x_i \sum_{l=1}^{n} x_l \left[\partial_{lj} f(x) - \delta_{lj} \sum_{k=1}^{n} x_k\, \partial_k f(x) - x_j\, \partial_l f(x) - x_j \sum_{k=1}^{n} x_k\, \partial_{lk} f(x)\right] \\ &= \partial_{ij} f(x) - \delta_{ij} \sum_{k=1}^{n} x_k\, \partial_k f(x) - x_j\, \partial_i f(x) - x_j \sum_{k=1}^{n} x_k\, \partial_{ik} f(x) \\ &\quad -x_i \sum_{l=1}^{n} x_l\, \partial_{lj} f(x) + 2x_i x_j \sum_{k=1}^{n} x_k\, \partial_k f(x) + x_i x_j \sum_{l=1}^{n}\sum_{k=1}^{n} x_l x_k\, \partial_{lk} f(x). \end{aligned}$$

In a slightly more compact form,

$$\begin{aligned} D_i u(x) = \partial_{ij} f(x) - x_j\, \partial_i f(x) - \delta_{ij}\, \langle \nabla f(x), x\rangle + 2x_i x_j\, \langle \nabla f(x), x\rangle \\ -x_j\, \langle f''(x)x, e_i\rangle - x_i\, \langle f''(x)x, e_j\rangle + x_i x_j\, \langle f''(x)x, x\rangle\,. \end{aligned}$$

It remains to restrict this function to the sphere. Proposition 9.3.2 is proved. □

It is well known (cf. e.g. [168]) that the Hilbert space $L^2(\mathbb{S}^{n-1}, \mathfrak{s}_{n-1})$ can be decomposed into the sum of orthogonal linear subspaces H_d, $d = 0, 1, 2, \dots$, consisting of all d-homogeneous harmonic polynomials (more precisely – restrictions of such polynomials to the sphere). In particular, H_0 is the space of constant functions, H_1 is the space of linear functions, H_2 consists of quadratic harmonics, and so on. Any element f_d of H_d represents an eigenfunction of the Laplacian with the eigenvalue $-d(n+d-2)$. That is, any function f in $L^2(\mathbb{S}^{n-1}, \mathfrak{s}_{n-1})$ admits a unique orthogonal (Fourier) expansion in spherical harmonics,

$$f = \sum_{d=0}^{\infty} f_d \quad (f_d \in H_d), \tag{9.7}$$

and $\Delta_{\mathbb{S}} f_d = -d(n+d-2)\, f_d$. Hence, if f is C^2-smooth, we get another representation for the Laplacian,

$$\Delta_{\mathbb{S}} f = -\sum_{d=1}^{\infty} d(n+d-2) f_d \tag{9.8}$$

which should be understood as equality in L^2. The Laplacian may also be introduced through the classical formula for spherical integration by parts,

$$\int \langle \nabla_{\mathbb{S}} f, \nabla_{\mathbb{S}} g\rangle \, d\mathfrak{s}_{n-1} = -\int f \Delta_{\mathbb{S}} g \, d\mathfrak{s}_{n-1}, \tag{9.9}$$

which may be taken as an equivalent definition of $\Delta_{\mathbb{S}}$. It yields the following characterization of the orthogonality to linear functions in terms of the spherical gradient.

Proposition 9.3.3 *For any C^1-smooth function f on $\mathbb{S}^{n-1}$,*

$$\int f(\theta)\theta \, d\mathfrak{s}_{n-1}(\theta) = \frac{1}{n-1} \int \nabla_{\mathbb{S}} f(\theta) \, d\mathfrak{s}_{n-1}(\theta).$$

In particular, f is orthogonal to all linear functions in $L^2(\mathbb{S}^{n-1})$ if and only if the linear forms $\langle \nabla_{\mathbb{S}} f(\theta), v\rangle$ have $\mathfrak{s}_{n-1}$-mean zero for any $v \in \mathbb{R}^n$.

Proof The linear function $g(\theta) = \langle v, \theta\rangle$ has spherical gradient and respectively spherical Laplacian

$$\nabla_{\mathbb{S}} g(\theta) = P_{\theta^\perp} v, \quad \Delta_{\mathbb{S}} g(\theta) = -(n-1)\langle v, \theta\rangle.$$

In this case, (9.9) becomes

$$\int \langle \nabla_{\mathbb{S}} f(\theta), P_{\theta^\perp} v\rangle \, d\mathfrak{s}_{n-1}(\theta) = (n-1) \int f(\theta) \langle v, \theta\rangle \, d\mathfrak{s}_{n-1}(\theta).$$

But $\nabla_{\mathbb{S}} f(\theta)$ is orthogonal to θ, so, $\langle \nabla_{\mathbb{S}} f(\theta), P_{\theta^\perp} v\rangle = \langle \nabla_{\mathbb{S}} f(\theta), v\rangle$. It follows that

$$\int \langle \nabla_{\mathbb{S}} f, v\rangle \, d\mathfrak{s}_{n-1} = (n-1) \int f(\theta) \langle v, \theta\rangle \, d\mathfrak{s}_{n-1}(\theta). \qquad \square$$

9.4 Poincaré and Logarithmic Sobolev Inequalities

Two fundamental tools in the analysis of functions on the sphere are the classical Poincaré inequality and the stronger logarithmic Sobolev inequality. Let us start with the weaker one.

Proposition 9.4.1 *For any Lipschitz function f on $\mathbb{S}^{n-1}$ with $\int f \, d\mathfrak{s}_{n-1} = 0$,*

$$\int |f|^2 \, d\mathfrak{s}_{n-1} \leq \frac{1}{n-1} \int |\nabla_{\mathbb{S}} f|^2 \, d\mathfrak{s}_{n-1}. \tag{9.10}$$

This inequality remains valid when $f = u + iv$ is complex-valued with the usual convention $|f|^2 = u^2 + v^2$ and $|\nabla_{\mathbb{S}} f|^2 = |\nabla_{\mathbb{S}} u|^2 + |\nabla_{\mathbb{S}} v|^2$ ($u = \mathrm{Re}(f), v = \mathrm{Im}(f)$).

For the (first) proof, one may use the orthogonal expansion (9.7) over spherical harmonics with $f_0 = 0$ and apply (9.8)–(9.9). Indeed, we have Parseval's identity

$$\|f\|^2_{L^2(\mathfrak{s}_{n-1})} = \sum_{d=1}^{\infty} \|f_d\|^2_{L^2(\mathfrak{s}_{n-1})},$$

while

$$\int |\nabla_{\mathbb{S}} f|^2 \, d\mathfrak{s}_{n-1} = -\int f \Delta_{\mathbb{S}} f \, d\mathfrak{s}_{n-1} = \sum_{d=1}^{\infty} d(n+d-2) \, \|f_d\|^2_{L^2(\mathfrak{s}_{n-1})}$$

in which $d(n+d-2) \geq n-1$. Equality in (9.10) is thus attained for all linear functions $f(\theta) = \langle \theta, v \rangle$, i.e., when $f_d = 0$ for all $d \geq 2$, and only in this case. Hence the constant $\frac{1}{n-1}$ is optimal, i.e. $\lambda_1(\mathfrak{s}_{n-1}) = n-1$.

This argument also shows that, if additionally f is orthogonal to all linear functions, then we get a slight improvement over (9.10), namely

$$\int f^2 \, d\mathfrak{s}_{n-1} \leq \frac{1}{2n} \int |\nabla_{\mathbb{S}} f|^2 \, d\mathfrak{s}_{n-1}. \tag{9.11}$$

In this case $f_0 = f_1 = 0$, while $d(n+d-2) \geq 2n$ for all $d \geq 2$.

Inequality (9.10) may also be derived from the Poincaré-type inequality

$$\mathrm{Var}_{\gamma_n}(u) \leq \int u^2 \, d\gamma_n$$

for the standard Gaussian measure γ_n on $\mathbb{R}^n$ with density $\frac{d\gamma_n(x)}{dx} = (2\pi)^{-\frac{n}{2}} \, e^{-|x|^2/2}$ (see item 5.7.4 and Section 5.4 about the more general Brascamp–Lieb inequality, Proposition 6.5.1). Namely, suppose that f is defined and smooth in some open neighborhood of $\mathbb{S}^{n-1}$ and has $\mathfrak{s}_{n-1}$-mean zero. We consider a smooth extension of f from the sphere to the whole space $\mathbb{R}^n$ using the equality

$$u(x) = r f(r^{-1}x), \quad \text{where } r = |x| = \sqrt{x_1^2 + \cdots + x_n^2}, \ \ x \in \mathbb{R}^n \ (x \neq 0),$$

and extending it to the origin as $u(0) = 0$. Denote by Z a standard normal random vector in $\mathbb{R}^n$, i.e., with distribution γ_n. Integration in polar coordinates gives

$$\int_{\mathbb{R}^n} u(x) \, d\gamma_n(x) = \mathbb{E}\,|Z| \int_{\mathbb{S}^{n-1}} f(\theta) \, d\mathfrak{s}_{n-1}(\theta) = 0,$$

$$\int_{\mathbb{R}^n} u(x)^2 \, d\gamma_n(x) = \mathbb{E}\,|Z|^2 \int_{\mathbb{S}^{n-1}} f(\theta)^2 \, d\mathfrak{s}_{n-1}(\theta) = n \int_{\mathbb{S}^{n-1}} f(\theta)^2 \, d\mathfrak{s}_{n-1}(\theta).$$

As is easy to check, we also have, for any $x \neq 0$,

$$\nabla u(x) \ = \ f(\theta)\,\theta + \nabla f(\theta) - \langle \nabla f(\theta), \theta \rangle \, \theta = f(\theta)\,\theta + \nabla_{\mathbb{S}} f(\theta),$$

so that

$$|\nabla u(x)|^2 = f(\theta)^2 + |\nabla_{\mathbb{S}} f(\theta)|^2 = f(\theta)^2 + |\nabla f(\theta)|^2 - \langle \nabla f(\theta), \theta \rangle^2 .$$

The last expression is bounded by a constant, which shows that f is Lipschitz on $\mathbb{R}^n$. Applying the Gaussian Poincaré-type inequality, we therefore obtain

$$\begin{aligned}\int_{\mathbb{S}^{n-1}} f(\theta)^2\,d\mathfrak{s}_{n-1}(\theta) + \int_{\mathbb{S}^{n-1}} |\nabla_{\mathbb{S}} f(\theta)|^2\,d\mathfrak{s}_{n-1}(\theta) &= \int_{\mathbb{R}^n} |\nabla u(x)|^2\,d\gamma_n(x)\\ &\geq \int_{\mathbb{R}^n} u(x)^2\,d\gamma_n(x) = n\int_{\mathbb{S}^{n-1}} f(\theta)^2\,d\mathfrak{s}_{n-1}(\theta),\end{aligned}$$

which is exactly (9.10).

A similar argument may also be used to derive a logarithmic Sobolev inequality on the sphere (although with a suboptimal constant) – on the basis of the logarithmic Sobolev inequality for the Euclidean space,

$$\mathrm{Ent}_{\gamma_n}(u^2) \leq 2\int |\nabla u|^2\,d\gamma_n.$$

It holds for any Lipschitz function u on $\mathbb{R}^n$, which may be weakened to the condition that u is Lipschitz outside any (small) neighborhood of the origin. Starting again with a smooth f on $\mathbb{S}^{n-1}$, we now consider the classical extension $u(x) = f(r^{-1}x)$, so that $\nabla u(x) = r^{-1}\,\nabla_{\mathbb{S}} f(\theta)$ for $x = r\theta$ $(r > 0)$. Integrating in polar coordinates, the Gaussian log-Sobolev inequality then leads to the relation

$$\mathrm{Ent}_{\mathfrak{s}_{n-1}}(f^2) \leq 2\,\mathbb{E}\,\frac{1}{|Z|^2}\int_{\mathbb{S}^{n-1}} |\nabla_{\mathbb{S}} f(\theta)|^2\,d\mathfrak{s}_{n-1}(\theta),$$

in which the expectation is finite for $n \geq 3$ and is asymptotically $1/n$. More precisely, the general formula

$$\mathbb{E}\,v(|Z|) = \frac{1}{(2\pi)^{\frac{n}{2}}}\int_{\mathbb{R}^n} v(|x|)\,e^{-|x|^2/2}\,dx = \frac{n\omega_n}{(2\pi)^{\frac{n}{2}}}\int_0^\infty v(r)\,r^{n-1}\,e^{-r^2/2}\,dr, \quad (9.12)$$

where ω_n denotes the volume of the unit ball in $\mathbb{R}^n$, gives, integrating by parts,

$$1 = \frac{n\omega_n}{(2\pi)^{\frac{n}{2}}}\int_0^\infty r^{n-1}\,e^{-r^2/2}\,dr = (n-2)\,\frac{n\omega_n}{(2\pi)^{\frac{n}{2}}}\int_0^\infty r^{n-3}\,e^{-r^2/2}\,dr = (n-2)\,\mathbb{E}\,\frac{1}{|Z|^2}.$$

Hence $\mathbb{E}\,\frac{1}{|Z|^2} = \frac{1}{n-2}$, and thus

$$\mathrm{Ent}_{\mathfrak{s}_{n-1}}(f^2) \leq \frac{2}{n-2}\int_{\mathbb{S}^{n-1}} |\nabla_{\mathbb{S}} f|^2\,d\mathfrak{s}_{n-1} \quad (n \geq 3).$$

This relation is sufficient for further applications, since it shows a correct asymptotic behavior of the constant with respect to the growing dimension n. Nevertheless, as we already mentioned in Section 7.2, let us recall a more precise and deeper result of Mueller and Weissler [149]: the (optimal) logarithmic Sobolev constant for the unit sphere coincides with the spectral gap.

Proposition 9.4.2 *For any Lipschitz function $f : \mathbb{S}^{n-1} \to \mathbb{R}$,*

$$\mathrm{Ent}_{\mathfrak{s}_{n-1}}(f^2) \le \frac{2}{n-1} \int |\nabla_{\mathbb{S}} f|^2 \, d\mathfrak{s}_{n-1}.$$

This result may be used to study the second order concentration phenomenon in the class of smooth functions f on the sphere, which we discuss in the next chapter. But, it is also important to control deviation bounds for f without smoothness and even without the continuity assumption. As we know, this can be done by virtue of the supremum- and infimum-convolution operators

$$P_t f(\theta) = \sup_{\theta' \in \mathbb{R}^n} \Big[f(\theta') - \frac{1}{2t} |\theta - \theta'|^2 \Big],$$

$$Q_t f(\theta) = \inf_{\theta' \in \mathbb{R}^n} \Big[f(\theta') + \frac{1}{2t} |\theta - \theta'|^2 \Big] \qquad (\theta \in \mathbb{R}^n, \ t > 0),$$

where for simplicity f is defined on the whole Euclidean space (cf. Definition 8.1.1). In particular, the general Proposition 8.4.1 in the form (8.11) yields:

Corollary 9.4.3 *Given an $\mathfrak{s}_{n-1}$-integrable function f on $\mathbb{R}^n$, for any $t > 0$,*

$$\int \exp\{tf\} \, d\mathfrak{s}_{n-1} \le \exp\Big\{ t \int P_{\frac{t}{n-1}} f \, d\mathfrak{s}_{n-1} \Big\},$$

$$\int \exp\big\{ tQ_{\frac{t}{n-1}} f \big\} \, d\mathfrak{s}_{n-1} \le \exp\Big\{ t \int f \, d\mathfrak{s}_{n-1} \Big\}.$$

9.5 Isoperimetric and Cheeger-type Inequalities

Up to an absolute factor, the Poincaré inequality on the sphere may also be sharpened by using the L^1-norms on both sides. In particular, we have:

Proposition 9.5.1 *For any Lipschitz function $f : \mathbb{S}^{n-1} \to \mathbb{R}$ with $\mathfrak{s}_{n-1}$-mean zero,*

$$\int |f| \, d\mathfrak{s}_{n-1} \le \frac{\pi}{\sqrt{2n}} \int |\nabla_{\mathbb{S}} f| \, d\mathfrak{s}_{n-1}. \tag{9.13}$$

Recall that inequalities such as (9.13) are in some sense stronger than Poincaré-type inequalities. For example, by Cheeger's inequality $\lambda_1 \ge \frac{1}{4} h^2$ and Corollary 5.6.2, (9.13) yields a spherical Poincaré inequality with an additional absolute factor. As we know, (9.13) is related to the isoperimetric problem. In case of the sphere, according to a theorem due to P. Lévy and E. Schmidt, the solution to this problem is provided by the balls (cups)

$$B = B(x) = \{\theta \in \mathbb{S}^{n-1} : \theta_1 \le x\}, \quad -1 < x < 1.$$

We accept this theorem without proof and refer the interested reader to [145], [89]. Note that any ε-neighborhood of $B(x)$ represents a ball on the sphere as well.

Proposition 9.5.2 *Given $t \in (0,1)$ and $\varepsilon > 0$, among all Borel sets $A \subset \mathbb{S}^{n-1}$ of measure $\mathfrak{s}_{n-1}(A) = t$, the measure $\mathfrak{s}_{n-1}(A_\varepsilon)$ is minimized for the ball $A = B$. Equivalently, if $\mathfrak{s}_{n-1}(A) = \mathfrak{s}_{n-1}(B)$, then*

$$\mathfrak{s}_{n-1}(A_\varepsilon) \geq \mathfrak{s}_{n-1}(B_\varepsilon), \quad \mathfrak{s}_{n-1}^+(A) \geq \mathfrak{s}_{n-1}^+(B).$$

By a simple integration argument, the last two inequalities are equivalent to each other. Here, it does not matter whether we consider ε-neighborhoods and $\mathfrak{s}_{n-1}$-perimeters of sets with respect to the geodesic or Euclidean distances on $\mathbb{S}^{n-1}$.

To restate Proposition 9.5.2 analytically (rather than geometrically by indicating extremal sets), one should introduce the distribution function $F_n(x) = \mathfrak{s}_{n-1}(B(x))$ of the first coordinate θ_1 of a point θ on the sphere, viewed as a random variable on the probability space $(\mathbb{S}^{n-1}, \mathfrak{s}_{n-1})$. By direct computation, it has density

$$p_n(x) = c_n \left(1 - x^2\right)_+^{\frac{n-3}{2}}, \quad x \in \mathbb{R},$$

with respect to the Lebesgue measure on the real line, where

$$c_n = \frac{\Gamma(\frac{n}{2})}{\sqrt{\pi}\,\Gamma(\frac{n-1}{2})}$$

is a normalizing constant. For example, if $n = 3$, then $c_3 = \frac{1}{2}$, and we deal with the uniform distribution on the interval $[-1, 1]$.

The equation $\mathfrak{s}_{n-1}(B(x)) = t$ is solved as $x = F_n^{-1}(t)$, where $F_n^{-1} : [0,1] \to [0,1]$ is the inverse of the distribution function. Since also $\mathfrak{s}_{n-1}^+(B(x)) = p_n(x)$, the associated isoperimetric profile is given by

$$I_{\mathfrak{s}_{n-1}}(t) = \mathfrak{s}_{n-1}^+(B(x)) = p_n(F_n^{-1}(t)), \quad 0 < t < 1.$$

Similarly to the Gaussian case, this function is concave on $(0,1)$ and symmetric around $t = \frac{1}{2}$. So, $I_{\mathfrak{s}_{n-1}}(t)/\min\{t, 1-t\}$ is minimized at this point, and thus the Cheeger isoperimetric constant for the sphere may be evaluated as

$$h(\mathfrak{s}_{n-1}) = 2I_{\mathfrak{s}_{n-1}}(1/2) = 2c_n.$$

One can check that the ratio $I_{\mathfrak{s}_{n-1}}(t)/(2t(1-t))$ is minimized for $t = 1/2$ as well (cf. [49]). Therefore, according to Corollary 5.6.2, the inequality

$$\int |f| \, d\mathfrak{s}_{n-1} \leq \frac{1}{2c_n} \int |\nabla_{\mathbb{S}} f| \, d\mathfrak{s}_{n-1} \tag{9.14}$$

holds true for any Lipschitz function f on $\mathbb{S}^{n-1}$ under the assumption that $m(f) = 0$ or $\int f \, d\mathfrak{s}_{n-1} = 0$. In both cases the constant $1/(2c_n)$ is optimal.

To simplify the dependence on n and prove (9.13), let us look at the sequence $c'_n = c_n/\sqrt{n}$. The first values are $c'_2 = \frac{1}{\pi\sqrt{2}} = 0.225\dots$, $c'_3 = \frac{1}{2\sqrt{3}} = 0.289\dots$, $c'_4 = \frac{1}{\pi} = 0.318\dots$ Clearly, $c'_n \to \frac{1}{\sqrt{2\pi}} = 0.399\dots$ as $n \to \infty$, and as a routine analysis shows, c'_n is minimized for $n = 2$, so that

$$\frac{1}{2c_n} \leq \frac{1}{c_2\sqrt{2n}} = \frac{\pi}{\sqrt{2n}}.$$

Thus, the inequality (9.14) leads to (9.13).

As an alternative approach to Proposition 9.5.1 not appealing to the isoperimetric theorem, one may also start with the L^1-Poincaré-type inequality for the Gaussian measure γ_n on $\mathbb{R}^n$, cf. Proposition 5.8.2, inequality (5.21), namely,

$$\int |\nabla u(x)|\,d\gamma_n(x) \leq \sqrt{\frac{\pi}{2}} \int |\nabla u(x)|\,d\gamma_n(x) \quad \Big(\int u(x)\,d\gamma_n(x) = 0\Big).$$

Let us repeat the argument used at the end of the previous section when deriving a logarithmic Sobolev inequality on the sphere with a suboptimal constant. Given a smooth function f on $\mathbb{S}^{n-1}$ with $\mathfrak{s}_{n-1}$-mean zero, define $u(x) = f(r^{-1}x)$, so that $\nabla u(x) = r^{-1}\,\nabla_{\mathbb{S}} f(\theta)$ for $x = r\theta$ ($r > 0$). Integrating in polar coordinates, we get

$$\int |f(\theta)|\,d\mathfrak{s}_{n-1}(\theta) \leq C_n \int |\nabla_{\mathbb{S}} f(\theta)|\,d\mathfrak{s}_{n-1}(\theta), \quad C_n = \sqrt{\frac{\pi}{2}}\,\mathbb{E}\,\frac{1}{|Z|},$$

where $Z \sim \gamma_n$. From (9.12), we find that $\mathbb{E}\,\frac{1}{|Z|} = \frac{1}{n-1}\,\mathbb{E}\,|Z| \leq \frac{1}{n-1}\sqrt{n}$, so that $C_n \leq \frac{1}{n-1}\sqrt{\frac{\pi n}{2}}$, which is better than the constant $\frac{\pi}{\sqrt{2n}}$ for $n \geq 3$. Also, in the case $n = 2$, we have $\mathbb{E}\,\frac{1}{|Z|} = \frac{1}{2}\sqrt{2\pi}$. Hence $C_2 = \frac{\pi}{2}$, the same as in (9.13).

9.6 Remarks

The material on spherical derivatives is rather standard. More details, including proofs, can be found in the extended ArXiv version of the paper [36].

Chapter 10
Second Order Spherical Concentration

In this chapter we discuss natural conditions imposed on the functions f on the unit sphere which guarantee smaller deviations of f from their means with respect to growing dimension n in comparison with deviations that are valid for the entire class of Lipschitz functions. These conditions involve derivatives of f of the second order, which may be considered both in the spherical and Euclidean setup.

10.1 Second Order Poincaré-type Inequalities

Let us return to the Poincaré inequality

$$\int f^2 \, d\mathfrak{s}_{n-1} \leq \frac{1}{n-1} \int |\nabla_{\mathbb{S}} f|^2 \, d\mathfrak{s}_{n-1} \quad \Big(\int f \, d\mathfrak{s}_{n-1} = 0\Big). \tag{10.1}$$

As we have already mentioned, if f is orthogonal to all linear functions in $L^2(\mathfrak{s}_{n-1})$, the constant $\frac{1}{n-1}$ can be improved to $\frac{1}{2n}$, cf. (9.11). Moreover, one may apply this inequality once more to bound the right-hand side of (10.1) by involving the second derivative. Indeed, one may write

$$\int |\nabla_{\mathbb{S}} f|^2 \, d\mathfrak{s}_{n-1} = n \int \bigg[\int \langle \nabla_{\mathbb{S}} f(\theta), v\rangle^2 \, d\mathfrak{s}_{n-1}(\theta)\bigg] \, d\mathfrak{s}_{n-1}(v)$$

and use the property that the function $g(\theta) = \langle \nabla_{\mathbb{S}} f(\theta), v\rangle$ has $\mathfrak{s}_{n-1}$-mean zero for any $v \in \mathbb{R}^n$ (Proposition 9.3.3). Using the identity (9.4) of Proposition 9.1.3 and applying (10.1) to g, we then get

S. Bobkov et al., *Concentration and Gaussian Approximation for Randomized Sums*, Probability Theory and Stochastic Modelling 104,
https://doi.org/10.1007/978-3-031-31149-9_10

$$
\begin{aligned}
(n-1)\int g^2\,d\mathfrak{s}_{n-1} &\le \int |f_{\mathbb{S}}''(\theta)\,v - \langle v,\theta\rangle\,\nabla_{\mathbb{S}} f(\theta)|^2\,d\mathfrak{s}_{n-1}(\theta)\\
&= \int |f_{\mathbb{S}}''(\theta)v|^2\,d\mathfrak{s}_{n-1}(\theta) + \int \langle v,\theta\rangle^2\,|\nabla_{\mathbb{S}} f(\theta)|^2\,d\mathfrak{s}_{n-1}(\theta)\\
&\quad - 2\int \big\langle f_{\mathbb{S}}''(\theta)\nabla_{\mathbb{S}} f(\theta), v\big\rangle\,\langle v,\theta\rangle\,d\mathfrak{s}_{n-1}(\theta).
\end{aligned}
$$

The next integration over $d\mathfrak{s}_{n-1}(v)$ cancels the last integral (due to $f_{\mathbb{S}}''(\theta)\theta = 0$), and we are led to

$$
(n-2)\int |\nabla_{\mathbb{S}} f(\theta)|^2\,d\mathfrak{s}_{n-1}(\theta) \le \int \|f_{\mathbb{S}}''(\theta)\|_{\mathrm{HS}}^2\,d\mathfrak{s}_{n-1}(\theta).
$$

If additionally f has mean zero, we arrive at

$$
\int f^2\,d\mathfrak{s}_{n-1} \le \frac{1}{2n(n-2)}\int \|f_{\mathbb{S}}''\|_{\mathrm{HS}}^2\,d\mathfrak{s}_{n-1}, \qquad n \ge 3.
$$

The constant in this second order Poincaré-type inequality is actually slightly worse than the optimal one, and in particular we lose information about extremal functions. To reach an optimal one, one may use the following identity, which can be obtained from the Bochner–Lichnerowicz formula in Riemannian Geometry and using the property that the unit sphere $\mathbb{S}^{n-1}$ in $\mathbb{R}^n$ has a constant curvature. (For a direct derivation, see also [36].)

Lemma 10.1.1 *For any C^4-smooth function f on $\mathbb{S}^{n-1}$,*

$$
\int \|f_{\mathbb{S}}''\|_{\mathrm{HS}}^2\,d\mathfrak{s}_{n-1} = \int f\left(\Delta_{\mathbb{S}}^2 f + (n-2)\Delta_{\mathbb{S}} f\right)d\mathfrak{s}_{n-1}, \tag{10.2}
$$

where $\Delta_{\mathbb{S}}^2 f = \Delta_{\mathbb{S}}(\Delta_{\mathbb{S}} f)$ is the squared Laplacian in the sense of superposition.

This identity can be used to properly bound the right-hand side of (10.1).

Proposition 10.1.2 *For any C^2-function f on $\mathbb{S}^{n-1}$,*

$$
\int |\nabla_{\mathbb{S}} f|^2\,d\mathfrak{s}_{n-1} \le \int \|f_{\mathbb{S}}''\|_{\mathrm{HS}}^2\,d\mathfrak{s}_{n-1},
$$

where equality is attained for all linear functions. Moreover, if f is orthogonal to all linear functions with respect to $\mathfrak{s}_{n-1}$, then

$$
\int |\nabla_{\mathbb{S}} f|^2\,d\mathfrak{s}_{n-1} \le \frac{1}{n+2}\int \|f_{\mathbb{S}}''\|_{\mathrm{HS}}^2\,d\mathfrak{s}_{n-1},
$$

with equality attainable for all quadratic harmonics.

Proof According to the integration by parts formula (9.9) and using (10.2), the relation of the form

$$c \int |\nabla_{\mathbb{S}} f|^2 \, d\mathfrak{s}_{n-1} \leq \int \|f''_{\mathbb{S}}\|_{\mathrm{HS}}^2 \, d\mathfrak{s}_{n-1} \quad (c > 0)$$

is equivalent to

$$\int f \, \Delta_{\mathbb{S}}^2 f \, d\mathfrak{s}_{n-1} \geq -(c + n - 2) \int f \, \Delta_{\mathbb{S}} f \, d\mathfrak{s}_{n-1}. \tag{10.3}$$

The latter can be studied by means of the orthogonal expansions (9.7) and (9.8) in spherical harmonics. Indeed, since $\Delta_{\mathbb{S}} f_d = -d(n + d - 2) \, f_d$ and hence $\Delta_{\mathbb{S}}^2 f_d = d^2(n + d - 2)^2 \, f_d$, we have the orthogonal expansions

$$\Delta_{\mathbb{S}} f = -\sum_{d=1}^{\infty} d(n + d - 2) f_d, \quad \Delta_{\mathbb{S}}^2 f = \sum_{d=1}^{\infty} d^2(n + d - 2)^2 f_d.$$

As a result, (10.3) is reduced to the comparison $d^2(n+d-2)^2 \geq (c+n-2) \cdot d(n+d-2)$, that is, to the requirement

$$c \leq d^2 + (d - 1)(n - 2).$$

Thus, if we want to involve all C^2-smooth functions f, the optimal value of c is described as the minimum of the right-hand side above over all $d \geq 1$. The minimum is achieved for $d = 1$, which leads to the optimal value $c = 1$ and thus proves the first relation of the proposition. However, if we require f to be orthogonal to all linear functions, then we can only allow the values $d \geq 2$, and the optimal value is $c = n + 2$. This gives the second relation of Proposition 10.1.2. □

Recalling the improved Poincaré inequality (9.11), from Proposition 10.1.2 we immediately obtain:

Proposition 10.1.3 *For any C^2-function f on $\mathbb{S}^{n-1}$ with mean zero,*

$$\int f^2 \, d\mathfrak{s}_{n-1} \leq \frac{1}{n-1} \int \|f''_S\|_{\mathrm{HS}}^2 \, d\mathfrak{s}_{n-1},$$

where equality is attained for all linear functions. Moreover, if f is orthogonal to all linear functions with respect to $\mathfrak{s}_{n-1}$, then

$$\int f^2 \, d\mathfrak{s}_{n-1} \leq \frac{1}{2n(n+2)} \int \|f''_S\|_{\mathrm{HS}}^2 \, d\mathfrak{s}_{n-1},$$

with equality attainable for all quadratic harmonics.

To get these relations, one can also apply (10.2) and the representation (9.7) in order to reduce them to spherical harmonics. But for $f = f_d$ in H_d, the inequality of the form $c \int f^2 \, d\mathfrak{s}_{n-1} \leq \int \|f''_{\mathbb{S}}\|_{\mathrm{HS}}^2 \, d\mathfrak{s}_{n-1}$ becomes

$$c \le d(n+d-2)\,\big(d(n+d-2)-(n-2)\big),$$

where the right-hand side is an increasing function of d.

An interesting consequence of the first relation of Proposition 10.1.3 is the statement that the equality $f''_{\mathbb{S}} = 0$ is only possible for constant functions (in contrast with the Euclidean Hessian).

10.2 Bounds on the L^2-norm in the Euclidean Setup

If f is defined and smooth in an open neighborhood of the sphere, and if we prefer to deal with usual Euclidean derivatives (rather than with spherical derivatives), one may use the general relation $|\nabla_{\mathbb{S}} f| \le |\nabla f|$. Then, from the Poincaré inequality (10.1) we have a formally weaker bound

$$\int f^2 \, d\mathfrak{s}_{n-1} \le \frac{1}{n-1} \int |\nabla f|^2 \, d\mathfrak{s}_{n-1}.$$

In fact, this inequality returns us to (10.1) when it is applied to the functions of the form $f(x/|x|)$. Our next step is to estimate the above integral on the right-hand side by involving the second derivative in analogy with Proposition 10.1.2.

Proposition 10.2.1 *Suppose that a function f is defined and C^2-smooth in some neighborhood of $\mathbb{S}^{n-1}$. If f is orthogonal to all linear functions with respect to $\mathfrak{s}_{n-1}$, then*

$$\int |\nabla f|^2 \, d\mathfrak{s}_{n-1} \le \frac{4}{n-1} \int \|f''\|_{\mathrm{HS}}^2 \, d\mathfrak{s}_{n-1}. \tag{10.4}$$

Proof By the Poincaré inequality, applied to the i-th partial derivative of f,

$$\int (\partial_i f(\theta))^2 \, d\mathfrak{s}_{n-1}(\theta) \le \Big(\int \partial_i f(\theta) \, d\mathfrak{s}_{n-1}(\theta)\Big)^2 + \frac{1}{n-1} \int \sum_{j=1}^{n} (\partial_{ij} f(\theta))^2 \, d\mathfrak{s}_{n-1}(\theta),$$

which after summation over all $i \le n$ yields

$$\int |\nabla f|^2 \, d\mathfrak{s}_{n-1} \le \Big|\int \nabla f \, d\mathfrak{s}_{n-1}\Big|^2 + \frac{1}{n-1} \int \|f''\|_{\mathrm{HS}}^2 \, d\mathfrak{s}_{n-1}. \tag{10.5}$$

In order to bound the intermediate Euclidean norm, we recall that the assumption that f is orthogonal to all linear functions is equivalent to the property that every function of the form

$$\langle \nabla_{\mathbb{S}} f(\theta), v\rangle = \langle \nabla f(\theta), v\rangle - \langle \nabla f(\theta), \theta\rangle \, \langle v, \theta\rangle$$

has $\mathfrak{s}_{n-1}$-mean zero (Proposition 9.3.3). In terms of $u(\theta) = \langle \nabla f(\theta), \theta \rangle$, this may be written as the vector equality

$$\int \nabla f \, d\mathfrak{s}_{n-1} = \int u(\theta)\, \theta \, d\mathfrak{s}_{n-1}(\theta).$$

Hence

$$\Big| \int \nabla f \, d\mathfrak{s}_{n-1} \Big|^2 = \int\!\!\int u(\theta) u(\theta') \, \langle \theta, \theta' \rangle \, d\mathfrak{s}_{n-1}(\theta) d\mathfrak{s}_{n-1}(\theta'),$$

which does not exceed

$$\Big(\int\!\!\int u(\theta)^2 u(\theta')^2 \, d\mathfrak{s}_{n-1}(\theta) d\mathfrak{s}_{n-1}(\theta') \Big)^{1/2} \Big(\int\!\!\int \langle \theta, \theta' \rangle^2 \, d\mathfrak{s}_{n-1}(\theta) d\mathfrak{s}_{n-1}(\theta') \Big)^{1/2}$$

$$= \frac{1}{\sqrt{n}} \int u^2 \, d\mathfrak{s}_{n-1} \, \le \, \frac{1}{\sqrt{n}} \int |\nabla f|^2 \, d\mathfrak{s}_{n-1},$$

where we used the Cauchy–Schwarz inequality together with $|u(\theta)| \le |\nabla f(\theta)|$. Inserting this bound into (10.5), we get a relation which is solved as

$$\int |\nabla f|^2 \, d\mathfrak{s}_{n-1} \le \frac{c_n}{n-1} \int \|f''\|_{\mathrm{HS}}^2 \, d\mathfrak{s}_{n-1}, \quad c_n = \frac{1}{1 - 1/\sqrt{n}}.$$

Clearly, $c_n \le c_2 = 2 + \sqrt{2} < 4$, thus proving (10.4). Note also that $c_n \to 1$ as $n \to \infty$. So, the constant 4 may be improved for large values of n. □

Combining (10.4) with the Poincaré inequality (9.11), we get a second order Poincaré-type inequality in the Euclidean setup,

$$\int (f - m)^2 \, d\mathfrak{s}_{n-1} \le \frac{2}{n(n-1)} \int \|f''\|_{\mathrm{HS}}^2 \, d\mathfrak{s}_{n-1},$$

assuming that f is orthogonal to all linear functions, and where m is the mean of f with respect to $\mathfrak{s}_{n-1}$. Here the left integral will not change when it is applied to $f_a(x) = f(x) - \frac{a}{2}\, |x|^2$ in place of f, while the right integral will depend on a. Hence

$$\int (f - m)^2 \, d\mathfrak{s}_{n-1} \le \frac{2}{n(n-1)} \int \|f'' - a\, \mathrm{I}_n\|_{\mathrm{HS}}^2 \, d\mathfrak{s}_{n-1},$$

where I_n denotes the $n \times n$ identity matrix. Thus, we arrive at:

Proposition 10.2.2 *If f is C^2-smooth in some neighborhood of $\mathbb{S}^{n-1}$ and is orthogonal to all affine functions with respect to $\mathfrak{s}_{n-1}$, then for any $a \in \mathbb{R}$,*

$$\int f^2 \, d\mathfrak{s}_{n-1} \le \frac{2}{n(n-1)} \int \|f'' - a\, \mathrm{I}_n\|_{\mathrm{HS}}^2 \, d\mathfrak{s}_{n-1}.$$

Let us finally mention that the constants in Propositions 10.2.1 and 10.2.2 are asymptotically correct with respect to the growing dimension n. This may be tested on the example of the quadratic functions $f(\theta) = \langle \theta, v\rangle^2 - \frac{1}{n}$ with unit vectors v.

10.3 First Order Concentration Inequalities

If a function f on $\mathbb{S}^{n-1}$ has $\mathfrak{s}_{n-1}$-mean zero and has Lipschitz semi-norm $\|f\|_{\mathrm{Lip}} \leq 1$ with respect to the geodesic distance, then, by the Poincaré inequality, its L^2-norm on the probability space $(\mathbb{S}^{n-1}, \mathfrak{s}_{n-1})$ satisfies

$$\|f\|_{L^2(\mathfrak{s}_{n-1})} \leq \frac{1}{\sqrt{n-1}}.$$

Hence, $|f|$ is of order $1/\sqrt{n}$ on a large subset of the sphere. This is part of the spherical concentration phenomenon, although expressed in a weak form, to which we refer as the first order concentration. In fact, as we know, Poincaré-type inequalities provide more – they are responsible for exponential integrability of Lipschitz functions, and, for example, Proposition 6.1.1 provides an exponential bound on the tails, namely

$$\mathfrak{s}_{n-1}\big\{|f| \geq r\big\} \leq 6\,\mathrm{e}^{-2\sqrt{n-1}\,r}, \quad r \geq 0.$$

Equivalently, modulo an absolute constant, we have the relation $\|f\|_{\psi_1} \leq \frac{c}{\sqrt{n-1}}$ for the Orlicz norm generated by the Young function $\psi_1(t) = \mathrm{e}^{|t|} - 1$.

Even more information is provided by the logarithmic Sobolev inequality on the unit sphere, cf. Proposition 9.4.2. Since the logarithmic Sobolev constant in this case is $\rho(\mathfrak{s}_{n-1}) = n - 1$, one may apply Corollary 7.3.3 and Corollary 7.5.3, which give an essential sharpening in terms of the tail behavior of Lipschitz functions.

Proposition 10.3.1 *For any $f : \mathbb{S}^{n-1} \to \mathbb{R}$ with $\|f\|_{\mathrm{Lip}} \leq 1$ and $\mathfrak{s}_{n-1}$-mean zero,*

$$\int \mathrm{e}^{tf}\, d\mathfrak{s}_{n-1} \leq \exp\Big\{\frac{t^2}{2(n-1)}\Big\}, \quad t \in \mathbb{R}.$$

In particular, for all $r > 0$,

$$\mathfrak{s}_{n-1}\big\{f \geq r\big\} \leq \mathrm{e}^{-(n-1)r^2/2}, \quad \mathfrak{s}_{n-1}\big\{|f| \geq r\big\} \leq 2\,\mathrm{e}^{-(n-1)r^2/2}. \tag{10.6}$$

When the Lipschitz semi-norm of the function f is too large, one may use the log-Sobolev inequality to bound the L^p-norms $\|f\|_p = \|f\|_{L^p(\mathfrak{s}_{n-1})}$ in terms of the L^p-norms of the modulus of the gradient. In particular, from Proposition 7.5.4 we obtain the following assertion (cf. also Remark 7.5.6).

Proposition 10.3.2 *Given a Lipschitz function $f : \mathbb{S}^{n-1} \to \mathbb{C}$ with $\mathfrak{s}_{n-1}$-mean zero, we have, for any $p \geq 2$,*

$$\|f\|_p \leq \frac{\sqrt{p-1}}{\sqrt{n-1}} \, \|\nabla_{\mathbb{S}} f\|_p.$$

When $p = 2$, here we return to the Poincaré inequality on the unit sphere.

As a simple example, we get the following bound for linear functions.

Corollary 10.3.3 *For any $v \in \mathbb{R}$ and $p \geq 2$,*

$$\Big(\int |\langle v, \theta\rangle|^p \, d\mathfrak{s}_{n-1}(\theta)\Big)^{1/p} \leq \sqrt{\frac{p}{n}} \, |v|.$$

Indeed, the function $f(\theta) = \langle v, \theta\rangle$ has a gradient satisfying $|\nabla_{\mathbb{S}} f\|_p \leq |\nabla f| = |v|$. Hence, by Proposition 10.3.2,

$$\Big(\int |\langle v, \theta\rangle|^p \, d\mathfrak{s}_{n-1}(\theta)\Big)^{1/p} \leq \frac{\sqrt{p-1}}{\sqrt{n-1}} \, |v| \leq \frac{\sqrt{p}}{\sqrt{n}} \, |v|, \quad 2 \leq p \leq n.$$

If $p > n$, the resulting bound holds true as well, since the L^p-norm on the left-hand side does not exceed $|v|$.

Similar concentration inequalities may also be derived from the isoperimetric theorem (Proposition 9.5.2), including the class of functions f that are defined and Lipschitz on $\mathbb{R}^n$ with power $\alpha \in (0, 1]$, i.e., satisfying

$$|f(\theta) - f(\theta')| \leq |\theta - \theta'|^\alpha \quad (\theta, \theta' \in \mathbb{R}^n). \tag{10.7}$$

In this case, we write $\|f\|_{\mathrm{Lip}(\alpha)} \leq 1$. Alternatively, one can use the supremum-convolution approach (based on the spherical logarithmic Sobolev inequality) to derive, for example, the following generalization of the exponential bound (10.6).

Proposition 10.3.4 *Given a function $f : \mathbb{R}^n \to \mathbb{R}$ with $\mathfrak{s}_{n-1}$-mean zero and $\|f\|_{\mathrm{Lip}(\alpha)} \leq 1$ $(0 < \alpha \leq 1)$, we have*

$$\mathfrak{s}_{n-1}\big\{f \geq r\big\} \leq \exp\Big\{-\frac{n-1}{2}\, r^{2/\alpha}\Big\}, \quad r \geq 0. \tag{10.8}$$

Proof By (10.7), for all $\theta, \theta' \in \mathbb{R}^n$ and $t > 0$,

$$f(\theta') - f(\theta) - \frac{1}{2t}\,|\theta - \theta'|^2 \leq |\theta - \theta'|^\alpha - \frac{1}{2t}\,|\theta - \theta'|^2 \leq \frac{2-\alpha}{2}\,(\alpha t)^{\frac{\alpha}{2-\alpha}},$$

so that for the supremum-convolution of f we have an upper bound

$$P_t f(\theta) = \sup_{\theta' \in \mathbb{R}^n} \Big[f(\theta') - \frac{1}{2t}\,|\theta - \theta'|^2\Big] \leq f(\theta) + \frac{2-\alpha}{2}\,(\alpha t)^{\frac{\alpha}{2-\alpha}}.$$

Hence, by Corollary 9.4.3, we obtain an upper bound on the Laplace transform,

$$\int e^{tf}\,d\mathfrak{s}_{n-1} \le \exp\Big\{t\int P_{\frac{t}{n-1}} f\,d\mathfrak{s}_{n-1}\Big\}$$
$$\le \exp\Big\{\frac{2-\alpha}{2}\,(n-1)^{-\frac{\alpha}{2-\alpha}}\,\alpha^{\frac{\alpha}{2-\alpha}}\,t^{\frac{2}{2-\alpha}}\Big\}.$$

By Markov's inequality, the latter yields

$$\mathfrak{s}_{n-1}\{f \ge r\} \le \exp\Big\{-rt + \frac{2-\alpha}{2}\,(n-1)^{-\frac{\alpha}{2-\alpha}}\,\alpha^{\frac{\alpha}{2-\alpha}}\,t^{\frac{2}{2-\alpha}}\Big\}.$$

It remains to optimize the right-hand side over all $t > 0$, and then we arrive at the required inequality (10.8). □

10.4 Second Order Concentration

Using a standard normal random variable Z, the Gaussian deviation bounds (10.6) may be stated informally as a stochastic dominance $|f| \le c\,\frac{|Z|}{\sqrt{n}}$. Up to an absolute constant $c > 0$, this may also be written equivalently as

$$\int e^{(n-1)\,f^2/c}\,d\mathfrak{s}_{n-1} \le 2.$$

In fact, under stronger assumptions involving the second derivative $f''_{\mathbb{S}}$, one can further strengthen this stochastic dominance with respect to the dimension n – for example, to get a comparison such as $|f| \le c\,(\frac{Z}{\sqrt{n}})^2$.

Given a real symmetric $n \times n$ matrix $A = (a_{ij})_{i,j=1}^n$ (e.g. when $A = f''_{\mathbb{S}}$) , we denote by $\|A\| = \max_{|\theta|\le 1} |A\theta|$ its operator norm, and by

$$\|A\|_{\rm HS} = \Big(\sum_{i,j=1}^n |a_{ij}|^2\Big)^{1/2}$$

its Hilbert–Schmidt norm. If the entries of A are complex numbers, the definition is similar, the only difference being that the supremum should be taken over all $\theta \in \mathbb{C}^n$ such that $|\theta| \le 1$ (that is, from the complex unit sphere).

Proposition 10.4.1 *Assume that a C^2-smooth function $f : \mathbb{S}^{n-1} \to \mathbb{R}$ is orthogonal to all affine functions. If $\|f''_{\mathbb{S}}\| \le 1$ on $\mathbb{S}^{n-1}$ and $\int \|f''_{\mathbb{S}}\|_{\rm HS}^2\,d\mathfrak{s}_{n-1} \le b$, then*

$$\int \exp\Big\{\frac{n-1}{2(1+b)}\,|f|\Big\}\,d\mathfrak{s}_{n-1} \le 2.$$

Proof By Proposition 9.2.1, $|\nabla^2_{\mathbb{S}} f(\theta)| \le \|f''_{\mathbb{S}}(\theta)\| \le 1$ for all points $\theta \in \mathbb{S}^{n-1}$, so that we may apply Proposition 7.6.1 on the space $\mathbb{M} = \mathbb{S}^{n-1}$ with $\mu = \mathfrak{s}_{n-1}$. It gives

$$\int e^{(n-1)f/2}\, d\mathfrak{s}_{n-1} \le \exp\Big\{\frac{n-1}{2}\int |\nabla_{\mathbb{S}} f|^2\, d\mathfrak{s}_{n-1}\Big\}.$$

To bound the last integral, we may apply the second relation of Proposition 10.1.2, and then we get

$$\log \int \exp\Big\{\frac{n-1}{2}\, f\Big\}\, d\mathfrak{s}_{n-1} \le \frac{n-1}{2(n+2)} \int \|f''_{\mathbb{S}}\|_{\rm HS}^2\, d\mathfrak{s}_{n-1} \le \frac{1}{2}\, b.$$

Using a similar inequality for the function $-f$, we have

$$\int e^{\frac{n-1}{2}\,|f|}\, d\mathfrak{s}_{n-1} \le \int e^{\frac{n-1}{2}\, f}\, d\mathfrak{s}_{n-1} + \int e^{-\frac{n-1}{2}\, f}\, d\mathfrak{s}_{n-1} \le 2e^{b/2}.$$

It follows that, for any $\lambda \ge 1$,

$$\int e^{\frac{n-1}{2}\,|f|/\lambda}\, d\mathfrak{s}_{n-1} \le \Big(\int e^{\frac{n-1}{2}\,|f|}\, d\mathfrak{s}_{n-1}\Big)^{1/\lambda} \le (2e^{b/2})^{1/\lambda}.$$

It remains to note that $(2e^{b/2})^{1/\lambda} = 2$ for $\lambda = 1 + \frac{b}{\log 4} \le 1 + b$.

Proposition 10.4.1 is proved. □

In the sequel, we will however work with Euclidean derivatives, and another version of Proposition 10.4.1 will be needed.

Proposition 10.4.2 *Let a real-valued function f be defined and C^2-smooth in some open neighborhood of $\mathbb{S}^{n-1}$. Assume that it is orthogonal to all affine functions and satisfies $\|f'' - a\,\mathbb{I}_n\| \le 1$ on $\mathbb{S}^{n-1}$ together with*

$$\int \|f'' - a\,\mathbb{I}_n\|_{\rm HS}^2\, d\mathfrak{s}_{n-1} \le b \tag{10.9}$$

for some $a \in \mathbb{R}$ and $b \ge 0$. Then

$$\int \exp\Big\{\frac{n-1}{2(1+3b)}\,|f|\Big\}\, d\mathfrak{s}_{n-1} \le 2. \tag{10.10}$$

Proof Suppose that f is orthogonal to all linear functions in $L^2(\mathfrak{s}_{n-1})$ and has mean m. We now apply Corollary 7.6.2 to the function $f - m$ together with bound (10.4) of Proposition 10.2.1, which gives

$$\log \int \exp\Big\{\frac{n-1}{2}\,(f-m)\Big\}\, d\mathfrak{s}_{n-1} \le 2 \int \|f''\|_{\rm HS}^2\, d\mathfrak{s}_{n-1}.$$

Applying it to $f_a(x) = f(x) - \frac{a}{2}\,|x|^2$ in place of f, we get

$$\log \int \exp\Big\{\frac{n-1}{2}\,(f-m)\Big\}\, d\mathfrak{s}_{n-1} \le 2 \int \|f'' - a\,\mathbb{I}_n\|_{\rm HS}^2\, d\mathfrak{s}_{n-1} \le 2b.$$

Now, assume that $m = 0$ and apply a similar inequality to the function $-f$. Then

$$\int e^{\frac{n-1}{2}\,|f|}\, d\mathfrak{s}_{n-1} \leq 2e^{2b}.$$

Hence, for any $\lambda \geq 1$,

$$\int e^{\frac{n-1}{2}\,|f|/\lambda}\, d\mathfrak{s}_{n-1} \leq \Big(\int e^{\frac{n-1}{2}\,|f|}\, d\mathfrak{s}_{n-1}\Big)^{1/\lambda} \leq (2e^{2b})^{1/\lambda}.$$

As before, it remains to note that $(2e^{2b})^{1/\lambda} = 2$ for $\lambda = 1 + \frac{2b}{\log 2} \leq 1 + 3b$.

Proposition 10.4.2 is proved. □

10.5 Second Order Concentration With Linear Parts

In Propositions 10.4.1–10.4.2 one may also start with an arbitrary C^2-smooth function f, but apply the hypotheses and the conclusions to the projection Tf of f onto the orthogonal complement of the space $\mathcal{H}$ of all affine functions on the sphere in $L^2(\mathfrak{s}_{n-1})$. This space has dimension $n+1$, and one may choose for the orthonormal basis in $\mathcal{H}$ the canonical functions

$$l_0(\theta) = 1, \qquad l_k(\theta) = \sqrt{n}\,\theta_k \ \ (k = 1, \dots, n, \ \ \theta = (\theta_1, \dots, \theta_n) \in \mathbb{S}^{n-1}).$$

Therefore, the "affine" part $l = Tf - f$ of f is described as the orthogonal projection in $L^2(\mathfrak{s}_{n-1})$ onto $\mathcal{H}$, namely

$$\begin{aligned} l(\theta) = \mathrm{Proj}_{\mathcal{H}}(f) &= \sum_{k=0}^{n} \langle f, l_k\rangle_{L^2(\mathfrak{s}_{n-1})}\, l_k(\theta) \\ &= \sum_{k=0}^{n} \Big(\int f(x) l_k(x)\, d\mathfrak{s}_{n-1}(x)\Big)\, l_k(\theta) \\ &= \int f(x)\, d\mathfrak{s}_{n-1}(x) + n \int \Big(f(x) \sum_{k=1}^{n} x_k \theta_k\Big)\, d\mathfrak{s}_{n-1}(x). \end{aligned}$$

In other words, $l(\theta) = m + \langle v, \theta\rangle$ with

$$m = \int f(x)\, d\mathfrak{s}_{n-1}(x), \qquad v = n \int x f(x)\, d\mathfrak{s}_{n-1}(x),$$

so, $Tf(\theta) = f(\theta) - l(\theta)$. The function $\langle v, \theta\rangle$ will be referred to as a linear part of f.

For example, if f is even, i.e. $f(-\theta) = f(\theta)$ for all $\theta \in \mathbb{S}^{n-1}$, then $Tf = f - m$.

In the setting of Proposition 10.4.2, the functions Tf and f have identical Euclidean second derivatives. Hence, if we want to obtain an inequality similar to (10.10) without the orthogonality assumption (still assuming conditions on the Euclidean second derivative), we need to verify that the affine part l is of order $\frac{1}{n}$. This may be achieved by estimating the L^2-norm of l and using the well-known fact that the linear functions on the sphere behave like Gaussian random variables. If, for definiteness, f has mean zero, then

$$\|l\|_{L^2}^2 = \frac{1}{n}\,|v|^2 = nI, \quad \text{where } I = \iint \langle x, y\rangle\, f(x) f(y)\, d\mathfrak{s}_{n-1}(x) d\mathfrak{s}_{n-1}(y).$$

Therefore, a natural requirement should be a bound $I \le b_0^2/n^3$ with b_0 of order 1. This leads to the following generalization of Proposition 10.4.2, which is more flexible in applications.

Proposition 10.5.1 *Let f be a real-valued, C^2-smooth function in some open neighborhood of $\mathbb{S}^{n-1}$. Assume that it has $\mathfrak{s}_{n-1}$-mean zero and $I \le b_0^2/n^3$ $(b_0 \ge 0)$. If $\|f'' - a\,\mathrm{I}_n\| \le 1$ on $\mathbb{S}^{n-1}$ for some $a \in \mathbb{R}$, and* (10.9) *holds, then*

$$\int \exp\Big\{\frac{n-1}{2(1+b_0+3b)}\,|f|\Big\}\, d\mathfrak{s}_{n-1} \le 2. \tag{10.11}$$

Proof Let $l(\theta) = \langle v, \theta\rangle$ be the linear part of f as above. First, we need a bound on the Orlicz norm corresponding to the Young function $\psi_1(t) = e^{|t|} - 1$ in the class of linear functions. Applying Proposition 10.3.1 to l and $-l$, we have

$$\int e^{t\,|l|}\, d\mathfrak{s}_{n-1} \le 2\,\exp\Big\{\frac{t^2}{2(n-1)}\,|v|^2\Big\}, \qquad t > 0.$$

Hence, for $\lambda > 1$,

$$\int e^{t\,|l|/\lambda}\, d\mathfrak{s}_{n-1} \le \Big(2\,\exp\Big\{\frac{t^2}{2(n-1)}\,|v|^2\Big\}\Big)^{1/\lambda} = 2,$$

where the last equality holds true for the value $\lambda = 1 + \frac{t^2\,|v|^2}{(n-1)\log 4}$. Equivalently, $\|l\|_{\psi_1} \le \frac{\lambda}{t} = \frac{1}{t} + \frac{t\,|v|^2}{(n-1)\log 4}$. Optimizing over $t > 0$, we arrive at

$$\|l\|_{\psi_1}^2 \le \frac{2\,|v|^2}{(n-1)\log 2}. \tag{10.12}$$

Since $|v| \le b_0/\sqrt{n}$, we thus have $\|l\|_{\psi_1} \le 2b_0/(n-1)$. On the other hand, by Proposition 10.4.2 with the same assumption on the second derivative of f,

$$\int \exp\Big\{\frac{n-1}{2(1+3b)}\,|Tf|\Big\}\, d\mathfrak{s}_{n-1} \le 2,$$

which means that $\|T\|_{\psi_1} \le \frac{2}{n-1}\,(1+3b)$. Using $|f| \le |Tf| + |l|$ and applying the triangle inequality in the Orlicz space $L^{\psi_1}(\mathfrak{s}_{n-1})$, we conclude that

$$\|f\|_{\psi_1} \le \frac{2b_0 + 2(1+3b)}{n-1}.$$

This is an equivalent form of (10.11), so, Proposition 10.5.1 is proved. □

Let us summarize and extend Propositions 10.2.2, 10.4.2 and 10.5.1 to the class of complex-valued functions.

Proposition 10.5.2 *Suppose that a complex-valued function f is C^2-smooth in some open neighborhood of $\mathbb{S}^{n-1}$ and has $\mathfrak{s}_{n-1}$-mean zero. For any $a \in \mathbb{C}$,*

$$\int |f|^2 \, d\mathfrak{s}_{n-1} \le \|l\|_{L^2}^2 + \frac{2}{n(n-1)} \int \|f'' - a\,\mathbb{I}_n\|_{\mathrm{HS}}^2 \, d\mathfrak{s}_{n-1}, \tag{10.13}$$

where l is a linear part of f. Moreover, if $\|f'' - a\,\mathbb{I}_n\| \le 1$ on $\mathbb{S}^{n-1}$, then

$$\|f\|_{\psi_1} \le 4\,\|l\|_{L^2} + \frac{6}{n-1} \int \|f'' - a\,\mathbb{I}_n\|_{\mathrm{HS}}^2 \, d\mathfrak{s}_{n-1} + \frac{4}{n-1}. \tag{10.14}$$

Proof Let $f_0 = \mathrm{Re}(f)$, $f_1 = \mathrm{Im}(f)$, $l_0 = \mathrm{Re}(l)$, $l_1 = \mathrm{Im}(l)$, and $a_0 = \mathrm{Re}(a)$, $a_1 = \mathrm{Im}(a)$. The inequality (10.13) follows from Proposition 10.2.2 applied to the functions $u_0 = f_0 - l_0$ and $u_1 = f_1 - l_1$ with a_0 and a_1 respectively.

Turning to the second assertion, first let us apply (10.12) to the linear part of f which gives

$$\|l_k\|_{\psi_1}^2 \le \frac{2n}{(n-1)\log 2} \|l_k\|_{L_2}^2 \le \frac{4}{\log 2} \|l_k\|_{L_2}^2, \quad k = 0, 1.$$

Using $|l| \le |l_0| + |l_1|$ and $\|l\|_{L^2}^2 = \|l_0\|_{L^2}^2 + \|l_1\|_{L^2}^2$, we then get

$$\begin{aligned} \|l\|_{\psi_1} &\le \|l_0\|_{\psi_1} + \|l_1\|_{\psi_1} \le \frac{2}{\sqrt{\log 2}} \Big(\|l_0\|_{L_2} + \|l_1\|_{L_2} \Big) \\ &\le \frac{2\sqrt{2}}{\sqrt{\log 2}} \sqrt{\|l_0\|_{L_2}^2 + \|l_1\|_{L_2}^2} \le 4\,\|l\|_{L_2}. \end{aligned}$$

The functions u_k ($k = 0, 1$) are orthogonal to all linear functions in $L^2(\mathfrak{s}_{n-1})$, and by the assumption,

$$|\langle (f''(\theta) - a\,\mathbb{I}_n)v, w\rangle| \le 1, \quad \theta \in \mathbb{S}^{n-1},$$

for all $v, w \in \mathbb{C}^n$ such that $|v| \le 1$ and $|w| \le 1$. Since $f''(\theta) = u_0''(\theta) + iu_1''(\theta)$, it follows that $\|u_k'' - a_k\,\mathbb{I}_n\| \le 1$ on $\mathbb{S}^{n-1}$. Hence, we may apply the inequality (10.10) of Proposition 10.4.2, which gives

$$\|u_k\|_{\psi_1} \le \frac{2}{n-1}(1 + 3b_k), \quad b_k = \int \|u_k'' - a_k\,\mathbb{I}_n\|_{\mathrm{HS}}^2 \, d\mathfrak{s}_{n-1},$$

so that

$$\begin{aligned}\|f - l\|_{\psi_1} &\le \|u_0\|_{\psi_1} + \|u_1\|_{\psi_1} \le \frac{2}{n-1}\,(2 + 3b_0 + 3b_1) \\ &= \frac{4}{n-1} + \frac{6}{n-1}\int \|f'' - a\,\mathbb{I}_n\|_{\rm HS}^2\,d\mathfrak{s}_{n-1}.\end{aligned}$$

It remains to combine the two bounds on the linear and non-linear parts by applying the triangle inequality $\|f\|_{\psi_1} \le \|f - l\|_{\psi_1} + \|l\|_{\psi_1}$. Proposition 10.5.2 is proved. □

10.6 Deviations for Some Elementary Polynomials

Let us illustrate the first and second order spherical concentration inequalities on the example of elementary polynomials such as

$$Q_p(\theta) = \sum_{k=1}^{n} a_k \theta_k^p, \quad \theta = (\theta_1, \dots, \theta_n) \in \mathbb{S}^{n-1}, \ p = 1, 2, \dots,$$

with real coefficients a_k. For applications to the randomized versions of the central limit theorem, we are mostly interested in the particular integer values $p = 3$ and $p = 4$.

To get an idea about the rate of fluctuations of these functions on the unit sphere with respect to $\mathfrak{s}_{n-1}$ for large values of n (with natural normalizations of the coefficients a_k), let us look first at their moments, including the moments of the first coordinate on the sphere. Since Q_p has a symmetric distribution under $\mathfrak{s}_{n-1}$ when p is odd, necessarily $\int Q_p\,d\mathfrak{s}_{n-1} = 0$. As for even integers, we have:

Proposition 10.6.1 *For any integer $p \ge 1$,*

$$\int Q_{2p}\,d\mathfrak{s}_{n-1} = \frac{(2p-1)!!}{n(n+2)\cdots(n+2p-2)}\sum_{k=1}^{n} a_k.$$

In particular,

$$\int \theta_1^{2p}\,d\mathfrak{s}_{n-1}(\theta) = \frac{(2p-1)!!}{n(n+2)\cdots(n+2p-2)} \le \frac{(2p-1)!!}{n^p}. \tag{10.15}$$

This formula follows from the fact that θ_1, viewed as a random variable on the probability space $(\mathbb{S}^{n-1}, \mathfrak{s}_{n-1})$, has density $c_n\,(1-x^2)^{\frac{n-3}{2}}$ supported on the interval $|x| < 1$. Using the beta function, this gives

$$\int \theta_1^{2p}\,d\mathfrak{s}_{n-1}(\theta) = \frac{B(p+\frac{1}{2}, \frac{n-1}{2})}{B(\frac{1}{2}, \frac{n-1}{2})} = \frac{\Gamma(p+\frac{1}{2})\,\Gamma(\frac{n}{2})}{\Gamma(p+\frac{n}{2})\,\Gamma(\frac{1}{2})},$$

which is further simplified to (10.15).

For computation of the mutual covariances between even powers of the spherical coordinates, one may use the identity

$$\int \theta_1^{2p}\, \theta_2^{2p}\, d\mathfrak{s}_{n-1}(\theta) = \frac{\big((2p-1)!!\big)^2}{n(n+2)\cdots(n+4p-2)}. \tag{10.16}$$

This follows from the relation $Z = |Z|\,\theta$ and similar formulas $\mathbb{E}Z_1^{2p} = (2p-1)!!$ and

$$\mathbb{E}\,|Z|^{2p} = 2^p\,\frac{\Gamma(p+\frac{n}{2})}{\Gamma(\frac{n}{2})} = n(n+2)\cdots(n+2p-2),$$

where $Z = (Z_1, \dots, Z_n)$ is a standard normal random vector in $\mathbb{R}^n$ independent of the random vector $\theta = (\theta_1, \dots, \theta_n)$ uniformly distributed over the unit sphere.

Let $p = 1$. The functional Q_1 is linear and has a Lipschitz semi-norm $\|Q_1\|_{\rm Lip} = (a_1^2 + \cdots + a_n^2)^{1/2}$. In this case, Proposition 10.3.1 with $f = Q_1$ correctly describes a subgaussian behavior of the distribution of Q_1: If $a_1^2 + \cdots + a_n^2 = 1$, then

$$\mathfrak{s}_{n-1}\big\{|Q_1| \geq r\big\} \leq 2\,e^{-(n-1)r^2/2}, \quad r > 0,$$

which corresponds to the short notation $|Q_1| \leq c\,\frac{|Z|}{\sqrt{n}}$ with $Z \sim N(0,1)$.

Let now $p = 3$. The functional Q_3 has mean zero, with Lipschitz semi-norm $\|Q_3\|_{\rm Lip} = 3\max_k |a_k|$. In this case, if $\max_k |a_k| \leq 1$, the first order concentration inequality (10.6) provides a similar relation

$$\mathfrak{s}_{n-1}\big\{|Q_3| \geq r\big\} \leq 2\,e^{-(n-1)r^2/18}.$$

However, this conclusion may be considerably strengthened with respect to the growing dimension n. Since Q_3 is not orthogonal to linear functions, we should appeal to the second order concentration inequality of Proposition 10.5.1, which involves functions with non-zero linear parts. To show that this part is sufficiently small for Q_3, consider the integral

$$\begin{aligned} I_{Q_3} &= \iint \langle x, y\rangle\, Q_3(x) Q_3(y)\, d\mathfrak{s}_{n-1}(x) d\mathfrak{s}_{n-1}(y) \\ &= \iint \sum_{i=1}^n x_i y_i \sum_{j=1}^n a_j x_j^3 \sum_{k=1}^n a_k y_k^3\, d\mathfrak{s}_{n-1}(x) d\mathfrak{s}_{n-1}(y) \\ &= \sum_{i=1}^n a_i^2 \Big(\int x_i^4\, d\mathfrak{s}_{n-1}(x)\Big)^2 = \frac{9}{n^2(n+2)^2} \sum_{i=1}^n a_i^2, \end{aligned}$$

which corresponds to (10.15) with $p = 2$. If $\max_k |a_k| \leq 1$, then $I_{Q_3} \leq \frac{9}{n^3}$. Now, the second derivative $Q_3''(\theta) = (6\,a_i \theta_i\, \delta_{ij})_{i,j=1}^n$ is a diagonal matrix, for which

$$\|Q_3''(\theta)\| = 6\max_k |a_k\theta_k|, \quad \|Q_3''(\theta)\|_{\rm HS}^2 = 36\sum_{k=1}^n a_k^2\theta_k^2.$$

Hence, the conditions of Proposition 10.5.1 are fulfilled for the function $f(\theta) = \frac{1}{6}\,Q_3(\theta)$ with parameters $a = 0$ and

$$b = \int \|f''\|_{\rm HS}^2\,d\mathfrak{s}_{n-1} = \frac{1}{36}\int \|Q_3''(\theta)\|_{\rm HS}^2\,d\mathfrak{s}_{n-1}(\theta) = \frac{1}{n}\sum_{k=1}^n a_k^2 \leq 1.$$

In addition,

$$I_f = \iint \langle x, y\rangle\, f(x)f(y)\,d\mathfrak{s}_{n-1}(x)\,d\mathfrak{s}_{n-1}(y) = \frac{1}{36}\,I_{Q_3} \leq \frac{1}{4n^3},$$

so, one may put $b_0 = \frac{1}{2}$. Since $\frac{1}{2(1+b_0+3b)} \geq \frac{1}{9}$, from the inequality (10.11), we get:

Corollary 10.6.2 *If* $\max_k |a_k| \leq 1$, *then*

$$\int \exp\Big\{\frac{n-1}{9}\,|Q_3|\Big\}\,d\mathfrak{s}_{n-1} \leq 2.$$

In particular, for all $r \geq 0$,

$$\mathfrak{s}_{n-1}\{(n-1)\,|Q_3| \geq r\} \leq 2\,e^{-r/9}. \tag{10.17}$$

Thus, Q_3 is of order $1/n$ with respect to n, and there is a subexponential behavior of tails of the distribution, i.e., $|Q_3| \preceq c\,(\frac{|Z|}{\sqrt{n}})^2$. In particular, we have a bound $\|Q_3\|_{L^2} \leq \frac{c}{n}$ for some absolute constant c. Note that, under our conditions, this $\frac{1}{n}$-rate cannot be improved, since by (10.15),

$$\mathrm{Var}_{\mathfrak{s}_{n-1}}(Q_3) = \|Q_3\|_{L^2}^2 = \int Q_3^2\,d\mathfrak{s}_{n-1} = \frac{15}{n(n+2)(n+4)}\sum_{k=1}^n a_k^2, \tag{10.18}$$

which dominates $(\frac{c}{n})^2$ when, for example, all $a_k = 1$.

In fact, at the expense of a worse decay in r on the right-hand side of (10.17), one may weaken the assumption on the coefficients in Corollary 10.6.2.

Proposition 10.6.3 *If* $\frac{1}{n}\sum_{k=1}^n a_k^2 = 1$, *then for all* $r > 0$,

$$\mathfrak{s}_{n-1}\{n\,|Q_3| \geq r\} \leq 2\,\exp\Big\{-\frac{1}{23}\,r^{2/3}\Big\}.$$

Proof We apply the L^p-Poincaré-type inequality of Proposition 10.3.2 to the function $f = nQ_3$. Using $\frac{p-1}{n-1} \leq \frac{2p}{n}$ and applying Jensen's inequality for the power function, we get, for any $p \geq 2$,

$$\begin{aligned}
\|f\|_p^p &\le n^p \left(\frac{2p}{n}\right)^{p/2} \|\nabla Q_3\|_p^p \\
&= n^{p/2}\,(2p)^{p/2}\,3^p \int \Big(\sum_{k=1}^n a_k^2\,\theta_k^4\Big)^{p/2} \,d\mathfrak{s}_{n-1}(\theta) \\
&\le n^p\,(2p)^{p/2}\,3^p \cdot \frac{1}{n}\sum_{k=1}^n a_k^2 \int |\theta_k|^{2p}\,d\mathfrak{s}_{n-1}(\theta) \\
&= n^p\,(18\,p)^{p/2} \int |\theta_1|^{2p}\,d\mathfrak{s}_{n-1}(\theta).
\end{aligned}$$

If $p = m$ is an integer, applying the relation (10.15), we get

$$\|f\|_m^m \le (18\,m)^{m/2}\,(2m-1)!! \le 18^{m/2}\,2^m\,m^{3m/2},$$

where we used the bound $(2m-1)!! \le (2m)^m$. Thus, $\|f\|_m \le 6\sqrt{2}\,m^{3/2}$. At the expense of a larger absolute factor, this inequality can be extended to all real $p \ge 2$ in place of m. Indeed, let m be the integer such that $m \le p < m+1$. Then

$$\begin{aligned}
\|f\|_p \le \|f\|_{m+1} &\le 6\sqrt{2}\,(m+1)^{3/2} \\
&\le 6\sqrt{2}\,(p+1)^{3/2} \le 9\sqrt{3}\,p^{3/2},
\end{aligned}$$

i.e. $\|f\|_p^p \le (bp)^{3p/2}$ with constant $b = (9\sqrt{3})^{2/3} = 3^{5/3}$. By Markov's inequality, choosing $p = \frac{1}{2^{1/3}\,b}\,r^{2/3}$ $(r > 0)$, we get

$$\mathfrak{s}_{n-1}\{|f| \ge r\} \le \frac{(bp)^{3p/2}}{r^p} = \exp\Big\{-\frac{p}{2}\,\log 2\Big\},$$

provided that $p \ge 2$. But, in the case $0 < p < 2$, the above right-hand side is greater than $1/2$, so that we have

$$\mathfrak{s}_{n-1}\{|f| \ge r\} \le 2\,\exp\Big\{-\frac{p}{2}\,\log 2\Big\}$$

for all $p > 0$ (hence for all $r > 0$). It remains to note that $\frac{p}{2}\,\log 2 = \frac{\log 2}{2^{4/3}\,3^{5/3}}\,r^{2/3} > \frac{1}{23}\,r^{2/3}$. Proposition 10.6.3 is proved. □

10.7 Polynomials of Fourth Degree

Similar observations can be made about the polynomials

$$Q_4(\theta) = \sum_{k=1}^n a_k\theta_k^4, \quad \theta = (\theta_1, \dots, \theta_n) \in \mathbb{S}^{n-1}.$$

They have $\mathfrak{s}_{n-1}$-means $\frac{3}{n+2}\,\bar{a}$, where

$$\bar{a} = \frac{1}{n}\,(a_1 + \cdots + a_n),$$

and Lipschitz semi-norms $\|Q_4\|_{\rm Lip} = 4\max_k |a_k|$. If $\max_k |a_k| \le 1$, the first order concentration inequality (10.6) provides the relation

$$\mathfrak{s}_{n-1}\Big\{\Big|Q_4 - \frac{3}{n+2}\,\bar{a}\Big| \ge r\Big\} \le 2\,{\rm e}^{-(n-1)r^2/32} \quad (r > 0).$$

This corresponds to the notation $|Q_4 - \frac{3}{n+2}\,\bar{a}| \le c\,\frac{|Z|}{\sqrt{n}}$, which also implies $|Q_4| \le c\,\frac{|Z|}{\sqrt{n}}$ (since $|\bar{a}| \le 1$). However, this conclusion may be strengthened with the help of Proposition 10.4.2. Indeed, Q_4 is orthogonal in $L^2(\mathfrak{s}_{n-1})$ to all linear functions, while the second derivative $Q_4''(\theta) = (12\,a_i\theta_i^2\,\delta_{ij})_{i,j=1}^n$ is a diagonal matrix, for which

$$\|Q_4''(\theta)\| = 12\max_k |a_k\theta_k^2|, \quad \|Q_4''(\theta)\|_{\rm HS}^2 = 144\sum_{k=1}^n a_k^2\theta_k^4.$$

Hence, the condition of Proposition 10.4.2 is fulfilled for the function $f(\theta) = \frac{1}{12}\,Q_4(\theta)$ with parameters $a = 0$ and

$$b = \frac{1}{144}\int \|Q_4''(\theta)\|_{\rm HS}^2\,{\rm d}\mathfrak{s}_{n-1}(\theta) = \frac{3}{n(n+2)}\sum_{k=1}^n a_k^2 \le \frac{3}{n+2} \le \frac{3}{4}.$$

As a result, from the inequality (10.10) we obtain that

$$\int \exp\Big\{\frac{n-1}{78}\,\Big|Q_4 - \frac{3\,\bar{a}}{n+2}\Big|\Big\}\,{\rm d}\mathfrak{s}_{n-1} \le 2,$$

which yields:

Corollary 10.7.1 *If* $\max_k |a_k| \le 1$, *then for all* $r \ge 0$,

$$\mathfrak{s}_{n-1}\Big\{(n-1)\,\Big|Q_4 - \frac{3\,\bar{a}}{n+2}\Big| \ge r\Big\} \le 2\,{\rm e}^{-r/78}.$$

In particular,

$$\mathfrak{s}_{n-1}\{(n-1)\,|Q_4| \ge r\} \le 3\,{\rm e}^{-r/78}. \tag{10.19}$$

Here the second bound follows from the first one, by noting that $|\bar{a}| \le 1$, so that $(n-1)\,|Q_4| \ge r$ implies $(n-1)\,|Q_4 - \frac{3\,\bar{a}}{n+2}| \ge r - \frac{3(n-1)}{n+2} > 0$ for $r \ge 3$. Hence, by Markov's inequality,

$$\mathfrak{s}_{n-1}\{(n-1)\,|Q_4| \ge r\} \le 2{\rm e}^{-\frac{1}{78}\,(r - \frac{3(n-1)}{n+2})} \le 2{\rm e}^{3/78}\,{\rm e}^{-r/78}.$$

In the case $r \le 3$, (10.19) is immediate, since the right-hand side is greater than 1.

Thus, Q_4 is also of order $1/n$ with respect to n, and we have a subexponential behavior of tails, i.e., $|Q_4| \le c\,(\frac{|Z|}{\sqrt{n}})^2$. The deviations of Q_4 around its mean are even smaller. To see this, let us look at the variance

$$\begin{aligned}
\mathrm{Var}_{\mathfrak{s}_{n-1}}(Q_4) &= \sum_{k=1}^n a_k^2\, \mathrm{Var}_{\mathfrak{s}_{n-1}}(\theta_k^4) + \sum_{i\neq j} a_i a_j\, \mathrm{Cov}_{\mathfrak{s}_{n-1}}(\theta_i^4, \theta_j^4) \\
&= \mathrm{Var}_{\mathfrak{s}_{n-1}}(\theta_1^4) \sum_{k=1}^n a_k^2 + \mathrm{Cov}_{\mathfrak{s}_{n-1}}(\theta_1^4, \theta_2^4) \sum_{i\neq j} a_i a_j \\
&= A_n \sum_{k=1}^n a_k^2 + B_n \Big(\sum_{k=1}^n a_k\Big)^2,
\end{aligned}$$

where we used the property that the measure $\mathfrak{s}_{n-1}$ is invariant under permutations of coordinates. Here, the coefficients are given by

$$\begin{aligned}
A_n &= \int \theta_1^8 \, d\mathfrak{s}_{n-1}(\theta) - \int \theta_1^4 \theta_2^4 \, d\mathfrak{s}_{n-1}(\theta), \\
B_n &= \int \theta_1^4 \theta_2^4 \, d\mathfrak{s}_{n-1}(\theta) - \Big(\int \theta_1^4 \, d\mathfrak{s}_{n-1}(\theta)\Big)^2.
\end{aligned}$$

To simplify, one may apply the formulas (10.15)–(10.16), from which we find that

$$\int \theta_1^4 \, d\mathfrak{s}_{n-1}(\theta) = \frac{3}{n(n+2)}, \quad \int \theta_1^4 \theta_2^4 \, d\mathfrak{s}_{n-1}(\theta) = \frac{9}{n(n+2)(n+4)(n+6)},$$

and

$$\int \theta_1^8 \, d\mathfrak{s}_{n-1}(\theta) = \frac{105}{n(n+2)(n+4)(n+6)},$$

so that

$$A_n = \frac{96}{n(n+2)(n+4)(n+6)}, \quad B_n = -\frac{72\,(n+3)}{n^2(n+2)^2(n+4)(n+6)}.$$

Since $B_n \le 0$, it follows that

$$\mathrm{Var}_{\mathfrak{s}_{n-1}}(Q_4) \le \frac{96}{n^3}\, \bar{a^2} \quad \text{where} \quad \bar{a^2} = \frac{1}{n} \sum_{k=1}^n a_k^2. \tag{10.20}$$

This inequality suggests that the function $Q_4 - \frac{3\,\bar{a}}{n+2}$ is of order $n^{-3/2}$, which should have some nice applications.

Proposition 10.7.2 *If* $\frac{1}{n}\sum_{k=1}^n a_k^2 = 1$, *then for all* $r \geq 0$,

$$\mathfrak{s}_{n-1}\Big\{n^{3/2}\,\Big|Q_4 - \frac{3\,\bar a}{n+2}\Big| \geq r\Big\} \leq 2\,\exp\Big\{-\frac{1}{38}\sqrt{r}\Big\}. \tag{10.21}$$

For short, this assertion may be written as the dominance $|Q_4 - \frac{3\,\bar a}{n+2}| \leq cZ^4\,n^{-3/2}$, where $Z \sim N(0,1)$.

Proof We use an argument which is similar to the one from the proof of Proposition 10.6.3. Applying Proposition 10.3.2 to the function $f = n^{3/2}\,(Q_4 - \frac{3\,\bar a}{n+2})$ together with Jensen's inequality, we have that, for any $p \geq 2$,

$$\begin{aligned}
\|f\|_p^p &\leq n^{3p/2}\Big(\frac{2p}{n}\Big)^{p/2}\,\|\nabla Q_4\|_p^p \\
&= n^{3p/2}\Big(\frac{2p}{n}\Big)^{p/2}\,4^p\int\Big(\sum_{k=1}^n a_k^2\,\theta_k^6\Big)^{p/2}\,d\mathfrak{s}_{n-1}(\theta) \\
&\leq n^{2p}\Big(\frac{2p}{n}\Big)^{p/2}\,4^p\,\frac{1}{n}\sum_{k=1}^n a_k^2\int|\theta_k|^{3p}\,d\mathfrak{s}_{n-1}(\theta) \\
&= n^{3p/2}\,(32\,p)^{p/2}\int|\theta_1|^{3p}\,d\mathfrak{s}_{n-1}(\theta).
\end{aligned}$$

Let us replace p with $2m$ assuming that $m \geq 1$ is an integer. By (10.15), we get

$$\|f\|_{2m}^{2m} \leq (64\,m)^m\,(6m-1)!! \leq (48\sqrt{6}\,m^2)^{2m},$$

where we used $(2k-1)!! \leq (2k)^k$. Hence $\|f\|_{2m} \leq 48\sqrt{6}\,m^2$. To extend this inequality to real $p \geq 2$ (although with a large absolute factor), pick up an integer m such that $2m \leq p < 2(m+1)$. Then

$$\begin{aligned}
\|f\|_p &\leq \|f\|_{2(m+1)} \leq 48\sqrt{6}\,(m+1)^2 \\
&\leq 48\sqrt{6}\,p^2 = (bp)^{2p}, \quad b = (48\sqrt{6})^{1/2}.
\end{aligned}$$

By Markov's inequality, given $r > 0$ and choosing $p = \frac{1}{2^{1/4}\,b}\sqrt{r}$, we get

$$\mathfrak{s}_{n-1}\{|f| \geq r\} \leq \frac{(bp)^{2p}}{r^p} = \exp\Big\{-\frac{p}{2}\,\log 2\Big\}$$

provided that $p \geq 2$. In the case $0 < p < 2$, the right-hand side is greater than $1/2$, so that

$$\mathfrak{s}_{n-1}\{|f| \geq r\} \leq 2\,\exp\Big\{-\frac{p}{2}\,\log 2\Big\}$$

for all $p > 0$. It remains to note that $\frac{p}{2}\,\log 2 > \frac{1}{38}\,r^{1/2}$, thus proving (10.18). □

10.8 Large Deviations for Weighted ℓ^p-norms

In the case where the coefficients a_k are bounded, the inequality (10.21) may be further sharpened for relatively small values of r. Consider the functionals

$$l_p(\theta) = \sum_{k=1}^{n} a_k\, |\theta_k|^p, \quad \theta = (\theta_1, \dots, \theta_n) \in \mathbb{S}^{n-1},$$

with $a_k \geq 0$, so that $Q_p = l_p$ when p is an even integer.

Proposition 10.8.1 *If all $a_k \leq 1$, then for all $r \geq 1$ and $p > 2$,*

$$\mathfrak{s}_{n-1}\Big\{n^{\frac{p-2}{2}}\, l_p \geq c_p\, r\Big\} \leq \exp\big\{-(rn)^{2/p}\big\} \tag{10.22}$$

where one may take $c_p = (\sqrt{p}+2)^p$. Correspondingly, if $a_1 + \dots + a_n \leq 1$, then

$$\mathfrak{s}_{n-1}\Big\{n^{\frac{p}{2}}\, l_p \geq c_p\, r\Big\} \leq 2\, \exp\big\{-r^{2/p}\big\} \tag{10.23}$$

with $c_p = (6p)^{p/2}$.

Note that (10.22) with $p = 4$ cannot be obtained on the basis of (10.21). Indeed, since $|\bar{a}| \leq 1$ and using $r - 3 \geq \frac{1}{4} r$ for $r \geq 4$, (10.21) only gives

$$\begin{aligned}\mathfrak{s}_{n-1}\{n\,|Q_4| \geq r\} &\leq \mathfrak{s}_{n-1}\Big\{n^{3/2}\,\Big|Q_4 - \frac{3\,\bar{a}}{n+2}\Big| \geq \frac{1}{4}\, r\sqrt{n}\Big\} \\ &\leq \exp\Big\{-c\sqrt{r}\, n^{1/4}\Big\}.\end{aligned}$$

Proof For simplicity, we assume that $p \geq 3$ is an integer. In the first assertion, one may also assume that all $a_k = 1$, in which case $l_p(\theta) = |\theta_1|^p + \dots + |\theta_n|^p$. We apply the spherical concentration inequality (10.6) to the ℓ_p^n-norm

$$f(x) = \|x\|_p = \Big(\sum_{k=1}^{n} |x_k|^p\Big)^{1/p}, \quad x = (x_1, \dots, x_n) \in \mathbb{R}^n.$$

Since $\|x\|_p \leq \|x\|_2 = |x|$, it has Lipschitz semi-norm 1, and thus

$$\mathfrak{s}_{n-1}\Big\{f \geq \int f\, d\mathfrak{s}_{n-1} + t\Big\} \leq e^{-nt^2/4}$$

for all $t \geq 0$, which implies that

$$\mathfrak{s}_{n-1}\Big\{l_p^{1/p} \geq \Big(\int l_p\, d\mathfrak{s}_{n-1}\Big)^{1/p} + t\Big\} \leq e^{-nt^2/4}. \tag{10.24}$$

Using the formula (10.15), one may derive the bound

$$\int l_p \, d\mathfrak{s}_{n-1} < A_p \, n^{-\frac{p-2}{2}}, \quad A_p = p^{p/2}. \tag{10.25}$$

Indeed, for even powers $p = 2k$, $k = 2, 3, \dots$, it implies

$$\int l_{2k} \, d\mathfrak{s}_{n-1} < \frac{(2k-1)!!}{n^{k-1}} < 2^k k! \, n^{-(k-1)} < (2k)^k \, n^{-(k-1)} = p^{p/2} \, n^{-\frac{p-2}{2}}.$$

Here, the resulting bound also holds for $p = 2k - 1$. This can be shown by using the property that the function $p \to l_p^{1/(p-2)}$ is non-decreasing in $p > 2$ (which is a particular case of Proposition 4.2.1 in view of $l_2 = 1$). It gives $l_{2k-1} \le l_{2k}^{\frac{2k-3}{2k-2}}$ and

$$\begin{aligned}\int l_{2k-1} \, d\mathfrak{s}_{n-1} &\le \Big(\int \theta_1^{2k} \, d\mathfrak{s}_{n-1}(\theta) \Big)^{\frac{2k-3}{2k-2}} \\ &< (2^k k! \, n^{-(k-1)})^{\frac{2k-3}{2k-2}} \; = \; (2^k k!)^{\frac{2k-3}{2k-2}} \, n^{-\frac{p-2}{2}} \\ &< (2k-1)^{\frac{k(2k-3)}{2k-2}} \, n^{-\frac{p-2}{2}} \; < \; (2k-1)^{\frac{2k-1}{2}} \, n^{-\frac{p-2}{2}},\end{aligned}$$

where we used the simple inequality $2^k k! < (2k-1)^k$.

Thus, (10.25) is derived, and returning to (10.24), we obtain a deviation inequality

$$\mathfrak{s}_{n-1}\Big\{ l_p^{1/p} \ge \sqrt{p} \, n^{-\frac{p-2}{2p}} + t \Big\} \le e^{-nt^2/4}.$$

Here, the choice $t = 2r^{1/p} \, n^{-\frac{p-2}{2p}}$ leads to

$$\mathfrak{s}_{n-1}\Big\{ n^{\frac{p-2}{2p}} \, l_p^{1/p} \ge \sqrt{p} + 2r^{1/p} \Big\} \le \exp\Big\{ -(rn)^{2/p} \Big\},$$

that is,

$$\mathfrak{s}_{n-1}\Big\{ n^{\frac{p-2}{2}} \, l_p \ge (\sqrt{p} + 2r^{1/p})^p \Big\} \le \exp\Big\{ -(rn)^{2/p} \Big\}.$$

Since $r \ge 1$, we have $(\sqrt{p} + 2r^{1/p})^p \le (\sqrt{p} + 2)^p \, r$, and the inequality (10.22) follows.

Turning to the second assertion, consider the function $f = n^{p/2} \, l_p$. We apply (10.25), which may be rewritten, using L^p-norms over the measure $\mathfrak{s}_{n-1}$, as the bound $\|\theta_1\|_p \le (\frac{p}{n})^{1/2}$. By the triangle inequality in L^m with an integer $m \ge 1$, this gives

$$\|f\|_m \le n^{p/2} \sum_{k=1}^n a_k \, \| \, |\theta_k|^p \|_m \le n^{p/2} \|\theta_1\|_{mp}^p \le (mp)^{p/2}.$$

Hence, by Markov's inequality, for any $t > 0$,

$$\mathfrak{s}_{n-1}\{f \ge t\} \le t^{-m} \, \|f\|_m^m \le \Big(\frac{(mp)^{p/2}}{t} \Big)^m.$$

Suppose that $t \geq 2\,(2p)^{p/2}$ and consider the largest integer m such that $\frac{(mp)^{p/2}}{t} \leq \frac{1}{2}$, i.e., $m \leq m_t = \frac{1}{p}\,(\frac{t}{2})^{2/p}$, so that

$$\mathfrak{s}_{n-1}\{f \geq t\} \leq 2^{-m}.$$

Since $m_t \geq 2$, necessarily $m \geq \frac{1}{2}\,m_t$, and the above large deviation bound yields

$$\begin{aligned}\mathfrak{s}_{n-1}\{f \geq t\} &\leq \exp\Big\{-\frac{\log 2}{2}\,m_t\Big\} \\ &= \exp\Big\{-\frac{\log 2}{2p \cdot 2^{2/p}}\,t^{2/p}\Big\} \leq \exp\Big\{-\frac{1}{6p}\,t^{2/p}\Big\}, \quad t \geq 2\,(2p)^{p/2},\end{aligned}$$

where we used the assumption $p > 2$.

For $t < 2\,(2p)^{p/2}$, the above right-hand side is larger than $\frac{1}{2}$. So, this inequality is fulfilled automatically with an additional factor 2, that is,

$$\mathfrak{s}_{n-1}\{f \geq t\} \leq 2\,\exp\Big\{-\frac{1}{6p}\,t^{2/p}\Big\}$$

for all $t > 0$. It remains to change the variable $t = (6p)^{p/2}\,r$, and then we arrive at the desired relation (10.23). Proposition 10.8.1 is proved. □

Using $(A_3^{1/3}+2)^3 < 33$ and $(A_4^{1/4}+2)^4 < 121$, we get in (10.22) for the particular cases $p = 3$ and $p = 4$ that, for all $r \geq 1$,

$$\mathfrak{s}_{n-1}\{n^{1/2}\,|l_3| \geq 33\,r\} \leq \exp\big\{-(rn)^{2/3}\big\}, \tag{10.26}$$

$$\mathfrak{s}_{n-1}\{n\,|l_4| \geq 121\,r\} \leq \exp\big\{-(rn)^{1/2}\big\}. \tag{10.27}$$

10.9 Remarks

The material of this chapter is mainly based on the papers [36] (Sections 10.1–10.5 and [35] (Sections 10.6–10.8).

An interesting question we did not discuss in this chapter concerns the asymptotic behavior of smooth symmetric functions on the n-sphere for large n. They may be asymptotically expressed in terms of polynomials of the functions Q_p of this chapter (with identical coefficients $a_k = 1$). If the symmetric functions are centered in the sense that they do not depend on Q_1, the approximation resembles Edgeworth-type expansions in the central limit theorem. A typical asymptotic expansion of first order of this type will be an affine function of Q_3 with an error term of order Q_4 to which exponential bounds apply. For details we refer the interested reader to [95], [96] (and there Theorem 2.2 with order $s = 2$). In these papers it is shown that this scheme provides a universal structure for all Edgeworth-type expansions.

Chapter 11
Linear Functionals on the Sphere

This chapter contains a more detailed analysis of distributions of linear functionals with respect to the normalized Lebesgue measure $\mathfrak{s}_{n-1}$ on the unit sphere $\mathbb{S}^{n-1} \subset \mathbb{R}^n$ ($n \geq 2$). The aim is in particular to quantify the asymptotic normality of these distributions and to include dimensional refinements of such approximation in analogy with Edgeworth expansions (which however we consider up to order 2).

11.1 First Order Normal Approximation

By the rotational invariance of the measure $\mathfrak{s}_{n-1}$, all linear functionals $f(\theta) = \langle \theta, v\rangle$ with $|v| = 1$ have equal distributions under $\mathfrak{s}_{n-1}$. Hence, it is sufficient to focus just on the first coordinate θ_1 of the vector $\theta \in \mathbb{S}^{n-1}$ viewed as a random variable on the probability space $(\mathbb{S}^{n-1}, \mathfrak{s}_{n-1})$. As already mentioned before, this random variable has density

$$c_n \left(1 - x^2\right)_+^{\frac{n-3}{2}}, \quad x \in \mathbb{R}, \ \ c_n = \frac{\Gamma(\frac{n}{2})}{\sqrt{\pi}\,\Gamma(\frac{n-1}{2})}$$

with respect to the Lebesgue measure on the real line, where c_n is a normalizing constant. We denote by φ_n the density of the normalized first coordinate $\sqrt{n}\,\theta_1$, i.e.,

$$\varphi_n(x) = c_n' \left(1 - \frac{x^2}{n}\right)_+^{\frac{n-3}{2}}, \quad c_n' = \frac{c_n}{\sqrt{n}}.$$

We have $c_n' \to \frac{1}{\sqrt{2\pi}}$, $c_n' < \frac{1}{\sqrt{2\pi}}$ (which will be explained in the end of this section), and

$$\varphi_n(x) \to \varphi(x) = \frac{1}{\sqrt{2\pi}}\, \mathrm{e}^{-x^2/2} \quad \text{as } n \to \infty.$$

We are interested in deviation bounds for $\varphi_n(x)$ from $\varphi(x)$ including deviations of the corresponding characteristic functions. First let us consider the asymptotic behavior of the functions

S. Bobkov et al., *Concentration and Gaussian Approximation for Randomized Sums*, Probability Theory and Stochastic Modelling 104,
https://doi.org/10.1007/978-3-031-31149-9_11

$$p_n(x) = \Big(1 - \frac{x^2}{n}\Big)_+^{\frac{n-3}{2}}, \quad x \in \mathbb{R}.$$

Clearly, $p_n(x) \to p(x) = e^{-x^2/2}$ for all x. These functions admit a Gaussian bound. Indeed, for $|x| < \sqrt{n}$ and $n \geq 4$,

$$-\log p_n(x) = -\frac{n-3}{2} \log\Big(1 - \frac{x^2}{n}\Big) \geq \frac{n-3}{2} \frac{x^2}{n}.$$

That is, we have:

Lemma 11.1.1 *If $n \geq 4$, then for all $x \in \mathbb{R}$,*

$$p_n(x) \leq e^{-\frac{n-3}{2n} x^2} \leq e^{-x^2/8}.$$

We also have $p_3(x) = 1$ in $|x| < \sqrt{3}$, while $p_2(x)$ is unbounded.

To study the rate of convergence of $p_n(x)$, assume that $|x| \leq \frac{1}{2}\sqrt{n}$ and $n \geq 4$. By Taylor's expansion, for some $0 \leq \varepsilon \leq 1$,

$$\begin{aligned} -\log p_n(x) &= -\frac{n-3}{2} \log\Big(1 - \frac{x^2}{n}\Big) \\ &= \frac{n-3}{2} \bigg(\frac{x^2}{n} + \Big(\frac{x^2}{n}\Big)^2 \sum_{k=2}^\infty \frac{1}{k} \Big(\frac{x^2}{n}\Big)^{k-2}\bigg) = \frac{n-3}{2} \Big(\frac{x^2}{n} + \frac{x^4}{n^2}\, \varepsilon\Big) \\ &= \frac{x^2}{2} - \frac{3x^2}{2n} + \frac{n-3}{2n^2} x^4 \varepsilon = \frac{x^2}{2} + \frac{x^2}{2n} \Big(-3 + \frac{n-3}{n} x^2 \varepsilon\Big). \end{aligned}$$

That is,

$$p_n(x) = e^{-x^2/2}\, e^{-\delta} \quad \text{with } \delta = \frac{x^2}{2n} \Big(-3 + \frac{n-3}{n} x^2 \varepsilon\Big).$$

Since $\delta \geq -\frac{3x^2}{2n} \geq -\frac{3}{32}$, we have $|e^{-\delta} - 1| \leq |\delta|\, e^{3/32} \leq 2\,|\delta|$. On the other hand,

$$\delta \leq \frac{x^2}{2n} \Big(-3 + \frac{n-3}{n} x^2\Big) \leq \frac{x^4}{2n}, \quad -\delta \leq \frac{3x^2}{2n},$$

giving $2\,|\delta| \leq \frac{1}{n}\,(3x^2 + x^4)$. As a result,

$$|p_n(x) - e^{-x^2/2}| \leq \frac{1}{n}\,(3x^2 + x^4)\, e^{-x^2/2}.$$

This implies in particular that, whenever $|x| \leq \frac{1}{2}\sqrt{n}$,

$$|p_n(x) - e^{-x^2/2}| \leq \frac{c}{n}\, e^{-x^2/4} \tag{11.1}$$

for some absolute constant c. Using the first bound of Lemma 11.1.1 with $n \geq 9$ and assuming that $|x| \geq \frac{1}{2}\sqrt{n}$, we also have $p_n(x) \leq \frac{c}{n}\, e^{-x^2/4}$. Hence, (11.1) holds for all x. In the case $3 \le n \le 8$, (11.1) holds as well, since p_n is bounded and supported on the finite interval $[-\sqrt{n}, \sqrt{n}]$.

Let us now integrate the inequality (11.1). Since

$$\int_{-\infty}^{\infty} p_n(x)\,\mathrm{d}x = \frac{1}{c_n'}, \qquad \int_{-\infty}^{\infty} \mathrm{e}^{-x^2/2}\,\mathrm{d}x = \sqrt{2\pi},$$

we get that $|\frac{1}{c_n'} - \sqrt{2\pi}| \le \frac{c}{n}$ and $|c_n' - \frac{1}{\sqrt{2\pi}}| \le \frac{c}{n}$ for some absolute constant c. Using (11.1), we then arrive at a similar conclusion about the density functions φ_n.

Proposition 11.1.2 *If $n \ge 3$, for all $x \in \mathbb{R}$, with some universal constant $c > 0$,*

$$|\varphi_n(x) - \varphi(x)| \le \frac{c}{n}\,\mathrm{e}^{-x^2/4}.$$

Moreover, in the interval $|x| \le \frac{1}{2}\sqrt{n}$,

$$|\varphi_n(x) - \varphi(x)| \le \frac{c}{n}\,(1 + x^4)\,\mathrm{e}^{-x^2/2}.$$

As for the normalizing constants in the definition of φ_n, let us show that $c_n' < \frac{1}{\sqrt{2\pi}}$. Equivalently, we need to prove that

$$\Gamma\Big(\frac{n}{2}\Big) < \Gamma\Big(\frac{n-1}{2}\Big)\sqrt{\frac{n}{2}}.$$

As we mentioned in the Remarks to Chapter 2 (Section 2.7), for any random variable $\xi > 0$ with a log-concave distribution, the function $u(x) = \log \mathbb{E}\,\xi^x - v(x)$ is concave in $x > 0$, where $v(x) = x \log x$. If ξ has a standard exponential distribution, we have $u(x) = \log \Gamma(x+1) - v(x)$, so, by Jensen's inequality,

$$\log \Gamma\Big(x + \frac{1}{2}\Big) - v\Big(x - \frac{1}{2}\Big) \ge \frac{1}{2}\,\big(\log \Gamma(x+1) - v(x)\big) + \frac{1}{2}\,\big(\log \Gamma(x) - v(x-1)\big)$$

for any $x \ge 1$. Hence, putting $\Delta v(x) = \frac{1}{2}\,v(x) + \frac{1}{2}\,v(x-1) - v(x - \frac{1}{2})$ and using $\Gamma(x) = \Gamma(x+1)/x$, we get

$$\log \Gamma(x+1) - \log \Gamma\Big(x + \frac{1}{2}\Big) - \frac{1}{2}\,\log(x+1) \;\le\; \Delta v(x) - \frac{1}{2}\,\log(1 + 1/x) \;\equiv\; \psi(x).$$

Let us show that $\psi(x)$ is negative in $x > 1$. We have $\psi(1) = 0$, while

$$\psi''(x) = \frac{P(x)}{4x^2(x-1)(x-\frac{1}{2})\,(x+1)^2}, \qquad P(x) = -3x^3 + 6x^2 + 2x - 1.$$

By the Taylor integral formula, $\Delta v(x) \to 0$ as $x \to \infty$, so, $\psi(\infty) = 0$.

The quadratic equation $P'(x) = 0$ has two roots $x_{1,2} = \frac{1}{3} \pm \frac{1}{3}\sqrt{3} < 1$, so that $P(x)$ is decreasing for $x > 1$. Since $P(1) = 4$ and $P(\infty) = -\infty$, there is a unique point $x_0 > 1$ such that $P(x) > 0$ in $1 \le x < x_0$ and $P(x) < 0$ in $x > x_0$. Hence, ψ is convex in $[1, x_0]$ and is concave in $[x_0, \infty)$, implying that $\psi(x) < 0$ for $x > 1$. In particular,

$$\Gamma(x+1) < \Gamma\Big(x + \frac{1}{2}\Big)\sqrt{x+1}, \quad x \geq 1,$$

which is only needed for $x = \frac{n}{2} - 1$.

11.2 Second Order Approximation

We now sharpen Proposition 11.1.2 by providing approximations for $\varphi_n(x)$ with errors of order $1/n^2$ by means of a suitable modification of the standard normal density. Assuming again that $|x| \leq \frac{1}{2}\sqrt{n}$ and $n \geq 3$, let us refine Taylor's expansion considered before. Namely, for $p_n(x) = (1 - \frac{x^2}{n})^{\frac{n-3}{2}}$ with some $0 \leq \varepsilon \leq 1$, we have

$$\begin{aligned}
-\log p_n(x) &= -\frac{n-3}{2} \log\Big(1 - \frac{x^2}{n}\Big) \\
&= \frac{n-3}{2}\Big(\frac{x^2}{n} + \frac{x^4}{2n^2} + \Big(\frac{x^2}{n}\Big)^3 \sum_{k=3}^{\infty} \frac{1}{k}\Big(\frac{x^2}{n}\Big)^{k-3}\Big) \\
&= \frac{n-3}{2}\Big(\frac{x^2}{n} + \frac{x^4}{2n^2} + \frac{x^6}{n^3}\frac{2\varepsilon}{3}\Big) = \frac{x^2}{2} - \frac{3x^2}{2n} + \frac{x^4}{4n} - \frac{1}{n^2}\Big(\frac{3}{4}x^4 - \frac{n-3}{3n}x^6\varepsilon\Big).
\end{aligned}$$

The remainder term

$$\delta = -\frac{3x^2}{2n} + \frac{x^4}{4n} - \frac{1}{n^2}\Big(\frac{3}{4}x^4 - \frac{n-3}{3n}x^6\varepsilon\Big)$$

satisfies $\delta \geq -\frac{3x^2}{2n} + \frac{x^4}{4n} - \frac{3x^4}{4n^2} > -\frac{27}{64}$. Hence

$$|e^{-\delta} - 1 + \delta| \leq \frac{\delta^2}{2}\, e^{27/64} \leq \delta^2.$$

Moreover, using $x^2 \leq \frac{1}{4}n$, we get

$$\begin{aligned}
|\delta| &\leq \frac{3x^2}{2n} + \frac{x^4}{4n} + \frac{1}{n^2}\Big(\frac{3}{4}x^4 + \frac{1}{3}x^6\Big) \\
&\leq \frac{3x^2}{2n} + \frac{x^4}{4n} + \frac{1}{4n}\Big(\frac{3}{4}x^2 + \frac{1}{3}x^4\Big) = \frac{x^2}{n}\Big(\frac{27}{16} + \frac{1}{3}x^2\Big).
\end{aligned}$$

Using $(a+b)^2 \leq 2a^2 + 2b^2$, we obtain that $\delta^2 \leq \frac{x^4}{n^2}\,(6 + \frac{2}{9}x^4)$. Hence,

$$\begin{aligned}
e^{x^2/2}\, p_n(x) &= e^{-\delta} = 1 - \delta + \varepsilon_1 \delta^2 \\
&= 1 + \frac{3x^2}{2n} - \frac{x^4}{4n} + \frac{1}{n^2}\Big(\frac{3}{4}x^4 - \frac{n-3}{3n}x^6\varepsilon\Big) + \varepsilon_2 \frac{x^4}{n^2}\Big(6 + \frac{2}{9}x^4\Big) \\
&= 1 + \frac{3x^2}{2n} - \frac{x^4}{4n} + \frac{A}{n^2},
\end{aligned}$$

for some $|\varepsilon_j| \leq 1$. Here

$$\begin{aligned}
|A| &\leq \Big|\frac{3}{4}x^4 - \frac{n-3}{3n}x^6\varepsilon\Big| + x^4\Big(6 + \frac{2}{9}x^4\Big) \\
&\leq \frac{3}{4}x^4 + \frac{1}{3}x^6 + x^4\Big(6 + \frac{2}{9}x^4\Big) \leq 7x^4 + \frac{1}{3}x^6 + \frac{2}{9}x^8 \leq 8x^4 + x^8.
\end{aligned}$$

One can summarize.

Lemma 11.2.1 *If $n \geq 3$, then in the interval $|x| \leq \frac{1}{2}\sqrt{n}$,*

$$p_n(x) = e^{-x^2/2}\Big[1 + \frac{6x^2 - x^4}{4n} + \frac{\varepsilon}{n^2}\,(8x^4 + x^8)\Big], \quad |\varepsilon| \leq 1.$$

We are now ready to derive a similar expansion for $\varphi_n(x)$, expressing its deviations from the standard normal density $\varphi(x)$ up to the quadratic term $1/n^2$. Denoting by Z a standard normal random variable, from Lemma 11.2.1 we obtain that

$$\frac{1}{\sqrt{2\pi}} \int_{-\infty}^{\infty} p_n(x)\,dx = 1 + \frac{1}{4n}\,(6\,\mathbb{E}Z^2 - \mathbb{E}Z^4) + O\Big(\frac{1}{n^2}\Big) = 1 + \frac{3}{4n} + O\Big(\frac{1}{n^2}\Big).$$

Here we used the property that $p_n(x)$ has a sufficiently fast decay for $|x| \geq \frac{1}{2}\sqrt{n}$, as indicated in Lemma 11.1.1. Since $\varphi_n(x) = c_n'\, p_n(x)$ is a density, we conclude that

$$1 = c_n'\sqrt{2\pi}\,\Big(1 + \frac{3}{4n} + O\Big(\frac{1}{n^2}\Big)\Big), \quad c_n'\sqrt{2\pi} = 1 - \frac{3}{4n} + O\Big(\frac{1}{n^2}\Big).$$

From this, by Lemma 11.2.1, for $|x| \leq \frac{1}{2}\sqrt{n}$,

$$\begin{aligned}
\sqrt{2\pi}\, e^{x^2/2}\, \varphi_n(x) &= \Big(1 - \frac{3}{4n} + O\Big(\frac{1}{n^2}\Big)\Big)\Big[1 + \frac{6x^2 - x^4}{4n} + \frac{\varepsilon}{n^2}\Big(8x^4 + x^8\Big)\Big] \\
&= 1 + \frac{6x^2 - x^4}{4n} - \frac{3}{4n} + O\Big(\frac{1 + x^8}{n^2}\Big).
\end{aligned}$$

Using the 4-th Chebyshev–Hermite polynomial $H_4(x) = x^4 - 6x^2 + 3$, we arrive at:

Proposition 11.2.2 *In the interval $|x| \leq \frac{1}{2}\sqrt{n}$, $n \geq 3$, the random variable $\theta_1\sqrt{n}$ has a density satisfying*

$$\varphi_n(x) = \varphi(x)\,\Big[1 - \frac{H_4(x)}{4n} + C\,\frac{1 + x^8}{n^2}\Big],$$

where the quantity C is bounded in absolute value by a universal constant.

Taking into account Lemma 11.1.1, we also have a non-uniform Gaussian bound on the whole real line.

Proposition 11.2.3 *For all $x \in \mathbb{R}$, $n \geq 3$, with some universal constant $C > 0$,*

$$\Big|\varphi_n(x) - \varphi(x)\Big(1 - \frac{H_4(x)}{4n}\Big)\Big| \leq \frac{C}{n^2}\, \mathrm{e}^{-x^2/4}.$$

11.3 Characteristic Function of the First Coordinate

In the sequel we use the following notation.

Definition. *Denote by $J_n = J_n(t)$ the characteristic function of the first coordinate θ_1 of a random vector $\theta = (\theta_1, \dots, \theta_n)$ which is uniformly distributed on $\mathbb{S}^{n-1}$.*

In a more explicit form, for any $t \in \mathbb{R}$,

$$\begin{aligned} J_n(t) &= c_n \int_{-\infty}^{\infty} \mathrm{e}^{itx}\, (1 - x^2)_+^{\frac{n-3}{2}}\, \mathrm{d}x \\ &= c_n' \int_{-\infty}^{\infty} \mathrm{e}^{itx/\sqrt{n}} \Big(1 - \frac{x^2}{n}\Big)_+^{\frac{n-3}{2}}\, \mathrm{d}x, \qquad c_n' = \frac{c_n}{\sqrt{n}}, \quad c_n = \frac{\Gamma(\frac{n}{2})}{\sqrt{\pi}\,\Gamma(\frac{n-1}{2})}. \end{aligned}$$

Note that the equality

$$\widetilde{J}_\nu(t) = \frac{1}{\sqrt{\pi}\,\Gamma(\nu + \frac{1}{2})} \Big(\frac{t}{2}\Big)^\nu \int_{-1}^{1} \mathrm{e}^{itx}\, (1 - x^2)^{\nu - \frac{1}{2}}\, \mathrm{d}x$$

defines the Bessel function of the first kind with index ν (cf. [11], p. 81). Therefore,

$$J_n(t) = \frac{1}{c_n} \sqrt{\pi}\, \Gamma\Big(\nu + \frac{1}{2}\Big) \Big(\frac{t}{2}\Big)^{-\nu} \widetilde{J}_\nu(t) \quad \text{with } \nu = \frac{n}{2} - 1.$$

Since this relationship will not be needed, we prefer to use the similar notation J_n as stated in Definition.

Thus, the characteristic function of the normalized first coordinate $\theta_1 \sqrt{n}$ is given by

$$\hat{\varphi}_n(t) = J_n\big(t\sqrt{n}\big) = \int_{-\infty}^{\infty} \mathrm{e}^{itx} \varphi_n(x)\, \mathrm{d}x,$$

which is the Fourier transform of the probability density φ_n.

Propositions 11.1.2 and 11.2.3 can be used, respectively, to compare $\hat{\varphi}_n(t)$ with the Fourier transform $\mathrm{e}^{-t^2/2}$ of the standard normal distribution and with the Fourier transform of the "corrected Gaussian measure", as well as to compare the derivatives of these transforms. Indeed, in general, if we have two integrable functions on the real line, say, p and q, their Fourier transforms

$$\hat{p}(t) = \int_{-\infty}^{\infty} \mathrm{e}^{itx} p(x)\,\mathrm{d}x, \quad \hat{q}(t) = \int_{-\infty}^{\infty} \mathrm{e}^{itx} q(x)\,\mathrm{d}x$$

satisfy, for all $t \in \mathbb{R}$,

$$|\hat{p}(t) - \hat{q}(t)| \le \int_{-\infty}^{\infty} |p(x) - q(x)|\,\mathrm{d}x.$$

In Propositions 11.1.2 and 11.2.3, the density $p = \varphi_n$ is respectively approximated by the functions

$$q(x) = \varphi(x) \quad \text{and} \quad q(x) = \varphi(x)\Big(1 - \frac{H_4(x)}{4n}\Big).$$

Since

$$\int_{-\infty}^{\infty} \mathrm{e}^{itx} \varphi(x)\,\mathrm{d}x = \mathrm{e}^{-t^2/2}, \quad \int_{-\infty}^{\infty} \mathrm{e}^{itx} H_4(x)\varphi(x)\,\mathrm{d}x = t^4 \mathrm{e}^{-t^2/2},$$

the associated Fourier transforms are described as

$$\hat{q}(t) = \mathrm{e}^{-t^2/2}, \quad \hat{q}(t) = \mathrm{e}^{-t^2/2}\Big(1 - \frac{t^4}{4n}\Big).$$

Therefore, applying the bounds in these propositions, we respectively get

$$|\hat{p}(t) - \hat{q}(t)| \le \frac{C}{n} \int_{-\infty}^{\infty} \mathrm{e}^{-x^2/4}\,\mathrm{d}x, \quad |\hat{p}(t) - \hat{q}(t)| \le \frac{C}{n^2} \int_{-\infty}^{\infty} \mathrm{e}^{-x^2/4}\,\mathrm{d}x.$$

Let us state this once more.

Proposition 11.3.1 *For all $t \in \mathbb{R}$, we have*

$$\Big|J_n(t\sqrt{n}) - \mathrm{e}^{-t^2/2}\Big| \le \frac{C}{n},$$

where C is an absolute constant. Moreover,

$$\Big|J_n(t\sqrt{n}) - \Big(1 - \frac{t^4}{4n}\Big)\mathrm{e}^{-t^2/2}\Big| \le \frac{C}{n^2}.$$

Returning to the general scheme, let us note that, when $\int_{-\infty}^{\infty} |x|^k\, p(x)\,\mathrm{d}x < \infty$ and $\int_{-\infty}^{\infty} |x|^k\, q(x)\,\mathrm{d}x < \infty$, the functions $\hat{p}$ and $\hat{q}$ are k times continuously differentiable, and their derivatives of the k-th order are given by

$$\frac{\mathrm{d}^k}{\mathrm{d}t^k}\,\hat{p}(t) = \int_{-\infty}^{\infty} \mathrm{e}^{itx}\,(ix)^k\,p(x)\,\mathrm{d}x$$

and similarly for q. Hence

$$\left|\frac{\mathrm{d}^k}{\mathrm{d}t^k}\,\hat{p}(t) - \frac{\mathrm{d}^k}{\mathrm{d}t^k}\,\hat{q}(t)\right| \leq \int_{-\infty}^{\infty} |x|^k\,|p(x) - q(x)|\,\mathrm{d}x.$$

Since

$$\int_{-\infty}^{\infty} |x|^k\,\mathrm{e}^{-x^2/4}\,\mathrm{d}x = 2^{k+1}\,\Gamma\Big(\frac{k+1}{2}\Big) \leq (ck)^{k/2}$$

for some absolute constant $c > 0$, the statement of Proposition 11.3.1 may be extended to all derivatives.

Proposition 11.3.2 *For all $k = 0, 1, \dots$ and $t \in \mathbb{R}$,*

$$\left|\frac{\mathrm{d}^k}{\mathrm{d}t^k}\,J_n\big(t\sqrt{n}\big) - \frac{\mathrm{d}^k}{\mathrm{d}t^k}\,\mathrm{e}^{-t^2/2}\right| \leq \frac{(Ck)^{k/2}}{n},$$

where C is an absolute constant. Moreover,

$$\left|\frac{\mathrm{d}^k}{\mathrm{d}t^k}\,J_n\big(t\sqrt{n}\big) - \frac{\mathrm{d}^k}{\mathrm{d}t^k}\,\Big(\Big(1 - \frac{t^4}{4n}\Big)\,\mathrm{e}^{-t^2/2}\Big)\right| \leq \frac{(Ck)^{k/2}}{n^2}.$$

In particular, taking $k = 1$, we have

$$\left|\big(J_n\big(t\sqrt{n}\big)\big)' + t\mathrm{e}^{-t^2/2}\right| \leq \frac{C}{n},$$

$$\left|\big(J_n\big(t\sqrt{n}\big)\big)' - \Big(\frac{t^5}{4n} - \frac{t^3}{n} - t\Big)\,\mathrm{e}^{-t^2/2}\right| \leq \frac{C}{n^2}.$$

One may also add a t-depending factor on the right-hand side. For t of order 1, this can be done just by virtue of Taylor's formula. Indeed, the first three derivatives of the functions

$$f_n(t) = J_n\big(t\sqrt{n}\big) = \mathbb{E}\,\mathrm{e}^{it\theta_1\sqrt{n}}, \quad g_n(t) = \Big(1 - \frac{t^4}{4n}\Big)\,\mathrm{e}^{-t^2/2}$$

are equal at zero. Since also $|f_n^{(4)}(t) - g_n^{(4)}(t)| \leq \frac{C}{n^2}$ according to Proposition 11.3.2, the Taylor formula refines this proposition for the interval $|t| \leq 1$.

Corollary 11.3.3 *With some absolute constant C we have, for all $t \in \mathbb{R}$,*

$$\left|J_n\big(t\sqrt{n}\big) - \Big(1 - \frac{t^4}{4n}\Big)\,\mathrm{e}^{-t^2/2}\right| \leq \frac{C}{n^2}\,\min\{1, t^4\},$$

$$\left|\big(J_n\big(t\sqrt{n}\big)\big)' + t\Big(1 + \frac{4t^2 - t^4}{4n}\Big)\,\mathrm{e}^{-t^2/2}\right| \leq \frac{C}{n^2}\,\min\{1, |t|^3\}.$$

The process of expanding the values $\varphi_n(x)$ and their Fourier transforms $J_n\big(t\sqrt{n}\big)$ in powers of $1/n$ can be done in analogy with Edgeworth expansions in the central limit theorem. Let us mention without a detailed derivation the next approximation, that is, the third order approximation (which however will not be needed for our applications). As before, $H_4(x) = x^4 - 6x^2 + 3$ denotes the 4-th Chebyshev–Hermite polynomial.

Proposition 11.3.4 *In the interval $|x| \le \frac{1}{2}\sqrt{n}$, $n \ge 3$, the random variable $\theta_1\sqrt{n}$ has a density satisfying*

$$\varphi_n(x) = \varphi(x)\left[1 - \frac{H_4(x)}{4n} + \frac{P(x)}{n^2} + C\,\frac{1 + x^{10}}{n^3}\right],$$

where the quantity C is bounded by a universal constant, and where P is a (universal) polynomial of degree 8.

Applying Lemma 11.1.1, we also have a non-uniform approximation for $\varphi_n(x)$ on the whole real line and a corresponding assertion about its Fourier transform.

Corollary 11.3.5 *For all $x \in \mathbb{R}$, with some universal constant C,*

$$\left|\varphi_n(x) - \varphi(x)\Big(1 - \frac{H_4(x)}{4n} + \frac{P(x)}{n^2}\Big)\right| \le \frac{C}{n^3}\,e^{-x^2/4}.$$

Therefore, with some universal polynomial Q of degree 8, for all $t \in \mathbb{R}$,

$$\left|J_n\big(t\sqrt{n}\big) - \Big(1 - \frac{t^4}{4n} + \frac{Q(t)}{n^2}\Big)\,e^{-t^2/2}\right| \le \frac{C}{n^3}.$$

11.4 Upper Bounds on the Characteristic Function

Although Proposition 11.3.1 provides good approximations for

$$\begin{aligned} J_n\big(t\sqrt{n}\big) &= \frac{c_n}{\sqrt{n}} \int_{-\sqrt{n}}^{\sqrt{n}} e^{itx}\Big(1 - \frac{x^2}{n}\Big)^{\frac{n-3}{2}}\,dx \\ &= c_n \int_{-1}^{1} e^{itx\sqrt{n}}\,\big(1 - x^2\big)^{\frac{n-3}{2}}\,dx, \qquad c_n = \frac{\Gamma(\frac{n}{2})}{\sqrt{\pi}\,\Gamma(\frac{n-1}{2})}, \end{aligned}$$

with respect to the growing dimension n, it does not say anything about the decay of these characteristic functions on large t-intervals such as $|t| \le c\sqrt{n}$ (which should be similar to the Gaussian characteristic function). To this aim, let us describe a simple approach based on contour integration, assuming temporarily that $n \ge 4$.

The function $z \to (1 - z^2)^{\frac{n-3}{2}}$ is analytic in the whole complex plane when n is odd and in the strip $z = x + iy$, $|x| < 1$, when n is even. Therefore, integrating along the boundary C of the rectangle $[-1, 1] \times [0, y]$ with $y > 0$ (slightly modifying the contour in a standard way near the points -1 and 1), we have

$$\int_C e^{itz\sqrt{n}}\,\big(1 - z^2\big)^{\frac{n-3}{2}}\,dz = 0.$$

Then we obtain a natural decomposition $J_n(t\sqrt{n}) = c_n\,(I_1(t) + I_2(t) + I_3(t))$, which we consider for $t > 0$, where

$$I_1(t) = \mathrm{e}^{-ty\sqrt{n}}\,(1+y^2)^{\frac{n-3}{2}} \int_{-1}^{1} \mathrm{e}^{itx\sqrt{n}} \Big(\frac{1-(x+iy)^2}{1+y^2}\Big)^{\frac{n-3}{2}}\,\mathrm{d}x,$$

$$I_2(t) = -\mathrm{e}^{it\sqrt{n}} \int_0^y \mathrm{e}^{-ts\sqrt{n}}\,\big(1-(1+is)^2\big)^{\frac{n-3}{2}}\,\mathrm{d}s,$$

$$I_3(t) = \mathrm{e}^{-it\sqrt{n}} \int_0^y \mathrm{e}^{-ts\sqrt{n}}\,\big(1-(1-is)^2\big)^{\frac{n-3}{2}}\,\mathrm{d}s.$$

For $0 \le s \le y \le \alpha$,

$$\big|1-(1+is)^2\big| = s\sqrt{s^2+4} \le \alpha\,\sqrt{\alpha^2+4} \equiv \beta.$$

Choosing $\alpha = \frac{1}{\sqrt{6}}$, we have $\beta = \frac{5}{6}$. Hence, for all $t > 0$,

$$|I_2(t)| \le \beta^{\frac{n-3}{2}} \int_0^y \mathrm{e}^{-ts\sqrt{n}}\,\mathrm{d}s \le \frac{1}{t\sqrt{n}}\,\beta^{\frac{n-3}{2}}, \quad |I_3(t)| \le \frac{1}{t\sqrt{n}}\,\beta^{\frac{n-3}{2}}. \tag{11.2}$$

In order to estimate $I_1(t)$, we use an elementary identity

$$\begin{aligned}\big|1-(x+iy)^2\big|^2 &= (1-x^2+y^2)^2 + 4x^2y^2 \\ &= (1-2x^2)(1+y^2)^2 + x^2\,(x^2+6y^2+2y^4) \quad (x,y \in \mathbb{R}),\end{aligned}$$

which for the region $|x| \le 1$ yields

$$\Big|\frac{1-(x+iy)^2}{1+y^2}\Big|^2 \le 1 - 2x^2 + x^2\,v(y^2), \quad v(z) = \frac{1+6z+2z^2}{(1+z)^2}.$$

Since $v'(z) = \frac{4-2z}{(1+z)^3} > 0$ in $0 \le z < 2$, this function increases in this interval, and assuming that $z = y^2 \le \frac{1}{6}$, we have $v(y^2) \le v(1/6) = \frac{74}{49}$. Hence

$$\Big|\frac{1-(x+iy)^2}{1+y^2}\Big|^2 \le 1 - \frac{24}{49}\,x^2 \le \exp\Big\{-\frac{24}{49}\,x^2\Big\}.$$

Using this estimate and assuming that n is sufficiently large, e.g. $n \ge 12$, so that $n-3 \ge \frac{3}{4}\,n$, we have

$$\begin{aligned}\int_{-1}^{1} \Big(\frac{|1-(x+iy)^2|}{1+y^2}\Big)^{\frac{n-3}{2}}\,\mathrm{d}x &\le \int_{-\infty}^{\infty} \mathrm{e}^{-\frac{12}{49}\,\frac{n-3}{2}\,x^2}\,\mathrm{d}x \\ &= \sqrt{\frac{49}{12}}\,\sqrt{\frac{2\pi}{n-3}} \le \frac{7}{3}\,\sqrt{\frac{2\pi}{n}}.\end{aligned}$$

This upper bound allows us to conclude that

$$|I_1(t)| \le \frac{7}{3}\sqrt{\frac{2\pi}{n}}\, e^{-ty\sqrt{n}}\, (1+y^2)^{\frac{n-3}{2}} \le \frac{7}{3}\sqrt{\frac{2\pi}{n}} \exp\Big\{-ty\sqrt{n} + \frac{n-3}{2}\, y^2\Big\}.$$

Choosing here $y = \frac{t}{\sqrt{n}}$, the expression in the exponent will be smaller than $t^2/2$. Recalling that $c_n = c_n'\sqrt{n} < \sqrt{n/(2\pi)}$, we then get

$$c_n\, |I_1(t)| \le \frac{7}{3}\, e^{-t^2/2}, \quad 0 \le t \le \sqrt{n/6}. \tag{11.3}$$

In the case $t > \sqrt{n/6}$, we choose $y = \frac{1}{\sqrt{6}}$ and then $-ty\sqrt{n} + \frac{n-3}{2}\, y^2 < -\frac{n}{12} - \frac{1}{4}$, so that

$$c_n\, |I_1(t)| \le \frac{7}{3}\, e^{-\frac{n}{12}-\frac{1}{4}}, \quad t \ge \sqrt{n/6}. \tag{11.4}$$

Let us collect these estimates, still assuming that $n \ge 12$. We use the bound $\beta^{\frac{n-3}{2}} \le C\, e^{-n/12}$. Here, since $\sqrt{\beta} < e^{-1/12}$, the optimal constant corresponds to $n = 12$, so, $\frac{C}{\sqrt{2\pi}} = \frac{1}{\sqrt{2\pi}}\, (\frac{5}{6})^{9/2}\, e < \frac{1}{2}$. Hence, by (11.2), we have, whenever $t \ge \sqrt{2}$,

$$c_n\, \big(|I_2(t)| + |I_3(t)|\big) \le \sqrt{2}\, \frac{c_n}{\sqrt{n}}\, \beta^{\frac{n-3}{2}} \le \frac{1}{\sqrt{2}}\, e^{-n/12}.$$

Combining this bound with (11.3), for the interval $\sqrt{2} \le t \le \sqrt{n/6}$ we obtain that

$$c_n\, \big(|I_1(t)| + |I_2(t)| + |I_3(t)|\big) \le \frac{1}{\sqrt{2}}\, e^{-n/12} + \frac{7}{3}\, e^{-t^2/2}.$$

Similarly, in the case $t > \sqrt{n/6}$, we use (11.4), leading to

$$c_n\, \big(|I_1(t)| + |I_2(t)| + |I_3(t)|\big) \le \frac{1}{\sqrt{2}}\, e^{-n/12} + \frac{7}{3}\, e^{-1/4}\, e^{-n/12} < 3\, e^{-n/12}.$$

Finally, if $0 \le t \le \sqrt{2}$, then $|J_n(t\sqrt{n})| \le 1 < 3\, e^{-t^2/2}$. One can summarize.

Proposition 11.4.1 *For all* $t \in \mathbb{R}$,

$$\big|J_n(t\sqrt{n})\big| \le 3\, e^{-t^2/2} + 3\, e^{-n/12}. \tag{11.5}$$

Here the assumption $n \ge 12$ may be removed, since $3\, e^{-n/12} > 1$ for $n < 12$, while $|J_n| \le 1$. Note that in the region $|t| \le \sqrt{n/6}$, the bound (11.5) yields

$$\big|J_n(t\sqrt{n})\big| \le 6\, e^{-t^2/2},$$

which complements the normal approximation given in Proposition 11.3.1.

11.5 Polynomial Decay at Infinity

Although the right-hand side of (11.5) is of an exponentially small order with respect to n for $|t| \geq \sqrt{n}$, this inequality does not reflect properties such as integrability of the characteristic function $J_n(t\sqrt{n})$ and polynomial decay in t at infinity. Indeed, for example, in the case $n = 3$, θ_1 has a uniform distribution on $[-1, 1]$, so that $J_3(t) = \frac{\sin t}{t}$, implying $|J_3(t)| \leq \frac{1}{|t|}$. More generally, for $n \geq 4$ and $t > 0$, let us integrate by parts to get

$$J_n(t) = \frac{c_n(n-3)}{it} \int_{-1}^{1} e^{itx}\, x(1-x^2)^{\frac{n-5}{2}}\, dx.$$

This implies that

$$\begin{aligned} |J_n(t)| &\leq \frac{2c_n(n-3)}{t} \int_{0}^{1} x(1-x^2)^{\frac{n-5}{2}}\, dx \\ &= \frac{c_n(n-3)}{t} \int_{0}^{1} (1-y)^{\frac{n-5}{2}}\, dy = \frac{2c_n}{t}. \end{aligned}$$

Since $c_n < \frac{1}{2}\sqrt{n}$, we may conclude (including the case $n = 3$).

Proposition 11.5.1 *If $n \geq 3$ and $t \neq 0$, then $|J_n(t\sqrt{n})| \leq 1/|t|$.*

Integrating by parts once more, we get, for $n \geq 6$ and $t \neq 0$,

$$\begin{aligned} J_n(t) &= \frac{c_n(n-3)}{(it)^2} \int_{-1}^{1} x(1-x^2)^{\frac{n-5}{2}}\, de^{itx} \\ &= \frac{c_n(n-3)}{t^2} \int_{-1}^{1} e^{itx}\Big((1-x^2)^{\frac{n-5}{2}} - (n-5)\,x^2(1-x^2)^{\frac{n-7}{2}}\Big)\, dx. \end{aligned}$$

This gives

$$|J_n(t)| \leq \frac{c_n(n-3)}{t^2} \bigg[\int_{-1}^{1} (1-x^2)^{\frac{n-5}{2}}\, dx + (n-5) \int_{-1}^{1} x^2\,(1-x^2)^{\frac{n-7}{2}}\, dx\bigg].$$

Here the first integral is equal to c_{n-2}^{-1}, and the second one with factor $n - 5$ is equal to

$$-\int_{-1}^{1} x\, d(1-x^2)^{\frac{n-5}{2}} = \int_{-1}^{1} (1-x^2)^{\frac{n-5}{2}}\, dx = c_{n-2}^{-1}.$$

Hence

$$|J_n(t)| \leq \frac{2(n-3)\, c_n c_{n-2}^{-1}}{t^2}.$$

On the other hand, $c_n\, c_{n-2}^{-1} = \frac{n-2}{n-3} < 1$. We can summarize.

Proposition 11.5.2 *If $n \geq 6$ and $t \neq 0$, then $|J_n(t\sqrt{n})| \leq 2/t^2$.*

One may continue this process to sharpen the decay in t and to properly reflect the dependence on the dimension n. Here, the Gegenbauer polynomials can be of help, since they form an orthogonal system in L^2 on the interval $[-1, 1]$ with weight function $(1-x^2)^{\alpha-\frac{1}{2}}$. However, we prefer a different route to the problem, which is based on a direct application of contour integration and Cauchy's theorem.

To describe the corresponding refinement, which is needed for the region $|t| \geq c\sqrt{n}$, let us modify the approach used in the proof of Proposition 11.4.1 and integrate the function $e^{itz}\,(1-z^2)^{\frac{n-3}{2}}$ along the contour containing the segment $[-1, 1]$ as before and the upper semi-circle $C^+ = \{e^{iu} : 0 \leq u \leq \pi\}$. In this case, by Cauchy's theorem,

$$J_n(t) = -c_n \int_{C^+} e^{itz}\,(1-z^2)^{\frac{n-3}{2}}\,dz.$$

Since $|1-z^2| = 2\,|\sin u|$ for $z = e^{iu}$, we get for $t > 0$

$$|J_n(t)| \leq c_n \int_0^\pi e^{-t\,\sin u}\,(2\sin u)^{\frac{n-3}{2}}\,du = 2c_n \int_0^{\pi/2} e^{-t\,\sin u}\,(2\sin u)^{\frac{n-3}{2}}\,du.$$

Using the bounds $u \geq \sin u \geq \frac{2}{\pi}u$ in the last integrand, the above is simplified to

$$\begin{aligned} |J_n(t)| &\leq 2c_n \int_0^\infty e^{-\frac{2t}{\pi}u}\,(2u)^{\frac{n-3}{2}}\,du \\ &= c_n \Big(\frac{\pi}{t}\Big)^{\frac{n-1}{2}} \Gamma\Big(\frac{n-1}{2}\Big) = \frac{1}{\sqrt{\pi}} \Big(\frac{\pi}{t}\Big)^{\frac{n-1}{2}} \Gamma\Big(\frac{n}{2}\Big). \end{aligned}$$

To further simplify, we recall the estimate $\Gamma(x+1) \leq x^x$ for $x \geq 1$ (cf. Section 1.3). This gives $\Gamma(\frac{n}{2}) \leq (\frac{n-2}{2})^{\frac{n-2}{2}} < (\frac{n}{2})^{\frac{n-1}{2}}$ for $n \geq 4$. The resulting bound is clearly valid for $n = 3$ as well. Thus, we arrive at:

Proposition 11.5.3 *If $n \geq 3$ and $t \neq 0$, then*

$$|J_n(t)| \leq \frac{1}{\sqrt{\pi}} \Big(\frac{\pi n}{2\,|t|}\Big)^{\frac{n-1}{2}}. \tag{11.6}$$

Thus, if n is fixed, $J_n(t) = O(t^{-\frac{n-1}{2}})$ as $t \to \infty$, and as is easy to check, the power $\frac{n-1}{2}$ is optimal, since it corresponds to the power $\frac{n-1}{2} - 1$ in the formula for the density of θ_1 under $\mathfrak{s}_{n-1}$.

In addition, (11.6) indicates exponential smallness of $|J_n(t\sqrt{n})|$ with respect to n in the region, say, $t \geq \pi\sqrt{n}$. Nevertheless, let us rewrite this bound (almost equivalently) to fit the form of the bound of Proposition 11.4.1. Since

$$\frac{\pi\sqrt{n}}{2t} \leq \Big(1 + \frac{3t^2}{\pi^2 n}\Big)^{-1/2}, \quad t \geq \pi\sqrt{n},$$

from (11.6) we get

$$|J_n(t\sqrt{n})| \leq \frac{1}{\sqrt{\pi}} \Big(1 + \frac{3t^2}{\pi^2 n}\Big)^{-\frac{n-1}{4}}. \tag{11.7}$$

This bound is as good as (11.5), with the additional feature that it continues to hold for the values $0 < t < \pi\sqrt{n}$ as well (modulo absolute constants). To see this, first note that in this region

$$\Big(1 + \frac{t^2}{3\pi^2 n}\Big)^{-\frac{n-1}{4}} \geq \Big(1 + \frac{1}{3}\Big)^{-\frac{n-1}{4}} \geq e^{-n/12}, \quad n \geq 3.$$

Also, $(1 + \frac{4\varepsilon}{n})^{-\frac{n-1}{4}} \geq e^{-\varepsilon}$ for all $\varepsilon \geq 0$. Hence, choosing $\varepsilon = \frac{1}{2}t^2$, we have

$$e^{-t^2/2} \leq \Big(1 + \frac{2t^2}{n}\Big)^{-\frac{n-1}{4}} \leq \Big(1 + \frac{t^2}{3\pi^2 n}\Big)^{-\frac{n-1}{4}}.$$

One may summarize, by combining (11.7) with the bound (11.5).

Proposition 11.5.4 *If $n \geq 3$, then for all $t \in \mathbb{R}$, with some universal constant,*

$$\big|J_n(t\sqrt{n})\big| \leq 6\Big(1 + \frac{ct^2}{n}\Big)^{-\frac{n-1}{4}}. \tag{11.8}$$

One may take $c = 1/(3\pi^2)$.

Here, the right-hand side describes the correct polynomial behavior, properly reflecting the role of the dimension n. In fact, (11.8) holds for $n = 2$ as well.

11.6 Remarks

The first order approximation as in Proposition 11.1.2 and the Gaussian bound of Proposition 11.4.1 (with somewhat worse absolute constants) can be found in [38]. A second order approximation was developed in [40]. Note that Proposition 11.5.3 seems to be new.

Part IV
First Applications to Randomized Sums

Chapter 12
Typical Distributions

In this part (Chapters 12–14) we describe various aspects of Sudakov's theorem. We start with the so-called *typical* distributions in his theorem and compare them with the standard normal law by means of the variance-type functionals including $\sigma_4^2(X)$.

12.1 Concentration Problems for Weighted Sums

Consider the model of random vectors $X = (X_1, \dots, X_n)$ in $\mathbb{R}^n$, $n \geq 2$, in full generality as introduced in Chapter 1. We denote by F_θ the distribution of the weighted sum

$$\langle X, \theta \rangle = \sum_{k=1}^{n} \theta_k X_k, \quad \theta = (\theta_1, \dots, \theta_n) \in \mathbb{S}^{n-1},$$

as well as the associated distribution function

$$F_\theta(x) = \mathbb{P}\{\langle X, \theta \rangle \leq x\}, \quad x \in \mathbb{R}.$$

Here we return to the striking observation due to Sudakov stating that, when n is large, most of F_θ's are concentrated about a certain probability distribution F with respect to the normalized Lebesgue measure $\mathfrak{s}_{n-1}$ on the unit sphere $\mathbb{S}^{n-1}$. Now we give a precise formulation as in [169] under the second moment assumption

$$\mathbb{E}\,\langle X, \theta \rangle^2 \leq M_2^2, \quad \theta \in \mathbb{S}^{n-1}, \tag{12.1}$$

and in terms of the (Kantorovich) L^1-distance between distribution functions.

S. Bobkov et al., *Concentration and Gaussian Approximation for Randomized Sums*, Probability Theory and Stochastic Modelling 104,
https://doi.org/10.1007/978-3-031-31149-9_12

Theorem 12.1.1 *For each $\delta > 0$, there is an integer n_δ depending on δ with the following property. If $n \ge n_\delta$, one can choose a set $\Theta \subset \mathbb{S}^{n-1}$ of measure $\mathfrak{s}_{n-1}(\Theta) \ge 1 - \delta$ and a distribution function F such that*

$$W(F_\theta, F) = \int_{-\infty}^{\infty} |F_\theta(x) - F(x)|\, dx \le M_2\,\delta \quad \textit{for all } \theta \in \Theta.$$

Thus, the concentration property of the family $\{F_\theta\}_{\theta\in S^{n-1}}$ has a surprisingly universal character, since no requirement on the distribution of X beyond the correlation-type condition (12.1) is needed.

Let us compare this assertion with the classical i.i.d. model assuming that the components X_k are independent, identically distributed, have mean zero and variance one (in particular, $M_2 = 1$). If the 3-rd absolute moment $\beta_3 = \mathbb{E}\,|X_1|^3$ is finite, then, by the non-uniform Berry–Esseen theorem, for all $\theta \in \mathbb{S}^{n-1}$,

$$W(F_\theta, \Phi) \le c\beta_3 \sum_{k=1}^n |\theta_k|^3,$$

where $c > 0$ is a universal constant (cf. e.g. [155]). This particular case is consistent with Theorem 12.1.1, which may be stated for all large $n \ge n(\delta, \beta_3)$ for a universal set, for example,

$$\Theta = \Big\{\theta \in \mathbb{S}^{n-1} : \sum_{k=1}^n |\theta_k|^3 \le 2/\sqrt{n}\Big\}.$$

Indeed, on the spaces $(\mathbb{S}^{n-1}, \mathfrak{s}_{n-1})$ we have convergence in probability

$$\sqrt{n}\sum_{k=1}^n |\theta_k|^3 \to \mathbb{E}\,|Z|^3 = \frac{4}{\sqrt{2\pi}} < 2 \quad (n \to \infty),$$

where $Z \sim N(0,1)$. Moreover, $W(F_\theta, \Phi) \le c\beta_3/\sqrt{n}$ for $\theta \in \Theta$, which quantifies the statement of Sudakov's theorem with $F = \Phi$ and provides a standard $\frac{1}{\sqrt{n}}$-rate for the normal approximation.

On the other hand, if we remove the 3-rd moment condition from this i.i.d. model, Theorem 12.1.1 will still be applicable, while the Berry–Esseen theorem is not available.

The distribution F in Theorem 12.1.1 is called *typical*. In general, it need not be normal and is determined in accordance with the distribution of X. In fact, it may always be chosen to be the spherical mean

$$F = \mathbb{E}_\theta F_\theta = \int_{S^{n-1}} F_\theta \, d\mathfrak{s}_{n-1}(\theta),$$

which we take as the definition of a typical distribution. Here and in the sequel, let us also agree to use $\mathbb{E}_\theta$ for the integral over the measure $\mathfrak{s}_{n-1}$.

The following general concentration problems have been raised in the literature after Sudakov's paper was published.

Problem 12.1.2 *Determine the rate of approximation of most F_θ by the typical distribution F, using standard metrics related to the topology of weak convergence in the space of probability measures on the line.*

As standard metrics, one usually considers the uniform or Kolmogorov distance, the Levý distance, the Kantorovich distance, etc. Of course, under stronger assumptions on the shape of the distribution of X (such as log-concavity, for example), it makes sense to involve stronger metrics, including total variation.

Problem 12.1.3 *How close is the typical distribution F to the normal distribution?*

In addition to these problems, another interesting and closely related issue is to determine the typical behavior of concentration functions Q_{F_θ} of F_θ.

12.2 The Structure of Typical Distributions

Theorem 12.1.1 and its possible quantification in Problem 12.1.2 are based on the spherical concentration phenomenon, which we discussed before. As for Problem 12.1.3, let us note that, since the measure $\mathfrak{s}_{n-1}$ is rotationally invariant, the typical distribution may be described as the distribution of the product ρZ_n, i.e.,

$$F = \mathcal{L}(\rho Z_n), \quad Z_n = \theta_1 \sqrt{n},$$

where

$$\rho^2 = \frac{1}{n}\,|X|^2 = \frac{1}{n}\,(X_1^2 + \cdots + X_n^2) \quad (\rho \geq 0),$$

and where θ_1, being the first coordinate of a point on the sphere, is treated as a random variable which is independent of X. Let us state this separately.

Proposition 12.2.1 *If ρ and Z_n are independent, then the distribution of ρZ_n is typical for the weighted sums of X_k's.*

Indeed, let π denote the Haar measure in the space $O(n)$ of all linear orthogonal transformations of $\mathbb{R}^n$, so that $\mathfrak{s}_{n-1}U^{-1} = \mathfrak{s}_{n-1}$ for any $U \in O(n)$. For any $x \in \mathbb{R}^n$, the distribution of Ux and $U^{-1}x$ under π are the same as the distribution of $|x|\theta$, where θ is uniformly distributed in $\mathbb{S}^{n-1}$. Hence, by the very definition of the typical distribution, for any bounded Borel measurable function u on the real line,

$$\begin{aligned}
\int_{-\infty}^{\infty} u\, \mathrm{d}F &= \mathbb{E}_\theta \int_{-\infty}^{\infty} u\, \mathrm{d}F_\theta = \mathbb{E}_\theta\, \mathbb{E}\, u(\langle X, \theta\rangle) \\
&= \int_{O(n)} \mathbb{E}_\theta\, \mathbb{E}\, u(\langle X, U\theta\rangle)\, \mathrm{d}\pi(U) = \int_{O(n)} \mathbb{E}_\theta\, \mathbb{E}\, u(\langle U^{-1}X, \theta\rangle)\, \mathrm{d}\pi(U) \\
&= \mathbb{E}_{\theta'}\, \mathbb{E}_\theta\, \mathbb{E}\, u(\langle |X|\theta', \theta\rangle) = \mathbb{E}_\theta\, \mathbb{E}\, u(|X|\,\theta_1),
\end{aligned}$$

where in the last step we made use of the property that $\langle \theta', \theta\rangle$ is distributed under $\mathfrak{s}_{n-1} \otimes \mathfrak{s}_{n-1}$ exactly as θ_1. This proves the claim.

Now, since the distribution of Z_n is nearly standard normal, one may conclude that the typical distribution F will be almost the same as the distribution of ρZ, where $Z \sim N(0,1)$ is independent of ρ. The distribution of ρZ represents a mixture of centered Gaussian measures on the line, and its associated distribution function is given by

$$\Phi_\rho(x) = \mathbb{P}\{\rho Z \le x\} = \mathbb{E}\,\Phi(x/\rho), \qquad x \in \mathbb{R},$$

where we use

$$\Phi(x) = \frac{1}{\sqrt{2\pi}} \int_{-\infty}^{x} e^{-y^2/2}\,dy$$

to denote the standard normal distribution function.

Thus, the typical distribution F is close to a mixture of centered Gaussian measures. Moreover, ρZ itself and therefore F are nearly standard normal if and only if the values of the random variable ρ^2 are concentrated about the point 1. This potential property may be quantified in terms of the variance functional

$$\sigma_4^2 = \frac{1}{n}\,\mathrm{Var}(|X|^2) = n\,\mathrm{Var}(\rho^2).$$

Note that the second moment of F is equal to $\frac{1}{n}\,\mathbb{E}\,|X|^2 = \mathbb{E}\rho^2$, so a normalization condition on this moment is desirable. For the beginning of our investigation, we examine closeness of the typical distribution to the normal law in terms of the variance of ρ. In the next section, the proximity is studied by means of the Kolmogorov distance, and then, in Section 12.4, we will derive the following bound, by using a much stronger weighted total variation distance.

Proposition 12.2.2 *Suppose that $\mathbb{E}\rho^2 = 1$. With some absolute constant $c > 0$ we have, for all $n \ge 1$,*

$$\int_{-\infty}^{\infty} (1+x^2)\,|F(dx) - \Phi(dx)| \le c\left(\frac{1}{n} + \mathrm{Var}(\rho)\right). \tag{12.2}$$

In particular,

$$\int_{-\infty}^{\infty} (1+x^2)\,|F(dx) - \Phi(dx)| \le c\,\frac{1+\sigma_4^2}{n},$$

$$\int_{-\infty}^{\infty} (1+x^2)\,|F(dx) - \Phi(dx)| \le c\,\frac{1+\sigma_2}{\sqrt{n}}.$$

Here the positive measure $|F(dx) - \Phi(dx)|$ denotes the variation in the sense of measure theory, and the left integral represents the weighted total variation of $F - \Phi$. In particular, we have a similar bound on the usual total variation distance between F and Φ.

The first assertion of the proposition is stronger than the second one because of the relation $\mathrm{Var}(\rho) \le \mathrm{Var}(\rho^2)$, cf. Proposition 1.5.1. However, in applications, it might be more convenient to use the second bound (although it requires finiteness of the 4-th moments of the X_k's). The last inequality uses the variance-type functional

$$\sigma_2 = \sigma_2(X) = \frac{1}{\sqrt{n}}\,\mathbb{E}\big|\,|X|^2 - n\big| = \sqrt{n}\,\mathbb{E}\,|\rho^2 - 1|.$$

It follows from the first one due to the relation $\mathrm{Var}(|X|) \le 2\sqrt{n}\,\sigma_2$, cf. Proposition 1.5.6. Although it provides worse dependence in n, it requires finiteness of the 2-nd moments of X_k and will be more preferable in obtaining the standard $\frac{1}{\sqrt{n}}$-rate (recall also that $\sigma_2 \le \sigma_4$).

12.3 Normal Approximation for Gaussian Mixtures

As a natural step, let us consider separately the question of the normal approximation for general mixtures of normal distributions Φ_t on the line with mean zero and standard deviations $t \ge 0$. Assuming that an arbitrary non-negative random variable $\rho = \rho(\omega)$ on a probability space $(\Omega, \mathcal{F}, \mathbb{P})$ is independent of $Z \sim N(0,1)$, introduce the distribution function of the random variable ρZ, i.e.,

$$\Phi_\rho(x) = \mathbb{E}\,\Phi_{\rho(\omega)}(x), \quad x \in \mathbb{R}. \tag{12.3}$$

In this section we focus on the Kolmogorov distance from Φ_ρ to Φ.

Proposition 12.3.1 *If* $\mathbb{E}\rho^2 = 1$, *then*

$$\sup_x |\Phi_\rho(x) - \Phi(x)| \le 2.1\,\mathrm{Var}(\rho^2).$$

Proof One may assume that $\rho > 0$ a.s. Given $0 < \varepsilon < 1$, we split the expectation in (12.3) into the events $A = \{\rho^2 \ge 1 - \varepsilon\}$ and $B = \{\rho^2 < 1 - \varepsilon\}$, so that

$$\Phi_\rho(x) - \Phi(x) = \mathbb{E}\,(\Phi_{\rho(\omega)}(x) - \Phi(x))\,1_A + \mathbb{E}\,(\Phi_{\rho(\omega)}(x) - \Phi(x))\,1_B. \tag{12.4}$$

Putting $\sigma^2 = \mathrm{Var}(\rho^2)$, $\sigma \ge 0$, we have, by Chebyshev's inequality,

$$\mathbb{P}(B) = \{1 - \rho^2 > \varepsilon\} \le \frac{\sigma^2}{\varepsilon^2}. \tag{12.5}$$

Using also $|\Phi_{\rho(\omega)}(x) - \Phi(x)| \le \frac{1}{2}$, which holds for all $x \in \mathbb{R}$ due to the symmetry of the involved distributions, we get the estimate

$$|\,\mathbb{E}\,(\Phi_{\rho(\omega)}(x) - \Phi(x))\,1_B| \le \frac{\sigma^2}{2\varepsilon^2} \tag{12.6}$$

for the second term on the right-hand side of (12.4). As for the first term, fix a value $\rho = \rho(\omega)$, $\omega \in \Omega$, and apply the inversion formula

$$\Phi_{\rho(\omega)}(x) - \Phi(x) = \frac{1}{2\pi} \int_{-\infty}^{\infty} e^{-itx} \, \frac{e^{-\rho^2 t^2/2} - e^{-t^2/2}}{-it} \, dt$$
$$= \frac{1}{\pi} \int_0^{\infty} \frac{\sin(tx)}{t} \left(e^{-\rho^2 t^2/2} - e^{-t^2/2}\right) dt,$$

which implies

$$\mathbb{E}\,(\Phi_\rho(x) - \Phi(x))\, 1_A = \frac{1}{\pi} \int_0^{\infty} \frac{\sin(tx)}{t} \, \mathbb{E}\left(e^{-\rho^2 t^2/2} - e^{-t^2/2}\right) 1_A \, dt.$$

Now write

$$e^{-\rho^2 t^2/2} - e^{-t^2/2} = e^{-t^2/2}\left(e^{-(\rho^2-1)t^2/2} - 1\right)$$

and apply the elementary bound

$$|e^{\delta} - 1 - \delta| \le \frac{1}{2}\,\delta^2 e^M, \quad -\infty < \delta \le M \ \ (M \ge 0).$$

Since $\delta \equiv -\frac{(\rho^2-1)t^2}{2} \le \frac{\varepsilon t^2}{2}$ on A, this gives that, for some random $|\gamma| \le 1$,

$$e^{-\rho^2 t^2/2} - e^{-t^2/2} = e^{-t^2/2}\Big(- \frac{(\rho^2-1)t^2}{2} + \gamma \, \frac{(\rho^2-1)^2\, t^4}{8} \, e^{\varepsilon t^2/2}\Big).$$

Hence, taking the expectation over the set A and using $1_A = 1 - 1_B$, we get for some constant $|c_1| \le 1$

$$\mathbb{E}\left(e^{-\rho^2 t^2/2} - e^{-t^2/2}\right) 1_A = \frac{t^2}{2}\, e^{-t^2/2}\, \mathbb{E}\,(\rho^2 - 1)\, 1_B + \frac{c_1 t^4}{8}\, e^{-(1-\varepsilon)t^2/2}\, \mathbb{E}\,(\rho^2-1)^2\, 1_A.$$

Returning to the inversion formula, this gives

$$\mathbb{E}\,(\Phi_\rho(x) - \Phi(x))\, 1_A = \alpha(x)\, \mathbb{E}\,(\rho^2 - 1)\, 1_B + c_2 \beta\, \mathbb{E}\,(\rho^2-1)^2\, 1_A$$

with a constant c_2 such that $|c_2| \le 1$, where

$$\beta = \frac{1}{8\pi} \int_0^{\infty} t^3\, e^{-(1-\varepsilon)t^2/2}\, dt = \frac{1}{4\pi\,(1-\varepsilon)^2}$$

and

$$\alpha(x) = \frac{1}{2\pi} \int_0^{\infty} t \sin(tx)\, e^{-t^2/2}\, dt = \frac{1}{2\sqrt{2\pi}}\, x\, e^{-x^2/2}.$$

Clearly, $|\alpha(x)| \le \frac{1}{2\sqrt{2\pi e}}$, which leads to the uniform bound

$$\big|\mathbb{E}\,(\Phi_\rho(x) - \Phi(x))\, 1_A\big| \le \frac{1}{2\sqrt{2\pi e}}\, |\mathbb{E}\,(\rho^2-1)\, 1_B| + \frac{1}{4\pi\,(1-\varepsilon)^2}\, \mathbb{E}\,(\rho^2-1)^2\, 1_A.$$

Next, by Cauchy's inequality, and recalling (12.5),

$$|\,\mathbb{E}\,(\rho^2 - 1)\, 1_B| \leq \sigma\sqrt{\mathbb{P}(B)} \leq \frac{\sigma^2}{\varepsilon},$$

while

$$\mathbb{E}\,(\rho^2 - 1)^2\, 1_A \leq \mathbb{E}\,(\rho^2 - 1)^2 = \sigma^2.$$

Using these bounds together with the bound (12.6) for the set B, we get the estimate

$$|\,\mathbb{E}\,(\Phi_\rho(x) - \Phi(x))| \leq C_\varepsilon \sigma^2$$

with constant

$$C_\varepsilon = \frac{1}{2\varepsilon^2} + \frac{1}{2\sqrt{2\pi e}\,\varepsilon} + \frac{1}{4\pi\,(1-\varepsilon)^2}.$$

Choosing $\varepsilon = 0.65$, we have $C_\varepsilon = 2.0191\ldots$ Thus, Proposition 12.3.1 is proved. □

12.4 Approximation in Total Variation

At the expense of a larger numerical constant, we now strengthen Proposition 12.3.1 by involving a much stronger distance. Keeping the same assumptions and notations as in the previous section, we prove:

Proposition 12.4.1 *If* $\mathbb{E}\rho^2 = 1$, *then for some absolute constant* $c > 0$,

$$\int_{-\infty}^{\infty} (1 + x^2)\, |\Phi_\rho(\mathrm{d}x) - \Phi(\mathrm{d}x)| \leq c\, \mathrm{Var}(\rho). \tag{12.7}$$

Proof The normal distribution function Φ_t with standard deviation $t > 0$ has density $\varphi_t(x) = \frac{1}{t}\,\varphi(x/t)$, where $\varphi(x) = \frac{1}{\sqrt{2\pi}}\,\mathrm{e}^{-x^2/2}$ is the standard normal density.

For a fixed number x, let us expand the function $u(t) = \varphi_t(x)$ according to the Taylor formula up to the quadratic term at the point $t_0 = 1$. We have $u(t_0) = \varphi(x)$,

$$u'(t) = (t^{-4}x^2 - t^{-2})\,\varphi(x/t), \quad u'(t_0) = (x^2 - 1)\,\varphi(x),$$

and

$$u''(t) = (t^{-7}x^4 - 5t^{-5}x^2 + 2t^{-3})\,\varphi(x/t) = t^{-3}\,\psi(x/t),$$

where $\psi(z) = (z^4 - 5z^2 + 2)\,\varphi(z)$. Therefore, using the integral Taylor formula

$$u(t) = u(t_0) + u'(t_0)(t - t_0) + (t - t_0)^2 \int_0^1 u''((1-s) + st)\,(1-s)\,\mathrm{d}s, \tag{12.8}$$

we get

$$\varphi_t(x) - \varphi(x) = (t-1)\,(x^2-1)\,\varphi(x) \\ + (t-1)^2 \int_0^1 \big((1-s)+st\big)^{-3}\,\psi\Big(\frac{x}{(1-s)+st}\Big)\,(1-s)\,ds.$$

We apply this representation with $t = \xi(\omega)$, where ξ is a positive random variable (on some probability space). Thus, Φ_ξ has density $\varphi_\xi = \mathbb{E}\,\varphi_{\xi(\omega)}$, which can be represented as

$$\varphi_\xi(x) - \varphi(x) = (x^2-1)\,\varphi(x)\,\mathbb{E}\,(\xi-1) \\ + \mathbb{E}\,(\xi-1)^2 \int_0^1 \big((1-s)+s\xi\big)^{-3}\,\psi\Big(\frac{x}{(1-s)+s\xi}\Big)\,(1-s)\,ds.$$

Putting

$$R_\xi(x) = \varphi_\xi(x) - \varphi(x) - (x^2-1)\,\varphi(x)\,\mathbb{E}\,(\xi-1)$$

and applying Fubini's theorem, we may conclude that the integral $\int_{-\infty}^\infty |R_\xi(x)|\,dx$ is bounded by

$$\mathbb{E}\int_{-\infty}^\infty (\xi-1)^2 \int_0^1 \big((1-s)+s\xi\big)^{-3}\,\Big|\psi\Big(\frac{x}{(1-s)+s\xi}\Big)\Big|\,(1-s)\,ds\,dx \\ = c_0\,\mathbb{E}\,(\xi-1)^2 \int_0^1 \big((1-s)+s\xi\big)^{-2}\,(1-s)\,ds$$

with $c_0 = \int_{-\infty}^\infty |\psi(x)|\,dx$. In particular, if $\xi \geq \frac{1}{2}$, the latter integral does not exceed $\int_0^1 \frac{1-s}{(1-\frac{s}{2})^2}\,ds = 4\log 2 - 2 < 1$, so that

$$\int_{-\infty}^\infty |R_\xi(x)|\,dx \leq c_0\,\mathbb{E}\,(\xi-1)^2.$$

This implies that for some $|\theta| \leq 1$,

$$\int_{-\infty}^\infty |\varphi_\xi(x) - \varphi(x)|\,dx = |\mathbb{E}\xi - 1| \int_{-\infty}^\infty |x^2-1|\,\varphi(x)\,dx + \theta c_0\,\mathbb{E}\,(\xi-1)^2.$$

Analogously, the integral $\int_{-\infty}^\infty x^2\,|R_\xi(x)|\,dx$ may be bounded from above by

$$\mathbb{E}\int_{-\infty}^\infty (\xi-1)^2 \int_0^1 \big((1-s)+s\xi\big)^{-3} x^2\,\Big|\psi\Big(\frac{x}{(1-s)+s\xi}\Big)\Big|(1-s)\,ds\,dx = c_1\,\mathbb{E}\,(\xi-1)^2,$$

where $c_1 = \frac{1}{2}\int_{-\infty}^\infty x^2\,|\psi(x)|\,dx$. Therefore (without any constraint on ξ),

$$\int_{-\infty}^\infty x^2\,|\varphi_\xi(x) - \varphi(x)|\,dx = |\mathbb{E}\xi - 1| \int_{-\infty}^\infty x^2\,|x^2-1|\,\varphi(x)\,dx + \theta c_1\,\mathbb{E}\,(\xi-1)^2.$$

The two representations can now be combined to

$$\int_{-\infty}^{\infty} (1+x^2)\,|\varphi_\xi(x) - \varphi(x)|\,dx \le a\,|\mathbb{E}\xi - 1| + b\,\mathbb{E}\,(\xi-1)^2,$$

where

$$a = \int_{-\infty}^{\infty} (1+x^2)\,|x^2-1|\,\varphi(x)\,dx, \quad b = \int_{-\infty}^{\infty} \Big(1+\frac{1}{2}x^2\Big)\,|\psi(x)|\,dx.$$

Let us derive numerical bounds on these absolute constants. If $Z \sim N(0,1)$, then using $(1+x^2)\,|x^2-1| \le 1+x^4$, we have $a \le 1+\mathbb{E}Z^4 = 4$. Since $|\psi(x)| \le (x^4+5x^2+2)\varphi(x)$, we also have

$$b \le \mathbb{E}\,(Z^4+5Z^2+2) + \frac{1}{2}\,\mathbb{E}\,(Z^6+5Z^4+2Z^2) = 26.$$

Thus,

$$\int_{-\infty}^{\infty} (1+x^2)\,|\varphi_\xi(x) - \varphi(x)|\,dx = 4\,|\mathbb{E}\xi - 1| + 26\,\mathbb{E}\,(\xi-1)^2, \tag{12.9}$$

provided that $\xi \ge 1/2$.

Now, consider the general case assuming without loss of generality that $\rho > 0$. Introduce the events $A_0 = \mathbb{P}\{\rho < 1/2\}$, $A_1 = \{\rho \ge 1/2\}$, and put $\alpha_0 = \mathbb{P}(A_0)$, $\alpha_1 = \mathbb{P}(A_1)$, assuming again without loss of generality that $\alpha_0 > 0$. Next we split the distribution Q of ρ into the two components supported on $(0,1/2)$ and $[1/2,\infty)$ and denote by ρ_0 and ρ_1 some random variables distributed respectively as the normalized restrictions of Q to these regions, so that $\rho_0 < 1/2$ and $\rho_1 \ge 1/2$. We thus represent the density of Φ_ρ as the convex mixture of two densities

$$\varphi_\rho = \alpha_0\varphi_{\rho_0} + \alpha_1\varphi_{\rho_1}, \tag{12.10}$$

where

$$\varphi_{\rho_0}(x) = \frac{1}{\alpha_0}\,\mathbb{E}\,\varphi(x/\rho)\,1_{\{\rho\in A_0\}}, \quad \varphi_{\rho_1}(x) = \frac{1}{\alpha_1}\,\mathbb{E}\,\varphi(x/\rho)\,1_{\{\rho\in A_1\}}.$$

Note that, since $\mathbb{E}\rho^2 = 1$, we necessarily have $\mathbb{E}\rho \le 1$. On the other hand,

$$\mathrm{Var}(\rho) = 1 - (\mathbb{E}\rho)^2 = (1-\mathbb{E}\rho)(1+\mathbb{E}\rho) \ge 1 - \mathbb{E}\rho.$$

Hence $|\mathbb{E}\rho - 1| \le \mathrm{Var}(\rho)$, and as a consequence,

$$\mathbb{E}\,(\rho-1)^2 = 2\,(1-\mathbb{E}\rho) \le 2\,\mathrm{Var}(\rho). \tag{12.11}$$

In particular, since $\rho < 1/2$ implies $(\rho-1)^2 > 1/4$, we have, by Chebyshev's inequality,

$$\alpha_0 = \mathbb{P}\{\rho < 1/2\} \le 8\,\mathrm{Var}(\rho). \tag{12.12}$$

Now, by the previous step (12.9) with $\xi = \rho_1$,

$$\int_{-\infty}^{\infty} (1 + x^2) \, |\varphi_{\rho_1}(x) - \varphi(x)| \, dx \le 4 \, |\mathbb{E}\rho_1 - 1| + 26 \, \mathbb{E} \, (\rho_1 - 1)^2. \tag{12.13}$$

On the other hand,

$$\int_{-\infty}^{\infty} (1 + x^2) \, \varphi_{\rho_0}(x) \, dx = 1 + \mathbb{E}\rho_0^2 \le \frac{5}{4},$$

so that

$$\int_{-\infty}^{\infty} (1 + x^2) \, |\varphi_{\rho_0}(x) - \varphi(x)| \, dx \le \frac{13}{4}.$$

From (12.10), (12.12) and (12.13), we then get that

$$\int_{-\infty}^{\infty} (1 + x^2) \, |\varphi_{\rho}(x) - \varphi(x)| \, dx \le 26 \, \mathrm{Var}(\rho) + 4 \, |\mathbb{E}\rho_1 - 1| + 26 \, \mathbb{E} \, (\rho_1 - 1)^2. \tag{12.14}$$

It remains to estimate the last two expectations. First suppose that $\mathrm{Var}(\rho) \le 1/16$, so that, by (12.12), $\alpha_0 \le \frac{1}{2}$ and $\alpha_1 \ge \frac{1}{2}$. By the definition,

$$\mathbb{E}\rho_1 = \frac{1}{\alpha_1} \, \mathbb{E}\rho \, 1_{\{\rho \in A_1\}} = \frac{1}{\alpha_1} \left(\mathbb{E}\rho - \mathbb{E}\rho \, 1_{\{\rho \in A_0\}}\right),$$

so

$$\mathbb{E}\rho_1 - 1 = \frac{1}{\alpha_1} \Big(\mathbb{E} \, (\rho - 1) - \mathbb{E} \, (\rho - 1) \, 1_{\{\rho \in A_0\}}\Big).$$

By Cauchy's inequality and applying (12.11) and (12.12),

$$\begin{aligned} \left|\mathbb{E} \, (\rho - 1) \, 1_{\{\rho \in A_0\}}\right| &\le \mathbb{E} \, |\rho - 1| \, 1_{\{\rho \in A_0\}} \\ &\le \left(\mathbb{E} \, (\rho - 1)^2\right)^{1/2} \alpha_0^{1/2} \le 4 \, \mathrm{Var}(\rho). \end{aligned} \tag{12.15}$$

Hence

$$|\mathbb{E}\rho_1 - 1| \le \frac{1}{\alpha_1} \Big(|\mathbb{E}\rho - 1| + |\mathbb{E} \, (\rho - 1) \, 1_{\{\rho \in A_0\}}|\Big) \le 10 \, \mathrm{Var}(\rho). \tag{12.16}$$

Similarly,

$$\mathbb{E}\rho_1^2 = \frac{1}{\alpha_1} \, \mathbb{E}\rho^2 \, 1_{\{\rho \in A_1\}} = \frac{1}{\alpha_1} \left(1 - \mathbb{E}\rho^2 \, 1_{\{\rho \in A_0\}}\right),$$

so, using $\rho \le 1/2$ on A_0 and applying (12.15), we have

$$\begin{aligned}\mathbb{E}\rho_1^2 - 1 &= \frac{1}{\alpha_1}\,\mathbb{E}\,(1-\rho^2)\,1_{\{\rho\in A_0\}} \\ &= \frac{1}{\alpha_1}\,\mathbb{E}\,(1-\rho)\,(1+\rho)\,1_{\{\rho\in A_0\}} \le \frac{3}{2\alpha_1}\,\mathbb{E}\,|1-\rho|\,1_{\{\rho\in A_0\}} \\ &\le \frac{3}{2\alpha_1}\cdot 4\,\mathrm{Var}(\rho) \le 12\,\mathrm{Var}(\rho).\end{aligned}$$

Writing $\mathbb{E}\,(\rho_1-1)^2 = (\mathbb{E}\rho_1^2-1) - 2\,\mathbb{E}\,(\rho_1-1)$ and applying (12.16), the above gives

$$\mathbb{E}\,(\rho_1-1)^2 \le 12\,\mathrm{Var}(\rho) + 20\,\mathrm{Var}(\rho) = 32\,\mathrm{Var}(\rho).$$

It remains to use this bound together with (12.16) in (12.14), which gives the desired estimate (12.7), i.e.,

$$\int_{-\infty}^{\infty} (1+x^2)\,|\varphi_\rho(x)-\varphi(x)|\,\mathrm{d}x \le c\,\mathrm{Var}(\rho)$$

with constant $c = 26 + 4\cdot 10 + 26\cdot 32 = 898$.

Finally, in the case $\mathrm{Var}(\rho) > 1/16$, one may just use

$$\begin{aligned}\int_{-\infty}^{\infty} (1+x^2)\,|\varphi_\rho(x)-\varphi(x)|\,\mathrm{d}x &\le \int_{-\infty}^{\infty} (1+x^2)\,\varphi_\rho(x)\,\mathrm{d}x + \int_{-\infty}^{\infty} (1+x^2)\,\varphi(x)\,\mathrm{d}x \\ &= (1+\mathbb{E}\,(\rho Z)^2) + (1+\mathbb{E}Z^2) = 4 < 64\,\mathrm{Var}(\rho).\end{aligned}$$

Proposition 12.4.1 is proved with $c = 898$. □

Proof (of Proposition 12.2.2) Assuming (without loss of generality) that $n \ge 3$, let Φ_n and φ_n denote respectively the distribution function and density of $Z_n = \theta_1\sqrt{n}$, where θ_1 is the first coordinate of a random point uniformly distributed on the unit sphere $\mathbb{S}^{n-1}$. If $\rho^2 = \frac{1}{n}\,|X|^2$ is independent of Z_n, then, by the definition of the typical distribution,

$$F(x) = \mathbb{P}\{\rho Z_n \le x\} = \mathbb{E}\,\Phi_n(x/\rho), \quad x \in \mathbb{R},$$

where we may also assume that $\rho > 0$ a.s. We know from Proposition 11.1.2 that

$$|\varphi_n(x) - \varphi(x)| \le \frac{c}{n}\,\mathrm{e}^{-x^2/4}$$

for all $x \in \mathbb{R}$ with some universal constant $c > 0$. Hence

$$\int_{-\infty}^{\infty} (1+x^2)\,|\Phi_n(\mathrm{d}x) - \Phi(\mathrm{d}x)| = \int_{-\infty}^{\infty} (1+x^2)\,|\varphi_n(x)-\varphi(x)|\,\mathrm{d}x \le \frac{c}{n} \tag{12.17}$$

(with some other constant). But, for any fixed value of ρ,

$$\int_{-\infty}^{\infty} (1+x^2)\,|\Phi_n(\mathrm{d}x/\rho) - \Phi(\mathrm{d}x/\rho)|(\mathrm{d}x) = \int_{-\infty}^{\infty} (1+\rho^2x^2)\,|\Phi_n(\mathrm{d}x) - \Phi(\mathrm{d}x)|.$$

So, taking the expectation with respect to ρ and using Jensen's inequality, we get

$$\int_{-\infty}^{\infty} (1+x^2)\,|F(\mathrm{d}x) - \Phi_\rho(\mathrm{d}x)| = \int_{-\infty}^{\infty} (1+x^2)\,|\mathbb{E}\,\Phi_n(\mathrm{d}x/\rho) - \mathbb{E}\,\Phi(\mathrm{d}x/\rho)|$$

$$\leq \mathbb{E} \int_{-\infty}^{\infty} (1+x^2)\,|\Phi_n(\mathrm{d}x/\rho) - \Phi(\mathrm{d}x/\rho)|$$

$$\leq \mathbb{E} \int_{-\infty}^{\infty} (1+\rho^2x^2)\,|\Phi_n(\mathrm{d}x) - \Phi(\mathrm{d}x)| = \int_{-\infty}^{\infty} (1+x^2)\,|\Phi_n(\mathrm{d}x) - \Phi(\mathrm{d}x)|.$$

It remains to apply (12.17) together with Proposition 12.4.1. □

12.5 L^p-distances to the Normal Law

Proposition 12.4.1 may be used to obtain the (a priori weaker) non-uniform bound

$$\sup_x \Big[(1+x^2)\,|\Phi_\rho(x) - \Phi(x)|\Big] \leq c\,\mathrm{Var}(\rho)$$

(cf. Lemma 12.5.4 below). The appearance of the weight $1+x^2$ on the left is important in order to control the L^p-distances between Φ_ρ and Φ, since we get

$$\Big(\int_{-\infty}^{\infty} |\Phi_\rho(x) - \Phi(x)|^p\,\mathrm{d}x\Big)^{1/p} \leq c\,\mathrm{Var}(\rho), \quad p \geq 1, \tag{12.18}$$

for some absolute constant $c > 0$. Here Φ_ρ denotes the distribution function of ρZ with an arbitrary random variable $\rho \geq 0$, independent of $Z \sim N(0,1)$ and such that $\mathbb{E}\rho^2 = 1$.

In the scheme of the weighted sums $\langle X, \theta\rangle$ with $\rho^2 = \frac{1}{n}\,|X|^2$, we obtain from Proposition 12.2.2 a similar bound on the normal approximation of the typical distribution function F by the standard normal distribution function Φ in all L^p-distances. Recall that $\sigma_2 = \frac{1}{\sqrt{n}}\,\mathbb{E}\,\big|\,|X|^2 - n\big|$ and $\sigma_4^2 = \frac{1}{n}\,\mathrm{Var}(|X|^2)$.

Proposition 12.5.1 *Suppose that* $\mathbb{E}\,|X|^2 = n$. *With some absolute constant* $c > 0$ *we have, for all* $p \geq 1$,

$$\Big(\int_{-\infty}^{\infty} |F(x) - \Phi(x)|^p\,\mathrm{d}x\Big)^{1/p} \leq c\,\Big(\frac{1}{n} + \mathrm{Var}(\rho)\Big).$$

In particular,

$$\Big(\int_{-\infty}^{\infty} |F(x) - \Phi(x)|^p\,\mathrm{d}x\Big)^{1/p} \leq c\,\frac{1+\sigma_2}{\sqrt{n}},$$

$$\Big(\int_{-\infty}^{\infty} |F(x) - \Phi(x)|^p\,\mathrm{d}x\Big)^{1/p} \leq c\,\frac{1+\sigma_4^2}{n}.$$

The particular value $p = 2$ will be considered later on in more detail (in Problem 12.1.2). In this case, we deal with a special important distance

$$\omega(F,G) = \Big(\int_{-\infty}^{\infty} |F(x) - G(x)|^2\,dx\Big)^{1/2},$$

which is finite, as long as the distribution functions F and G have finite first absolute moments. Returning to (12.18), we may thus write

$$\omega(\Phi_\rho, \Phi) \le c\,\mathrm{Var}(\rho). \tag{12.19}$$

In fact, here the distance may be evaluated explicitly in terms of ρ, by using the following elementary identity (which will be needed in the sequel as well).

Lemma 12.5.2 *For all* $\alpha, \alpha_0 \ge 0$,

$$\int_{-\infty}^{\infty} \frac{e^{-\alpha t^2} - e^{-\alpha_0 t^2}}{t^2}\,dt = 2\sqrt{\pi}\,\big(\sqrt{\alpha_0} - \sqrt{\alpha}\big).$$

Indeed, denoting the left integral by $\psi(\alpha)$, we obtain a smooth function on $(0,\infty)$ with derivative

$$\psi'(\alpha) = -\int_{-\infty}^{\infty} e^{-\alpha t^2}\,dt = -\frac{\sqrt{\pi}}{\sqrt{\alpha}}.$$

Since also $\psi(\alpha_0) = 0$, after integration of the above equality we get the assertion.

Now, let ρ' be an independent copy of ρ, so as to represent the square of the characteristic function of Φ_ρ as

$$|\mathbb{E}\,e^{it\rho Z}|^2 = |\mathbb{E}\,e^{-\rho^2 t^2/2}|^2 = \mathbb{E}\,e^{-(\rho^2+\rho'^2)\,t^2/2}.$$

Hence, by Plancherel's theorem,

$$\begin{aligned}\int_{-\infty}^{\infty} (\Phi_\rho(x) - \Phi(x))^2\,dx &= \frac{1}{2\pi}\int_{-\infty}^{\infty} \Big|\frac{\mathbb{E}\,e^{it\rho Z} - e^{-t^2/2}}{t}\Big|^2\,dt \\ &= \frac{1}{2\pi}\int_{-\infty}^{\infty} \frac{\mathbb{E}\,e^{-(\rho^2+\rho'^2)\,t^2/2} - 2\,\mathbb{E}\,e^{-(\rho^2+1)\,t^2/2} + e^{-t^2}}{t^2}\,dt.\end{aligned}$$

Applying Lemma 12.5.2 with $\alpha_0 = 0$, we arrive at the identity

$$\omega^2(\Phi_\rho, \Phi) = \frac{1}{\sqrt{\pi}}\Big[2\,\mathbb{E}\sqrt{\frac{\rho^2+1}{2}} - \mathbb{E}\sqrt{\frac{\rho^2+\rho'^2}{2}} - 1\Big].$$

As a result, (12.19) can be restated as the relation

$$2\,\mathbb{E}\sqrt{\frac{\rho^2+1}{2}} - \mathbb{E}\sqrt{\frac{\rho^2+\rho'^2}{2}} - 1 \le c\,(\mathrm{Var}(\rho))^2,$$

where c is an absolute constant. It is unclear how to obtain such an estimate by a different argument (which would not be based on Proposition 12.4.1).

Finally, let us sharpen Proposition 12.5.1 under the stronger hypothesis that the distribution of X lies on the sphere of radius $\sqrt{n}$ (that is, when $\sigma_2 = \sigma_4 = 0$).

Proposition 12.5.3 *Let X be a random vector in $\mathbb{R}^n$ with $|X|^2 = n$ a.s. Then*

$$\omega^2(F, \Phi) = \frac{15}{256\sqrt{\pi}} \frac{1}{n^2} + O(n^{-3}).$$

Proof The characteristic function of F is described as

$$f(t) = \mathbb{E}_\theta f_\theta(t) = \mathbb{E}_\theta \, \mathbb{E} \, e^{it\langle X, \theta\rangle} = \mathbb{E} \, J_n(t|X|) = J_n(t\sqrt{n}), \quad t \in \mathbb{R},$$

where J_n is the characteristic function of the first coordinate θ_1 under the uniform measure $\mathfrak{s}_{n-1}$ on the unit sphere $\mathbb{S}^{n-1}$. We apply the Plancherel formula and the expansion for J_n from Corollary 11.3.3,

$$J_n\big(t\sqrt{n}\big) = \Big(1 - \frac{t^4}{4n}\Big) \, e^{-t^2/2} + O\Big(\frac{1}{n^2} \min(1, t^4)\Big).$$

This gives

$$\begin{aligned}
\omega^2(F, \Phi) &= \frac{1}{2\pi} \int_{-\infty}^{\infty} |J_n\big(t\sqrt{n}\,\big) - e^{-t^2/2}|^2 \, \frac{dt}{t^2} \\
&= \frac{1}{2\pi} \int_{-\infty}^{\infty} \Big|\frac{t^4}{4n} \, e^{-t^2/2} + O(n^{-2}) \min(1, t^4)\Big|^2 \, \frac{dt}{t^2} \\
&= \frac{1}{2\pi} \int_{-\infty}^{\infty} \frac{t^6}{16\,n^2} \, e^{-t^2} dt + O(n^{-3}) \, = \, \frac{1}{2\pi} \, \frac{15\sqrt{2\pi}}{16n^2 \cdot 8\sqrt{2}} + O(n^{-3}),
\end{aligned}$$

thus proving the proposition. □

Finally, let us explain the non-uniform bound mentioned in the beginning of this section. It is a consequence of Proposition 12.4.1, using the following general observation.

Lemma 12.5.4 *Suppose that the distribution functions F and G have densities p and q such that the weighted total variation distance*

$$L = \int_{-\infty}^{\infty} (1 + x^2) \, |p(x) - q(x)| \, dx$$

is finite. Then

$$\sup_x \Big[(1 + x^2) \, |F(x) - G(x)|\Big] \le 2L.$$

Proof By Fubini's theorem,

$$\begin{aligned}\int_0^\infty |x\,(F(x)-G(x))|\,\mathrm{d}x &= \int_0^\infty x\left|\int_x^\infty (p(y)-q(y))\,\mathrm{d}y\right|\mathrm{d}x\\ &\le \int_0^\infty x\left[\int_x^\infty |p(y)-q(y)|\,\mathrm{d}y\right]\mathrm{d}x\\ &= \frac{1}{2}\int_0^\infty y^2\,|p(y)-q(y)|\,\mathrm{d}y.\end{aligned}$$

A similar bound also holds for the integral over the negative half-axis, so

$$\int_{-\infty}^\infty |x\,(F(x)-G(x))|\,\mathrm{d}x \le \frac{1}{2}\,L.$$

The function $H(x) = (1+x^2)\,(F(x)-G(x))$ has a Radon–Nikodym derivative

$$h(x) = 2x\,(F(x)-G(x)) + (1+x^2)\,(p(x)-q(x)),$$

implying that H has a total variation norm satisfying

$$\begin{aligned}\|H\|_{\mathrm{TV}} &= \int_{-\infty}^\infty |h(x)|\,\mathrm{d}x\\ &\le 2\int_{-\infty}^\infty |x\,(F(x)-G(x))|\,\mathrm{d}x + \int_{-\infty}^\infty (1+x^2)\,|p(x)-q(x)|\,\mathrm{d}x \le 2L.\end{aligned}$$

Since $|H(x)| \le \|H\|_{\mathrm{TV}}$ for all x, the required inequality follows. □

As its consequence, we obtain that, for any $p \ge 1$,

$$\left(\int_{-\infty}^\infty |F(x)-G(x)|^p\,\mathrm{d}x\right)^{1/p} \le 2L\left(\int_{-\infty}^\infty \frac{1}{(1+x^2)^p}\,\mathrm{d}x\right)^{1/p}.$$

The last integral is smaller than 4, so that the L^p-norm on the left is bounded by $8L$. This explains the bound (12.18), as well as the first bound in Proposition 12.5.1 via an application of Proposition 12.2.2.

12.6 Lower Bounds

In a certain sense, the upper bound of Proposition 12.3.1 on the Kolmogorov distance is optimal with respect to the variance of ρ^2. At least, this is the case when ρ is bounded, as the following assertion shows.

Proposition 12.6.1 *If* $\mathbb{E}\rho^2 = 1$ *and* $0 \le \rho \le M$ *a.s., then for the distribution function* Φ_ρ *of the random variable* ρZ*, where* $Z \sim N(0,1)$ *is independent of* ρ*, we have*

$$\sup_x |\Phi_\rho(x) - \Phi(x)| \ge \frac{c}{M^5}\,\mathrm{Var}(\rho^2), \tag{12.20}$$

where $c > 0$ *is an absolute constant.*

Note that the left-hand side is dominated by the total variation $\|\Phi_\rho - \Phi\|_{\mathrm{TV}}$, while $\mathrm{Var}(\rho^2) \ge \mathrm{Var}(\rho)$. Also, the assumption $\mathbb{E}\rho^2 = 1$ ensures that $M \ge 1$.

Proof We will need a general lower bound on the Kolmogorov distance

$$\Delta = \sup_x |F(x) - G(x)|$$

between distribution functions F and G in terms of their characteristic functions f and g. Recall that according to Proposition 3.4.1,

$$\Delta \ge \frac{1}{2\sqrt{2\pi}}\bigg|\int_{-\infty}^{\infty} (f(t) - g(t))\, e^{-t^2/2}\,dt\bigg|.$$

We apply this bound with $F = \Phi_\rho$ and $G = \Phi$, in which case, by Jensen's inequality,

$$f(t) = \mathbb{E}\, e^{-\rho^2 t^2/2} \ge e^{-t^2/2} = g(t).$$

Hence

$$\begin{aligned}\sup_x |\Phi_\rho(x) - \Phi(x)| &\ge \frac{1}{2\sqrt{2\pi}}\int_{-\infty}^{\infty} \Big(\mathbb{E}\, e^{-\rho^2 t^2/2} - e^{-t^2/2}\Big)\, e^{-t^2/2}\,dt \\ &= \frac{1}{2}\Big(\mathbb{E}\,\frac{1}{\sqrt{1+\rho^2}} - \frac{1}{\sqrt{2}}\Big).\end{aligned} \tag{12.21}$$

One can now expand the function $u(t) = (1+t)^{-1/2}$ near the point $t_0 = 1$ according to the integral Taylor formula (12.8) up to the quadratic term. Since $u''(t) = \frac{3}{4}\,(1+t)^{-5/2}$, this gives

$$\begin{aligned}\mathbb{E}\, u(\rho^2) - u(1) &= \frac{3}{4}\,\mathbb{E}\,(\rho^2 - 1)^2 \int_0^1 \Big(1 + ((1-s) + s\rho^2)\Big)^{-5/2} (1-s)\,ds \\ &\ge \frac{3}{4}\,\mathrm{Var}(\rho^2) \int_0^1 \Big(1 + ((1-s) + sM^2)\Big)^{-5/2} (1-s)\,ds \\ &\ge \frac{3}{8}\,\mathrm{Var}(\rho^2)\,(1 + M^2)^{-5/2},\end{aligned}$$

where we used $M \ge 1$ in the last step. It remains to apply (12.21) and then we arrive at (12.20). Proposition 12.6.1 follows. □

Finally, returning to the scheme of the weighted sums $\langle X, \theta \rangle$ ($\theta \in \mathbb{S}^{n-1}$) with typical distribution function F, let us give an asymptotic analog of Proposition 12.5.3 for the Kolmogorov distance under the same hypothesis that the distribution of the random vector X in $\mathbb{R}^n$ lies on the sphere of radius $\sqrt{n}$.

Proposition 12.6.2 *If* $|X| = \sqrt{n}$ *a.s., then*

$$\frac{c_0}{n} \leq \sup_x |F(x) - \Phi(x)| \leq \frac{c_1}{n}$$

for some numerical constants $c_1 > c_0 > 0$.

Proof Since $\sigma_4^2(X) = 0$, the upper bound follows from Proposition 12.2.2. For the lower bound, we use the second lower bound of Proposition 3.4.1,

$$\Delta \geq \frac{1}{3T} \left| \int_0^T (f(t) - g(t)) \left(1 - \frac{t}{T}\right) \mathrm{d}t \right|,$$

and once more Corollary 11.3.3 with its asymptotic formula for the characteristic function f of F, namely,

$$f(t) = \left(1 - \frac{t^4}{4n}\right) \mathrm{e}^{-t^2/2} + O(n^{-2} \min(1, t^4)).$$

Putting $g(t) = \mathrm{e}^{-t^2/2}$ and choosing $T = 1$, the above lower bound then yields

$$\Delta = \sup_x |F(x) - \Phi(x)| \geq \frac{c_1}{n} - \frac{c_1'}{n^2} \geq \frac{c_2}{n}, \qquad n \geq n_0,$$

for some numerical constants $c_1, c_2 > 0$, $c_1' \in \mathbb{R}$ and some integer $n_0 \geq 1$. In the case $n < n_0$, the statement is obvious, since the distribution F is supported on the interval $[-\sqrt{n_0}, \sqrt{n_0}]$, so that Δ is bounded away from zero for all $n < n_0$.

Proposition 12.6.2 follows. □

12.7 Remarks

Propositions 12.2.2 and 12.4.1 were shown in [37].

A closely related property in the form of a local limit theorem was studied in [51], which contains the following observation. Suppose that a random vector X is uniformly distributed in a convex body K in $\mathbb{R}^n$ with volume $\mathrm{Vol}_n(K) = 1$ and satisfies

$$\mathbb{E} \langle X, \theta \rangle^2 = \int_K \langle x, \theta \rangle^2 \, \mathrm{d}x = L_K^2, \quad \theta \in \mathbb{S}^{n-1}.$$

That is, the random vector $Y = \frac{1}{L_K} X$ is isotropic. Then, for the density p of the typical distribution F, we have

$$\left|p(x) - \frac{1}{L_K\sqrt{2\pi}}\, \mathrm{e}^{-x^2/(2L_K^2)}\right| \le C\left(\frac{\sigma_4 L_K}{x^2\sqrt{n}} + \frac{1}{n}\right), \quad 0 < x \le c\sqrt{n},$$

for some absolute constants C and c, where

$$\sigma_4^2 = \sigma_4^2(Y) = \frac{1}{n}\,\mathrm{Var}(|Y|^2) = \frac{1}{nL_K^4}\,\mathrm{Var}(|X|^2).$$

Note that the typical density may also be described as

$$p(x) = \int_{S^{n-1}} \mathrm{Vol}_{n-1}\Big(K \cap H_\theta(|x|)\Big)\, \mathrm{d}\mathfrak{s}_{n-1}(\theta), \quad x \in \mathbb{R},$$

which expresses the average $(n-1)$-dimensional volume of the section of K by the hyperplanes $H_\theta(|x|) = \{y \in \mathbb{R}^n : \langle y, \theta\rangle = |x|\}$ perpendicular to θ at distance $|x|$ from the origin.

For some special bodies the normal approximation as described above has been studied in [71] and [126], cf. also [70].

Chapter 13
Characteristic Functions of Weighted Sums

In order to study deviations of the distribution functions F_θ from the typical distribution F by means of the Kolmogorov distance, Berry–Esseen-type inequalities, which we discussed in Chapter 3, will be used. To this end we need to focus first on the behavior of characteristic functions of F_θ.

13.1 Upper Bounds on Characteristic Functions

As before, let $X = (X_1, \dots .X_n)$ be a random vector in $\mathbb{R}^n$, $n \geq 2$. Problem 12.1.2 asking for the concentration of the distribution functions F_θ of the weighted sums $\langle X, \theta \rangle$ around the typical distribution $F = \mathbb{E}_\theta F_\theta$ may be attacked by the study of the characteristic functions

$$f_\theta(t) = \mathbb{E}\, e^{it\langle X,\theta\rangle}, \quad t \in \mathbb{R}. \tag{13.1}$$

Therefore, we aim to quantify the concentration of f_θ around the characteristic function f of the typical distribution F for most directions θ in terms of the correlation-type functionals introduced in Chapter 1. Recall that, in view of Proposition 12.2.1, the characteristic function of F is described as

$$f(t) = \mathbb{E}_\theta f_\theta(t) = \mathbb{E}_\theta\, \mathbb{E}\, e^{it\langle X,\theta\rangle} = \mathbb{E}\, J_n(t|X|), \quad t \in \mathbb{R},$$

where J_n is the characteristic function of the first coordinate θ_1 under the uniform measure $\mathfrak{s}_{n-1}$ on the unit sphere $\mathbb{S}^{n-1}$.

First let us look at the decay of $f_\theta(t)$ at infinity on average. From (13.1),

$$\mathbb{E}_\theta\, |f_\theta(t)|^2 = \mathbb{E}_\theta\, \mathbb{E}\, e^{it\langle X-Y,\theta\rangle} = \mathbb{E}\, J_n(t|X-Y|),$$

where Y is an independent copy of X. To proceed, let us rewrite the Gaussian-type bound (11.5) of Proposition 11.4.1 as

S. Bobkov et al., *Concentration and Gaussian Approximation for Randomized Sums*, Probability Theory and Stochastic Modelling 104,
https://doi.org/10.1007/978-3-031-31149-9_13

$$|J_n(t)| \leq 3\,\mathrm{e}^{-t^2/2n} + 3\,\mathrm{e}^{-n/12} \tag{13.2}$$

which gives

$$\mathbb{E}_\theta\,|f_\theta(t)|^2 \leq 3\,\mathbb{E}\,\mathrm{e}^{-t^2|X-Y|^2/2n} + 3\,\mathrm{e}^{-n/12}.$$

Splitting the latter expectation into the event $B = \{|X-Y|^2 \leq \lambda n\}$ and its complement, we get the following general statement.

Lemma 13.1.1 *The characteristic functions f_θ satisfy, for all $t \in \mathbb{R}$ and $\lambda > 0$,*

$$\frac{1}{3}\,\mathbb{E}_\theta\,|f_\theta(t)|^2 \leq \mathrm{e}^{-\lambda t^2/2} + \mathrm{e}^{-n/12} + \mathbb{P}\{|X-Y|^2 \leq \lambda n\},$$

where Y is an independent copy of X. In particular,

$$\frac{1}{\sqrt{3}}\,\mathbb{E}_\theta\,|f_\theta(t)| \leq \mathrm{e}^{-\lambda t^2/4} + \mathrm{e}^{-n/24} + \sqrt{\mathbb{P}\{|X-Y|^2 \leq \lambda n\}}.$$

If $\mathbb{E}\,|X|^2 = n$, the right-hand side of these bounds can be further developed by involving the moment and variance-type functionals

$$m_p = \frac{1}{\sqrt{n}}\,(\mathbb{E}\,|\langle X, Y\rangle|^p)^{1/p}, \qquad \sigma_{2p} = \sqrt{n}\left(\mathbb{E}\left|\frac{|X|^2}{n} - 1\right|^p\right)^{1/p}.$$

Both are non-decreasing functions in $p \geq 1$. Recall that, by Proposition 1.1.3, $m_2 \geq 1$ due to $\mathbb{E}\,|X|^2 = n$, so that $m_p \geq 1$ for all $p \geq 1$. The most interesting special values are $2p = 2, 3, 4$, in which cases

$$\sigma_2 = \frac{1}{n^{1/2}}\,\mathbb{E}\,\big|\,|X|^2 - n\big|, \quad \sigma_3^{3/2} = \frac{1}{n^{3/4}}\,\mathbb{E}\,\big|\,|X|^2 - n\big|^{3/2}, \quad \sigma_4^2 = \frac{1}{n}\,\mathrm{Var}(|X|^2).$$

In order to estimate the probability of the event B, let us choose $\lambda = 1/4$. Then, by Proposition 1.6.2, $\mathbb{P}(B) \leq Cn^{-p}$ with constant $C = 4^{2p}\,(m_{2p}^{2p} + \sigma_{2p}^{2p})$. From Lemma 13.1.1 we therefore deduce:

Lemma 13.1.2 *Let the random vector X in $\mathbb{R}^n$ satisfy $\mathbb{E}\,|X|^2 = n$. If the moment m_{2p} is finite for $p \geq 1$, then for some constant $c > 0$,*

$$c^p\,\mathbb{E}_\theta\,|f_\theta(t)|^2 \leq \frac{m_{2p}^{2p} + \sigma_{2p}^{2p}}{n^p} + \mathrm{e}^{-t^2/8},$$

$$c^p\,\mathbb{E}_\theta\,|f_\theta(t)| \leq \frac{m_{2p}^{p} + \sigma_{2p}^{p}}{n^{p/2}} + \mathrm{e}^{-t^2/16}.$$

Note that, by Jensen's inequality, $|f(t)| \leq \mathbb{E}_\theta\,|f_\theta(t)|$. Hence, the characteristic function of the typical distribution shares the same bounds. In fact, the parameter m_{2p} is not needed. Indeed, by Proposition 1.5.1 with $\lambda = \frac{1}{2}$, if $\mathbb{E}\,|X|^2 = n$, then

$$\mathbb{P}\Big\{|X|^2 \leq \frac{1}{2}\,n\Big\} \leq 2^p\,\frac{\sigma_{2p}^p}{n^{p/2}}.$$

Hence, by Lemma 13.1.1, and using (13.2),

$$\begin{aligned}|f(t)| &\leq \mathbb{E}\,|J_n(t|X|)|\,1_{\{|X|\leq\sqrt{n/2}\}} + \mathbb{E}\,|J_n(t|X|)|\,1_{\{|X|>\sqrt{n/2}\}} \\ &\leq 2^p\,\frac{\sigma_{2p}^p}{n^{p/2}} + 3\,\big(\mathrm{e}^{-t^2/4} + \mathrm{e}^{-n/12}\big).\end{aligned}$$

Thus, we get:

Lemma 13.1.3 *Let the random vector X in $\mathbb{R}^n$ satisfy $\mathbb{E}\,|X|^2 = n$. Then with some constant $c_p > 0$ depending on $p \geq 1$, for all $t \in \mathbb{R}$,*

$$c_p\,|f(t)| \leq \frac{1+\sigma_{2p}^p}{n^{p/2}} + \mathrm{e}^{-t^2/4}.$$

Therefore, for all $T > 0$,

$$\frac{c_p}{T}\int_0^T |f(t)|\,\mathrm{d}t \leq \frac{1+\sigma_{2p}^p}{n^{p/2}} + \frac{1}{T}.$$

These upper bounds may be considerably sharpened in the case of independent random variables. Define the average 4-th moment

$$\bar{\beta}_4 = \frac{1}{n}\sum_{k=1}^n \mathbb{E}X_k^4.$$

Lemma 13.1.4 *If $X_1, \dots, X_n$ are independent, have mean zero, variance one, and finite moment $\bar{\beta}_4$, then*

$$\mathbb{E}_\theta\,|f_\theta(t)|^2 \leq 3\mathrm{e}^{-t^2/2} + 6\mathrm{e}^{-n/(16\bar{\beta}_4)}.$$

As a consequence,

$$\mathbb{E}_\theta\,|f_\theta(t)| \leq \sqrt{3}\,\mathrm{e}^{-t^2/4} + \sqrt{6}\,\mathrm{e}^{-n/(32\bar{\beta}_4)}.$$

Proof Let $Y = (Y_1, \dots, Y_n)$ be an independent copy of X. The random vector $\widetilde{X} = (X-Y)/\sqrt{2}$ is isotropic and has mean zero, so, $M_2(\widetilde{X}) = 1$, while

$$\sigma_4^2(\widetilde{X}) = \frac{1}{n}\,\mathrm{Var}\big(|\widetilde{X}|^2\big) = \frac{1}{4n}\sum_{k=1}^n \mathrm{Var}\big((X_k - Y_k)^2\big) = \frac{1}{2}\,(\bar{\beta}_4 + 1).$$

Since $\bar{\beta}_4 \geq 1$, we get

$$\sigma_4^2(\widetilde{X}) + M_2^4(\widetilde{X}) = \frac{1}{2}\,(\bar{\beta}_4 + 1) + 1 \leq 2\bar{\beta}_4.$$

Applying Proposition 2.1.4 to $\widetilde{X}$ with $\lambda = \frac{1}{2}$, we may therefore conclude that

$$\begin{aligned}\mathbb{P}\{|X-Y|^2 \le n\} &= \mathbb{P}\Big\{|\widetilde{X}|^2 \le \lambda n\Big\} \\ &\le \exp\Big\{-\frac{(1-\lambda)^2}{2\,(\sigma_4^2(\widetilde{X}) + M_2^4(\widetilde{X}))}\,n\Big\} \le \exp\Big\{-n/(16\bar{\beta}_4)\Big\}.\end{aligned}$$

Hence, Lemma 13.1.1, now with $\lambda = 1$, gives

$$\mathbb{E}_\theta\,|f_\theta(t)|^2 \le 3\mathrm{e}^{-t^2/2} + 3\mathrm{e}^{-n/12} + 3\mathrm{e}^{-n/(16\bar{\beta}_4)} \le 3\mathrm{e}^{-t^2/2} + 6\mathrm{e}^{-n/(16\bar{\beta}_4)}.$$

It remains to apply Cauchy's inequality. Lemma 13.1.4 is proved. □

13.2 Concentration Functions of Weighted Sums

The integral appearing in Lemma 13.1.3 provides a well-known upper bound on the concentration function, which in general is defined by

$$Q_\xi(h) = Q_F(h) = \sup_x \mathbb{P}\{x \le \xi \le x+h\}, \quad h \ge 0.$$

Here, ξ is an arbitrary random variable with distribution function F. This quantity gives important information about the behavior of F – in particular, about possible concentration of the "mass" around a point. But this coincidence in terminology is irrelevant to the spherical concentration and its consequences.

According to Lemma 3.2.2,

$$Q_F(h) \le 2\Big(\frac{96}{95}\Big)^2 h \int_0^{1/h} |f(t)|\,\mathrm{d}t, \quad h > 0,$$

where f is the characteristic function of ξ. We will use this bound to estimate the concentration function of the weighted sums $\langle X, \theta\rangle$, which thus gives

$$Q_{F_\theta}(h) \le 3h \int_0^{1/h} |f_\theta(t)|\,\mathrm{d}t.$$

Applying Lemma 13.1.2 with $p = 2$, we immediately obtain an upper bound on the concentration functions on average, namely

$$\mathbb{E}_\theta\, Q_{F_\theta}(1/T) \le c\,(m_4^2 + \sigma_4^2)\,\frac{1}{n} + \frac{c}{T}, \quad T > 0.$$

Here we may choose $T = n/(m_4^2 + \sigma_4^2)$, and then we get:

Proposition 13.2.1 *Given a random vector X in $\mathbb{R}^n$ with finite 4-th moment and such that $\mathbb{E}\,|X|^2 = n$, put $\gamma = m_4^2(X) + \sigma_4^2(X)$. For some absolute constant $c > 0$ we have*

$$\mathbb{E}_\theta\, Q_{F_\theta}\Big(\frac{\gamma}{n}\Big) \le \frac{c\gamma}{n}.$$

By Markov's inequality, given $r > 0$, on the part of the unit sphere $\mathbb{S}^{n-1}$ of $\mathfrak{s}_{n-1}$-measure at least $1 - \frac{2c}{r}$, we have that

$$Q_{F_\theta}\Big(\frac{\gamma}{n}\Big) \le \frac{r}{2}\,\frac{\gamma}{n}.$$

Let us note that all concentration functions are subadditive, that is, $Q(h_1 + h_2) \le Q(h_1) + Q(h_2)$ for all $h_1, h_2 \ge 0$. In particular, $Q(\lambda h) \le (\lambda + 1)\, Q(h) \le 2\lambda\, Q(h)$, whenever $\lambda \ge 1$, $h \ge 0$. Hence, the above inequality implies

$$Q_{F_\theta}\Big(\lambda\,\frac{\gamma}{n}\Big) \le \lambda r\,\frac{\gamma}{n}, \quad \lambda \ge 1.$$

Substituting $h = \lambda\,\frac{\gamma}{n}$, we obtain a partial answer to the question raised after the formulation of Problem 12.1.3.

Corollary 13.2.2 *Let $\mathbb{E}\,|X|^2 = n$. Given $r > 0$, on the part of the unit sphere of $\mathfrak{s}_{n-1}$-measure at least $1 - \frac{c}{r}$ with an absolute constant $c > 0$, we have*

$$Q_{F_\theta}(h) \le rh \quad \textit{for all}\ \ h \ge \frac{m_4^2(X) + \sigma_4^2(X)}{n}.$$

This statement shows that, when m_4 and σ_4 are bounded, most of the distribution functions F_θ behave like Lipschitz functions whose "almost" Lipschitz semi-norm

$$\|F_\theta\|_{\mathrm{Lip}_\varepsilon} \equiv \sup_{x-y\ge\varepsilon} \frac{F_\theta(x) - F_\theta(y)}{x-y} \quad \text{with } \varepsilon = \frac{m_4^2(X) + \sigma_4^2(X)}{n}$$

is of order 1.

13.3 Deviations of Characteristic Functions

Now we are ready to apply the first order concentration on the sphere to study Problem 12.1.2 in terms of characteristic functions $f_\theta(t)$ with fixed $t \in \mathbb{R}$, rather than directly for the distributions F_θ. This can be done using the moment functionals

$$M_p(X) = \sup_{\theta\in\mathbb{S}^{n-1}} \Big(\mathbb{E}\,|\langle X, \theta\rangle|^p\Big)^{1/p}.$$

As before, let $f(t) = \mathbb{E}_\theta f_\theta(t)$ denote the characteristic function of the typical distribution F. The complex-valued function $u_t(\theta) = f_\theta(t) = \mathbb{E}\, e^{it\langle X,\theta\rangle}$ is smooth on the whole space $\mathbb{R}^n$ and has partial derivatives

$$\frac{\partial_k u_t(\theta)}{\partial \theta_k} = it\,\mathbb{E} X_k\, e^{it\langle X,\theta\rangle}, \quad k = 1,\dots,n,$$

or in vector form

$$\langle \nabla u_t(\theta), v\rangle = it\,\mathbb{E}\,\langle X, v\rangle\, e^{it\langle X,\theta\rangle}, \quad v \in \mathbb{C}^n,$$

where we use a standard inner product in the complex Euclidean n-space. Hence, writing $v = a + bi$, $a, b \in \mathbb{R}^n$, we have in terms of $M_1 = M_1(X)$

$$\begin{aligned} |\langle \nabla u_t(\theta), v\rangle| &\le |t|\,\mathbb{E}\,|\langle X, v\rangle| \\ &\le |t|\,(\mathbb{E}\,|\langle X, a\rangle| + \mathbb{E}\,|\langle X, b\rangle|) \le M_1 |t|\,(|a| + |b|). \end{aligned}$$

This readily implies uniform bounds on the modulus of the gradient, namely

$$|\nabla \operatorname{Re}(u_t(\theta))| \le M_1|t|, \quad |\nabla \operatorname{Im}(u_t(\theta))| \le M_1|t|,$$

so that $|\nabla u_t(\theta)| \le \sqrt{2}\,M_1|t|$. By Cauchy's inequality, we also have

$$|\langle \nabla u_t(\theta), v\rangle|^2 \le t^2\,\mathbb{E}\,|\langle X, v\rangle|^2 \le M_2\,t^2\,(|a|^2 + |b|^2) = M_2\,t^2\,|v|^2,$$

so that $|\nabla u_t(\theta)| \le M_2|t|$ (where $M_2 = M_2(X)$).

It is also interesting to get a similar bound holding on average. With this aim, let us square the vector representation and write

$$\langle \nabla u_t(\theta), v\rangle^2 = t^2\,\mathbb{E}\,\langle X, v\rangle\,\langle Y, v\rangle\, e^{it\langle X-Y,\theta\rangle}, \quad v \in \mathbb{S}^{n-1},$$

where Y is an independent copy of X. Integrating over v with respect to $\mathfrak{s}_{n-1}$, we get

$$|\nabla u_t(\theta)|^2 = t^2\,\mathbb{E}\,\langle X, Y\rangle\, e^{it\langle X-Y,\theta\rangle}.$$

One can now integrate over $\theta \in \mathbb{S}^{n-1}$ and summarize.

Lemma 13.3.1 *Given $t \in \mathbb{R}$, the function $u_t(\theta) = \mathbb{E}\, e^{it\langle X,\theta\rangle}$ satisfies*

$$|\nabla u_t(\theta)| \le \sqrt{2} M_1(X)\,|t|, \quad |\nabla u_t(\theta)| \le M_2(X)\,|t|$$

for all $\theta \in \mathbb{R}^n$, Moreover, using an independent copy Y of X, we have

$$\mathbb{E}_\theta\,|\nabla u_t(\theta)|^2 = t^2\,\mathbb{E}\,\langle X, Y\rangle\, J_n(t(X-Y)).$$

Recall that $J_n(t)$ denotes the characteristic function of the first coordinate of a point drawn from the unit sphere $\mathbb{S}^{n-1}$ according to the uniform distribution $\mathfrak{s}_{n-1}$.

Thus, the Lipschitz property of the functions $\theta \to f_\theta(t)$ can be expressed in terms of the first absolute moment $M_1(X)$ of the random vector X in $\mathbb{R}^n$. Applying Proposition 10.3.1 separately to the real and imaginary parts of the function

$$u(\theta) = \frac{1}{|t| M_1(X)} \left(f_\theta(t) - f(t)\right) \quad (t \neq 0),$$

we then get the subgaussian deviation inequality

$$\mathfrak{s}_{n-1}\left\{(n-1)\,|\mathrm{Re}(u(\theta))| \geq r\right\} \leq 2\mathrm{e}^{-r^2/2},$$

and similarly for $\mathrm{Im}(u(\theta))$, valid for any $r \geq 0$. To unite both, note that $|u(\theta)| \geq r$ implies

$$|\mathrm{Re}(u(\theta))| \geq \frac{r}{\sqrt{2}} \ \text{ or } \ |\mathrm{Im}(u(\theta))| \geq \frac{r}{\sqrt{2}}.$$

As a result, we arrive at:

Proposition 13.3.2 *If $M_1(X) \leq M_1$, then for all $t \in \mathbb{R}$ and $r \geq 0$,*

$$\mathfrak{s}_{n-1}\left\{|f_\theta(t) - f(t)| \geq r\right\} \leq 4\,\mathrm{e}^{-(n-1)\,r^2/(4M_1^2 t^2)}.$$

We also have corresponding bounds on the L^p-norms $\|f_\theta(t) - f(t)\|_p$ with respect to the measure $\mathfrak{s}_{n-1}$. In the cases $p = 1$ and $p = 2$ such bounds can also be derived from the Cheeger-type (Proposition 9.5.1) and Poincaré inequalities on the sphere.

Proposition 13.3.3 *If $M_1(X) \leq M_1$, then for all $t \in \mathbb{R}$,*

$$\mathbb{E}_\theta\, |f_\theta(t) - f(t)|^2 \leq \frac{2}{n-1}\,(M_1 t)^2. \tag{13.3}$$

More generally

$$\mathbb{E}_\theta\, |f_\theta(t) - f(t)|^p \leq \left(\frac{2(p-1)}{n-1}\right)^{p/2} (M_1|t|)^p, \quad p \geq 2.$$

Here, the L^p-bounds are based on Proposition 10.3.2.

If we choose to use $M_2(X)$ rather than $M_1(X)$, the constant 2 may be removed from both inequalities. Moreover, under the additional assumption that the functions $u_t(\theta) = f_\theta(t)$ are orthogonal to all linear functions in $L^2(\mathfrak{s}_{n-1})$, the bound (13.3) may be improved by virtue of the Poincaré-type inequality (9.11), which gives

$$\mathbb{E}_\theta\, |f_\theta(t) - f(t)|^2 \leq \frac{1}{2n}\, M_2^2(X)\, t^2. \tag{13.4}$$

Thus, the first order moment condition guarantees the standard $\frac{1}{\sqrt{n}}$-rate of deviations for $f_\theta(t)$ around $f(t)$ and hence for the distribution functions F_θ around F, at least in a weak sense (by applying tools from Fourier Analysis). Sudakov formulated this property under a (stronger) second order moment condition $M_2(X) \leq M_2$, which includes all isotropic random vectors. Let us state Proposition 13.3.2 once more for this important particular case.

Corollary 13.3.4 *If the random vector X in $\mathbb{R}^n$ is isotropic, then, for all $t \in \mathbb{R}$ and $r \geq 0$,*

$$\mathfrak{s}_{n-1}\big\{(n-1)\,|f_\theta(t) - f(t)| \geq r\big\} \leq 4\,\mathrm{e}^{-r^2/(4t^2)}.$$

Moreover,

$$\mathbb{E}_\theta\,|f_\theta(t) - f(t)|^p \leq \Big(\frac{p-1}{n-1}\Big)^{p/2}\,|t|^p, \quad p \geq 2.$$

13.4 Deviations in the Symmetric Case

In order to improve the standard rate in the bounds of Proposition 13.3.3 and Corollary 13.3.4 with respect to the dimension for deviations of

$$u_t(\theta) = f_\theta(t) = \mathbb{E}\,\mathrm{e}^{it\langle X,\theta\rangle}, \quad \theta = (\theta_1,\dots,\theta_n) \in \mathbb{R}^n, \tag{13.5}$$

about the mean characteristic function $f(t)$ (under the measure $\mathfrak{s}_{n-1}$), we employ the second order concentration phenomenon on the sphere. This requires us to impose a corresponding orthogonality condition and to consider the second partial derivatives of $u_t(\theta)$ with t being a real parameter. In turn, the study of the Hessian of u_t inspires us to impose a second order correlation condition on X.

Thus, let $X = (X_1,\dots,X_n)$ be an isotropic random vector in $\mathbb{R}^n$ satisfying a second order correlation condition with finite constant $\Lambda = \Lambda(X)$, that is,

$$\mathbb{E}\Big|\sum_{j,k=1}^n z_{jk}\,(X_jX_k - \delta_{jk})\Big|^2 \leq \Lambda \sum_{j,k=1}^n |z_{jk}|^2 \tag{13.6}$$

for all complex numbers z_{jk}. To apply Proposition 10.2.2 (the second order Poincaré-type inequality) and Proposition 10.4.2 (the second order concentration), let us also assume that X has a symmetric distribution. In this case, $u_t(\theta)$ are even with respect to θ, i.e., $u_t(-\theta) = u_t(\theta)$, and hence these functions are orthogonal to all linear functions in $L^2(\mathfrak{s}_{n-1})$ for any $t \in \mathbb{R}$.

In order to show that the deviations of u_t are of order $1/n$, we need to choose a suitable value of $a \in \mathbb{C}$ and develop the operator norm $\|u_t'' - a\,\mathbb{I}_n\|$ and the Hilbert–Schmidt norm $\|u_t'' - a\,\mathbb{I}_n\|_{\mathrm{HS}}$. First note that, by direct differentiation in (13.5),

$$[u_t''(\theta)]_{jk} = \frac{\partial^2}{\partial\theta_j\partial\theta_k}\,f_\theta(t) = -t^2\,\mathbb{E}\,X_jX_k\,\mathrm{e}^{it\langle X,\theta\rangle}.$$

Hence, it makes sense to take $a = -t^2 f(t)$ to balance the diagonal elements in the matrix of second derivatives of u_t. For all vectors $v, w \in \mathbb{C}^n$ with complex components, using the canonical inner product in the complex n-space, we have

$$\big\langle u_t''(\theta)v, w\big\rangle = -t^2\,\mathbb{E}\,\langle X,v\rangle\,\overline{\langle X,w\rangle}\,\mathrm{e}^{it\langle X,\theta\rangle}.$$

The isotropy assumption ensures that $\mathbb{E}\,|\,\langle X, v\rangle\,|^2 = |v|^2$ and similarly for w. Hence, with this choice of a, by Cauchy's inequality,

$$|\,\big\langle (u_t''(\theta) - a\,\mathbb{I}_n)v, w\big\rangle\,| \,\le\, t^2\,\mathbb{E}\,|\,\langle X, v\rangle\,|\,|\,\langle X, w\rangle\,| + |a|\,|\,\langle v, w\rangle\,| \le 2t^2,$$

whenever $|v| = 1$ and $|w| = 1$. This bound ensures that

$$\|u_t''(\theta) - a\,\mathbb{I}_n\| \le 2t^2 \quad (\theta \in \mathbb{R}^n,\ t \in \mathbb{R}). \tag{13.7}$$

Now, putting $a(\theta) = -t^2 f_\theta(t)$, we have

$$\begin{aligned}
\|u_t''(\theta) - a(\theta)\,\mathbb{I}_n\|_{\mathrm{HS}}^2 &= \sum_{j,k=1}^n \big|u_t''(\theta)_{jk} - a(\theta)\delta_{jk}\big|^2 \\
&= \sup\Big|\sum_{j,k=1}^n z_{jk}\,\Big(u_t''(\theta)_{jk} - a(\theta)\delta_{jk}\Big)\Big|^2 \\
&= t^4 \sup\Big|\,\mathbb{E}\sum_{j,k=1}^n z_{jk}\,\big(X_jX_k - \delta_{jk}\big)\,e^{it\langle X,\theta\rangle}\Big|^2 \\
&\le t^4\,\sup\,\mathbb{E}\Big|\sum_{j,k=1}^n z_{jk}\,(X_jX_k - \delta_{jk})\Big|^2,
\end{aligned}$$

where the supremum runs over all complex numbers z_{jk} such that $\sum_{j,k=1}^n |z_{jk}|^2 = 1$. But, under this constraint (with complex coefficients), the last expectation does not exceed Λ, according to the hypothesis (13.6). Thus,

$$\|u_t''(\theta) - a(\theta)\,\mathbb{I}_n\|_{\mathrm{HS}}^2 \le \Lambda t^4 \tag{13.8}$$

for all θ. On the other hand, by Proposition 13.3.3 in the improved form (13.4), and using the isotropy assumption, we have

$$\mathbb{E}_\theta\,\|(a(\theta) - a)\,\mathbb{I}_n\|_{\mathrm{HS}}^2 = nt^4\,\mathbb{E}_\theta\,|f_\theta(t) - f(t)|^2 \le \frac{1}{2}\,t^6. \tag{13.9}$$

Combining the last two bounds on the Hilbert–Schmidt norm, we arrive at

$$\mathbb{E}_\theta\|u_t''(\theta) - a\,\mathbb{I}_n\|_{\mathrm{HS}}^2 \le 2\Lambda t^4 + t^6. \tag{13.10}$$

One can now involve the second order Poincaré-type inequality of Proposition 10.2.2, which yields

$$\begin{aligned}
\mathbb{E}_\theta|f_t(\theta) - f(t)|^2 &\le \frac{2}{n(n-1)}\,\mathbb{E}_\theta\|u_t''(\theta) - a\,\mathbb{I}_n\|_{\mathrm{HS}}^2 \\
&\le \frac{2}{n(n-1)}\,(2\Lambda t^4 + t^6).
\end{aligned} \tag{13.11}$$

This inequality already produces the desired $\frac{1}{n}$-rate with respect to n, and in fact, one may control large deviations. In terms of the function

$$u(\theta) = \frac{1}{2t^2}\,(f_\theta(t) - f(t)), \quad \theta \in \mathbb{R}^n,\ t \neq 0,$$

the inequalities (13.7) and (13.10) take the form

$$\Big\|u''(\theta) + \frac{1}{2}\,f(t)\,\mathbb{I}_n\Big\| \leq 1, \quad \mathbb{E}_\theta\,\Big\|u''(\theta) + \frac{1}{2}\,f(t)\,\mathbb{I}_n\Big\|_{\mathrm{HS}}^2 \leq \frac{\Lambda}{2} + \frac{t^2}{4},$$

which means that the conditions of Proposition 10.4.2 are fulfilled for the real and imaginary parts of u with parameter $b = \frac{1}{2}\Lambda + \frac{1}{4}\,t^2$. Applying the exponential bound (10.10) to both real and imaginary parts, we then get

$$\|\operatorname{Re}(f_\theta(t) - f(t))\|_{\psi_1} \leq \frac{4t^2}{n-1}\,(1+3b),$$

and similarly for $\operatorname{Im}(f_\theta(t) - f(t))$. Hence, by the triangle inequality for the ψ_1-norm,

$$\begin{aligned}\|f_\theta(t) - f(t)\|_{\psi_1} &\leq \frac{8t^2}{n-1}\,(1+3b)\\ &= \frac{2t^2}{n-1}\,(4 + 6\Lambda + 3t^2) \;\leq\; \frac{2t^2}{n-1}\,(14\,\Lambda + 3t^2),\end{aligned}$$

where we used that $\Lambda \geq \frac{1}{2}$. As a result, we arrive at the following conclusion.

Proposition 13.4.1 *Let X be an isotropic random vector in $\mathbb{R}^n$ with a symmetric distribution and finite constant Λ. Then, the characteristic functions $f_\theta(t) = \mathbb{E}\,e^{it\langle X,\theta\rangle}$ satisfy, for any $t \neq 0$,*

$$\mathbb{E}_\theta\,\exp\Big\{\frac{n-1}{2t^2\,(14\Lambda + 3t^2)}\,|f_\theta(t) - f(t)|\Big\} \leq 2.$$

However, in further applications it is desirable to improve the dependence on t on the left-hand side of this exponential bound, at least, for not too large values of $|t|$. With this aim, let us apply (13.11) in (13.9) which leads to the improved relation

$$\|(a(\theta) - a)\,\mathbb{I}_n\|_{\mathrm{HS}}^2 \leq \frac{2}{n-1}\,(2\Lambda t^8 + t^{10}).$$

Together with (13.8) this gives

$$\mathbb{E}_\theta\|u_t''(\theta) - a\,\mathbb{I}_n\|_{\mathrm{HS}}^2 \leq 2\Lambda t^4 + \frac{4}{n-1}\,(2\Lambda t^8 + t^{10}).$$

This quantity is bounded by $c\Lambda t^4$ in the region $|t| \leq n^{1/6}$. But, one may enlarge it by repeating the argument. Indeed, using the above upper bound, one may apply Proposition 10.2.2 once more, which gives

$$\mathbb{E}_\theta |f_t(\theta) - f(t)|^2 \le \frac{2}{n(n-1)}\Big(2\Lambda t^4 + \frac{4}{n-1}\,(2\Lambda t^8 + t^{10})\Big).$$

Using this in (13.9), we get

$$\|(a(\theta) - a)\,\mathbb{I}_n\|_{\mathrm{HS}}^2 \le \frac{2}{n-1}\Big(2\Lambda t^8 + \frac{4}{n-1}\,(2\Lambda t^{12} + t^{14})\Big),$$

which together with (13.8) yields

$$\mathbb{E}_\theta\,\|u_t''(\theta) - a\,\mathbb{I}_n\|_{\mathrm{HS}}^2 \le 2\Lambda t^4 + \frac{4}{n-1}\Big(2\Lambda t^8 + \frac{4}{n-1}\,(2\Lambda t^{12} + t^{14})\Big).$$

The expression on the right-hand side is bounded by multiples of Λt^4 in larger regions such as $|t| \le An^{1/5}$. By (13.7), the conditions of Proposition 10.4.2 are therefore fulfilled for the real and imaginary parts of the same function u as above with parameter $b = c\Lambda/2$. As a result, Proposition 10.4.2 yields:

Proposition 13.4.2 *Let X be an isotropic random vector in $\mathbb{R}^n$ with a symmetric distribution and finite constant Λ. Then, the characteristic functions $f_\theta(t) = \mathbb{E}\,e^{it\langle X,\theta\rangle}$ satisfy, for any $t \neq 0$, $|t| \le An^{1/5}$,*

$$\mathbb{E}_\theta\,|f_\theta(t) - f(t)|^2 \le \frac{c}{n^2}\,\Lambda t^4$$

with some constant $c > 0$ depending on the parameter $A > 0$ only. Moreover,

$$\mathbb{E}_\theta\,\exp\Big\{\frac{n}{c\Lambda t^2}\,|f_\theta(t) - f(t)|\Big\} \le 2.$$

Continuing the process of repeated applications of (13.9), one can obtain Proposition 13.4.2 for the regions $|t| \le An^\alpha$ with any fixed $\alpha < \frac{1}{4}$.

13.5 Deviations in the Non-symmetric Case

We now drop the symmetry assumption and extend the bounds of Proposition 13.4.2 to the general case of an isotropic random vector X in $\mathbb{R}^n$ with a finite constant Λ as in (13.6). This is possible by involving an additional parameter, namely, the L^2-norm of the linear part of characteristic functions

$$u_t(\theta) = f_\theta(t) = \mathbb{E}\,e^{it\langle X,\theta\rangle}, \quad \theta \in \mathbb{R}^n.$$

More precisely, consider an orthogonal decomposition

$$u_t(\theta) = f(t) + l_t(\theta) + v_t(\theta), \tag{13.12}$$

where

$$l_t(\theta) = a_1(t)\,\theta_1 + \cdots + a_n(t)\,\theta_n, \quad \theta = (\theta_1, \dots, \theta_n) \in \mathbb{R}^n,$$

is the orthogonal projection of u_t in $L^2(\mathbb{R}^n, \mathfrak{s}_{n-1})$ onto the space of all linear functions (the linear part), and

$$v_t(\theta) = u_t(\theta) - f(t) - l_t(\theta)$$

is the non-linear part of u_t. By the orthogonality,

$$\mathbb{E}_\theta\, |f_\theta(t) - f(t)|^2 = \mathbb{E}_\theta\, |l_t(\theta)|^2 + \mathbb{E}_\theta\, |v_t(\theta)|^2. \tag{13.13}$$

To bound the last L^2-norm, first note that u_t and v_t have equal second order partial derivatives on $\mathbb{R}^n$. Choosing $a = -t^2 f(t)$ and putting $a(\theta) = -t^2 f_\theta(t)$ as before, the inequalities (13.7)–(13.8) do not change, so that, for all $\theta \in \mathbb{R}^n$ and $t \in \mathbb{R}$,

$$\|v_t''(\theta) - a\,\mathbb{I}_n\| \le 2t^2, \quad \|v_t''(\theta) - a(\theta)\,\mathbb{I}_n\|_{\rm HS}^2 \le \Lambda t^4. \tag{13.14}$$

On the other hand, by the isotropy assumption, we have $|\nabla u_t| \le |t|$ (Lemma 13.3.1), so that, by the Poincaré inequality,

$$\mathbb{E}_\theta\, |f_\theta(t) - f(t)|^2 \le \frac{n}{n-1}\, t^2 \le 2t^2. \tag{13.15}$$

Hence

$$\mathbb{E}_\theta\, \|(a(\theta) - a)\,\mathbb{I}_n\|_{\rm HS}^2 = nt^4\, \mathbb{E}_\theta\, |f_\theta(t) - f(t)|^2 \le 2t^6. \tag{13.16}$$

Combining the last two bounds leads to

$$\mathbb{E}_\theta \|v_t''(\theta) - a\,\mathbb{I}_n\|_{\rm HS}^2 \le 2\Lambda t^4 + 4t^6,$$

which, by Proposition 10.2.2, gives

$$\mathbb{E}_\theta |v_t(\theta)|^2 \le \frac{2}{n(n-1)}\, \mathbb{E}_\theta \|v_t''(\theta) - a\,\mathbb{I}_n\|_{\rm HS}^2 \le \frac{2}{n(n-1)}\,(2\Lambda t^4 + 4t^6).$$

Next, we need to remove the last term $4t^6$ for the range $|t| \le An^{1/5}$ (by adding a smaller quantity). Using the above bound in (13.13), we have

$$\mathbb{E}_\theta\, |f_\theta(t) - f(t)|^2 \le \mathbb{E}_\theta\, |l_t(\theta)|^2 + \frac{2}{n(n-1)}\,(2\Lambda t^4 + 4t^6),$$

which, according to the identity in (13.16), gives

$$\mathbb{E}_\theta \|(a(\theta) - a)\,\mathbb{I}_n\|_{\rm HS}^2 \le nt^4\, \mathbb{E}_\theta\, |l_t(\theta)|^2 + \frac{2}{n-1}\,(2\Lambda t^8 + 4t^{10}).$$

Combining this with the second inequality in (13.14), we get

$$\mathbb{E}_\theta \|v_t''(\theta) - a\,\mathbb{I}_n\|_{\rm HS}^2 \le 2nt^4\, \mathbb{E}_\theta\, |l_t(\theta)|^2 + 2\Lambda t^4 + \frac{4}{n-1}\,(2\Lambda t^8 + 4t^{10}).$$

Hence, by Proposition 10.2.2 once more,

$$\begin{aligned}\mathbb{E}_\theta |v_t(\theta)|^2 &\le \frac{2}{n(n-1)}\Big(2nt^4\,\mathbb{E}_\theta\,|l_t(\theta)|^2 + 2\Lambda t^4 + \frac{4}{n-1}\,(2\Lambda t^8 + 4t^{10})\Big)\\ &= \frac{4t^4}{n-1}\,\mathbb{E}_\theta\,|l_t(\theta)|^2 + \frac{4}{n(n-1)}\,\Lambda t^4 + \frac{16}{n(n-1)^2}\,(\Lambda t^8 + 2t^{10}),\end{aligned}$$

so that, by (13.13),

$$\begin{aligned}\mathbb{E}_\theta\,|f_\theta(t) - f(t)|^2 &\le \Big(1 + \frac{4t^4}{n-1}\Big)\,\mathbb{E}_\theta\,|l_t(\theta)|^2\\ &\quad + \frac{4}{n(n-1)}\,\Lambda t^4 + \frac{16}{n(n-1)^2}\,(\Lambda t^8 + 2t^{10}).\end{aligned}$$

According to the identity (13.16) once more, this gives

$$\begin{aligned}\mathbb{E}_\theta \|(a(\theta) - a)\,\mathbb{I}_n\|_{\mathrm{HS}}^2 &\le nt^4\,\Big(1 + \frac{4t^4}{n-1}\Big)\,\mathbb{E}_\theta\,|l_t(\theta)|^2\\ &\quad + \frac{4}{n-1}\,\Lambda t^8 + \frac{16}{(n-1)^2}\,(\Lambda t^{12} + 2t^{14}).\end{aligned}$$

Combining this inequality with the second inequality in (13.14), we get

$$\begin{aligned}\mathbb{E}_\theta \|v_t''(\theta) - a\,\mathbb{I}_n\|_{\mathrm{HS}}^2 &\le 2nt^4\,\Big(1 + \frac{4t^4}{n-1}\Big)\,\mathbb{E}_\theta\,|l_t(\theta)|^2\\ &\quad + 2\Lambda t^4 + \frac{8}{n-1}\,\Lambda t^8 + \frac{32}{(n-1)^2}\,(\Lambda t^{12} + 2t^{14}).\end{aligned}$$

Now, if $|t| \le An^{1/5}$, the coefficient in front of $\mathbb{E}_\theta\,|l_t(\theta)|^2$ does not exceed a multiple of nt^4. Similarly, in this region the last three terms can be bounded by Λt^4 up to a numerical factor. Hence, this bound simplifies to

$$c\,\mathbb{E}_\theta \|v_t''(\theta) - a\,\mathbb{I}_n\|_{\mathrm{HS}}^2 \le nt^4\,\mathbb{E}_\theta\,|l_t(\theta)|^2 + \Lambda t^4$$

for some constant $c > 0$ depending on the parameter A, only. Once more by Proposition 10.2.2, it follows that

$$c\,\mathbb{E}_\theta |v_t(\theta)|^2 \le \mathbb{E}_\theta\,|l_t(\theta)|^2 + \frac{2}{n(n-1)}\,\Lambda t^4.$$

It remains to use this inequality in (13.13) to get a similar upper bound

$$c\,\mathbb{E}_\theta\,|f_\theta(t) - f(t)|^2 \le 2\,\mathbb{E}_\theta\,|l_t(\theta)|^2 + \frac{2}{n(n-1)}\,\Lambda t^4.$$

To get a stronger deviation inequality in terms of the Orlicz ψ_1-norm, note that, by (13.7), the conditions of Proposition 10.5.2 are fulfilled for the same function

$$u(\theta) = \frac{1}{2t^2}\,(f_\theta(t) - f(t))$$

as in the symmetric case, with $-\frac{1}{2}\,f(t)$ in place of a. Since (13.16) holds for u_t as well, this inequality may be rewritten as

$$c\,\mathbb{E}_\theta\,\Big\|u''(\theta) + \frac{1}{2}\,f(t)\,\mathbb{I}_n\Big\|_{\rm HS}^2 \le n\,\|l_t\|_{L^2}^2 + \Lambda.$$

Note that the linear part of u is $\frac{1}{2t^2}\,l_t$. Hence, the inequality (13.14) of Proposition 10.5.2 applied to u yields

$$c\,\Big\|\frac{1}{2t^2}\,(f_\theta(t) - f(t))\Big\|_{\psi_1} \le \frac{1+\Lambda}{n} + \|l_t\|_{L^2}^2 + \frac{1}{2t^2}\,\|l_t\|_{L^2}.$$

Using once more $\Lambda \ge \frac{1}{2}$, the above is simplified to

$$c\,\|f_\theta(t) - f(t))\|_{\psi_1} \le \frac{\Lambda t^2}{n} + \|l_t\|_{L^2} + \|l_t\|_{L^2}^2\,t^2.$$

Here, the last term on the right-hand side is dominated by the second to last term in the smaller interval $|t| \le An^{1/6}$. Indeed, according to the L^2-bound (13.15),

$$\|l_t\|_{L^2}\,t^2 \le \|f_\theta(t) - f(t)\|_{L^2}\,t^2 \le \frac{2\,|t|^3}{\sqrt{n}} \le 2A^3.$$

Hence $\|l_t\|_{L^2}^2\,t^2 \le 2A^3\,\|l_t\|_{L^2}$. As a result, we arrive at:

Proposition 13.5.1 *Given an isotropic random vector X in $\mathbb{R}^n$ with finite constant Λ, the characteristic functions $f_\theta(t) = \mathbb{E}\,e^{it\langle X,\theta\rangle}$ satisfy, for any $|t| \le An^{1/5}$,*

$$c\,\mathbb{E}_\theta\,|f_\theta(t) - f(t)|^2 \le \mathbb{E}_\theta\,|l_t(\theta)|^2 + \frac{\Lambda t^4}{n^2} \tag{13.17}$$

with a constant $c > 0$ depending on $A > 0$ only, where $l_t(\theta)$ is the linear part of $f_\theta(t)$ from the orthogonal decomposition (13.12). *Moreover, in the interval $|t| \le An^{1/6}$,*

$$c\,\|f_\theta(t) - f(t)\|_{\psi_1} \le \|l_t\|_{L^2} + \frac{\Lambda t^2}{n}. \tag{13.18}$$

If the distribution of X is symmetric about the origin, then $l_t(\theta) = 0$, and we return in (13.17) to the Poincaré-type inequality of Proposition 13.4.2, while (13.18) leads to the large deviation bound.

The linear part $l_t(\theta)$ is also vanishing when X has mean zero and a constant Euclidean norm, i.e. when $|X| = \sqrt{n}$ a.s. This will be discussed in the next section.

13.6 The Linear Part of Characteristic Functions

In order to make the bounds (13.17)–(13.18) effective, we need to properly estimate the L^2-norm of the linear part $l_t(\theta)$ of $f_\theta(t)$, that is, the integrals

$$I(t) = \|l_t(\theta)\|^2_{L^2(\mathfrak{s}_{n-1})} = n\,\mathbb{E}_\theta\,\mathbb{E}_{\theta'}\,\langle\theta,\theta'\rangle\, f_\theta(t)\bar{f}_{\theta'}(t).$$

Thus, we are looking for conditions and formulas which yield bounds $I(t) = O(\frac{1}{n^2})$.

As an equivalent formula, write

$$I(t) = n\sum_{k=1}^n |\,\mathbb{E}_\theta\,\theta_k f_\theta(t)|^2 = n\sum_{k=1}^n \mathbb{E}\,\mathbb{E}_\theta\,\mathbb{E}_{\theta'}\Big[\theta_k\theta'_k\, e^{it\langle X,\theta\rangle - it\langle Y,\theta'\rangle}\Big],$$

where Y is an independent copy of X. To compute the inner expectations, introduce the function

$$K_n(t) = J_n\big(\sqrt{tn}\big), \quad t \ge 0,$$

where, as before, J_n denotes the characteristic function of the first coordinate of a point on the unit sphere $\mathbb{S}^{n-1}$ under $\mathfrak{s}_{n-1}$. By the definition,

$$\mathbb{E}_\theta\, e^{i\langle v,\theta\rangle} = J_n(|v|) = K_n\Big(\frac{|v|^2}{n}\Big), \quad v = (v_1,\dots,v_n) \in \mathbb{R}^n.$$

Differentiating this equality with respect to the variable v_k, we obtain that

$$i\,\mathbb{E}_\theta\,\theta_k e^{i\langle v,\theta\rangle} = \frac{2v_k}{n}\,K'_n\Big(\frac{|v|^2}{n}\Big).$$

Let us multiply this by a similar equality

$$-i\,\mathbb{E}_\theta\,\theta_k e^{-i\langle w,\theta\rangle} = \frac{2w_k}{n}\,K'_n\Big(\frac{|w|^2}{n}\Big),$$

and conclude that, for all $v, w \in \mathbb{R}^n$,

$$\mathbb{E}_\theta\,\mathbb{E}_{\theta'}\Big[\theta_k\theta'_k\, e^{i\langle v,\theta\rangle - i\langle w,\theta'\rangle}\Big] = \frac{4v_k w_k}{n^2}\,K'_n\Big(\frac{|v|^2}{n}\Big)K'_n\Big(\frac{|w|^2}{n}\Big).$$

Hence, summing over all $k \le n$, we get

$$\sum_{k=1}^n \mathbb{E}_\theta\,\mathbb{E}_{\theta'}\Big[\theta_k\theta'_k\, e^{i\langle v,\theta\rangle - i\langle w,\theta'\rangle}\Big] = \frac{4\,\langle v,w\rangle}{n^2}\,K'_n\Big(\frac{|v|^2}{n}\Big)\,K'_n\Big(\frac{|w|^2}{n}\Big).$$

It remains to make the substitution $v = tX$, $w = tY$, and to take the expectation over (X,Y). Then we arrive at the following expression for the integrals $I(t)$.

Lemma 13.6.1 *For all $t \in \mathbb{R}$, the characteristic function $f_\theta(t)$ as a function of θ on the sphere $\mathbb{S}^{n-1}$ has a linear part whose squared $L^2(\mathfrak{s}_{n-1})$-norm is equal to*

$$I(t) = \frac{4t^2}{n}\,\mathbb{E}\,\langle X, Y\rangle\, K_n'\Big(\frac{t^2|X|^2}{n}\Big) K_n'\Big(\frac{t^2|Y|^2}{n}\Big),$$

where $K_n(t) = J_n\big(\sqrt{tn}\big)$, $t \geq 0$, and where Y is an independent copy of X.

In particular, if $|X| = \sqrt{n}$ a.s., then

$$I(t) = \frac{4t^2}{n}\,K_n'^2(t^2)\,\mathbb{E}\,\langle X, Y\rangle\,,$$

which is vanishing as long as X has mean zero. Note that the property $I(t) = 0$ continues to hold for more general random vectors. In particular, this is the case when the conditional distribution of X given that $|X| = r$ has mean zero for any $r > 0$.

Now, let us write down an asymptotic formula for the function K_n and its derivative. We know from Corollary 11.3.3 that

$$\frac{d}{dt}\,J_n(t\sqrt{n}) = -t\,\Big(1 - \frac{t^4 - 4t^2}{4n}\Big)\,e^{-t^2/2} + O\Big(n^{-2}\min(1, |t|^3)\Big).$$

Since $J_n(t\sqrt{n}) = K_n(t^2)$, after differentiation we find that

$$2tK_n'(t^2) = \frac{d}{dt}\,K_n(t^2) = -t\,\Big(1 - \frac{t^4 - 4t^2}{4n}\Big)\,e^{-t^2/2} + O\Big(n^{-2}\min(1, |t|^3)\Big).$$

Changing the variable, we arrive at

$$K_n'(t) = -\frac{1}{2}\,\Big(1 - \frac{t^2 - 4t}{4n}\Big)\,e^{-t/2} + O\Big(n^{-2}\min(1, t)\Big), \quad t \geq 0.$$

From this

$$K_n'(t)K_n'(s) = \frac{1}{4}\,\Big(1 - \frac{(t^2 + s^2) - 4(t+s)}{4n}\Big)\,e^{-(t+s)/2} + O\big(n^{-2}\big)$$

uniformly over all $t, s \geq 0$, so,

$$4K_n'\Big(\frac{t^2|X|^2}{n}\Big)K_n'\Big(\frac{t^2|Y|^2}{n}\Big) = \left(1 - \frac{t^4\,\big(\frac{|X|^4}{n^2} + \frac{|Y|^4}{n^2}\big) - 4t^2\,\big(\frac{|X|^2}{n} + \frac{|Y|^2}{n}\big)}{4n}\right) e^{-\frac{t^2(|X|^2+|Y|^2)}{2n}} + \varepsilon$$

with a remainder term satisfying $|\varepsilon| \leq \frac{c}{n^2}$ up to some absolute constant c.

If X is isotropic, we get

$$\frac{t^2}{n}\,\mathbb{E}\,|\langle X, Y\rangle|\,|\varepsilon| \leq \frac{ct^2}{n^3}\,\sqrt{\mathbb{E}\,\langle X, Y\rangle^2} = \frac{ct^2}{n^{5/2}}.$$

But, even assuming that $\mathbb{E}\,|X|^2 = n$, one may use $\mathbb{E}\,|\langle X, Y\rangle| \le \mathbb{E}\,\frac{|X|^2+|Y|^2}{2} = n$, which leads to a bound of order at most $\frac{t^2}{n^2}$. Hence, from Lemma 13.6.1 we obtain:

Lemma 13.6.2 *Let X be a random vector in $\mathbb{R}^n$ such that $\mathbb{E}\,|X|^2 = n$. For all $t \in \mathbb{R}$, the characteristic function $f_\theta(t)$ as a function of θ on the unit sphere has a linear part whose squared $L^2(\mathfrak{s}_{n-1})$-norm is represented as*

$$I(t) = \frac{t^2}{n}\,\mathbb{E}\,\langle X, Y\rangle \left(1 - \frac{t^4\big(\frac{|X|^4}{n^2} + \frac{|Y|^4}{n^2}\big) - 4t^2\big(\frac{|X|^2}{n} + \frac{|Y|^2}{n}\big)}{4n}\right) e^{-\frac{t^2(|X|^2+|Y|^2)}{2n}} + O(t^2 n^{-2}),$$

where Y is an independent copy of X. Moreover, if X is isotropic, then the error term may be improved to $O(t^2 n^{-5/2})$.

Let us apply this result to the inequalities of Proposition 13.5.1, still assuming that Y is an independent copy of X.

Proposition 13.6.3 *Let X be an isotropic random vector in $\mathbb{R}^n$ with finite constant $\Lambda = \Lambda(X)$. For any $|t| \le An^{1/5}$, the characteristic functions $f_\theta(t) = \mathbb{E}\,e^{it\langle X,\theta\rangle}$ satisfy*

$$c\,\mathbb{E}_\theta\,|f_\theta(t) - f(t)|^2 \le \frac{\Lambda t^4}{n^2} + I(t)$$

with some constant $c > 0$ depending on the parameter $A > 0$ only, where

$$I(t) = \frac{t^2}{n}\,\mathbb{E}\,\langle X, Y\rangle \left(1 - \frac{(U^2+V^2)\,t^4 - 8R^2t^2}{4n}\right) e^{-R^2t^2} + O\big(t^2 n^{-5/2}\big)$$

with $R^2 = \frac{|X|^2+|Y|^2}{2n}$ and $U = \frac{|X|^2}{n}$, $V = \frac{|Y|^2}{n}$. Moreover, in the interval $|t| \le An^{1/6}$,

$$c\,\|f_\theta(t) - f(t)\|_{\psi_1} \le \frac{\Lambda t^2}{n} + \sqrt{I(t)}.$$

13.7 Remarks

First order concentration bounds on deviations of characteristic functions like the one of Proposition 13.3.2 were derived in [25]. Proposition 13.4.1 is proved in [39], and its extension to the non-linear case has been developed in [40].

Chapter 14
Fluctuations of Distributions

The main theme in this chapter is the various bounds on the deviations of the distribution functions $F_\theta(x)$ from the typical distribution function $F(x)$ and from the standard normal distribution function $\Phi(x)$ at individual points x and in several weak metrics, including the L^1-distance (the Kantorovich metric), the Lévy and Kolmogorov distances.

14.1 The Kantorovich Transport Distance

In dimension one (i.e., on the real line), the Kantorovich transport distance between two probability distributions, say F and G, takes the simple form of the L^1-distance

$$W(F,G) = \int_{-\infty}^{\infty} |F(x) - G(x)|\, dx,$$

where we use the same symbols to denote the distribution functions associated with given measures. This quantity, which defines a metric in the space of all probability measures on the real line with finite first absolute moments, is closely related to the topology of weak convergence (cf. e.g. [174]).

As before, given a random vector X in $\mathbb{R}^n$ ($n \geq 2$), we denote by

$$F_\theta(x) = \mathbb{P}\{\langle X, \theta\rangle \leq x\}, \quad x \in \mathbb{R},$$

the distribution functions of the weighted sums $\langle X, \theta\rangle$ with $\theta \in \mathbb{S}^{n-1}$ and by $F(x) = \mathbb{E}_\theta F_\theta(x)$ the average (typical) distribution function.

In order to deal with the main Problem 12.1.2, we start with the Kantorovich distance for bounding possible fluctuations of F_θ around F on average. As a main parameter of the underlying distribution of X, we use the moment functionals

$$M_p = M_p(X) = \sup_{\theta \in \mathbb{S}^{n-1}} \Big(\mathbb{E}\,|\,\langle X, \theta\rangle\,|^p\Big)^{1/p}, \quad p \geq 1.$$

S. Bobkov et al., *Concentration and Gaussian Approximation for Randomized Sums*, Probability Theory and Stochastic Modelling 104,
https://doi.org/10.1007/978-3-031-31149-9_14

Denote by

$$q = \frac{p}{p-1}$$

the conjugate power. Our aim is to prove the following statement quantifying and generalizing Theorem 12.1.1.

Proposition 14.1.1 *If $p > 1$, then*

$$\mathbb{E}_\theta\, W(F_\theta, F) \le 7qM_p\, \frac{1}{(2n)^{1/(2q)}}. \tag{14.1}$$

For example, in the case $p = 2$, we get

$$\mathbb{E}_\theta\, W(F_\theta, F) \le 14\, M_2\, (2n)^{-\frac{1}{4}}.$$

By Chebyshev's inequality, it follows that, for any $\delta > 0$,

$$\mathfrak{s}_{n-1}\big\{W(F_\theta, F) \ge M_2\, \delta\big\} \le \frac{14}{\delta}\, (2n)^{-\frac{1}{4}}.$$

The expression on the right does not exceed δ as soon as $n \ge 14^4\, \delta^{-8}$, so that Theorem 12.1.1 holds true with $n_\delta = [14^4\, \delta^{-8}] + 1$.

For the proof of Proposition 14.1.1 we need one elementary lemma.

Lemma 14.1.2 *Given distribution functions F and G, for all real numbers $a < b$ and any integer $N \ge 1$,*

$$\int_a^b |F(x) - G(x)|\,dx \le \sum_{k=1}^N \bigg|\int_{a_{k-1}}^{a_k} (F(x) - G(x))\,dx\bigg| + \frac{2(b-a)}{N},$$

where $a_k = a + (b-a)\frac{k}{N}$.

Proof Denote by I the collection of all indexes $k = 1, \dots, N$ such that in the k-th interval $\Delta_k = (a_{k-1}, a_k)$ the function $\psi(x) = F(x) - G(x)$ does not change the sign. The remaining indexes form a complementary set $J \subset \{1, \dots, N\}$. Then

$$\int_{\Delta_k} |F(x) - G(x)|\,dx = \bigg|\int_{\Delta_k} (F(x) - G(x))\,dx\bigg|$$

for all $k \in I$. In the case $k \in J$, we obviously have

$$\sup_{x \in \Delta_k} |\psi(x)| \le \sup_{x,y \in \Delta_k} (\psi(x) - \psi(y)) \le F(\Delta_k) + G(\Delta_k),$$

where F and G are treated as probability measures. In this case,

$$\int_{\Delta_k} |F(x) - G(x)|\,dx \le (F(\Delta_k) + G(\Delta_k))|\Delta_k|, \qquad |\Delta_k| = \frac{b-a}{N}.$$

Combining these estimates, we can bound the integral $\int_a^b |F(x) - G(x)|\, dx$ by

$$\sum_{k\in I} \Big| \int_{\Delta_k} (F(x) - G(x))\, dx \Big| + \sum_{k\in J} (F(\Delta_k) + G(\Delta_k))\, |\Delta_k|$$

$$\leq \sum_{k=1}^{N} \Big| \int_{\Delta_k} (F(x) - G(x))\, dx \Big| + \frac{b-a}{N} \sum_{k=1}^{N} (F(\Delta_k) + G(\Delta_k)).$$

The lemma is proved. □

As a preliminary step towards Proposition 14.1.1, let us also derive:

Lemma 14.1.3 *If M_p is finite, then, for any bounded, Borel measurable function $w : \mathbb{R} \to \mathbb{R}$,*

$$\mathbb{E}_\theta \Big| \int_{-\infty}^{\infty} w(x)\, (F_\theta(x) - F(x))\, dx \Big| \leq \frac{\pi M_p}{\sqrt{2n}} \Big(\int_{-\infty}^{\infty} |w(x)|^q\, dF(x) \Big)^{1/q}. \tag{14.2}$$

In particular, for all $a < b$,

$$\mathbb{E}_\theta \Big| \int_a^b (F_\theta(x) - F(x))\, dx \Big| \leq \frac{\pi M_p}{\sqrt{2n}}\, (F(b) - F(a))^{1/q}. \tag{14.3}$$

Proof Note that the left integral in (14.2) is well-defined and is finite, once M_1 is finite (which is fulfilled under the p-th moment assumption $M_p < \infty$).

Given a function $g : \mathbb{R} \to \mathbb{R}$, having a bounded continuous derivative g', put

$$u(\theta) = \int_{-\infty}^{\infty} g\, dF_\theta - \int_{-\infty}^{\infty} g\, dF = \mathbb{E}\, g(\langle X, \theta\rangle) - \int_{-\infty}^{\infty} g\, dF, \quad \theta \in \mathbb{R}^n.$$

This function is C^1-smooth on $\mathbb{R}^n$, has $\mathfrak{s}_{n-1}$-mean zero on the unit sphere, and we have the identity

$$\langle \nabla u(\theta), v\rangle = \mathbb{E}\, \langle X, v\rangle\, g'(\langle X, \theta\rangle).$$

Using $(\mathbb{E}\, |\, \langle X, v\rangle\, |^p)^{1/p} \leq M_p$ for $v \in \mathbb{S}^{n-1}$ and applying Hölder's inequality, we get

$$|\, \langle \nabla u(\theta), v\rangle\, | \leq M_p \Big(\mathbb{E}\, |g'(\langle X, \theta\rangle)|^q \Big)^{1/q},$$

which implies

$$|\nabla u(\theta)| \leq M_p \Big(\mathbb{E}\, |g'(\langle X, \theta\rangle)|^q \Big)^{1/q}. \tag{14.4}$$

We now integrate this inequality over the unit sphere and apply the L^1-Poincaré-type inequality of Proposition 9.5.1, which yields

$$\mathbb{E}_\theta\, |u| \leq \frac{\pi}{\sqrt{2n}}\, \mathbb{E}_\theta\, |\nabla u|. \tag{14.5}$$

Since, by Hölder's inequality,

$$\begin{aligned}\mathbb{E}_\theta \left(\mathbb{E}\,|g'(\langle X,\theta\rangle)|^q\right)^{1/q} &\le \left(\mathbb{E}_\theta\,\mathbb{E}\,|g'(\langle X,\theta\rangle)|^q\right)^{1/q} \\ &= \left(\mathbb{E}_\theta \int_{-\infty}^{\infty} |g'|^q\,\mathrm{d}F_\theta\right)^{1/q} = \left(\int_{-\infty}^{\infty} |g'|^q\,\mathrm{d}F\right)^{1/q},\end{aligned}$$

from (14.4)–(14.5) we obtain that

$$\mathbb{E}_\theta \left|\int_{-\infty}^{\infty} g\,\mathrm{d}(F_\theta - F)\right| \le \frac{\pi M_p}{\sqrt{2n}} \left(\int_{-\infty}^{\infty} |g'|^q\,\mathrm{d}F\right)^{1/q}.$$

Integrating by parts on the left and replacing g' with w, the above relation becomes

$$\mathbb{E}_\theta \left|\int_{-\infty}^{\infty} w(x)\,(F_\theta(x) - F(x))\,\mathrm{d}x\right| \le \frac{\pi M_p}{\sqrt{2n}} \left(\int_{-\infty}^{\infty} |w|^q\,\mathrm{d}F\right)^{1/q},$$

which is the required inequality (14.2) with an additional assumption that w is continuous. But this can easily be relaxed to the condition that w is measurable. The lemma is thus proved. □

Proof (of Proposition 14.1.1) Since the inequality (14.1) is homogeneous with respect to X, we may assume that $M_p = 1$.

Applying (14.3) with the intervals $\Delta_k = (a_{k-1}, a_k)$ as in Lemma 14.1.2, we get

$$\mathbb{E}_\theta \left|\int_{\Delta_k} (F_\theta(x) - F(x))\,\mathrm{d}x\right| \le \frac{\pi}{\sqrt{2n}}\, F(\Delta_k)^{1/q}.$$

Therefore, by Lemma 14.1.2, for any integer $N \ge 1$,

$$\mathbb{E}_\theta \int_a^b |F_\theta(x) - F(x)|\,\mathrm{d}x \le \frac{2(b-a)}{N} + \frac{\pi}{\sqrt{2n}} \sum_{k=1}^{N} F(\Delta_k)^{1/q}.$$

By Hölder's inequality, the last sum does not exceed $N^{1/p}\left(\sum_{k=1}^N F(\Delta_k)\right)^{1/q} \le N^{1/p}$, and we arrive at the estimate

$$\mathbb{E}_\theta \int_a^b |F_\theta(x) - F(x)|\,\mathrm{d}x \le \frac{2(b-a)}{N} + \frac{\pi}{\sqrt{2n}}\, N^{1/p}.$$

In particular, putting $a = -b$, $b > 0$, we have

$$\mathbb{E}_\theta \int_{-b}^b |F_\theta(x) - F(x)|\,\mathrm{d}x \le \frac{4b}{N} + \frac{\pi}{\sqrt{2n}}\, N^{1/p}. \tag{14.6}$$

To extend this integral to the whole real line, we use the assumption $M_p = 1$, which, by Markov's inequality, implies that $F_\theta\{y : |y| \geq x\} \leq x^{-p}$ for all $x > 0$ together with a similar bound for F after averaging over θ. This gives

$$\int_{|x|\geq b} |F_\theta(x) - F(x)|\,dx \leq \int_{|x|\geq b} |x|^{-p}\,dx = \frac{2b^{1-p}}{p-1}.$$

Combining this estimate with (14.6), we get

$$\mathbb{E}_\theta \int_{-\infty}^{\infty} |F_\theta(x) - F(x)|\,dx \leq \frac{2b^{1-p}}{p-1} + \frac{4b}{N} + \frac{\pi}{\sqrt{2n}}\,N^{1/p}.$$

Here, the right-hand side is minimized at $b = (\frac{N}{2})^{1/p}$, for which

$$\frac{2b^{1-p}}{p-1} + \frac{4b}{N} = \frac{4N^{(1-p)/p}}{2^{1/p}\,(p-1)} + \frac{4N^{(1-p)/p}}{2^{1/p}} = q\,2^{1+1/q}\,N^{-1/q}.$$

Thus,

$$\mathbb{E}_\theta\, W(F_\theta, F) \leq q\,2^{1+1/q}\,N^{-1/q} + \frac{\pi}{\sqrt{2n}}\,N^{1/p}. \tag{14.7}$$

Now, one should optimize the right-hand side of this inequality over all integers $N \geq 1$. Consider the function of the real variable $x > 0$

$$\psi(x) = \alpha x^{-1/q} + \beta x^{1/p}, \quad \alpha = q\,2^{1+1/q},\ \beta = \frac{\pi}{\sqrt{2n}}.$$

It has a minimum at

$$x_0 = \frac{\alpha}{\beta}\,\frac{p}{q} = \frac{1}{\pi}\,2^{1+1/q}\,p\sqrt{2n}.$$

Note that $x_0 > \frac{2}{\pi}\sqrt{2n} > 1$. Therefore, choosing $N = [x_0] + 1$, we have $x_0 < N \leq 2x_0$. For this value

$$\psi(N) \leq \alpha\, x_0^{-1/q} + \beta\,(2x_0)^{1/p} = c_q\,(2n)^{-1/(2q)}$$

with constant

$$c_q = \pi^{1/q}\,2^{-1/q^2}\,(2p)^{1/p}\,(2 + 2^{1/q}\,(q-1)).$$

Hence, by (14.7),

$$\mathbb{E}_\theta\, W(F_\theta, F) \leq c_q\,(2n)^{-1/(2q)}.$$

It remains to bound the expression for c_q. First, one may use $2 + 2^{1/q}\,(q-1) \leq 2q$ and $(2p)^{1/p} \leq e^{2/e}$, so that $c_q \leq 2e^{2/e}\,u(q)q$, where $u(q) = \pi^{1/q}\,2^{-1/q^2}$. This function attains its maximum at $q_0 = (\log 4)/(\log \pi)$, which gives $c_q \leq 2e^{2/e}\,u(q_0)\,q < 6.7\,q$, and we arrive at the desired inequality (14.1). □

14.2 Large Deviations for the Kantorovich Distance

Since Proposition 14.1.1 only deals with the average approximation of F_θ by F by means of the Kantorovich distance W, one may wonder how to control deviations for most of the directions θ. To this end one can employ again the first order concentration property on the sphere. We keep the same notations as in the previous section, assuming in particular that $p > 1$ and $q = \frac{p}{p-1}$.

Proposition 14.2.1 *If M_1 is finite, then for any $r > 0$,*

$$\mathfrak{s}_{n-1}\{W(F_\theta, F) \geq m + r\} \leq e^{-nr^2/(4M_1^2)},$$

$$\mathfrak{s}_{n-1}\{W(F_\theta, F) \leq m - r\} \leq e^{-nr^2/(4M_1^2)},$$

where $m = \mathbb{E}_\theta W(F_\theta, F)$. In particular, for any $p \geq 2$,

$$\Big(\mathbb{E}_\theta \, W(F_\theta, F)^p\Big)^{1/p} \leq m + M_1 \frac{\sqrt{2p}}{\sqrt{n}}.$$

Involving other M_p-functionals, one can eliminate the quantity m in the bound for deviations above the mean.

Proposition 14.2.2 *If M_p is finite $(p > 1)$, then for all $r \geq 0$,*

$$\mathfrak{s}_{n-1}\Big\{W(F_\theta, F) \geq 7qM_p \, (2n)^{-1/(2q)} + M_p \, r\Big\} \leq e^{-nr^2/4}. \tag{14.8}$$

In particular,

$$\mathfrak{s}_{n-1}\Big\{W(F_\theta, F) \geq 8qM_p \, (2n)^{-1/(2q)}\Big\} \leq \exp\Big\{-\frac{1}{8}\, n^{1/p}\Big\}. \tag{14.9}$$

Thus, the distances $W(F_\theta, F)$ may exceed $CqM_p \, (2n)^{-1/(2q)}$ only on a very small part of the unit sphere (when the dimension n is large).

Note also that the bound (14.8) implies (14.1) up to an absolute factor (in place of the constant 7). Indeed, the non-negative random variable

$$\xi(\theta) = \Big(\frac{1}{M_p}\, W(F_\theta, F) - 7q \, (2n)^{-\frac{1}{2q}}\Big)^+$$

has subgaussian tails on $(\mathbb{S}^{n-1}, \mathfrak{s}_{n-1})$. More precisely, (14.8) implies that $\mathbb{E}_\theta \xi \leq \frac{c}{\sqrt{n}}$, so that

$$\frac{1}{M_p}\, \mathbb{E}\, W(F_\theta, F) \leq 7q \, (2n)^{-1/(2q)} + \mathbb{E}_\theta \xi \leq c' q n^{-1/(2q)}$$

for some positive absolute constants c and c'. In this sense, Proposition 14.2.2 sharpens Proposition 14.1.1.

Proof (of Propositions 14.2.1 and 14.2.2) The distance function

$$u(\theta) = W(F_\theta, F) = \int_{-\infty}^{\infty} |F_\theta(x) - F(x)|\,\mathrm{d}x$$

is Lipschitz on $\mathbb{R}^n$. Indeed, according to the classical Kantorovich–Rubinstein theorem, it admits the representation

$$u(\theta) = \sup\Big[\int_{-\infty}^{\infty} g\,\mathrm{d}F_\theta - \int_{-\infty}^{\infty} g\,\mathrm{d}F\Big] = \sup\Big[\mathbb{E}\,g(\langle X,\theta\rangle) - \int_{-\infty}^{\infty} g\,\mathrm{d}F\Big],$$

where the supremum runs over all functions $g : \mathbb{R} \to \mathbb{R}$ with $\|g\|_{\mathrm{Lip}} \le 1$, cf. e.g. [80] (note that in dimension one as above this representation is elementary). It follows that, for all $\theta, \theta' \in \mathbb{R}^n$,

$$u(\theta) - u(\theta') \le \sup\Big[\mathbb{E}\,g(\langle X,\theta\rangle) - \mathbb{E}\,g(\langle X,\theta'\rangle)\Big].$$

But, using the Lipschitz property, we have

$$|\mathbb{E}\,g(\langle X,\theta\rangle) - \mathbb{E}\,g(\langle X,\theta'\rangle)| \le \mathbb{E}\,|\,\langle X,\theta\rangle - \langle X,\theta'\rangle\,| \le M_1\,|\theta - \theta'|.$$

Therefore, $\|u\|_{\mathrm{Lip}} \le M_1$, so that, by Proposition 10.3.1, for all $r \ge 0$,

$$\mathfrak{s}_{n-1}\big\{u - \mathbb{E}_\theta\,u \ge r\big\} \le \mathrm{e}^{-nr^2/(4M_1^2)},$$
$$\mathfrak{s}_{n-1}\big\{u - \mathbb{E}_\theta\,u \le -r\big\} \le \mathrm{e}^{-(nr^2/(4M_1^2)},$$

which yield the large deviation bounds of Proposition 14.2.1, as well as (14.8), by involving the bound (14.1) on the mean $\mathbb{E}_\theta\,u$ and noting that $M_1 \le M_p$. The second bound (14.9) follows from the first one when applying (14.8) with $r = q\,(2n)^{-1/(2q)}$.

It remains to explain the moment inequality. Putting $\xi = (W(F_\theta, F) - m)^+$ and using $\Gamma(x+1) \le x^x$ with $x = p/2 \ge 1$, we have

$$\begin{aligned}\mathbb{E}_\theta\,\xi^p &= \int_0^\infty \mathfrak{s}_{n-1}\{\xi \ge r\}\,\mathrm{d}r^p \le \int_0^\infty \mathrm{e}^{-nr^2/(4M_1^2)}\,\mathrm{d}r^p \\ &= \Big(\frac{2M_1}{\sqrt{n}}\Big)^p\,\Gamma\Big(\frac{p}{2}+1\Big) \le \Big(\frac{2M_1}{\sqrt{n}}\Big)^p\Big(\frac{p}{2}\Big)^{p/2} = \Big(\frac{M_1\sqrt{2p}}{\sqrt{n}}\Big)^p \equiv A.\end{aligned}$$

Thus, $\|\xi\|_p = (\mathbb{E}_\theta\,\xi^p)^{1/p} \le A$. Since $W(F_\theta, F) \le \xi + m$, we conclude, by the triangle inequality, that $\|W(F_\theta, F)\|_p \le \|\xi\|_p + m \le A + m$. Both propositions are now proved. □

Note that for growing values of p the rate of approximation of F_θ by F in the Kantorovich distance W as in (14.1) with respect to the growing dimension n gets better and approaches the standard $\frac{1}{\sqrt{n}}$-rate (like in the classical central limit theorem with i.i.d. summands). On the other hand, the moment functionals M_p grow as well. To balance both effects, we need to require an appropriate hypothesis on the rate of growth of the moments as in the following statement, which provides a nearly standard rate of approximation.

Proposition 14.2.3 *Suppose that* $\mathbb{E}\, e^{|\langle X,\theta\rangle|/\lambda} \leq 2$ *for all* $\theta \in \mathbb{S}^{n-1}$ *with some* $\lambda > 0$. *Then*

$$\mathbb{E}_\theta\, W(F_\theta, F) \leq c\lambda\, \frac{\log n}{\sqrt{n}}$$

for some absolute constant c. Moreover, for all $r > 0$,

$$\mathfrak{s}_{n-1}\Big\{W(F_\theta, F) \geq c\lambda\, \frac{\log n}{\sqrt{n}} + \lambda r\Big\} \leq e^{-nr^2/4}.$$

The assumption about the moments is expressed here in terms of the Orlicz norms $\|\,\langle X,\theta\rangle\,\|_\psi$ generated by the Young function $\psi(r) = e^{|r|} - 1$. The finiteness of this norm is equivalent to the property that the L^p-norms $\|\,\langle X,\theta\rangle\,\|_p$ grow at worst at a linear rate as p grows to infinity.

Proof One may assume that $\lambda = 1$. In view of the bound $r^p \leq p^p e^{-p} e^r$ $(r \geq 0)$, the moment hypothesis of the proposition implies that

$$\|\,\langle X,\theta\rangle\,\|_p^p \leq p^p e^{-p}\, \mathbb{E}\, e^{|\langle X,\theta\rangle|} \leq 2p^p e^{-p},$$

so that $M_1 \leq M_p \leq 2p/e$. Applying Proposition 14.1.1, we obtain that

$$\mathbb{E}_\theta\, W(F_\theta, F) \leq \frac{14}{e}\, pq\,(2n)^{1/(2p)}\, \frac{1}{\sqrt{2n}}.$$

It is sufficient to choose here $p = 2 + \log n$, in which case the right-hand side will be bounded by $(c \log n)/\sqrt{n}$.

The second assertion of the proposition follows from the first bound of Proposition 14.2.1, by noting that $\mathbb{E}\, e^{|\langle X,\theta\rangle|} \geq 1 + \mathbb{E}\,|\,\langle X,\theta\rangle\,|$, which implies $M_1 \leq 1$. □

14.3 Pointwise Fluctuations

Let X be a random vector in $\mathbb{R}^n$ $(n \geq 2)$. We now turn to other distances that give more information about possible fluctuations of the distribution functions

$$F_\theta(x) = \mathbb{P}\{\langle X,\theta\rangle \leq x\}$$

about the typical distribution function $F(x) = \mathbb{E}_\theta\, F_\theta(x)$. Here, we consider deviations for individual points $x \in \mathbb{R}$ and use only the moment functional $M_2 = M_2(X)$.

As a natural approach to the problem, one may employ the supremum-convolution inequality on the unit sphere described in Corollary 9.4.3. It implies that

$$\mathbb{E}_\theta\, \exp\{tu(\theta)\} \leq \exp\Big\{t\,\mathbb{E}_\theta P_{\frac{2t}{n}}\, u(\theta)\Big\}, \quad t > 0. \tag{14.10}$$

where u may be an arbitrary $\mathfrak{s}_{n-1}$-integrable function on $\mathbb{R}^n$, and

$$P_s u(\theta) = \sup_{\theta' \in \mathbb{R}^n} \Big[u(\theta') - \frac{|\theta - \theta'|^2}{2s}\Big], \quad s > 0,$$

is the supremum-convolution operator with the quadratic cost function.

We apply the inequality (14.10) to the functions of the form $u(\theta) = F_\theta(x)$ with fixed $x \in \mathbb{R}$. By Markov's inequality, for any $h > 0$,

$$\begin{aligned} F_{\theta'}(x) &= \mathbb{P}\{\langle X, \theta'\rangle \le x\} = \mathbb{P}\{\langle X, \theta\rangle \le x + \langle X, \theta - \theta'\rangle\} \\ &\le \mathbb{P}\{\langle X, \theta\rangle \le x + h\} + \mathbb{P}\{\langle X, \theta - \theta'\rangle \ge h\} \le F_\theta(x+h) + \frac{M_2^2}{h^2}\,|\theta - \theta'|^2 \end{aligned}$$

and similarly

$$F_{\theta'}(x) \ge F_\theta(x-h) - \frac{M_2^2}{h^2}\,|\theta - \theta'|^2.$$

Putting $h = 2M_2\sqrt{\frac{t}{n}}$, we then get

$$F_{\theta'}(x) \le F_\theta(x+h) + \frac{n}{4t}\,|\theta - \theta'|^2, \quad F_{\theta'}(x) \ge F_\theta(x-h) - \frac{n}{4t}\,|\theta - \theta'|^2.$$

Hence, by the definition of the supremum-convolution operators,

$$P_{\frac{2t}{n}} u(\theta) = \sup_{\theta' \in \mathbb{R}^n} \Big[\,F_{\theta'}(x) - \frac{n}{4t}\,|\theta - \theta'|^2\Big] \le F_\theta(x+h),$$

$$P_{\frac{2t}{n}} (-u)(\theta) = \sup_{\theta' \in \mathbb{R}^n} \Big[-F_{\theta'}(x) - \frac{n}{4t}\,|\theta - \theta'|^2\Big] \le -F_\theta(x-h).$$

Let us now rewrite (14.10) in the form

$$\mathbb{E}_\theta \exp\{t\,(u - \mathbb{E}_\theta\, u)\} \le \exp\Big\{t\,\Big(\mathbb{E}_\theta\, P_{\frac{2t}{n}}\, u - \mathbb{E}_\theta\, u\Big)\Big\}.$$

Replacing u with $-u$, we also have

$$\mathbb{E}_\theta \exp\{t\,(\mathbb{E}_\theta\, u - u)\} \le \exp\Big\{t\,\Big(\mathbb{E}_\theta\, P_{\frac{2t}{n}}\,(-u) + \mathbb{E}_\theta\, u\Big)\Big\}.$$

For $u(\theta) = F_\theta(x)$, we have $\mathbb{E}_\theta\, u = F(x)$, and similarly, by the above pointwise estimates on the sup-convolutions,

$$\mathbb{E}_\theta\, P_{\frac{2t}{n}}\, u(\theta) \le F(x+h), \quad \mathbb{E}_\theta\, P_{\frac{t}{2n}}\,(-u)(\theta) \le -F(x-h).$$

Thus, we have obtained the following estimates on the exponential moments for deviations of $F_\theta(x)$ around $F(x)$.

Proposition 14.3.1 *For any $x \in \mathbb{R}$ and $t > 0$, with $h = 2M_2\sqrt{\frac{t}{n-1}}$,*

$$\mathbb{E}_\theta \exp\{t\,(F_\theta(x) - F(x))\} \le \exp\{t\,(F(x+h) - F(x))\},$$
$$\mathbb{E}_\theta \exp\{t\,(F(x) - F_\theta(x))\} \le \exp\{t\,(F(x) - F(x-h))\}.$$

The two bounds can be united in one relation: For any $t \in \mathbb{R}$,

$$\mathbb{E}_\theta\, e^{t\,(F_\theta(x)-F(x))} \le \exp\Big\{|t|(F(x+h) - F(x-h))\Big\}, \quad h = 2M_2\sqrt{\frac{|t|}{n}}.$$

This already shows that the local behavior of the distribution function F near a given point x turns out to be responsible for the large deviation behavior at this point of the distribution function F_θ around its mean (as a function of $\theta \in \mathbb{S}^{n-1}$).

For a quantitative statement, let us recall that F represents the distribution of $\theta_1|X|$, where θ_1 is the first coordinate of the vector θ on the unit sphere viewed as a random variable independent of X. Clearly, F is absolutely continuous with respect to the Lebesgue measure on the real line, symmetric about the origin (as a measure) and unimodal. Denoting by φ_n the density of the random variable $\theta_1\sqrt{n}$, the density of F may be written as

$$p(x) = \mathbb{E}\,\frac{\sqrt{n}}{|X|}\,\varphi_n\Big(\frac{x\sqrt{n}}{|X|}\Big), \quad x \in \mathbb{R}.$$

Hence, p is even, smooth and non-decreasing on $(0,\infty)$. Also, F has a finite Lipschitz semi-norm $\ell = \|F\|_{\mathrm{Lip}}$ if and only if the expectation $\mathbb{E}\,\frac{\sqrt{n}}{|X|}$ is finite, and then

$$c_0\,\mathbb{E}\,\frac{\sqrt{n}}{|X|} \le \ell \le c_1\,\mathbb{E}\,\frac{\sqrt{n}}{|X|}$$

for some absolute constants $c_1 > c_0 > 0$.

In addition, using $F(x+h) - F(x-h) \le 2\ell h$, the exponential bound on deviations of $F_\theta(x)$ yields

$$\mathbb{E}_\theta\, e^{t\,(F_\theta(x)-F(x))} \le \exp\{2\ell h\,|t|\} = \exp\Big\{4M_2\ell\,\frac{|t|^{3/2}}{\sqrt{n}}\Big\}.$$

Equivalently, making the substitution $t = \frac{n^{1/3}}{(4M_2\ell)^{2/3}}\,\lambda$, for the function

$$\xi(\theta) = \frac{n^{1/3}}{(4M_2\ell)^{2/3}}\,(F_\theta(x) - F(x))$$

(viewed as a random variable on the sphere), we get a bound on the Laplace transform

$$\mathbb{E}_\theta \exp\{\lambda\xi\} \le \exp\{|\lambda|^{3/2}\}, \quad \lambda \in \mathbb{R}.$$

By Markov's inequality, for any $r > 0$,

$$\mathfrak{s}_{n-1}\{\xi \geq r\} \leq \exp\{\lambda^{3/2} - \lambda r\} = \exp\{-4r^3/27\} \quad \text{for } \lambda = \frac{4}{9}\,r^2.$$

Similarly, $\mathfrak{s}_{n-1}\{\xi \leq -r\} \leq \exp\{-4r^3/27\}$. Therefore, we arrive at:

Corollary 14.3.2 *Assume that the density of the typical distribution F is bounded by a number ℓ. Then, for any $x \in \mathbb{R}$ and $r > 0$,*

$$\mathfrak{s}_{n-1}\Big\{\frac{n^{1/3}}{(4M_2\ell)^{2/3}}\,|F_\theta(x) - F(x)| \geq r\Big\} \leq 2\,\exp\{-4r^3/27\}.$$

In particular, for some absolute constant c,

$$\mathbb{E}_\theta\,|F_\theta(x) - F(x)| \leq c\,(M_2\ell)^{2/3}\,\frac{1}{n^{1/3}}.$$

14.4 The Lévy Distance

Recall that the Lévy distance $L(F_\theta, F)$ between the distribution functions F_θ and F is defined as the minimum over all $h \in [0,1]$ such that the inequality

$$F(x-h) - h \leq F_\theta(x) \leq F(x+h) + h$$

holds for all $x \in \mathbb{R}$. In general, there is an elementary relation $L^2 \leq W$ connecting it with the Kantorovich distance. Indeed, if $h > L(F_\theta, F)$, one can find x such that

$$F_\theta(x) > F(x+h) + h \quad \text{or} \quad F(x) > F_\theta(x+h) + h.$$

In the first case (for definiteness), we then have

$$W(F_\theta, F) \geq \int_x^{x+h} (F_\theta(y) - F(y))\,dy \geq \int_x^{x+h} (F_\theta(x) - F(x+h))\,dy > h^2.$$

Letting $h \to L(F_\theta, F)$ yields the required inequality

$$L(F_\theta, F) \leq \sqrt{W(F_\theta, F)}.$$

Now, using Proposition 14.2.1, we may therefore conclude that, for all $r \geq 0$,

$$\mathfrak{s}_{n-1}\Big\{L(F_\theta, F) \geq \sqrt{7qM_p}\,\frac{1}{(2n)^{1/(4q)}} + \sqrt{M_p r}\Big\} \leq e^{-(n-1)r^2/4}.$$

In particular (which also follows from Proposition 14.1.1),

$$\mathbb{E}\, L(F_\theta, F) \le \sqrt{7qM_p}\,\frac{1}{n^{1/(4q)}}.$$

With respect to n, the latter bound behaves worse than $n^{-1/4}$ for any value of $p > 1$. In fact, in order to reach this rate (modulo a logarithmically growing factor), only the functional $M_1 = M_1(X)$ is needed.

Proposition 14.4.1 *For any random vector X in $\mathbb{R}^n$ with finite first absolute moment,*

$$\mathbb{E}_\theta\, L(F_\theta, F) \le \frac{M_1 + 2\log n}{n^{1/4}}.$$

Proof Introduce the characteristic functions $f_\theta(t) = \mathbb{E}\, e^{it\langle X,\theta\rangle}$ and $f(t) = \mathbb{E}_\theta f_\theta(t)$ ($t \in \mathbb{R}$), corresponding to the distribution functions F_θ of the weighted sums $\langle X,\theta\rangle$ and the typical distribution F. We apply Zolotarev's estimate from Proposition 3.3.1: For any $T > 0$,

$$\mathbb{E}_\theta\, L(F_\theta, F) \le \frac{1}{\pi}\int_{-T}^{T} \frac{\mathbb{E}_\theta\,|f_\theta(t) - f(t)|}{|t|}\, dt + \frac{4\log(1+T)}{T}.$$

Here, according to Proposition 13.3.3, $\mathbb{E}_\theta\,|f_\theta(t) - f(t)| \le \frac{\sqrt{2}\,M_1}{\sqrt{n-1}}\,|t|$, so that

$$\mathbb{E}_\theta\, L(F_\theta, F) \le \frac{2\sqrt{2}}{\pi}\,\frac{M_1 T}{\sqrt{n-1}} + \frac{4\log(1+T)}{T}.$$

To bound the above right-hand side, choose $T = n^{1/4}$ for $n \ge 16$. Then $\frac{4}{T}\log(1+T) < 2n^{-1/4}\log n$ and $\frac{2\sqrt{2}\,T}{\pi\sqrt{n-1}} < n^{-1/4}$, which implies the desired inequality. If $n < 16$, it holds automatically, since then $\frac{2\log n}{n^{1/4}} > 1$, while $L \le 1$. Proposition 14.4.1 follows.□

The bound of Corollary 14.3.2 suggests, however, that, by involving the moment functional M_2, one can further improve the rate in the bound of Proposition 14.4.1. Indeed, this turns out to be possible by virtue of the supremum-convolution approach, as can be seen from the following:

Proposition 14.4.2 *For any $r > 0$,*

$$\mathfrak{s}_{n-1}\big\{L(F_\theta, F) \ge r\big\} \le 12\,\big(M_2\, r^{-3/2} + 1\big)\, \exp\Big\{-\frac{n}{32\,M_2^2}\, r^3\Big\}. \tag{14.11}$$

In particular, for some absolute constant c,

$$\mathbb{E}_\theta\, L(F_\theta, F) \le c\,\Big(\frac{\log n}{n}\Big)^{1/3} M_2^{2/3}. \tag{14.12}$$

Proof We rewrite the exponential bounds of Proposition 14.3.1 as

$$\mathbb{E}_\theta\, e^{t\,(F_\theta(x)-F(x+h))} \le 1, \quad \mathbb{E}_\theta\, e^{t\,(F(x-h)-F_\theta(x))} \le 1,$$

where $h = 2M_2\sqrt{\frac{t}{n}}$ with arbitrary $t > 0$. By Markov's inequality, for any $r \ge 0$,

$$\begin{aligned} \mathfrak{s}_{n-1}\big\{F_\theta(x) - F(x+h) \ge r/2\big\} &\le e^{-tr}, \\ \mathfrak{s}_{n-1}\big\{F(x-h) - F_\theta(x) \ge r/2\big\} &\le e^{-tr/2}. \end{aligned}$$

Let us recall that the typical distribution is symmetric about the origin and has a unimodal density (in particular, F is continuous).

Fix an integer $N \ge 1$ large enough so that $1 - F(Nh) = F(-Nh) \le \varepsilon$ for a given $\varepsilon > 0$. For the points $x_k = kh$, $k = 0, \pm 1, \pm(N+1)$, the above bounds yield

$$\begin{aligned} \mathfrak{s}_{n-1}\Big\{\max_{|k|\le N+1}\,[F_\theta(x_k) - F(x_k+h)] \ge \frac{r}{2}\Big\} &\le (2N+3)\, e^{-tr/2}, \\ \mathfrak{s}_{n-1}\Big\{\max_{|k|\le N+1}\,[F(x_k-h) - F_\theta(x_k)] \ge \frac{r}{2}\Big\} &\le (2N+3)\, e^{-tr/2}. \end{aligned}$$

Hence, for the random variable

$$\xi_N = \max\Big\{\max_{|k|\le N+1}\,[F_\theta(x_k) - F(x_k+h)],\ \max_{|k|\le N+1}\,[F(x_k-h) - F_\theta(x_k)]\Big\}$$

we have

$$\mathfrak{s}_{n-1}\{\xi_N \ge r/2\} \le 2(2N+3)\, e^{-tr/2}.$$

Now, take any point $x \in [-(N+1)h, (N+1)h]$ different from all x_k's and choose k such that $x_{k-1} < x < x_k$ $(-N \le k \le N+1)$. Then

$$\begin{aligned} F_\theta(x) - F(x+2h) &\le F_\theta(x_k) - F(x_{k-1}+2h) \\ &= F_\theta(x_k) - F(x_k+h) \le \xi_N, \end{aligned}$$

and similarly

$$\begin{aligned} F(x-2h) - F_\theta(x) &\le F(x_k-2h) - F_\theta(x_{k-1}) \\ &= F(x_{k-1}-h) - F_\theta(x_{k-1}) \le \xi_N. \end{aligned}$$

In addition, if $x > x_{N+1} = (N+1)h$, then $F_\theta(x) - F(x+2h) \le 1 - F(Nh) \le \varepsilon$ and

$$\begin{aligned} F(x-2h) - F_\theta(x) &\le 1 - F_\theta((N+1)h) \\ &= (1 - F(x_N)) + (F(x_{N+1}-h) - F_\theta(x_{N+1})) \le \xi_N + \varepsilon. \end{aligned}$$

Using similar bounds for $x < -(N+1)h$, we conclude that the random variable

$$\xi = \max\Big\{\sup_x\,[F_\theta(x) - F(x+2h)],\ \sup_x\,[F(x-2h) - F_\theta(x)]\Big\}$$

satisfies $\xi \leq \xi_N + \varepsilon$. Hence, for all $r > 0$,

$$\mathfrak{s}_{n-1}\Big\{\xi \geq \frac{r}{2} + \varepsilon\Big\} \leq 2(2N+3)\, e^{-tr/2}.$$

Let us now choose a suitable N by using the functional M_2. Since

$$\int_{-\infty}^{\infty} x^2\, dF_\theta(x) = \mathbb{E}\,\langle X, \theta\rangle^2 \leq M_2^2$$

for all $\theta \in \mathbb{S}^{n-1}$, the same is true for F as well. But if η is a random variable distributed according to F, then, by Markov's inequality,

$$\mathbb{P}\{|\eta| \geq Nh\} \leq \frac{M_2^2}{(Nh)^2} \leq \varepsilon,$$

where the last inequality holds true, as long as $N \geq \frac{M_2}{h}\, \varepsilon^{-1/2}$. Let N be the smallest such integer, so that $N \leq \frac{M_2}{h}\, \varepsilon^{-1/2} + 1$. Inserting $t = \frac{n}{4M_2^2}\, h^2$ into the deviation bound for ξ, we then get

$$\mathfrak{s}_{n-1}\Big\{\xi \geq \frac{r}{2} + \varepsilon\Big\} \leq \Big(\frac{4M_2}{h}\, \varepsilon^{-1/2} + 10\Big) \exp\Big\{-\frac{n}{8M_2^2}\, rh^2\Big\}.$$

Now we choose $\varepsilon = \frac{1}{2}\, r$, in which case the above inequality becomes

$$\mathfrak{s}_{n-1}\{\xi \geq r\} \leq \Big(\frac{4M_2}{h}\, r^{-1/2}\sqrt{2} + 10\Big) \exp\Big\{-\frac{n}{8M_2^2}\, rh^2\Big\}.$$

Let us replace here h with $h/2$ and formulate this inequality once more in terms of the functional

$$L_h(F_\theta, F) = \max\Big\{\sup_x\, [F_\theta(x) - F(x+h)],\ \sup_x\, [F(x-h) - F_\theta(x)]\Big\}.$$

Let us also use $8\sqrt{2} < 12$ to simplify the constants. Then, for any $r, h > 0$, we have

$$\mathfrak{s}_{n-1}\Big\{L_h(F_\theta, F) \geq r\Big\} \leq 12\,\Big(\frac{M_2}{h}\, r^{-1/2} + 1\Big) \exp\Big\{-\frac{n}{32\, M_2^2}\, rh^2\Big\}.$$

It is interesting that the probabilities on the right-hand side decay exponentially fast with respect to n. In order to relate this inequality to the Lévy distance, we recall that $L(F_\theta, F) \leq h \iff L_h(F_\theta, F) \leq h$. Hence, in the case $h = r$, the above yields the required inequality (14.11).

Now, rewrite this inequality in terms of the function $\eta = M_2^{-2/3} L(F_\theta, F)$ as

$$\mathfrak{s}_{n-1}\{\eta \geq r\} \leq 12\,\big(r^{-3/2} + 1\big) \exp\Big\{-\frac{n}{32}\, r^3\Big\}.$$

If $r \geq 6a_n$, $a_n = (\frac{\log n}{n})^{1/3}$, it follows that $\exp\{-\frac{n}{32}r^3\} \leq \frac{1}{n^3}\exp\{-\frac{1}{64}r^3\}$, while

$$r^{-3/2} + 1 \leq (6a_n)^{-3/2} + 1 = \frac{n^{1/2}}{6^{3/2}(\log n)^{1/2}} + 1 < \sqrt{n}.$$

Hence

$$\int_{6a_n}^{\infty} \mathfrak{s}_{n-1}\{\eta \geq r\}\,\mathrm{d}r \leq \int_{6a_n}^{\infty} 12\,(r^{-3/2}+1)\,\mathrm{e}^{-\frac{n}{32}r^3}\,\mathrm{d}r \leq \frac{12}{n^2}\int_0^{\infty} \mathrm{e}^{-\frac{1}{64}r^3}\,\mathrm{d}r.$$

This shows that $\mathbb{E}\eta \leq ca_n$ for some absolute constant c. Proposition 14.4.2 is proved. □

Substituting in (14.11) r with $cM_2^{2/3}(\frac{r\log n}{n})^{1/3}$, we get a somewhat simpler estimate for large deviations.

Corollary 14.4.3 *With some absolute constant $c > 0$, for any $r \geq 1$,*

$$\mathfrak{s}_{n-1}\Big\{L(F_\theta, F) \geq cr\Big(\frac{\log n}{n}\Big)^{1/3} M_2^{2/3}\Big\} \leq n^{-r}.$$

The dependence on r in the exponent on the right-hand side of (14.11) can be sharpened by involving the functionals M_p with $p > 2$. This may be achieved using the Lipschitz property as in the following:

Lemma 14.4.4 *For all $\theta, \theta' \in \mathbb{R}^n$ and $p \geq 1$,*

$$L(F_\theta, F_{\theta'}) \leq M_p^{\frac{p}{p+1}}\,|\theta - \theta'|^{\frac{p}{p+1}}.$$

In particular, for any distribution function G,

$$|L(F_\theta, G) - L(F_{\theta'}, G)| \leq M_p^{\frac{p}{p+1}}\,|\theta - \theta'|^{\frac{p}{p+1}}.$$

Thus, the distance functions $\theta \rightarrow L(F_\theta, G)$ belong to the class Lip(α) on the unit sphere for $\alpha = \frac{p}{p+1}$ and have Lip(α)-seminorms at most $M_p^{p/(p+1)}$.

Proof Let $r = |\theta - \theta'| > 0$ and put $v = \frac{1}{r}(\theta - \theta')$. By Markov's inequality,

$$\mathbb{P}\{|\langle X, \theta\rangle - \langle X, \theta'\rangle| > rt\} = \mathbb{P}\{|\langle X, v\rangle| > t\} \leq \frac{M_p^p}{t^p}$$

for any $t > 0$. Hence, for any $x \in \mathbb{R}$,

$$\begin{aligned} F_{\theta'}(x) &= \mathbb{P}\{\langle X, \theta'\rangle \leq x\} = \mathbb{P}\{\langle X, \theta\rangle \leq x + (\langle X, \theta\rangle - \langle X, \theta'\rangle)\} \\ &\leq \mathbb{P}\{\langle X, \theta\rangle \leq x + rt\} + \mathbb{P}\{\langle X, \theta\rangle - \langle X, \theta'\rangle > rt\} \leq F_\theta(x + rt) + \frac{M_p^p}{t^p}. \end{aligned}$$

Choose $t = M_p^{p/(p+1)}\, r^{-1/(p+1)}$ so that for $h = rt = \frac{M_p^p}{t^p}$, we obtain that

$$F_{\theta'}(x) \leq F_\theta(x+h) + h$$

with arbitrary x. This implies that $L(F_\theta, F_{\theta'}) \leq rt$, which is the first assertion. The second assertion follows from the triangle inequality

$$|L(F_\theta, G) - L(F_{\theta'}, G)| \leq L(F_\theta, F_{\theta'}),$$

thus proving the lemma. □

We can now combine Proposition 10.3.4 about the deviations of Lip(α) functions and Lemma 14.4.4, where we choose $G = F$.

Proposition 14.4.5 *If M_p is finite $(p \geq 1)$, then for any $r > 0$,*

$$\mathfrak{s}_{n-1}\Big\{|L(F_\theta, F) - \mathbb{E}_\theta\, L(F_\theta, F)| \geq r\Big\} \leq 2\, \exp\Big\{-(n-1)\,\frac{r^{2(p+1)/p}}{2M_p^2}\Big\}.$$

The particular case $p = 2$ leads to the following estimate, which is consistent with (14.11):

$$\mathfrak{s}_{n-1}\Big\{|L(F_\theta, F) - \mathbb{E}_\theta L(F_\theta, F)| \geq r\Big\} \leq 2\, \exp\Big\{-(n-1)\,\frac{r^3}{2M_2^2}\Big\}.$$

14.5 Berry–Esseen-type Bounds

The moment functionals $M_p = M_p(X)$ may also be used to control the Kolmogorov distance

$$\rho(F_\theta, F) = \sup_x\, |F_\theta(x) - F(x)|.$$

However, in general it is stronger than the Lévy distance, and the associated topology is stronger than the weak topology in the space of probability distributions on the line. So, in order to get good bounds on $\rho(F_\theta, F)$, we will need to involve some other characteristics of the underlying distribution of the random vector X. We are mostly interested in bounding the first and the second moments, $\mathbb{E}_\theta\, \rho(F_\theta, F)$ and $\mathbb{E}_\theta\, \rho^2(F_\theta, F)$, in the sense of the uniform distribution $\mathfrak{s}_{n-1}$ on the unit sphere $\mathbb{S}^{n-1}$.

Our approach to the main Problem 12.1.2 will be based on Fourier analysis, namely, on general Berry–Esseen-type bounds, which allow one to transfer bounds on the characteristic functions to bounds on the Kolmogorov distance. First let us state separately one general relation which is adapted to the characteristic functions of linear forms of the random vector X in $\mathbb{R}^n$,

$$f_\theta(t) = \mathbb{E}\, e^{it\langle X, \theta\rangle} = \int_{-\infty}^{\infty} e^{itx}\, dF_\theta(x), \quad t \in \mathbb{R},\ \theta \in \mathbb{S}^{n-1}.$$

As before,

$$f(t) = \mathbb{E}_\theta f_\theta(t) = \int_{-\infty}^{\infty} e^{itx}\, dF(x)$$

denotes the characteristic function of the typical distribution $F = \mathbb{E}_\theta F$.

Lemma 14.5.1 *Given a random vector X in $\mathbb{R}^n$, for all $T \geq T_0 > 0$,*

$$c\,\mathbb{E}_\theta\, \rho(F_\theta, F) \leq \int_0^{T_0} \frac{\mathbb{E}_\theta\, |f_\theta(t) - f(t)|}{t}\, dt + \int_{T_0}^{T} \frac{\mathbb{E}_\theta\, |f_\theta(t)|}{t}\, dt + \frac{1}{T}\int_0^T |f(t)|\, dt, \tag{14.13}$$

where $c > 0$ is an absolute constant.

Lemma 14.5.2 *Moreover, if $T \geq T_0 \geq 1$, then*

$$c\,\mathbb{E}_\theta\, \rho^2(F_\theta, F) \leq \int_0^1 \frac{\mathbb{E}_\theta\, |f_\theta(t) - f(t)|^2}{t^2}\, dt + \log T \int_0^{T_0} \frac{\mathbb{E}_\theta\, |f_\theta(t) - f(t)|^2}{t}\, dt + \log T \int_{T_0}^{T} \frac{\mathbb{E}_\theta\, |f_\theta(t)|^2}{t}\, dt + \frac{1}{T^2}\Big(\int_0^T |f(t)|\, dt\Big)^2. \tag{14.14}$$

Proof By Proposition 3.2.3, for any $\theta \in \mathbb{S}^{n-1}$,

$$c\,\rho(F_\theta, F) \leq \int_0^T \frac{|f_\theta(t) - f(t)|}{t}\, dt + \frac{1}{T}\int_0^T |f(t)|\, dt, \tag{14.15}$$

where one may take $c = 1/\pi^3$. After taking the expectation of both sides, we get

$$c\,\mathbb{E}_\theta\, \rho(F_\theta, F) \leq \int_0^T \frac{\mathbb{E}_\theta\, |f_\theta(t) - f(t)|}{t}\, dt + \frac{1}{T}\int_0^T |f(t)|\, dt.$$

In further estimations, when T is chosen to be large, it makes sense to divide the second to last integral into two intervals, $[0, T_0]$ and $[T_0, T]$ with T_0 being much smaller than T. In that case, in the integral over the second interval one may use the bounds

$$\mathbb{E}_\theta\, |f_\theta(t) - f(t)| \leq \mathbb{E}_\theta\, |f_\theta(t)| + |f(t)| \leq 2\,\mathbb{E}_\theta\, |f_\theta(t)|,$$

which leads to (14.13).

By a similar argument, one may estimate the L^2-norm of the Kolmogorov distance. Squaring (14.15), we get

$$\frac{c^2}{2}\,\rho^2(F_\theta, F) \leq \Big(\int_0^T \frac{|f_\theta(t) - f(t)|}{t}\, dt\Big)^2 + \frac{1}{T^2}\Big(\int_0^T |f(t)|\, dt\Big)^2.$$

Assuming that $T > 1$, we split integration in the first integral into the intervals $[0, 1]$ and $[1, T]$. For the first interval, by Cauchy's inequality,

$$\Big(\int_0^1 \frac{|f_\theta(t) - f(t)|}{t}\,\mathrm{d}t\Big)^2 \le \int_0^1 \frac{|f_\theta(t) - f(t)|^2}{t^2}\,\mathrm{d}t.$$

To treat the second interval, consider the probability measure μ on $[1, T]$ with density $\frac{\mathrm{d}\mu(t)}{\mathrm{d}t} = \frac{1}{t \log T}$. Then, by Cauchy's inequality,

$$\begin{aligned}
\int_1^T \frac{|f_\theta(t) - f(t)|}{t}\,\mathrm{d}t &= \log T \int_1^T |f_\theta(t) - f(t)|\,\mathrm{d}\mu(t) \\
&\le \log T \bigg(\int_1^T |f_\theta(t) - f(t)|^2\,\mathrm{d}\mu(t)\bigg)^{1/2} \\
&= \bigg(\log T \int_1^T \frac{|f_\theta(t) - f(t)|^2}{t}\,\mathrm{d}t\bigg)^{1/2}.
\end{aligned}$$

Hence, for some absolute constant $c > 0$,

$$\begin{aligned}
c\,\rho^2(F_\theta, F) \le{}& \int_0^1 \frac{|f_\theta(t) - f(t)|^2}{t^2}\,\mathrm{d}t \\
&+ \log T \int_1^T \frac{|f_\theta(t) - f(t)|^2}{t}\,\mathrm{d}t + \frac{1}{T^2}\Big(\int_0^T |f(t)|\,\mathrm{d}t\Big)^2.
\end{aligned}$$

Without any essential loss, one may extend integration in the second integral to the larger interval $[0, T]$. Moreover, taking the expectation over θ, we then get

$$\begin{aligned}
c\,\mathbb{E}_\theta\,\rho^2(F_\theta, F) \le{}& \int_0^1 \frac{\mathbb{E}_\theta\,|f_\theta(t) - f(t)|^2}{t^2}\,\mathrm{d}t \\
&+ \log T \int_0^T \frac{\mathbb{E}_\theta\,|f_\theta(t) - f(t)|^2}{t}\,\mathrm{d}t + \frac{1}{T^2}\Big(\int_0^T |f(t)|\,\mathrm{d}t\Big)^2.
\end{aligned}$$

Again, one may split the integration in the second integral into two intervals $[0, T_0]$ and $[T_0, T]$ with $1 \le T_0 \le T$, so as to consider separately sufficiently large values of t for which $|f_\theta(t)|$ is small enough (with high probability). More precisely, since

$$f(t) = \mathbb{E}_\theta\, f_\theta(t) \quad \text{and} \quad |f_\theta(t) - f(t)|^2 \le 2\,|f_\theta(t)|^2 + 2\,|f(t)|^2,$$

we have $\mathbb{E}_\theta\,|f_\theta(t) - f(t)|^2 \le 4\,\mathbb{E}_\theta\,|f_\theta(t)|^2$. This leads to the bound (14.14). □

One may simplify the bounds of Lemmas 14.5.1–14.5.2 by involving M_p together with the moment and normalized variance-type functionals

$$m_p = \frac{1}{\sqrt{n}}\,(\mathbb{E}\,|\langle X, Y\rangle|^p)^{1/p}, \qquad \sigma_{2p} = \sqrt{n}\,\bigg(\mathbb{E}\,\Big|\frac{|X|^2}{n} - 1\Big|^p\bigg)^{1/p}$$

(where Y is an independent copy of X). Recall that $m_p \le M_p^2$ (Proposition 1.4.2). In addition, if $\mathbb{E}\,|X|^2 = n$, then, by Proposition 1.1.3, for any $p \ge 2$,

$$m_p \ge m_2 \ge \frac{1}{n}\,\mathbb{E}\,|X|^2 = 1.$$

From Lemma 14.5.1 we obtain:

Lemma 14.5.3 *Let X be a random vector in $\mathbb{R}^n$ with finite moment of order $2p$ ($p \ge 1$), such that $\mathbb{E}\,|X|^2 = n$. Then with some constant c_p depending on p only, for all $T \ge T_0 \ge 1$,*

$$c_p\,\mathbb{E}_\theta\,\rho(F_\theta, F) \le \int_0^{T_0} \mathbb{E}_\theta\,|f_\theta(t) - f(t)|\,\frac{dt}{t} + \frac{m_{2p}^p + \sigma_{2p}^p}{n^{p/2}}\Big(1 + \log\frac{T}{T_0}\Big) + \frac{1}{T} + e^{-T_0^2/16}.$$

Proof According to Lemma 13.1.3, for all $T > 0$,

$$\frac{c_p}{T}\int_0^T |f(t)|\,dt \le \frac{1 + \sigma_{2p}^p}{n^{p/2}} + \frac{1}{T}.$$

This allows us to estimate the last integral in (14.13). Moreover, by Lemma 13.1.2,

$$c_p \int_{T_0}^T \frac{\mathbb{E}_\theta\,|f_\theta(t)|}{t}\,dt \le \int_{T_0}^T \Big(\frac{m_{2p}^p + \sigma_{2p}^p}{n^{p/2}} + e^{-T_0^2/16}\Big)\frac{dt}{t} \le \frac{m_{2p}^p + \sigma_{2p}^p}{n^{p/2}}\,\log\frac{T}{T_0} + \frac{8}{T_0^2}\,e^{-T_0^2/16}$$

for the second to last integral in (14.13). Since $m_2 \ge 1$ and $T_0 \ge 1$, Lemma 14.5.3 follows. □

In the spirit of Berry–Esseen-type bounds, one may also develop lower bounds. In particular, we have the follow general estimate.

Lemma 14.5.4 *Given a function G of bounded variation such that $G(-\infty) = 0$ and $G(\infty) = 1$, for any $T > 0$,*

$$\mathbb{E}_\theta\,\sup_x |F_\theta(x) - G(x)| \ge \frac{1}{6T}\,\mathbb{E}_\theta\Big|\int_0^T (f_\theta(t) - f(t))\Big(1 - \frac{t}{T}\Big)\,dt\Big|.$$

Proof We need the following assertion: Given a complex-valued random variable ξ with finite first absolute moment, we have $\mathbb{E}\,|\xi - b| \ge \frac{1}{2}\,\mathbb{E}\,|\xi - \mathbb{E}\xi|$ for any complex number b. Indeed, by the triangle inequality,

$$|\xi - \mathbb{E}\xi| \le |\xi - b| + |b - \mathbb{E}\xi| \le |\xi - b| + \mathbb{E}\,|b - \xi|.$$

It remains to take the expectations of both sides. We use this inequality with

$$\xi = \int_0^T f_\theta(t)\left(1 - \frac{t}{T}\right)\mathrm{d}t, \quad b = \int_0^T g(t)\left(1 - \frac{t}{T}\right)\mathrm{d}t,$$

where g is the Fourier–Stieltjes transform of G. Applying the bound (3.16) of Proposition 3.4.1 on the probability space $(\mathbb{S}^{n-1}, \mathfrak{s}_{n-1})$, we then get

$$\begin{aligned}\mathbb{E}_\theta \sup_x |F_\theta(x) - G(x)| &\geq \frac{1}{3T}\,\mathbb{E}_\theta \Big| \int_0^T (f_\theta(t) - g(t))\left(1 - \frac{t}{T}\right)\mathrm{d}t \Big| \\ &\geq \frac{1}{6T}\,\mathbb{E}_\theta \Big| \int_0^T (f_\theta(t) - f(t))\left(1 - \frac{t}{T}\right)\mathrm{d}t \Big|.\end{aligned}$$

Lemma 14.5.4 is proved. □

14.6 Preliminary Bounds on the Kolmogorov Distance

One can use the bounds of Lemmas 14.5.1–14.5.2 to obtain the rate $1/\sqrt{n}$ modulo a logarithmic factor. In addition to the moment functionals $M_p = M_p(X)$, we involve the variance-type functional

$$\sigma_2 = \sigma_2(X) = \frac{1}{\sqrt{n}}\,\mathbb{E}\,\big|\,|X|^2 - n\big|.$$

Proposition 14.6.1 *If a random vector X in $\mathbb{R}^n$ has finite first moment M_1, then for some absolute constant $c > 0$ we have*

$$c\,\mathbb{E}_\theta\,\rho^2(F_\theta, F) \leq \frac{M_1^2\,(\log n)^2}{n} + \mathbb{P}\big\{|X - Y|^2 \leq n/4\big\}\,(\log n)^2 + \frac{1}{n^2},$$

where Y is an independent copy of X. As a consequence, if X has finite 2-nd moment M_2 and $\mathbb{E}\,|X|^2 = n$, then

$$c\,\mathbb{E}_\theta\,\rho^2(F_\theta, F) \leq (M_2^4 + \sigma_2^2)\,\frac{(\log n)^2}{n}.$$

A similar bound also holds for the normal distribution function Φ in place of F.

Proof We apply Lemma 14.5.2 with $T_0 = 5\sqrt{\log n}$ and $T = 5n$. The first two integrals in (14.14) can be bounded by using the spherical Poincaré inequality, i.e., using the first bound of Proposition 13.3.3,

$$\mathbb{E}_\theta\,|f_\theta(t) - f(t)|^2 \leq \frac{2M_1^2}{n-1}\,t^2.$$

This gives

$$\int_0^1 \frac{\mathbb{E}_\theta\, |f_\theta(t) - f(t)|^2}{t^2}\, dt + \log T \int_0^{T_0} \frac{\mathbb{E}_\theta\, |f_\theta(t) - f(t)|^2}{t}\, dt$$
$$\leq \frac{2M_1^2}{n-1}\Big(1 + \frac{25}{2}\,(\log(5n))^2\Big).$$

To bound the third and fourth integrals in (14.14), we involve Lemma 13.1.1 with $\lambda = 1/4$, which gives

$$\frac{1}{3}\,\mathbb{E}_\theta\, |f_\theta(t)|^2 \leq e^{-t^2/8} + e^{-n/12} + \mathbb{P}\{|X - Y|^2 \leq n/4\}.$$

This implies that, for some constant C,

$$\frac{1}{T}\int_0^T |f(t)|\, dt \leq \frac{1}{T}\int_0^T \mathbb{E}_\theta\, |f_\theta(t)|\, dt \;\leq\; \frac{1}{T}\int_0^T \Big(\mathbb{E}_\theta\, |f_\theta(t)|^2\Big)^{1/2} dt$$
$$\leq C\Big(\frac{1}{T} + \sqrt{\mathbb{P}\{|X - Y|^2 \leq n/4\}}\,\Big),$$

so that

$$\frac{1}{T^2}\Big(\int_0^T |f(t)|\, dt\Big)^2 \leq 2\,C^2\Big(\frac{1}{n^2} + \mathbb{P}\{|X - Y|^2 \leq n/4\}\Big).$$

Similarly, for some absolute constants $c > 0$ and $C > 0$,

$$c\int_{T_0}^T \frac{\mathbb{E}_\theta\, |f_\theta(t)|^2}{t}\, dt \leq \Big(e^{-n/12} + \mathbb{P}\{|X - Y|^2 \leq n/4\}\Big) \log \frac{T}{T_0} + e^{-T_0^2/8}$$
$$\leq C\Big(\frac{1}{n^3} + \mathbb{P}\{|X - Y|^2 \leq n/4\}\, \log n\Big).$$

These bounds prove the first assertion of the proposition.

For the second assertion, note that $M_2^2 \geq m_2 \geq 1$, so that the last term $1/n^2$ is dominated by M_2^4/n. It remains to recall that, by Proposition 1.6.2,

$$\mathbb{P}\{|X - Y|^2 \leq n/4\} \leq 16\,\frac{m_2^2 + \sigma_2^2}{n} \leq 16\,\frac{M_2^4 + \sigma_2^2}{n}.$$

This leads to the second bound.

Here, F may be replaced with the standard normal distribution function Φ due to Proposition 12.2.2. Indeed, it provides the bound on the weighted total variation distance, which implies that $\rho(F, \Phi) \leq \frac{C}{\sqrt{n}}\,(1+\sigma_2)$. Hence, by the triangle inequality,

$$\rho^2(F_\theta,\Phi) \le \big(\rho(F_\theta,F)+\rho(F,\Phi)\big)^2$$
$$\le 2\rho^2(F_\theta,F)+2\rho^2(F,\Phi) \le 2\rho^2(F_\theta,F)+\frac{4C^2}{n}\,\big(1+\sigma_2^2\big).$$

After averaging over θ, Proposition 14.6.1 is thus proved. □

Thus, in a rather general situation (when M_2 and σ_2^2 are of order one), the Kolmogorov distances $\rho(F_\theta,F)$ and $\rho(F_\theta,\Phi)$ are at most of order $1/\sqrt{n}$ on average, modulo a logarithmic factor. In particular,

$$\mathbb{E}_\theta\,\rho(F_\theta,\Phi) \le C(M_2^2+\sigma_2)\,\frac{\log n}{\sqrt{n}}$$

for some absolute constant C. It is interesting that for the rate $\frac{\log n}{\sqrt{n}}$ in this central limit theorem, we do not need moments higher than 2.

As for the first inequality of Proposition 14.6.1, it implies that

$$c\,\mathbb{E}_\theta\,\rho(F_\theta,F) \le \frac{M_1\log n}{\sqrt{n}}+\sqrt{\mathbb{P}\{|X-Y|^2\le n/4\}}\,\log n+\frac{1}{n}.$$

In fact, this assertion may slightly be refined by using Lemma 14.5.1 (rather than 14.5.2). Namely, with similar arguments as in the proof of Proposition 14.6.1, we then arrive at:

Proposition 14.6.2 *If a random vector X in $\mathbb{R}^n$ has finite moment M_1, then*

$$c\,\mathbb{E}_\theta\,\rho(F_\theta,F) \le \frac{M_1\sqrt{\log n}}{\sqrt{n}}+\sqrt{\mathbb{P}\{|X-Y|^2\le n/4\}}\,\log n+\frac{1}{n},$$

where $c>0$ is an absolute constant and Y is an independent copy of X.

Proof Using the same parameters $T_0=5\sqrt{\log n}$ and $T=5n$, we now apply Lemma 14.5.1. Again, the first integral in (14.13) can be bounded by virtue of the the first bound of Proposition 13.3.3, which gives

$$\int_0^{T_0}\frac{\mathbb{E}_\theta\,|f_\theta(t)-f(t)|}{t}\,dt \le \frac{5\sqrt{2}\,M_1}{\sqrt{n-1}}\sqrt{\log n}.$$

To bound the second and third integrals in (14.13), Lemma 13.1.1 with $\lambda=1/4$ gives

$$c\,\mathbb{E}_\theta\,|f_\theta(t)| \le e^{-t^2/16}+e^{-n/24}+\sqrt{\mathbb{P}\{|X-Y|^2\le n/4\}},$$

which implies that, for some constant C,

$$\frac{1}{T}\int_0^T|f(t)|\,dt \le \frac{1}{T}\int_0^T\mathbb{E}_\theta\,|f_\theta(t)|\,dt \le C\Big(\frac{1}{n}+\sqrt{\mathbb{P}\{|X-Y|^2\le n/4\}}\Big)$$

and

$$\int_{T_0}^{T} \frac{\mathbb{E}_\theta\,|f_\theta(t)|}{t}\,dt \leq C\left(e^{-n/24} + \sqrt{\mathbb{P}\{|X-Y|^2 \leq n/4\}}\right)\log\frac{T}{T_0} + Ce^{-T_0^2/16}.$$

These bounds prove Proposition 14.6.2. □

As an illustration of Propositions 14.6.1–14.6.2, let us mention the following consequence for the i.i.d. case.

Corollary 14.6.3 *If the components of the random vector X in $\mathbb{R}^n$ are independent, identically distributed, have mean zero and finite second moment, then*

$$\mathbb{E}_\theta\,\rho(F_\theta, F) \leq c\sqrt{\frac{\log n}{n}}, \quad \mathbb{E}_\theta\,\rho^2(F_\theta, F) \leq c\,\frac{(\log n)^2}{n},$$

where the constant c depends on the distribution of X_1 only.

Indeed, since the Kolmogorov distance is invariant under linear transformations, without loss of generality one may assume that $\mathbb{E}\,(X_1 - Y_1)^2 = 1$, where $Y = (Y_1, \dots, Y_n)$ is an independent copy of $X = (X_1, \dots, X_n)$. But then, by Proposition 2.1.5, applied to the random variables $X_k - Y_k$, we have

$$\mathbb{P}\{|X-Y|^2 \leq n/4\} \leq e^{-cn},$$

for some constant $c > 0$ depending on the distribution of X_1 only. It remains to apply Propositions 14.6.1–14.6.2.

However, one cannot draw a similar conclusion to Corollary 14.6.3 by replacing F with the normal distribution function Φ. If $\mathbb{E}X_1^2 = 1$, the second bound in Proposition 14.6.1 is applicable, but it contains the parameter

$$\sigma_2 = \frac{1}{\sqrt{n}}\,\mathbb{E}\,\big|\xi_1 + \dots + \xi_n\big|, \quad \xi_k = X_k^2 - 1,$$

which might grow to infinity as $n \to \infty$, when ξ_1 does not have a finite second moment, i.e., when $\mathbb{E}X_1^4 = \infty$.

As for the typical distribution, let us recall that it is described as the law $F = \mathcal{L}(|X|\theta_1)$, where the first coordinate θ_1 of a point θ on the unit sphere is viewed as being independent of X. The distribution of $\theta_1\sqrt{n}$ is close to the standard normal law Φ at the rate of order $1/n$ (by Proposition 11.1.2). Hence, in the bounds of Corollary 14.6.3, F may be replaced with the law $G = \mathcal{L}(|X|\frac{Z}{\sqrt{n}})$, i.e., we still have

$$\mathbb{E}_\theta\,\rho(F_\theta, G) \leq c\frac{\sqrt{\log n}}{\sqrt{n}}, \quad \mathbb{E}_\theta\,\rho^2(F_\theta, G) \leq c\frac{\log^2 n}{n}.$$

Recall that G represents the mixture of centered Gaussian measures on the real line with characteristic function

$$\int_{-\infty}^{\infty} \mathrm{e}^{itx}\, \mathrm{d}G(x) = \mathbb{E}\, \mathrm{e}^{-t^2|X|^2/2n} = \left(\mathbb{E}\, \mathrm{e}^{-t^2 X_1^2/2n}\right)^n$$
$$= \left(\int_{-\infty}^{\infty} \mathrm{e}^{-t^2 r^2/2n}\, \mathrm{d}\nu(r)\right)^n, \quad t \in \mathbb{R},$$

where the last two expressions correspond to the i.i.d. case as in Corollary 14.6.3 with a common distribution ν of the components X_k. Equivalently, in this case, G serves as the distribution of the normalized sum

$$\Sigma_n = \frac{X_1 Z_1 + \cdots + X_n Z_n}{\sqrt{n}}$$

with independent random variables $Z_k \sim N(0,1)$ that are also independent of X.

Remark 14.6.4 Under the assumptions of Corollary 14.6.3, even if additionally $\mathbb{E}X_1 = 0$, $\mathbb{E}X_1^2 = 1$, the normalized sums $\frac{1}{\sqrt{n}}\,(X_1 + \cdots + X_n)$ (that is, the weighted sums $\langle X, \theta\rangle$ with equal coefficients) have distributions F_n which do not need be close to the standard normal law Φ at the standard rate. Moreover, as was shown by Matskyavichyus [135], the distances $\rho(F_n, \Phi)$ may tend to zero at an arbitrarily slow rate. Namely, for any sequence $\varepsilon_n \to 0$, X_1 may have a distribution with mean zero and variance one, such that $\rho(F_n, \Phi) \geq \varepsilon_n$ for all n large enough. In the construction of [135], this distribution has characteristic function of the form

$$\mathbb{E}\, \mathrm{e}^{itX_1} = \int_0^{\infty} \mathrm{e}^{-t^2 r^2/2}\, \mathrm{d}\nu(r), \quad t \in \mathbb{R},$$

for some discrete probability measure ν on $(0, \infty)$. Hence, the distribution of Z_n is the same as the distribution of Σ_n when X_1^2 is distributed according to ν. This means that the concentration of the F_θ around their mean $F = \mathbb{E}_\theta F_\theta$ may be essentially stronger than their concentration around the standard normal law.

14.7 Bounds With a Standard Rate

To get rid of the logarithmic term in the bounds of Propositions 14.6.1–14.6.2, one may involve the 3-rd moment assumptions in terms of the moment and variance-type functionals m_p, M_p, and σ_p of index $p = 3$. They are defined by

$$m_3 = \frac{1}{\sqrt{n}}\, \left(\mathbb{E}\,|\,\langle X, Y\rangle\,|^3\right)^{1/3}, \quad M_3 = \sup_{|\theta|=1}\, \left(\mathbb{E}\,|\,\langle X, \theta\rangle\,|^3\right)^{1/3},$$
$$\sigma_3 = \sqrt{n}\Big(\mathbb{E}\,\Big|\frac{|X|^2}{n} - 1\Big|^{\frac{3}{2}}\Big)^{\frac{2}{3}} = \frac{1}{\sqrt{n}}\Big(\mathbb{E}\,\big|\,|X|^2 - n\big|^{\frac{3}{2}}\Big)^{\frac{2}{3}},$$

where Y is an independent copy of X. Let us recall that $m_3 \le M_3^2$ (Proposition 1.4.2), while $\sigma_3^2 \le \sigma_4^2 = \frac{1}{n}\,\mathrm{Var}(|X|^2)$.

Proposition 14.7.1 *Let X be a random vector in $\mathbb{R}^n$ with mean $a = \mathbb{E}X$, finite 3-rd moment, and such that $\mathbb{E}\,|X|^2 = n$. Then for some absolute constant c,*

$$\mathbb{E}_\theta\, \rho(F_\theta, \Phi) \le c\,(m_3^{3/2} + \sigma_3^{3/2} + |a|)\,\frac{1}{\sqrt{n}}. \tag{14.16}$$

In particular,

$$\mathbb{E}_\theta\, \rho(F_\theta, \Phi) \le c\,(M_3^3 + \sigma_3^{3/2})\,\frac{1}{\sqrt{n}}. \tag{14.17}$$

Let us recall that $|a| \le M_2$, while $m_3 \ge m_2 \ge 1$ due to the assumption $\mathbb{E}\,|X|^2 = n$ (Proposition 1.1.3 and Proposition 1.2.2). Hence, in the isotropic case, the term $|a|$ may be removed from (14.16). As for (14.17), this inequality follows from (14.16), since $m_3 \le M_3^2$, while $|a| \le M_2 \le M_3$.

In general, the term $|a|$ may not be removed from (14.16). See Remark 14.7.2, where we discuss the sharpness of this inequality with respect to the expectation of X.

Proof (of Proposition 14.7.1) By Lemma 14.5.3 with $p = 3/2$, we have for all $T \ge T_0 \ge 1$,

$$\begin{aligned} c\,\mathbb{E}_\theta\, \rho(F_\theta, F) \le &\int_0^{T_0} \frac{\mathbb{E}_\theta\,|f_\theta(t) - f(t)|}{t}\,\mathrm{d}t \\ &+ \frac{m_3^{3/2} + \sigma_3^{3/2}}{n^{3/4}}\Big(1 + \log\frac{T}{T_0}\Big) + \frac{1}{T} + \mathrm{e}^{-T_0^2/16}, \end{aligned}$$

where $c > 0$ is an absolute constant. Let us take here $T = 4n$ and $T_0 = 4\sqrt{\log n}$. Since $m_3 \ge 1$, the last two terms are negligible, and we get the bound

$$c\,\mathbb{E}_\theta\, \rho(F_\theta, F) \le \int_0^{T_0} \frac{\mathbb{E}_\theta\,|f_\theta(t) - f(t)|}{t}\,\mathrm{d}t + (m_3^{3/2} + \sigma_3^{3/2})\,\frac{\log n}{n^{3/4}}.$$

To analyze the above integral over the interval $[0, T_0]$, we apply the first order Poincaré-type inequality on the unit sphere via the equality of Lemma 13.3.1,

$$\mathbb{E}_\theta\,|\nabla f_\theta(t)|^2 = t^2\,\mathbb{E}\,\langle X, Y\rangle\, J_n(t(X - Y)),$$

which gives

$$\begin{aligned} c\,\mathbb{E}_\theta\, \rho(F_\theta, F) \le &\frac{1}{\sqrt{n}}\int_0^{T_0} \sqrt{\mathbb{E}\,\langle X, Y\rangle\, J_n(t(X - Y))}\,\mathrm{d}t \\ &+ (m_3^{3/2} + \sigma_3^{3/2})\,\frac{\log n}{n^{3/4}}. \end{aligned} \tag{14.18}$$

Next, we apply the bound of Proposition 11.3.1

$$\left|J_n(t\sqrt{n}) - \mathrm{e}^{-t^2/2}\right| \leq \frac{C}{n}$$

which allows one to replace the J_n-term with $\mathrm{e}^{-t^2|X-Y|^2/2n}$ at the expense of an error not exceeding, up to an absolute constant,

$$\frac{1}{n}\,T_0\sqrt{\mathbb{E}\,|\langle X,Y\rangle|} \leq \frac{\sqrt{m_2}}{n^{3/4}}\,T_0 \leq \frac{\sqrt{m_3}}{n^{3/4}}\,T_0 \leq \frac{m_3^{3/2}}{n^{3/4}}\,T_0.$$

As a result, the bound (14.18) simplifies to

$$c\,\mathbb{E}_\theta\,\rho(F_\theta,F) \leq \frac{1}{\sqrt{n}}\int_0^{T_0}\sqrt{I(t)}\,dt + (m_3^{3/2}+\sigma_3^{3/2})\,\frac{\log n}{n^{3/4}} \tag{14.19}$$

with

$$I(t) = \mathbb{E}\,\langle X,Y\rangle\,\mathrm{e}^{-t^2|X-Y|^2/2n} = \mathbb{E}_Z\big|\,\mathbb{E}\,X\mathrm{e}^{it\langle X,Z\rangle/\sqrt{n}}\,|^2.$$

Note that $I(t) \geq 0$, which follows from the second representation, in which the random vector Z is independent of X and has a standard normal distribution on $\mathbb{R}^n$.

Now, focusing on $I(t)$, consider the events

$$A = \Big\{|X-Y|^2 \leq \frac{1}{4}\,n\Big\}, \quad B = \Big\{|X-Y|^2 > \frac{1}{4}\,n\Big\}$$

and split the expectation in the definition of $I(t)$ into the sets A and B, so that $I(t) = I_1(t) + I_2(t)$, where

$$I_1(t) = \mathbb{E}\,\langle X,Y\rangle\,\mathrm{e}^{-t^2|X-Y|^2/2n}\,1_A, \quad I_2(t) = \mathbb{E}\,\langle X,Y\rangle\,\mathrm{e}^{-t^2|X-Y|^2/2n}\,1_B.$$

As we know (cf. Proposition 1.6.2),

$$\mathbb{P}(A) \leq 64\,\frac{m_3^3+\sigma_3^3}{n^{3/2}}.$$

Hence, applying Hölder's inequality, as well as the simple inequality $xy^2 \leq x^3 + y^3$ $(x,y \geq 0)$ with $x = m_3$ and $y = \sigma_3$, we have

$$\begin{aligned}|I_1(t)| &\leq \big(\mathbb{E}\,|\langle X,Y\rangle|^3\big)^{1/3}\,(\mathbb{P}(A))^{2/3}\\ &\leq m_3\sqrt{n}\cdot 16\,\frac{m_3^2+\sigma_3^2}{n} \leq \frac{32}{\sqrt{n}}\,(m_3^3+\sigma_3^3).\end{aligned}$$

Now, using $\mathbb{E}\langle X, Y\rangle = |a|^2$, we represent the second expectation as

$$\begin{aligned} I_2(t) &= e^{-t^2}\, \mathbb{E}\,\langle X, Y\rangle\, e^{-t^2\,(\frac{|X-Y|^2}{2n}-1)}\, 1_B \\ &= e^{-t^2}\, \mathbb{E}\,\langle X, Y\rangle \left(e^{-t^2\,(\frac{|X-Y|^2}{2n}-1)} - 1\right) 1_B - e^{-t^2}\, \mathbb{E}\,\langle X, Y\rangle\, 1_A + e^{-t^2}|a|^2. \end{aligned}$$

Here, the last expectation has already been bounded in absolute value by $32\,(m_3^3+\sigma_3^3)/\sqrt{n}$. To estimate the first one, we use the inequality $|e^{-x}-1| \le |x|\, e^{x_0}$ ($x_0 \ge 0$, $x \ge -x_0$). Since on the set B there is a uniform bound

$$t^2\left(\frac{|X-Y|^2}{2n} - 1\right) \ge -\frac{7}{8}\,t^2,$$

we conclude by virtue of Hölder's inequality that

$$\begin{aligned} \mathbb{E}\,|\langle X, Y\rangle|\, \left|e^{-t^2\,(\frac{|X-Y|^2}{2n}-1)} - 1\right| 1_B &\le t^2 e^{7t^2/8}\, \mathbb{E}\,|\langle X, Y\rangle|\, \left|\frac{|X-Y|^2}{2n} - 1\right| \\ &\le\ t^2 e^{7t^2/8} \left(\mathbb{E}\,|\langle X, Y\rangle|^3\right)^{\frac{1}{3}} \left(\mathbb{E}\left|\frac{|X-Y|^2}{2n} - 1\right|^{\frac{3}{2}}\right)^{\frac{2}{3}}. \end{aligned}$$

The first expectation on the right-hand side is $\mathbb{E}\,|\langle X, Y\rangle|^3 = (m_3\sqrt{n})^3$. As for the second one, first write

$$\frac{|X-Y|^2}{2n} - 1\ =\ \frac{1}{2}\left(\frac{|X|^2}{n} - 1\right) + \frac{1}{2}\left(\frac{|Y|^2}{n} - 1\right) - \frac{1}{n}\,\langle X, Y\rangle.$$

Using the inequality $(\frac{1}{2}x + \frac{1}{2}y + z)^{\frac{3}{2}} \le x^{\frac{3}{2}} + y^{\frac{3}{2}} + 2z^{\frac{3}{2}}$ ($x, y, z \ge 0$), we then also have

$$\mathbb{E}\left|\frac{|X-Y|^2}{2n} - 1\right|^{\frac{3}{2}} \le 2\,\mathbb{E}\left|\frac{|X|^2}{n} - 1\right|^{\frac{3}{2}} + \frac{2}{n^{3/2}}\left(\mathbb{E}\,|\langle X, Y\rangle|^3\right)^{1/2} = \frac{2}{n^{3/4}}\,(\sigma_3^{3/2} + m_3^{3/2}),$$

which gives

$$\left(\mathbb{E}\left|\frac{|X-Y|^2}{2n} - 1\right|^{\frac{3}{2}}\right)^{\frac{2}{3}} \le \frac{2}{\sqrt{n}}\,(\sigma_3 + m_3).$$

Therefore,

$$\begin{aligned} \mathbb{E}\,|\langle X, Y\rangle|\, \left|e^{-t^2\,(\frac{|X-Y|^2}{2n}-1)} - 1\right| 1_B &\le 2t^2 e^{7t^2/8}\, m_3\,(m_3 + \sigma_3) \\ &\le 4t^2 e^{7t^2/8}\,(m_3^2 + \sigma_3^2), \end{aligned}$$

and, as a result,

$$I_2(t) \le 32\,\frac{e^{-t^2}}{\sqrt{n}}\,(m_3^3 + \sigma_3^3) + 4t^2 e^{-t^2/8}\,(m_3^2 + \sigma_3^2) + e^{-t^2}|a|^2,$$

where the factor e^{-t^2} in the first term may be removed without loss of strength.

Together with the estimate on $I_1(t)$, we get

$$I(t) \leq \frac{64}{\sqrt{n}}\,(m_3^3 + \sigma_3^3) + 4t^2 e^{-t^2/8}\,(m_3^2 + \sigma_3^2) + e^{-t^2}\,|a|^2,$$

so

$$\sqrt{I(t)} \leq \frac{8}{n^{1/4}}\,(m_3^{3/2} + \sigma_3^{3/2}) + 2\,|t|\,e^{-t^2/16}\,(m_3 + \sigma_3) + e^{-t^2/2}\,|a|$$

and

$$\frac{1}{\sqrt{n}} \int_0^{T_0} \sqrt{I(t)}\,dt \;\leq\; \frac{8T_0}{n^{3/4}}\,(m_3^{3/2} + \sigma_3^{3/2}) + \frac{C}{\sqrt{n}}\,(m_3 + \sigma_3 + |a|)$$

for some absolute constant C.

Returning to the bound (14.19), we thus obtain that

$$c\,\mathbb{E}_\theta\,\rho(F_\theta, F) \leq \frac{\log n}{n^{3/4}}\,(m_3^{3/2} + \sigma_3^{3/2}) + \frac{C}{\sqrt{n}}\,(m_3 + \sigma_3 + |a|).$$

To simplify it, one may use again the property $m_3 \geq 1$, which implies that $m_3 + \sigma_3 \leq 2(m_3^{3/2} + \sigma_3^{3/2})$ for all values of σ_3. Thus, for some absolute constant $c > 0$,

$$c\,\mathbb{E}_\theta\,\rho(F_\theta, F) \leq \frac{C}{\sqrt{n}}\,(m_3^{3/2} + \sigma_3^{3/2} + |a|).$$

To obtain a similar bound with Φ in place of F, it remains to recall Proposition 12.2.2 with its estimate $\rho(F, \Phi) \leq C\,\frac{1+\sigma_2}{\sqrt{n}}$, where $1 + \sigma_2$ may further be bounded by $2(m_3^{3/2} + \sigma_3^{3/2})$. The inequality (14.16) has now been proved. □

Remark 14.7.2 Let us illustrate the inequality (14.16) on the example where the random vector X has a normal distribution with a large mean value. Given a standard normal random vector Z in $\mathbb{R}^{n-1}$ (which we identify with the space of all points in $\mathbb{R}^n$ with zero last coordinate), define

$$X = \alpha Z + \lambda e_n \quad \text{with } 1 \leq \lambda \leq n^{1/4}, \;\; \alpha^2(n-1) + \lambda^2 = n,$$

where $e_n = (0, \dots, 0, 1)$ is the last unit vector in the canonical basis of $\mathbb{R}^n$ (necessarily, $|\alpha| < 1$). Since Z is orthogonal to e_n, we have $|X|^2 = \alpha^2\,|Z|^2 + \lambda^2$ and $\mathbb{E}\,|X|^2 = n$. In addition,

$$\sigma_3^2 \leq \sigma_4^2 = \frac{\alpha^4}{n}\,\mathrm{Var}(|Z|^2) = \frac{2\alpha^4\,(n-1)}{n} \leq 2.$$

Let Z' be an independent copy of Z, so that $Y = \alpha Z' + \lambda e_n$ is an independent copy of X. From this, by Jensen's inequality,

$$\begin{aligned} m_3^4 \leq m_4^4 &= \frac{1}{n^2}\,\mathbb{E}\,\langle X, Y\rangle^4 = \frac{1}{n^2}\,\mathbb{E}\,(\alpha^2\,\langle Z, Z'\rangle + \lambda^2)^4 \\ &\leq \frac{8}{n^2}\,(\alpha^8\,\mathbb{E}\,\langle Z, Z'\rangle^4 + \lambda^8) = \frac{8}{n^2}\,(3\alpha^8\,(n^2 + 2n) + \lambda^8) \leq 80. \end{aligned}$$

Thus, both m_3 and σ_3 are bounded, while the mean $a = \mathbb{E}X = \lambda e_n$ has the Euclidean length $|a| = \lambda \geq 1$. Hence, the inequality (14.16) simplifies to

$$\mathbb{E}_\theta\, \rho(F_\theta, \Phi) \leq \frac{c\lambda}{\sqrt{n}}.$$

Let us see that this bound may be reversed (which would imply that $|a|$ may not be removed from (14.16)). For any unit vector $\theta = (\theta_1, \dots, \theta_n)$, the linear form $S_\theta = \langle X, \theta\rangle$ has a normal distribution on the real line with mean $\mathbb{E}S_\theta = \lambda\theta_n$ and variance $\mathrm{Var}(S_\theta) = \alpha^2\,(1-\theta_n)^2 < 1$. Note that, for the normal distribution function $\Phi_{\mu,\sigma^2}(x) = \Phi(\frac{x-\mu}{\sigma})$ with parameters $|\mu| \leq 1$, $\frac{1}{2} \leq \sigma^2 \leq 1$, we have

$$\begin{aligned}\rho(\Phi_{\mu,\sigma^2}, \Phi) &\geq |\Phi_{\mu,\sigma^2}(0) - \Phi(0)| \\ &= |\Phi(-\mu/\sigma) - \Phi(0)| \geq \frac{1}{\sqrt{2\pi}}\, e^{-\mu^2/2\sigma^2}\, \frac{|\mu|}{\sigma} \geq \frac{1}{\sqrt{2\pi e}}\, |\mu|.\end{aligned}$$

In our case, since $\lambda \leq n^{1/4}$ and

$$\alpha^2 = \frac{n-\lambda^2}{n-1} \geq \frac{n-\sqrt{n}}{n-1} \geq 1 - \frac{1}{\sqrt{n}},$$

we have $|\mathbb{E}S_\theta| \leq 1$ and $\mathrm{Var}(S_\theta) \geq \frac{1}{2}$ on the set $\Omega = \{\theta : |\theta_n| < \frac{\log n}{\sqrt{n}}\}$ with n large enough. It then easily follows that for some constant $c > 0$

$$\mathbb{E}_\theta\, \rho(F_\theta, \Phi) \geq \frac{\lambda}{\sqrt{2\pi e}}\, \mathbb{E}\,|\theta_n|\, 1_{\{\theta\in\Omega\}} \geq \frac{c\lambda}{\sqrt{n}}.$$

14.8 Deviation Bounds for the Kolmogorov Distance

In order to see that the bound (14.17) is valid not only on average, but also for all directions θ from a large part of the unit sphere, it is natural to complement or strengthen this bound in terms of large deviations. Note that, in general

$$L(F_\theta, \Phi) \leq \rho(F_\theta, \Phi) \leq \Big(1 + \frac{1}{\sqrt{2\pi}}\Big)\, L(F_\theta, \Phi),$$

so, it does not matter which of the two distances is used to measure the closeness to the standard normal law.

Let us return to Proposition 14.4.5 and combine it with Proposition 14.7.1. Then we immediately obtain:

Proposition 14.8.1 *Suppose that the random vector X in $\mathbb{R}^n$ satisfies $\mathbb{E}\,|X|^2 = n$, and let M_3 be finite. Then, for all $p > 0$ and $r > 0$,*

$$\mathfrak{s}_{n-1}\Big\{L(F_\theta,\Phi) \geq \frac{A}{\sqrt{n}} + r\Big\} \leq \exp\Big\{-n\,\frac{r^{2(p+1)/p}}{4M_p^2}\Big\},$$

where A depends on M_3 and σ_3, only.

One may take $A = c\,(M_3^3 + \sigma_3^{3/2})$ with some absolute constant $c > 0$. For example, in the isotropic case, that is, with $p = 2$ and $M_2 = 1$,

$$\mathfrak{s}_{n-1}\Big\{L(F_\theta,\Phi) \geq \frac{A}{\sqrt{n}} + r\Big\} \leq \mathrm{e}^{-nr^3/4},$$

so that with high $\mathfrak{s}_{n-1}$-probability

$$\rho(F_\theta,\Phi) \leq 2A\,\frac{(\log n)^{1/3}}{n^{1/3}}.$$

But, this "typical" bound can be improved by choosing larger values of p, as long as the moments M_p do not grow too fast. To this end, assume, for example, that

$$\mathbb{E}\,\mathrm{e}^{|\langle X,\theta\rangle|/\lambda} \leq 2$$

for all $\theta \in \mathbb{S}^{n-1}$ with some $\lambda > 0$. Using $x^p\,\mathrm{e}^{-x} \leq (\frac{p}{\mathrm{e}})^p$ $(x, p > 0)$, we then get

$$\Big|\frac{\langle X,\theta\rangle}{\lambda}\Big|^p \leq \Big(\frac{p}{\mathrm{e}}\Big)^p\,\mathrm{e}^{|\langle X,\theta\rangle|/\lambda}$$

and hence $M_p^p \leq 2\,(\frac{\lambda p}{\mathrm{e}})^p$, i.e.,

$$M_p \leq 2^{1/p}\,\frac{\lambda p}{\mathrm{e}} = 2^s\,\frac{\lambda}{\mathrm{e}s}, \quad s = \frac{1}{p}.$$

From this, if $p \geq 1$,

$$\begin{aligned}\frac{r^{2(p+1)/p}}{2M_p^2} &= r^2\,\frac{r^{2/p}}{2M_p^2} \geq r^2\,\frac{\mathrm{e}^2}{2\cdot 4^s\lambda^2}\,s^2 r^{2s} \\ &\geq r^2\,\frac{\mathrm{e}^2}{8\lambda^2}\,s^2\,\exp\big\{-2s\,\log(1/r)\big\}.\end{aligned}$$

Assuming that $0 < r \leq \mathrm{e}^{-1}$, one may choose $s = 1/\log(1/r)$, and then the latter expression is greater than or equal to $r^2/(8\lambda^2\,\log^2(1/r))$. One may now use this lower bound in the inequality of Proposition 14.8.1, to get the following corollary.

Corollary 14.8.2 *Suppose that* $\mathbb{E}\,|X|^2 = n$ *and* $\mathbb{E}\,e^{|\langle X,\theta\rangle|/\lambda} \le 2$ *for all* $\theta \in \mathbb{S}^{n-1}$ *with some* $\lambda > 0$. *Then*

$$\mathfrak{s}_{n-1}\Big\{L(F_\theta,\Phi) \ge \frac{A}{\sqrt{n}} + r\Big\} \le \exp\Big\{-\frac{n}{16\lambda^2}\,\frac{r^2}{\log^2(1/r)}\Big\}$$

for all $0 < r \le e^{-1}$, *where* $A > 0$ *depends on* λ *and* σ_3 *only.*

In the inequality $e^{|\langle X,\theta\rangle|/\lambda} \ge \frac{1}{2\lambda^2}\,\langle X,\theta\rangle^2$ one may take the expectation, which leads to $2 \ge \frac{1}{2\lambda^2}\,\mathbb{E}\,\langle X,\theta\rangle^2$. Taking another expectation with respect to $d\mathfrak{s}_{n-1}(\theta)$, we see that the condition $\mathbb{E}\,|X|^2 = n$ forces λ to be bounded away from zero: $\lambda \ge \frac{1}{2}$. Let us choose

$$r = 4\lambda\,\frac{(\log n)^{3/2}}{\sqrt{n}}.$$

It follows that $r \ge 2\,\frac{(\log 2)^{3/2}}{\sqrt{n}} > \frac{1}{\sqrt{n}}$, so that $\log(1/r) < \frac{1}{2}\log n$ and

$$\frac{r}{\log(1/r)} \ge \frac{2r}{\log n} = 8\lambda\,\frac{(\log n)^{1/2}}{\sqrt{n}}.$$

Hence

$$\frac{n}{16\lambda^2}\,\frac{r^2}{\log^2(1/r)} \ge 4\log n.$$

If $r < 1/e$, one may apply Corollary 14.8.2, which gives

$$\mathfrak{s}_{n-1}\Big\{L(F_\theta,\Phi) > \frac{A}{\sqrt{n}} + er\Big\} \le \mathfrak{s}_{n-1}\Big\{L(F_\theta,\Phi) \ge \frac{A}{\sqrt{n}} + r\Big\} \le \frac{1}{n^4}.$$

In the case $r \ge 1/e$, the left $\mathfrak{s}_{n-1}$-probability is zero, that is, we arrive at the following:

Corollary 14.8.3 *Under the assumptions of Corollary 14.8.2, with some constant A depending on* λ *and* σ_3, *the inequality*

$$\rho(F_\theta,\Phi) \le A\,\frac{(\log n)^{3/2}}{\sqrt{n}}$$

holds true for all $\theta \in \mathbb{S}^{n-1}$ *except for a set of directions of* $\mathfrak{s}_{n-1}$*-measure* $\le n^{-4}$.

14.9 The Log-concave Case

Most of the previous observations may be refined under an additional log-concavity assumption on the distribution of $X = (X_1,\dots,X_n)$. More precisely, we assume here that the random vector X is isotropic, has mean zero, and a log-concave distribution on $\mathbb{R}^n$ $(n \ge 2)$. The aim of this section is to derive the following bounds on the fluctuations of the distribution functions F_θ about the typical distribution $F = \mathbb{E}_\theta F_\theta$. First, we remove the $\log n$ term from Proposition 14.2.3 when speaking about the Kantorovich distance.

Proposition 14.9.1 *For some absolute constant $c > 0$, we have*

$$\mathbb{E}_\theta W(F_\theta, F) \le \frac{c}{\sqrt{n}}.$$

Moreover, for any $r > 0$,

$$\mathfrak{s}_{n-1}\Big\{W(F_\theta, F) \ge \frac{c}{\sqrt{n}} + r\Big\} \le \mathrm{e}^{-nr^2/4}.$$

Turning to the Kolmogorov distance, let us note that all weighted sums $\langle X, \theta\rangle$ have a log-concave distribution on the real line with mean zero and variance one. Consequently, they have a bounded L^{ψ_1}-norm, i.e., $\mathbb{E}\, \mathrm{e}^{|\langle X,\theta\rangle|/\lambda} \le 2$ for some absolute constant $\lambda > 0$ (cf. Corollary 2.5.2). Corollary 14.8.3 is thus applicable and provides the rate of fluctuation of the F_θ from the standard normal law Φ of order at most $\frac{A}{\sqrt{n}}\, (\log n)^{3/2}$. However, the constant A in that statement depends upon the variance functional $\sigma_3(X)$, about which we cannot say whether or not it is bounded by a universal constant (in the general log-concave isotropic situation). As it turns out, when replacing Φ with F, there is no need to involve this functional.

Proposition 14.9.2 *For all $r > 0$,*

$$\mathfrak{s}_{n-1}\Big\{\sup_x \Big[\mathrm{e}^{c|x|}|F_\theta(x) - F(x)|\Big] \ge r\Big\} \le C\sqrt{n}\log n\, \mathrm{e}^{-cnr^2}$$

with some absolute constants $C > 0$ and $c > 0$. In particular,

$$\mathbb{E}_\theta\, \rho(F_\theta, F) \le \frac{c\sqrt{\log n}}{\sqrt{n}}.$$

By the Borell characterization theorem (Proposition 2.4.1), the distribution μ of X has a log-concave density $p(x)$ (which is also due to the assumption that X is isotropic, so that μ is full-dimensional).

Let us write the points in $\mathbb{R}^n$ in the form $x = (y, t)$ with $x \in \mathbb{R}^{n-1}$ and $t \in \mathbb{R}$. As a first step of the proof, we need:

Lemma 14.9.3 *For every $t \in \mathbb{R}$ and every unit vector l in $\mathbb{R}^{n-1}$,*

$$\int_{\mathbb{R}^{n-1}} |\langle l, y\rangle|\, p(y,t)\, \mathrm{d}y \le C\, \mathrm{e}^{-c|t|},$$

where C and c are positive universal constants.

Proof The function $l^+(y) = \max\{\langle l, y\rangle\, , 0\}$ is log-concave on $\mathbb{R}^{n-1}$, and so is the function $l^+(y)\, p(y,t)$ on $\mathbb{R}^n$. Hence, by the Prékopa theorem (Corollary 2.4.3), the function

$$u(t) = \int_{\mathbb{R}^{n-1}} l^+(y)\, p(y,t)\, dy$$

is log-concave on $\mathbb{R}$. We consider the linear functions $\xi(y,t) = \langle l, y\rangle$ and $\eta(y,t) = t$ as log-concave random variables on the probability space $(\mathbb{R}^n, \mu)$. By the assumptions, $\mathbb{E}\xi = \mathbb{E}\eta = 0$ and $\mathbb{E}\xi^2 = \mathbb{E}\eta^2 = 1$. Thus, using $\xi^+ = \max\{\xi, 0\}$, we have

$$\int_{-\infty}^{\infty} u(t)\, dt = \mathbb{E}\xi^+, \quad \int_{-\infty}^{\infty} tu(t)\, dt = \mathbb{E}\xi^+\eta, \quad \int_{-\infty}^{\infty} t^2 u(t)\, dt = \mathbb{E}\xi^+\eta^2.$$

Introduce the log-concave probability density

$$q(t) = \frac{u(t)}{\mathbb{E}\xi^+}, \quad t \in \mathbb{R}.$$

Since we need to estimate the function $u(t)$ from above, we apply Proposition 2.6.4 to q. Let ζ be a random variable with this density, so that

$$\mathbb{E}\zeta = \frac{\mathbb{E}\xi^+\eta}{\mathbb{E}\xi^+}, \quad \mathrm{Var}(\zeta) = \frac{\mathbb{E}\xi^+\eta^2\, \mathbb{E}\xi^+ - (\mathbb{E}\xi^+\eta)^2}{(\mathbb{E}\xi^+)^2}.$$

Let ν denote the normalized restriction of μ to the half-space $A = \{\xi > 0\}$ in $\mathbb{R}^n$. Then the above variance can also be written as

$$\begin{aligned}\mathrm{Var}(\zeta) &= \frac{1}{\mathbb{E}\xi^+} \int_{-\infty}^{\infty} (t-a)^2\, u(t)\, dt \\ &= \frac{\mu(A)}{\mathbb{E}\xi^+} \int (\eta - a)^2\, \xi\, d\nu = \frac{\mu(A)}{\mathbb{E}\xi^+} \|(\eta - a)^2\, \xi\|_1, \end{aligned} \tag{14.20}$$

where $a = \mathbb{E}\zeta$ and where we use the notation $\|\psi\|_\alpha = (\int |\psi|^\alpha\, d\nu)^{1/\alpha}$ with the usual convention $\|\psi\|_0 = \exp\{\int \log|\psi|\, d\nu\}$. We have

$$\|(\eta - a)^2\, \xi\|_1 \geq \|(\eta - a)^2\, \xi\|_0 = \|\, |\eta - a|\, \|_0^2\, \|\xi\|_0,$$

while, since the measure ν is log-concave, by Propositions 2.5.1 and 2.5.3, we have $\|\psi\|_0 \geq c_\alpha \|\psi\|_\alpha$ for every norm ψ up to some constants $c_\alpha > 0$ depending on α only. Hence

$$\|(\eta - a)^2\, \xi\|_1 \geq c\, \|\eta - a\|_2^2\, \|\xi\|_2 \tag{14.21}$$

for some absolute constant $c > 0$.

Recall that

$$\|\eta - a\|_2^2 = \frac{1}{\mu(A)} \int_A (\eta - a)^2\, d\mu \quad \text{and} \quad \|\xi\|_2^2 = \frac{1}{\mu(A)} \int_A \xi^2\, d\mu.$$

Applying Proposition 2.6.3 to the random variables $\eta - a$ and ξ, and using $\mathrm{Var}(\eta) = \mathrm{Var}(\xi) = 1$, we obtain that

$$\|\eta - a\|_2^2 \ge \frac{1}{24}\,\mu(A)^2, \quad \|\xi\|_2^2 \ge \frac{1}{24}\,\mu(A)^3.$$

Using these estimates in (14.21), the identity (14.20) then yields

$$\mathrm{Var}(\zeta) \ge \frac{c}{24\sqrt{24}}\,\frac{\mu(A)^{7/2}}{\mathbb{E}\xi^+}.$$

In addition, $\mathbb{E}\xi^+ \le \sqrt{\mathbb{E}\xi^2} = 1$, while, by Proposition 2.6.2, $\mu(A) \ge \frac{1}{\mathrm{e}}$. Hence $\sigma^2 \equiv \mathrm{Var}(\zeta) \ge c$ for some other constant $c > 0$.

Now, in order to bound the expectation $a = \mathbb{E}\zeta = \frac{\mathbb{E}\xi^+\eta}{\mathbb{E}\xi^+}$, just note that $|\mathbb{E}\xi^+\eta|^2 \le \mathbb{E}\xi^2\,\mathbb{E}\eta^2 = 1$, and once more by Propositions 2.6.2–2.6.3,

$$\mathbb{E}\xi^+ = \int_A \xi\,\mathrm{d}\mu \ge \frac{1}{4\sqrt{2}}\,\mu(A)^2 \ge \frac{1}{4\sqrt{2}\,\mathrm{e}^2}.$$

Thus, $|a| \le 4\sqrt{2}\,\mathrm{e}^2$. Applying now Proposition 2.6.4, we finally get

$$q(t) \le \frac{C}{\sigma}\,\exp\Big\{-\frac{|t-a|}{\sigma\sqrt{12}}\Big\} \le C'\mathrm{e}^{-c|t|}$$

for some positive absolute constants c, C, C' (since we have universal bounds for σ and a). It remains to note that $u(t) = q(t)\,\mathbb{E}\xi^+ \le q(t)$.

At last, replacing l with $-l$, we get a similar estimate for $\int_{\mathbb{R}^{n-1}} l^-(y)\,p(y,t)\,\mathrm{d}y$, where $l^-(y) = \max\{-\langle l, y\rangle, 0\}$. Lemma 14.9.3 is proved. □

Lemma 14.9.4 *For every $t \in \mathbb{R}$, the function*

$$u_t(\theta) = F_\theta(t) = \mathbb{P}\{\langle X, \theta\rangle \le t\}$$

has a Lipschitz semi-norm on $\mathbb{S}^{n-1}$ satisfying, with some universal constants $C > 0$ and $c > 0$,

$$\|u_t\|_{\mathrm{Lip}} \le C\,\mathrm{e}^{-c|t|}.$$

Proof Without loss of generality, we may assume that the density p of X is positive on $\mathbb{R}^n$ (hence everywhere continuous). It is readily verified that u_t is differentiable at all points of $\mathbb{R}^n$ except for the origin, and we need to show that the spherical gradient of u_t satisfies $|\nabla_{\mathbb{S}}\, u_t(\theta)| \le C\,\mathrm{e}^{-c|t|}$ for all $\theta \in \mathbb{S}^{n-1}$. Moreover, by the isotropy assumption, it is sufficient to consider the point $\theta = e_n = (0, \dots, 0, 1)$. Note that

$$|\nabla_{\mathbb{S}}\, u_t(e_n)| = \bigg(\sum_{k=1}^{n-1}\Big|\frac{\partial u_t(e_n)}{\partial\theta_k}\Big|^2\bigg)^{1/2} = \sup_{|l|=1}\,|\langle \nabla u_t(e_n), l\rangle|,$$

where the supremum is taken over all unit vectors $l = (l_1, \dots, l_{n-1}, 0)$ in $\mathbb{R}^n$ with the last coordinate $l_n = 0$.

Fix $k = 1, \dots, n-1$. The two-dimensional random vector (X_k, X_n) has a positive (continuous) log-concave density $p_k = p_k(x_k, x_n)$ on $\mathbb{R}^2$. Therefore, using the canonical orthonormal basis $e_1, \dots, e_{n-1}, e_n$ in $\mathbb{R}^n$, for any $\varepsilon > 0$, the difference

$$u_t(e_n + \varepsilon e_k) - u_t(e_n) = \mathbb{P}\{X_n + \varepsilon X_k \le t\} - \mathbb{P}\{X_n \le t\}$$

may be written in terms of the density as

$$\begin{aligned}
&\int_{-\infty}^{0} \bigg[\int_{t}^{t-\varepsilon x_k} p_k(x_k, x_n)\,\mathrm{d}x_n\bigg]\mathrm{d}x_k - \int_{0}^{\infty} \bigg[\int_{t-\varepsilon x_k}^{t} p_k(x_k, x_n)\,\mathrm{d}x_n\bigg]\mathrm{d}x_k \\
&= \varepsilon \int_{-\infty}^{0} \bigg[\int_{0}^{-x_k} p_k(x_k, t+\varepsilon u)\,\mathrm{d}u\bigg]\mathrm{d}x_k - \varepsilon \int_{0}^{\infty} \bigg[\int_{-x_k}^{0} p_k(x_k, t+\varepsilon u)\,\mathrm{d}u\bigg]\mathrm{d}x_k \\
&= (\varepsilon + o(\varepsilon)) \int_{-\infty}^{0} (-x_k)\, p_k(x_k, t)\,\mathrm{d}x_k - (\varepsilon + o(\varepsilon)) \int_{0}^{\infty} x_k\, p_k(x_k, t)\,\mathrm{d}x_k.
\end{aligned}$$

Note that all the integrals are well-defined and finite, since log-concave densities decrease exponentially fast at infinity and, more precisely, they admit exponential upper bounds such as $C\,\mathrm{e}^{-c|x|}$. Thus, letting $\varepsilon \to 0$, we find that

$$\frac{\partial u_t(e_n)}{\partial \theta_k} = -\int_{-\infty}^{\infty} x_k\, p_k(x_k, t)\,\mathrm{d}x_k.$$

Since, by Fubini's theorem, $p_k(x_k, x_n) = \int_{\mathbb{R}^{n-2}} p(x)\,\frac{\mathrm{d}x}{\mathrm{d}x_k \mathrm{d}x_n}$, we arrive at the representation

$$\frac{\partial u_t(e_n)}{\partial \theta_k} = -\int_{\mathbb{R}^{n-1}} x_k\, p(x_1, \dots, x_{n-1}, t)\,\mathrm{d}x_1 \dots \mathrm{d}x_{n-1}.$$

Hence, in a vector form, given a unit vector l in $\mathbb{R}^{n-1}$,

$$\langle \nabla u_t(e_n), l\rangle = -\int_{\mathbb{R}^{n-1}} \langle l, y\rangle\, p(y, t)\, dy,$$

where we write $y = (x_1, \dots, x_{n-1})$. It remains to apply Lemma 14.9.3. □

Proof (of Proposition 14.9.1) By Proposition 9.5.1 (L^1-Poincaré-type inequality on the unit sphere), applied to the function $u_t(\theta) = F_\theta(t)$, we have

$$\mathbb{E}_\theta\, |F_\theta(t) - F(t)| \le \frac{\pi}{\sqrt{2n}}\, |\nabla_S\, u_t(\theta)| \le \frac{C}{\sqrt{n}}\, \mathrm{e}^{-c|t|}$$

for some absolute positive constants C and c, using Lemma 14.9.4 in the last step. Integrating this inequality over the variable t, we then arrive at

$$\mathbb{E}_\theta\, W(F_\theta, F) = \mathbb{E}_\theta \int_{-\infty}^{\infty} |F_\theta(t) - F(t)|\,\mathrm{d}t \le \frac{C'}{\sqrt{n}},$$

which is the first assertion of the proposition. To prove the second one, it remains to apply Proposition 14.2.1. □

Proof (of Proposition 14.9.2) Since every weighted sum $\langle X, \theta \rangle$ with $\theta \in \mathbb{S}^{n-1}$ represents a log-concave random variable with mean $a = 0$ and standard deviation $\sigma = 1$, we know from Proposition 2.6.4 that the density $q_\theta(x) = F_\theta'(x)$ of $\langle X, \theta \rangle$ satisfies, for all $x \in \mathbb{R}$,

$$q_\theta(x) \le C_1 e^{-|x|/4}, \tag{14.22}$$

where C_1 is some positive universal constant. Integrating this inequality over θ, we have a similar bound

$$q(x) \le C_1 e^{-|x|/4} \tag{14.23}$$

for the density q of F (although q does not need to be log-concave). Hence

$$\max\{F_\theta(-x), 1 - F_\theta(x)\} \le 4C_1 \, e^{-x/4}, \quad x \ge 0,$$

and similarly for F as well. Equivalently, we have

$$e^{x/8} \max\{F_\theta(-x), 1 - F_\theta(x)\} \le 4C_1 \, e^{-x/8}$$

and the same for F, so,

$$e^{|x|/8} \, |F_\theta(x) - F(x)| \le 4C_1 \, r \quad \text{for all } |x| \ge 8 \log(1/r), \tag{14.24}$$

where $r \in (0, \frac{1}{2}]$ is a fixed number.

Put

$$\Omega(x) = \{\theta \in \mathbb{S}^{n-1} : e^{c|x|} \, |F_\theta(x) - F(x)| \le Cr\}, \quad x \in \mathbb{R},$$

where c and C are constants from Lemma 14.9.3 (we may assume that $c \le 1/8$). By this lemma, one may apply the spherical concentration inequality (10.6) of Proposition 10.3.1 to the Lipschitz function $\theta \to \frac{1}{C} \, e^{c|x|} \, (F_\theta(x) - F(x))$, which gives

$$\mathfrak{s}_{n-1}(\Omega(x)) \ge 1 - 2 \, e^{-nr^2/4}.$$

We apply this inequality to a sequence of points $x = x_1, \dots, x_N$ increasing with step r and such that $x_1 = -x_N$ (the number $N = N_r$ will grow as $c' \, \frac{\log(1/r)}{r}$ when $r \downarrow 0$, where the constant c' has to be later specified). As a result, for the set $\Omega = \cap_{k=1}^N \Omega(x_k)$, we have

$$\mathfrak{s}_{n-1}(\Omega(t)) \ge 1 - 2N e^{-nr^2/4}. \tag{14.25}$$

Now, take $\theta \in \Omega$ so that, for all $1 \le k \le N$,

$$e^{c|x_k|} \, |F_\theta(x_k) - F(x_k)| \le Cr. \tag{14.26}$$

To extend this estimate to all points $x \in [x_1, x_N]$, assume that $x_k < x < x_{k+1}$ for some $k = 1, \dots, N-1$. Since $x_{k+1} - x_k = r$, from (14.22)–(14.24) and (14.26) it follows that

$$\begin{aligned} |F_\theta(x) - F(x)| &\le |F_\theta(x_k) - F(x_k)| + |F_\theta(x) - F_\theta(x_k)| + |F(x) - F(x_k)| \\ &\le Cre^{-c|x_k|} + r \sup_{x_k<y<x_{k+1}} q_\theta(y) + r \sup_{x_k<y<x_{k+1}} q(y) \\ &\le Cre^{c/2}\, \mathrm{e}^{-c|x|} + 8C_1 re^{1/8}\, \mathrm{e}^{-|x|/4}. \end{aligned}$$

Together with (14.24), we conclude that the inequality

$$\mathrm{e}^{c|x|}\, |F_\theta(x) - F(x)| \le C_2 r$$

holds true for all $x \in \mathbb{R}$ as long as $x_N \ge 8 \log(1/r)$ with some absolute constant $C_2 \ge 1$. By the construction, $(N-1)r = x_N - x_1 = 2x_N$, so, the condition

$$N - 1 \ge \frac{16}{r} \log(1/r)$$

has to be fulfilled. The smallest such number satisfies $N \le 2 + \frac{16}{r} \log(1/r) \le \frac{18}{r} \log(1/r)$, so that, by (14.25),

$$\mathfrak{s}_{n-1}\Big\{ \sup_x \mathrm{e}^{c|x|} |F_\theta(x) - F(x)| \ge C_2 r \Big\} \le \frac{36}{r} \log(1/r)\, \mathrm{e}^{-nr^2/4}.$$

If $nr^2 \ge 1$, we arrive at the desired relation

$$\mathfrak{s}_{n-1}\Big\{ \sup_t \mathrm{e}^{c|x|} |F_\theta(x) - F(x)| \ge C_2 r \Big\} \le 18\sqrt{n}\, \log n\, \mathrm{e}^{-nr^2/4}.$$

If $nr^2 \le 1$, it also holds, since the right-hand side is greater than 1. It remains to note that, for values $r > \frac{1}{2}$, this inequality is fulfilled automatically by using a suitable value of C_2, since $\sup_x \mathrm{e}^{|x|/4} |F_\theta(x) - F(x)|$ is bounded from above by a universal constant.

Thus, after a change of the variable, we get

$$\mathfrak{s}_{n-1}\big\{ \rho(F_\theta, F) \ge r \big\} \le C\sqrt{n}\, \log n\, \mathrm{e}^{-cnr^2}, \quad r > 0,$$

for some absolute constants $C > 0$ and $c > 0$. Using the property $\rho(F_\theta, F) \le 1$, this implies

$$\begin{aligned} \mathbb{E}_\theta\, \rho(F_\theta, F) &= \mathbb{E}_\theta\, \rho(F_\theta, F)\, 1_{\{\rho(F_\theta,F)<2\sqrt{\log n}/\sqrt{cn}\}} \\ &\quad + \mathbb{E}_\theta\, \rho(F_\theta, F)\, 1_{\{\rho(F_\theta,F)\ge 2\sqrt{\log n}/\sqrt{cn}\}} \\ &\le \frac{2\sqrt{\log n}}{\sqrt{cn}} + \mathfrak{s}_{n-1}\Big\{ \rho(F_\theta, F) \ge \frac{2\sqrt{\log n}}{\sqrt{cn}} \Big\} \\ &\le \frac{2\sqrt{\log n}}{\sqrt{cn}} + \frac{C\sqrt{n}\, \log n}{n^4}. \end{aligned}$$

Proposition 14.9.2 is proved. □

14.10 Remarks

Propositions 14.1.1 and 14.2.1 are proved in [31], where Lemma 14.1.2 is derived as well. Large deviations for the Lévy distance $L(F_\theta, F)$ were considered in [25], where the following result is obtained for isotropic random vectors X in $\mathbb{R}^n$: For any $r > 0$,

$$\mathfrak{s}_{n-1}\{L(F_\theta, F) \geq r\} \leq 4n^{3/8}\, e^{-nr^4/8}. \tag{14.27}$$

This bound makes sense for $r \geq c\,(\frac{\log n}{n})^{1/4}$. Hence, Proposition 14.4.2 and Corollary 14.4.3 provide an improvement of the rate from $n^{-1/4}$ to $n^{-1/3}$ (modulo logarithmically growing terms with respect to n). Proposition 14.9.2 was proved in [25] as well. The key Lemma 14.9.3 was actually discovered before in [4] in the case where X is uniformly distributed over a convex body (using arguments from Convex Geometry). Proposition 14.9.1, as a consequence of Lemma 14.9.3, was mentioned in [31].

The material of Sections 14.6 and 14.7 is based on [38].

The closeness of most F_θ to Φ may also be studied for stronger metrics, or in stronger senses. An important result in this direction is due to S. Sodin ([167], Theorem 3). Let X be a random vector in $\mathbb{R}^n$ with mean zero and ψ_1-marginals, that is, satisfying $\mathbb{E}\, e^{|\langle X, \theta\rangle|/\lambda} \leq 2$ for all $\theta \in \mathbb{S}^{n-1}$ with some constant $\lambda > 0$. Suppose that the typical distribution F is almost standard normal in the sense that, for some $\varepsilon \in (0, \varepsilon_0)$,

$$1 - \varepsilon \leq \frac{1 - F(x)}{1 - \Phi(x)} \leq 1 + \varepsilon, \quad 0 \leq x \leq T.$$

Then

$$\mathfrak{s}_{n-1}\Big\{\sup_{0 \leq x \leq T} \Big|\frac{1 - F_\theta(x)}{1 - \Phi(x)} - 1\Big| \geq 10\,\varepsilon\Big\} \leq \frac{CT^8}{n\varepsilon^4}\, \exp\Big\{- cn\varepsilon^2/T^6\Big\}.$$

Here the positive constants ε_0, c, C do not depend on n and the distribution of X.

Part V
Refined Bounds and Rates

Chapter 15
L^2 Expansions and Estimates

We now consider more precise assertions about fluctuations of the distribution functions $F_\theta(x)$ of the weighted sums $\langle X, \theta \rangle$ around their means $F(x)$. This is possible for the L^2-distance defined for arbitrary distribution functions F and G by

$$\omega(F, G) = \left(\int_{-\infty}^{\infty} (F(x) - G(x))^2 \, dx \right)^{1/2},$$

which is always finite as long as these distributions have finite first absolute moments. The main tool in this investigation will be the Plancherel theorem, which relates this distance to the difference of the corresponding characteristic functions f and g by virtue of the explicit formula

$$\omega(F, G) = \left(\frac{1}{2\pi} \int_{-\infty}^{\infty} \left| \frac{f(t) - g(t)}{t} \right|^2 dt \right)^{1/2}.$$

15.1 General Approximations

For an arbitrary random vector X in $\mathbb{R}^n$ and a weighted sum of its components, we therefore have the identity

$$\mathbb{E}_\theta \, \omega^2(F_\theta, F) = \frac{1}{2\pi} \int_{-\infty}^{\infty} \frac{\mathrm{Var}_\theta(f_\theta(t))}{t^2} \, dt, \tag{15.1}$$

where as before F_θ ($\theta \in \mathbb{S}^{n-1}$) denotes the distribution function of the weighted sum $\langle X, \theta \rangle$ with characteristic function f_θ, and where $F = \mathbb{E}_\theta F_\theta$ denotes the typical distribution with characteristic function $f(t) = \mathbb{E}_\theta f_\theta(t)$ ($\mathbb{E}_\theta$ and Var_θ mean integration and taking the variance over the uniform probability measure $\mathfrak{s}_{n-1}$ on the unit sphere). Recall that

$$\mathrm{Var}_\theta(f_\theta(t)) = \mathbb{E}_\theta \, |f_\theta(t) - f(t)|^2 = \mathbb{E}_\theta \, |f_\theta(t)|^2 - |f(t)|^2,$$

S. Bobkov et al., *Concentration and Gaussian Approximation for Randomized Sums*, Probability Theory and Stochastic Modelling 104,
https://doi.org/10.1007/978-3-031-31149-9_15

and

$$f(t) = \mathbb{E}_\theta \, \mathbb{E}\, e^{it\langle X,\theta\rangle} = \mathbb{E} J_n(t|X|),$$

where J_n is the characteristic function of the first coordinate θ_1 of θ under $\mathfrak{s}_{n-1}$. Similarly, if Y is an independent copy of X,

$$\begin{aligned}\mathbb{E}_\theta \, |f_\theta(t)|^2 &= \mathbb{E}_\theta \left(\mathbb{E}\, e^{it\langle X,\theta\rangle} \, \mathbb{E}\, e^{-it\langle Y,\theta\rangle}\right) \\ &= \mathbb{E}_\theta \, \mathbb{E}\, e^{it\langle X-Y,\theta\rangle} = \mathbb{E} J_n(t|X-Y|).\end{aligned}$$

Hence, the Plancherel formula (15.1) becomes

$$\mathbb{E}_\theta \, \omega^2(F_\theta, F) = \frac{1}{2\pi} \int_{-\infty}^{\infty} \left(\mathbb{E} J_n(t|X-Y|) - |\mathbb{E} J_n(t|X|)|^2\right) \frac{dt}{t^2}. \tag{15.2}$$

To further simplify this expression (asymptotically with respect to the growing dimension n), let us assume that $\mathbb{E}\,|X| \le b\sqrt{n}$ and apply the normal approximation

$$\left|J_n\big(t\sqrt{n}\big) - e^{-t^2/2}\right| \le \frac{C}{n} \, \min\{1, t^2\},$$

where C is an absolute constant (cf. Section 11.3). Note that, for any $\eta > 0$,

$$\delta(\eta) \equiv \int_{-\infty}^{\infty} \frac{\min\{1, t^2\eta^2\}}{t^2} \, dt = 2\int_0^{1/\eta} \eta^2 \, dt + 2\int_{1/\eta}^{\infty} \frac{1}{t^2} \, dt = 4\eta, \tag{15.3}$$

with $\delta(0) = 0$. Hence

$$\begin{aligned}\mathbb{E} \int_{-\infty}^{\infty} \left|J_n(t|X-Y|) - e^{-t^2|X-Y|^2/2n}\right| \frac{dt}{t^2} &\le \frac{C}{n} \, \mathbb{E} \int_{-\infty}^{\infty} \min\left\{1, \frac{t^2|X-Y|^2}{n}\right\} \frac{dt}{t^2} \\ &= \frac{4C}{n} \, \mathbb{E} \, \frac{|X-Y|}{\sqrt{n}} \le \frac{8Cb}{n}.\end{aligned}$$

Similarly,

$$\begin{aligned}\int_{-\infty}^{\infty} \left|(\mathbb{E} J_n(t|X|))^2 - (\mathbb{E}\, e^{-t^2|X|^2/2n})^2\right| \frac{dt}{t^2} &\le 2\,\mathbb{E} \int_{-\infty}^{\infty} \left|J_n(t|X|) - e^{-t^2|X|^2/2n}\right| \frac{dt}{t^2} \\ &\le \frac{2C}{n} \, \mathbb{E} \int_{-\infty}^{\infty} \min\left\{1, \frac{t^2|X|^2}{n}\right\} \frac{dt}{t^2} = \frac{8C}{n} \, \mathbb{E} \, \frac{|X|}{\sqrt{n}} \le \frac{8Cb}{n}.\end{aligned}$$

Using this bound in (15.2), we arrive at the following general approximation.

Proposition 15.1.1 *Let Y be an independent copy of the random vector X in $\mathbb{R}^n$ satisfying $\mathbb{E}\,|X| \le b\sqrt{n}$ a.s. Then*

$$\mathbb{E}_\theta \, \omega^2(F_\theta, F) = \frac{1}{2\pi} \int_{-\infty}^{\infty} \left(\mathbb{E}\, e^{-t^2|X-Y|^2/2n} - \left(\mathbb{E}\, e^{-t^2|X|^2/2n}\right)^2\right) \frac{dt}{t^2} + \frac{Cb}{n},$$

where C is a quantity bounded by an absolute constant (in absolute value).

In order to obtain more precise formulas with a remainder term of a smaller order than $1/n$, we need more precise approximations of the function J_n, such as

$$J_n(t\sqrt{n}) = \left(1 - \frac{t^4}{4n}\right) e^{-t^2/2} + O\Big(n^{-2} \min\{1, t^2\}\Big), \tag{15.4}$$

which holds uniformly on the real line with a universal constant in O (note that the remainder term can be made even smaller for small values of t, cf. Corollary 11.3.3). With this approximation, as before, we have

$$\mathbb{E} \int_{-\infty}^{\infty} \left| J_n(t|X-Y|) - \left(1 - \frac{t^4|X-Y|^4}{4n^3}\right) e^{-\frac{t^2|X-Y|^2}{2n}} \right| \frac{dt}{t^2} \leq \frac{8Cb}{n^2}.$$

Next we note that the main term $(1 - \frac{t^4}{4n})\, e^{-t^2/2}$ in (15.4) is bounded by an absolute constant, which implies that

$$\begin{aligned} J_n(t\sqrt{n})\, J_n(s\sqrt{n}) &= \left(1 - \frac{t^4}{4n}\right)\left(1 - \frac{s^4}{4n}\right) e^{-(t^2+s^2)/2} + O\Big(n^{-2} \min\{1, t^2 + s^2\}\Big) \\ &= \left(1 - \frac{t^4 + s^4}{4n}\right) e^{-(t^2+s^2)/2} + O\Big(n^{-2} \min\{1, t^2 + s^2\}\Big). \end{aligned}$$

Hence

$$\begin{aligned} |\mathbb{E} J_n(t|X|)|^2 = \mathbb{E}\, J_n(t|X|)\, J_n(t|Y|) &= \mathbb{E}\left(1 - \frac{t^4\,(|X|^4 + |Y|^4)}{4n^3}\right) e^{-\frac{t^2\,(|X|^2+|Y|^2)}{2n}} \\ &\quad + O\Big(n^{-2}\, \mathbb{E} \min\Big\{1, \frac{t^2\,(|X|^2 + |Y|^2)}{n}\Big\}\Big). \end{aligned}$$

Integrating over t according to (15.2) and applying (15.3), we obtain the following relation. As before, Y denotes an independent copy of X.

Proposition 15.1.2 *If X is a random vector in $\mathbb{R}^n$ satisfying $\mathbb{E}\,|X| \leq b\sqrt{n}$, then with some quantity C which is bounded by an absolute constant, we have*

$$\mathbb{E}_\theta\, \omega^2(F_\theta, F) = \frac{1}{2\pi} I + \frac{Cb}{n^2},$$

where

$$I = \int_{-\infty}^{\infty} \left[\mathbb{E}\left(1 - \frac{t^4|X-Y|^4}{4n^3}\right) e^{-\frac{t^2|X-Y|^2}{2n}} - \mathbb{E}\left(1 - \frac{t^4\,(|X|^4 + |Y|^4)}{4n^3}\right) e^{-\frac{t^2\,(|X|^2+|Y|^2)}{2n}}\right] \frac{dt}{t^2}.$$

15.2 Bounds for L^2-distance With a Standard Rate

In this section our aim is to show that the mean distances

$$\mathbb{E}_\theta\, \omega^2(F_\theta, F) \quad \text{and} \quad \mathbb{E}_\theta\, \omega^2(F_\theta, \Phi)$$

are of order at most $O(1/n)$ with involved constants depending only on

$$m_2^2 = m_2^2(X) = \frac{1}{n}\, \mathbb{E}\, \langle X, Y\rangle^2 \quad \text{and} \quad \sigma_4^2 = \sigma_4^2(X) = \frac{1}{n}\, \mathrm{Var}(|X|^2)$$

(where Y is an independent copy of X). In particular, we do not need to involve the functional M_4.

Proposition 15.2.1 *Let X be a random vector in $\mathbb{R}^n$ with $\mathbb{E}X = a$ and $\mathbb{E}\,|X|^2 = n$. For some absolute constant $c \geq 0$,*

$$\mathbb{E}_\theta\, \omega^2(F_\theta, F) \leq \frac{cA}{n}, \tag{15.5}$$

where $A = |a|^2 + m_2^2 + \sigma_4^2$. In particular, if X is isotropic, then

$$\mathbb{E}_\theta\, \omega^2(F_\theta, F) \leq \frac{c\,(1 + \sigma_4^2)}{n}.$$

Similar inequalities continue to hold with the standard normal distribution function Φ in place of F.

If X is isotropic, then $m_2 = 1$, while $|a| \leq 1$ (by Bessel's inequality). Hence, both characteristics m_2 and a may be removed from the parameter A in this case (more precisely, one may put $A = 1 + \sigma_4^2$).

If X is not necessarily isotropic, one may involve the functional M_2 by using $|a| \leq M_2$ and $m_2 \leq M_2^2$ (Propositions 1.4.2 and 1.2.2). Hence, in (15.5) one may take $A = M_2^4 + \sigma_4^2$.

In general, it may however happen that m_2 and σ_4 are bounded, while $|a|$ is large. The example in Remark 15.2.2 shows that this parameter cannot be removed.

Proof To derive (15.5), introduce the random variable

$$\rho^2 = \frac{|X - Y|^2}{2n} \quad (\rho \geq 0).$$

By Jensen's inequality, $\mathbb{E}\, e^{-t^2|X|^2/2n} \geq e^{-t^2/2}$, so that

$$\mathbb{E}\, e^{-t^2|X-Y|^2/2n} - \big(\mathbb{E}\, e^{-t^2|X|^2/2n}\big)^2 \leq \mathbb{E}\, e^{-t^2|X-Y|^2/2n} - e^{-t^2}.$$

By Proposition 15.1.1,

$$\mathbb{E}_\theta\, \omega^2(F_\theta, F) \le \frac{1}{2\pi}\, \mathbb{E} \int_{-\infty}^{\infty} \frac{e^{-\rho^2 t^2} - e^{-t^2}}{t^2}\, dt + \frac{c}{n}$$

for some absolute constant $c > 0$. As we know, the above integral may be evaluated according to the identity of Lemma 12.5.2,

$$\int_{-\infty}^{\infty} \frac{e^{-\alpha t^2} - e^{-\alpha_0 t^2}}{t^2}\, dt = 2\sqrt{\pi}\left(\sqrt{\alpha_0} - \sqrt{\alpha}\right), \quad \alpha, \alpha_0 \ge 0. \tag{15.6}$$

Applying it with $\alpha_0 = 1$ and $\alpha = \rho^2$, we arrive at the bound

$$\sqrt{\pi}\, \mathbb{E}_\theta\, \omega^2(F_\theta, F) \le (1 - \mathbb{E}\rho) + \frac{c}{n}$$

for some absolute constant c. Furthermore, let us recall the bound of Lemma 1.5.4,

$$1 - \mathbb{E}\rho \le \frac{1}{2}\, \mathbb{E}\,(1-\rho^2) + \mathbb{E}\,(1-\rho^2)^2.$$

Then we get

$$\sqrt{\pi}\, \mathbb{E}_\theta\, \omega^2(F_\theta, F) \le \frac{1}{2}\, \mathbb{E}\,(1-\rho^2) + \mathbb{E}\,(1-\rho^2)^2 + \frac{c}{n}. \tag{15.7}$$

Since $|X-Y|^2 = |X|^2 + |Y|^2 - 2\langle X, Y\rangle$, we may write

$$1 - \rho^2 = \frac{n - |X|^2}{2n} + \frac{n - |Y|^2}{2n} + \frac{\langle X, Y\rangle}{n},$$

which implies that

$$1 - \mathbb{E}\rho^2 = \frac{1}{n}\, \mathbb{E}\,\langle X, Y\rangle = \frac{1}{n}\, |a|^2, \quad a = \mathbb{E}X.$$

In addition,

$$(1-\rho^2)^2 \le 2\left(\frac{n - |X|^2}{2n} + \frac{n - |Y|^2}{2n}\right)^2 + 2\, \frac{\langle X, Y\rangle^2}{n^2},$$

which gives

$$\mathbb{E}\,(1-\rho^2)^2 \le \frac{\mathrm{Var}(|X|^2)}{n^2} + 2\, \frac{\mathbb{E}\,\langle X, Y\rangle^2}{n^2} = \frac{\sigma_4^2 + 2m_2^2}{n}.$$

It remains to apply these bounds in (15.7), which leads to

$$\sqrt{\pi}\, \mathbb{E}_\theta\, \omega^2(F_\theta, F) \le \frac{\frac{1}{2}\, |a|^2 + \sigma_4^2 + 2m_2^2 + c}{n}.$$

Due to the assumption $\mathbb{E}\,|X|^2 = n$, we necessarily have $m_2 \geq 1$ (cf. Proposition 1.1.3), and the inequality (15.5) follows.

Finally, in order to replace F with Φ, let us recall that $\omega(F,\Phi) \leq c\,(1+\sigma_2)/\sqrt{n}$ up to an absolute positive constant c (Proposition 12.5.1). Squaring this inequality and using $\sigma_2 \leq \sigma_4$, we get $\omega^2(F,\Phi) \leq 2c^2\,(1+\sigma_4^2)/n$. Proposition 15.2.1 is proved. □

Returning to the last step, let us explain how to estimate $\omega^2(F,\Phi)$ with arguments developed in the previous and current sections. By the Plancherel formula,

$$\omega^2(F,\Phi) = \frac{1}{2\pi}\int_{-\infty}^{\infty} |\,\mathbb{E}J_n(t|X|) - e^{-t^2/2}|^2\,\frac{dt}{t^2}. \tag{15.8}$$

Applying the approximation $|J_n(t\sqrt{n}) - e^{-t^2/2}| \leq \frac{c}{n}\min\{1,|t|\}$ together with the identity (15.3), we conclude that

$$\begin{aligned}\int_{-\infty}^{\infty} \Big|\mathbb{E}J_n(t|X|) - \mathbb{E}\,e^{-t^2|X|^2/2n}\Big|^2\,\frac{dt}{t^2} &\leq \mathbb{E}\int_{-\infty}^{\infty} \Big|J_n(t|X|) - e^{-t^2|X|^2/2n}\Big|^2\,\frac{dt}{t^2}\\ &\leq \frac{2c^2}{n^2}\,\mathbb{E}\int_{-\infty}^{\infty}\min\Big\{1,\frac{t^2|X|^2}{n}\Big\}\,\frac{dt}{t^2} = \frac{8c^2}{n^2}\,\mathbb{E}\,\frac{|X|}{\sqrt{n}} \leq \frac{8c^2}{n^2}.\end{aligned}$$

Hence, up to an error of order at most $O(n^{-2})$, one may replace the characteristic function $f(t) = \mathbb{E}J_n(t|X|)$ of F with the characteristic function $g(t) = \mathbb{E}\,e^{-t^2|X|^2/2n}$ of the corresponding mixture of centered Gaussian mixtures on the real line. More precisely, subtracting $g(t)$ from $f(t)$ and adding in (15.8), it follows that

$$\omega^2(F,\Phi) \leq \frac{1}{\pi}\int_{-\infty}^{\infty} \frac{(\mathbb{E}\,e^{-t^2|X|^2/2n} - e^{-t^2/2})^2}{t^2}\,dt + \frac{8c^2}{\pi n^2}.$$

Let us remove the square and apply (15.6) with $\alpha = \frac{1}{2n}\,|X|^2$ and $\alpha_0 = \frac{1}{2}$. Again, by Lemma 1.5.4 with $\xi = \frac{1}{\sqrt{n}}\,|X|$, we then see that the above integral does not exceed

$$\begin{aligned}\int_{-\infty}^{\infty} \frac{\mathbb{E}\,e^{-t^2|X|^2/2n} - e^{-t^2/2}}{t^2}\,dt &= \sqrt{2\pi}\,\mathbb{E}\,(1 - |X|/\sqrt{n})\\ &\leq \sqrt{2\pi}\,\mathrm{Var}\Big(\frac{1}{n}\,|X|^2\Big) = \frac{\sqrt{2\pi}}{n}\,\sigma_4^2.\end{aligned}$$

Thus,

$$\omega^2(F,\Phi) \leq \frac{\sqrt{2}\,\sigma_4^2}{\sqrt{\pi}\,n} + O\Big(\frac{1}{n^2}\Big).$$

Remark 15.2.2 To see that the presence of the parameter a in (15.5) is necessary under the condition $\mathbb{E}\,|X|^2 = n$, let us return to the example of the Gaussian random vector described in Remark 14.7.2,

$$X = \alpha Z + \lambda e_n \quad \text{with } 1 \leq \lambda \leq n^{1/4},\ \ \alpha^2(n-1) + \lambda^2 = n,$$

where Z is a standard normal random vector in $\mathbb{R}^{n-1}$. As was noticed, $\sigma_4^2 \le 2$ and $m_2^2 \le m_4^2 \le 9$, while the mean $a = \mathbb{E}X$ has length $|a| = \lambda \ge 1$. Hence, the inequality (15.5), being stated for the normal distribution function in place of F, simplifies to

$$\mathbb{E}_\theta\, \omega^2(F_\theta, \Phi) \le \frac{c\lambda^2}{n}.$$

Let us show that this bound may be reversed up to an absolute factor. Note that, for any unit vector θ, the linear form $S_\theta = \langle X, \theta\rangle$ has a normal distribution on the line with mean $\lambda\theta_n$ and variance $\alpha^2(1-\theta_n^2)$. Consider the normal distribution function $\Phi_{\mu,\sigma^2}(x) = \Phi(\frac{x-\mu}{\sigma})$ with parameters $0 \le \mu \le 1$ and $\frac{1}{2} \le \sigma^2 \le 1$ ($\sigma > 0$). If $x \le \frac{\mu}{1+\sigma}$, then $\frac{x-\mu}{\sigma} \le x$, while $\frac{|x-\mu|}{\sigma} \ge |x|$. Hence, the interval $[\frac{x-\mu}{\sigma}, x]$ has length

$$x - \frac{x-\mu}{\sigma} = \frac{\mu - (1-\sigma)x}{\sigma} \ge \frac{2}{1+\sigma}\,\mu \ge \mu,$$

and the standard normal density $\varphi(y)$ attains a minimum on it at the left endpoint. It follows that

$$|\Phi_{\mu,\sigma^2}(x) - \Phi(x)| = \int_{\frac{x-\mu}{\sigma}}^{x} \varphi(y)\, dy \ge \mu\, \varphi\Big(\frac{x-\mu}{\sigma}\Big),$$

so that

$$\begin{aligned}\omega^2(\Phi_{\mu,\sigma^2}, \Phi) &\ge \mu^2 \int_{-\infty}^{\frac{\mu}{1+\sigma}} \varphi\Big(\frac{x-\mu}{\sigma}\Big)^2 dx \\ &= \frac{\sigma\mu^2}{2\pi} \int_{-\infty}^{-\frac{\mu}{1+\sigma}} e^{-y^2}\, dy \ge \frac{\sigma\mu^2}{2\pi} \int_{-\infty}^{-1} e^{-y^2}\, dy \ge c\mu^2.\end{aligned}$$

In our case, $\mu = \mathbb{E}S_\theta = \lambda\theta_n$ and $\sigma^2 = \mathrm{Var}(S_\theta) = \alpha^2(1-\theta_n^2)$. Since $\lambda \le n^{1/4}$ and $\alpha^2 = \frac{n-\lambda^2}{n-1}$, we have that, for all n large enough, $0 \le \mu \le 1$ and $\sigma^2 \ge \frac{1}{2}$ on the set $\Omega_n = \{\theta \in \mathbb{S}^{n-1} : 0 \le \theta_n < \frac{\log n}{\sqrt{n}}\}$. It then follows that

$$\mathbb{E}_\theta\, \omega^2(F_\theta, \Phi) \ge c\lambda^2\, \mathbb{E}_\theta\, \theta_n^2\, 1_{\{\theta\in\Omega_n\}} \ge \frac{c'\lambda^2}{n}.$$

15.3 Expansion With Error of Order n^{-1}

Let us now sharpen Proposition 15.2.1 by using a more precise asymptotic expansion for the functions J_n. With this aim, we strengthen the condition $\mathbb{E}\,|X|^2 = n$ to the pointwise assumption $|X|^2 = n$ a.s. In this case,

$$\rho^2 = \frac{|X-Y|^2}{2n} = 1 - \xi, \quad \xi = \frac{\langle X, Y\rangle}{n},$$

where Y is an independent copy of X. Thus, ξ may take values in $[-1, 1]$, and in addition, $\mathbb{E}\xi^2 = \frac{1}{n}\, m_2^2$.

Proposition 15.3.1 *Let X be a random vector in $\mathbb{R}^n$ such that $|X|^2 = n$ a.s. Then*

$$\sqrt{\pi}\, \mathbb{E}_\theta\, \omega^2(F_\theta, F) = \Big(1 + \frac{1}{4n}\Big)\, \mathbb{E}\, \Big(1 - (1-\xi)^{1/2}\Big) - \frac{1}{8n} + O(n^{-2}).$$

A similar expression holds true for Φ in place of the typical distribution F.

Proof Let us return to the identity of Proposition 15.1.2,

$$\mathbb{E}_\theta\, \omega^2(F_\theta, F) = \frac{1}{2\pi}\, I + O(n^{-2}),$$

in which the integral I may be written in terms of ξ as

$$I = \mathbb{E} \int_{-\infty}^{\infty} \bigg[\Big(1 - \frac{(1-\xi)^2\, t^4}{n}\Big)\, e^{-(1-\xi)\, t^2} - \Big(1 - \frac{t^4}{2n}\Big)\, e^{-t^2}\bigg]\, \frac{dt}{t^2}.$$

To evaluate this integral, consider the function

$$\psi(\alpha) = \int_{-\infty}^{\infty} \bigg[\Big(1 - \frac{\alpha^2 t^4}{n}\Big)\, e^{-\alpha t^2} - \Big(1 - \frac{t^4}{2n}\Big)\, e^{-t^2}\bigg]\, \frac{dt}{t^2}, \qquad \alpha \geq 0.$$

We have

$$\begin{aligned}
\psi'(\alpha) &= -\int_{-\infty}^{\infty} \Big(1 + \frac{2\alpha t^2 - \alpha^2 t^4}{n}\Big)\, e^{-\alpha t^2}\, dt \\
&= -\frac{1}{\sqrt{2\alpha}} \int_{-\infty}^{\infty} \Big(1 + \frac{s^2 - \frac{1}{4}\, s^4}{n}\Big)\, e^{-s^2/2}\, ds = -\sqrt{\pi}\, \Big(1 + \frac{1}{4n}\Big)\, \alpha^{-1/2},
\end{aligned}$$

and after integration

$$\psi(\alpha) = \psi(1) - 2\sqrt{\pi}\, \Big(1 + \frac{1}{4n}\Big)\, (\alpha^{1/2} - 1).$$

Also,

$$\psi(1) = -\int_{-\infty}^{\infty} \frac{t^2}{2n}\, e^{-t^2}\, dt = -\frac{\sqrt{\pi}}{4n},$$

so

$$\psi(\alpha) = 2\sqrt{\pi}\, \bigg[\Big(1 + \frac{1}{4n}\Big)\, (1 - \alpha^{1/2}) - \frac{1}{8n}\bigg].$$

It remains to insert the value $\alpha = 1 - \xi$, since $I = \psi(1 - \xi)$.

The last assertion follows from the asymptotic expansion for $\omega^2(F, \Phi)$ given in Proposition 12.5.3. Proposition 15.3.1 is thus proved. □

Note that in the case $|X| = \sqrt{n}$ a.s., the random variable $\rho^2 = \frac{|X-Y|^2}{2n}$ satisfies $\rho^2 \leq 2$ a.s., while $\mathbb{E}\rho^2 = 1 - \frac{1}{n}\,|\mathbb{E}X|^2 \leq 1$. Hence, we may apply the lower bound of Lemma 1.5.4, which gives

$$1 - \mathbb{E}\rho \geq \frac{1}{2}\,\mathbb{E}\,(1-\rho^2) + \frac{1}{16}\,\mathbb{E}\,(1-\rho^2)^2 \geq \frac{1}{16\,n^2}\,\mathbb{E}\,\langle X, Y\rangle^2.$$

Applying the asymptotic expansion of Proposition 15.3.1 (with Φ in place of F), this leads to the following lower bound on the L^2-distance on average.

Corollary 15.3.2 *Let X be a random vector in $\mathbb{R}^n$ such that $|X|^2 = n$ a.s. Then*

$$\sqrt{\pi}\,\mathbb{E}_\theta\,\omega^2(F_\theta, \Phi) \geq \frac{1}{16\,n^2}\,\mathbb{E}\,\langle X, Y\rangle^2 - \frac{1}{8n} + O(n^{-2}),$$

where Y is an independent copy of X.

This lower bound shows the importance of isotropy for obtaining bounds on $\mathbb{E}_\theta\,\omega^2(F_\theta, \Phi)$ that are better than $\frac{1}{n}$. Note that the boundedness of the moment M_2 does not guarantee better rates. In the isotropic case, we have $\mathbb{E}\,\langle X, Y\rangle^2 = n$, but if, for example $\mathbb{E}\,\langle X, Y\rangle^2 \geq 4n$, Corollary 15.3.2 yields

$$\sqrt{\pi}\,\mathbb{E}_\theta\,\omega^2(F_\theta, \Phi) \geq \frac{1}{8n} - \frac{c}{n^2}$$

for some absolute constant c.

Example 15.3.3 Let $\xi_1, \dots, \xi_{n/4}$ ($n = 4, 8, 12, \dots$) be independent Bernoulli random variables taking the values ± 1 with probability $\frac{1}{2}$, and let $X = (X_1, \dots, X_n)$ be the random vector with components $X_k = 2\xi_k$ for $k \leq \frac{n}{4}$ and $X_k = 0$ for $k > \frac{n}{4}$. In this case, $|X|^2 = n$ and $M_2(X) = 2$. Also,

$$\mathbb{E}\,\langle X, Y\rangle^2 = \sum_{k=1}^n (\mathbb{E}X_k^2)^2 = 4n.$$

15.4 Two-sided Bounds

We will now simplify the statement of Proposition 15.3.1 by involving the 4-th moment condition and the (non-negative) moment functionals

$$\alpha_p = \frac{1}{\sqrt{n}}\Big(\mathbb{E}\,\langle X, Y\rangle^p\Big)^{1/p},$$

where Y is an independent copy of X. We considered these functionals in Section 1.4 for positive integers p, and now need them for the particular values $p = 3$ and $p = 4$. Note that $\alpha_4 = m_4$ and $\alpha_3 \leq m_3$.

In the case $|X|^2 = n$ a.s., m_4 is finite, but in general it may be of order 1, as well as being large in comparison with the dimension n. On the other hand, if additionally X is isotropic, then using $\langle X,Y\rangle^4 \le n^2 \langle X,Y\rangle^2$, we have $\mathbb{E}\,\langle X,Y\rangle^4 \le n^3$, i.e., $m_4 \le n^{1/4}$.

Proposition 15.4.1 *Let X be an isotropic random vector in $\mathbb{R}^n$ such that $|X|^2 = n$ a.s. and $\mathbb{E}X = 0$. Then*

$$\sqrt{\pi}\,\mathbb{E}_\theta\,\omega^2(F_\theta,F) = \frac{1}{16\,n^3}\,\mathbb{E}\,\langle X,Y\rangle^3 + \frac{c}{n^4}\,\mathbb{E}\,\langle X,Y\rangle^4 + O(n^{-2})$$

with $\frac{1}{100} \le c \le 3$. A similar expansion also holds for F replaced with the normal distribution function Φ. In particular,

$$\mathbb{E}_\theta\,\omega^2(F_\theta,\Phi) = \frac{c_1}{n^{3/2}}\,\alpha_3 + \frac{c_2}{n^2}\,\alpha_4 + O(n^{-2})$$

with $c_1 = \frac{1}{16\sqrt{\pi}}$ and $\frac{1}{200} < c_2 < 2$.

Since $\alpha_4 \ge 1$, we immediately obtain:

Corollary 15.4.2 *Under the same assumptions,*

$$\mathbb{E}_\theta\,\omega^2(F_\theta,F) \le \frac{1}{16\sqrt{\pi}\,n^3}\,\mathbb{E}\,\langle X,Y\rangle^3 + \frac{c}{n^4}\,\mathbb{E}\,\langle X,Y\rangle^4, \tag{15.9}$$

where c is an absolute constant. A similar bound also holds for F replaced with the normal distribution function Φ.

In view of the pointwise bound $\langle X,Y\rangle^3 \le n\,\langle X,Y\rangle^2$, we have $\mathbb{E}\,\langle X,Y\rangle^3 \le n^2$. Therefore, the inequality (15.9) implies

$$\mathbb{E}_\theta\,\omega^2(F_\theta,F) \le \frac{c}{n}.$$

But this has been already obtained in the bound (15.5) of Proposition 15.2.1 with $\sigma_4 = 0$ and $M_2 = 1$ (and where the assumption about the mean was not used).

If the distribution of X is symmetric about the origin, then the cubic term α_3 is vanishing, and the upper bound (15.9) is further simplified. In fact, since $\alpha_3 \ge 0$ in general, in this case a similar lower bound holds as well (regardless of the symmetry property).

Corollary 15.4.3 *Let X be an isotropic random vector in $\mathbb{R}^n$ with mean zero and such that $|X|^2 = n$ a.s. Then for some positive absolute constants $c_j > 0$,*

$$\mathbb{E}_\theta\,\omega^2(F_\theta,F) \ge \frac{c_1}{n^4}\,\mathbb{E}\,\langle X,Y\rangle^4 - \frac{c_2}{n^2}.$$

Moreover, if the distribution of X is symmetric about the origin, then

$$\mathbb{E}_\theta\, \omega^2(F_\theta, F) \le \frac{c_3}{n^4}\, \mathbb{E}\,\langle X, Y\rangle^4.$$

Similar bounds also hold for F replaced with Φ.

The last lower bound makes sense when $\mathbb{E}\,\langle X, Y\rangle^4$ has an order larger than n^2. In this case we obtain two-sided bounds, so that

$$\mathbb{E}_\theta\, \omega^2(F_\theta, F) \sim \frac{1}{n^4}\, \mathbb{E}\,\langle X, Y\rangle^4.$$

Let us also recall that, by Proposition 1.4.2 with $p = 4$, $(\mathbb{E}\,\langle X, Y\rangle^4)^{1/4} \le \sqrt{n}\, M_4(X)^2$. Hence the upper bound of Corollary 14.4.3 implies

$$\mathbb{E}_\theta\, \omega^2(F_\theta, F) \le \frac{cM_4^8}{n^2},$$

where the right hand-side is thus of order n^{-2} when M_4 is bounded.

For the proof of Proposition 15.4.1, first let us further develop the expansion for the function $w(\varepsilon) = 1 - \sqrt{1-\varepsilon}$, which was considered in the proof of Lemma 1.5.4, up to the term ε^4.

Lemma 15.4.4 *For all* $|\varepsilon| \le 1$,

$$1 - \sqrt{1-\varepsilon} \le \frac{1}{2}\,\varepsilon + \frac{1}{8}\,\varepsilon^2 + \frac{1}{16}\,\varepsilon^3 + 3\varepsilon^4.$$

In addition,

$$1 - \sqrt{1-\varepsilon} \ge \frac{1}{2}\,\varepsilon + \frac{1}{8}\,\varepsilon^2 + \frac{1}{16}\,\varepsilon^3 + 0.01\,\varepsilon^4.$$

Proof By Taylor's formula for the function $w(\varepsilon)$ around zero on the half-axis $\varepsilon < 1$, we have

$$w(\varepsilon) = \frac{1}{2}\,\varepsilon + \frac{1}{8}\,\varepsilon^2 + \frac{1}{16}\,\varepsilon^3 + \frac{5}{128}\,\varepsilon^4 + \frac{w^{(5)}(\varepsilon_1)}{120}\,\varepsilon^5$$

for some point ε_1 between zero and ε. Since $w^{(5)}(\varepsilon) = \frac{105}{32}\,(1-\varepsilon)^{-9/2} \ge 0$, we have

$$w(\varepsilon) \le \frac{1}{2}\,\varepsilon + \frac{1}{8}\,\varepsilon^2 + \frac{1}{16}\,\varepsilon^3 + \frac{5}{128}\,\varepsilon^4, \quad \varepsilon \le 0.$$

Also, $w^{(5)}(\varepsilon) \le \frac{105}{32}\, 3^{9/2} < 461$ for $0 \le \varepsilon \le \frac{2}{3}$, so, in this interval

$$\frac{5}{128}\,\varepsilon^4 + \frac{w^{(5)}(\varepsilon_1)}{120}\,\varepsilon^5 \le 3\varepsilon^4.$$

Thus, in both cases,

$$w(\varepsilon) \le \frac{1}{2}\,\varepsilon + \frac{1}{8}\,\varepsilon^2 + \frac{1}{16}\,\varepsilon^3 + 3\varepsilon^4.$$

This inequality also holds for the remaining values $\frac{2}{3} \le \varepsilon \le 1$, since the right-hand side is greater than or equal to 1, while $w(\varepsilon) \le 1$. The upper bound of the lemma is thus proved. Now, from Taylor's formula we also get that

$$1 - \sqrt{1-\varepsilon} \ge \frac{1}{2}\,\varepsilon + \frac{1}{8}\,\varepsilon^2 + \frac{1}{16}\,\varepsilon^3 + \frac{5}{128}\,\varepsilon^4, \qquad \varepsilon \ge 0.$$

In addition, if $-1 \le \varepsilon \le 0$, then $w^{(5)}(\varepsilon) \le \frac{105}{32}$, so

$$\begin{aligned} 1 - \sqrt{1-\varepsilon} &= \frac{1}{2}\,\varepsilon + \frac{1}{8}\,\varepsilon^2 + \frac{1}{16}\,\varepsilon^3 + \frac{5}{128}\,\varepsilon^4 \left(1 + \frac{w^{(5)}(\varepsilon_1)}{120}\,\varepsilon\right) \\ &\ge \frac{1}{2}\,\varepsilon + \frac{1}{8}\,\varepsilon^2 + \frac{1}{16}\,\varepsilon^3 + \varepsilon^4 \left(\frac{5}{128} - \frac{\frac{105}{32}}{120}\right) \ge \frac{1}{2}\,\varepsilon + \frac{1}{8}\,\varepsilon^2 + \frac{1}{16}\,\varepsilon^3 + 0.01\,\varepsilon^4. \end{aligned}$$

Lemma 15.4.4 is proved. □

Proof (of Proposition 15.4.1) Using the lemma with $\varepsilon = \xi = \frac{\langle X,Y\rangle}{n}$ and applying Proposition 15.3.1, we get an asymptotic representation

$$\sqrt{\pi}\,\mathbb{E}_\theta\,\omega^2(F_\theta, F) = \left(1 + \frac{1}{4n}\right)\left(\frac{1}{8}\,\mathbb{E}\xi^2 + \frac{1}{16}\,\mathbb{E}\xi^3 + c\,\mathbb{E}\xi^4\right) - \frac{1}{8n} + O(n^{-2})$$

with $0.01 \le c \le 3$. If additionally X is isotropic, then $\mathbb{E}\,\langle X, Y\rangle^2 = n$, i.e. $\mathbb{E}\xi^2 = \frac{1}{n}$, and the representation is simplified to

$$\sqrt{\pi}\,\mathbb{E}_\theta\,\omega^2(F_\theta, F) = \left(1 + \frac{1}{4n}\right)\left(\frac{1}{16}\,\mathbb{E}\xi^3 + c\,\mathbb{E}\xi^4\right) + O(n^{-2}),$$

thus removing the term of order $1/n$. As was already noted, $\mathbb{E}\xi^3 \le \frac{1}{n}$ and $\mathbb{E}\xi^4 \le \frac{1}{n}$. Hence, the above is further simplified to

$$\sqrt{\pi}\,\mathbb{E}_\theta\,\omega^2(F_\theta, F) = \frac{1}{16}\,\mathbb{E}\xi^3 + c\,\mathbb{E}\xi^4 + O(n^{-2}),$$

which is exactly the expansion of Proposition 15.4.1. □

15.5 Asymptotic Formulas in the General Case

Let us now return to the Plancherel formula in the general case,

$$\mathbb{E}_\theta\,\omega^2(F_\theta, F) = \frac{1}{2\pi}\int_{-\infty}^{\infty} \frac{\mathbb{E}J_n(t|X-Y|) - (\mathbb{E}J_n(t|X|))^2}{t^2}\,dt.$$

If $\mathbb{E}\,|X| \le b\sqrt{n}$, then, according to Proposition 15.1.2,

$$\mathbb{E}_\theta\,\omega^2(F_\theta, F) = \frac{1}{2\pi} I + \frac{Cb}{n^2},$$

where C is bounded by an absolute constant and

$$I = \mathbb{E} \int_{-\infty}^{\infty} \left[\left(1 - \frac{t^4 |X-Y|^4}{4n^3}\right) e^{-\frac{t^2 |X-Y|^2}{2n}} - \left(1 - \frac{t^4\,(|X|^4 + |Y|^4)}{4n^3}\right) e^{-\frac{t^2\,(|X|^2+|Y|^2)}{2n}} \right] \frac{dt}{t^2}.$$

To evaluate such integrals, consider the functions of the form

$$\psi_r(\alpha) = \frac{1}{\sqrt{2\pi}} \int_{-\infty}^{\infty} \left((1 - rt^4)\, e^{-\alpha t^2/2} - e^{-t^2/2} \right) \frac{dt}{t^2} \qquad (\alpha > 0,\ r \in \mathbb{R}).$$

Clearly,

$$\psi_r(1) = -\frac{1}{\sqrt{2\pi}} \int_{-\infty}^{\infty} rt^2\, e^{-t^2/2}\, dt = -r$$

and

$$\psi_r'(\alpha) = -\frac{1}{2}\, \frac{1}{\sqrt{2\pi}} \int_{-\infty}^{\infty} (1 - rt^4)\, e^{-\alpha t^2/2}\, dt = -\frac{1}{2\sqrt{\alpha}} \left(1 - \frac{3r}{\alpha^2}\right).$$

Hence

$$\psi_r(\alpha) - \psi_r(1) = \int_1^{\alpha} \left(-\frac{1}{2}\, z^{-1/2} + \frac{3r}{2}\, z^{-5/2} \right) dz = (1 + r) - (\alpha^{1/2} + r\alpha^{-3/2}),$$

and we get

$$\psi_r(\alpha) = 1 - (\alpha^{1/2} + r\alpha^{-3/2}).$$

Here, when α and r approach zero such that $r = O(\alpha^2)$, in the limit it follows that $\psi_0(0) = 1$. Note also that for $r = 0$ the above formula returns us to the identity of Lemma 12.5.2 in the particular case $\alpha_0 = 1/2$ and with $\alpha/2$ in place of α. From this,

$$\frac{1}{\sqrt{2\pi}}\, I = \mathbb{E}\,(\psi_{r_1}(\alpha_1) - \psi_{r_2}(\alpha_2)) = \mathbb{E}\,(\alpha_2^{1/2} + r_2 \alpha_2^{-3/2}) - \mathbb{E}\,(\alpha_1^{1/2} + r_1 \alpha_1^{-3/2}),$$

which we need with

$$\alpha_1 = \frac{|X-Y|^2}{n}, \quad r_1 = \frac{|X-Y|^4}{4n^3}, \quad \alpha_2 = \frac{|X|^2 + |Y|^2}{n}, \quad r_2 = \frac{|X|^4 + |Y|^4}{4n^3}.$$

It follows that

$$\begin{aligned}
\alpha_2^{1/2} + r_2 \alpha_2^{-3/2} &= \left(\frac{|X|^2 + |Y|^2}{n}\right)^{1/2} \left(1 + \frac{1}{4n}\, \frac{|X|^4 + |Y|^4}{(|X|^2 + |Y|^2)^2}\right), \\
\alpha_1^{1/2} + r_1 \alpha_1^{-3/2} &= \left(\frac{|X-Y|^2}{n}\right)^{1/2} \left(1 + \frac{1}{4n}\right)
\end{aligned}$$

with the assumption that both expressions are equal to zero in the case $X = Y = 0$. Therefore, we have the following general, although preliminary, observation on the basis of Proposition 15.1.2.

Lemma 15.5.1 *Let X be a random vector in $\mathbb{R}^n$ satisfying $\mathbb{E}\,|X| \le b\sqrt{n}$. Then with some quantity C, which is bounded by an absolute constant, we have*

$$\mathbb{E}_\theta\, \omega^2(F_\theta, F) = \frac{1}{\sqrt{2\pi}}\, \mathbb{E}R + \frac{Cb}{n^2},$$

where

$$R = \frac{(|X|^2 + |Y|^2)^{1/2}}{\sqrt{n}} \left(1 + \frac{1}{4n}\, \frac{|X|^4 + |Y|^4}{(|X|^2 + |Y|^2)^2}\right) - \frac{|X - Y|}{\sqrt{n}} \left(1 + \frac{1}{4n}\right)$$

with the assumption that $R = 0$ in the case $X = Y = 0$.

When $|X|^2 = |Y|^2 = n$, this lemma returns us to Proposition 15.3.1.

As for the general case, note that $R \le 2\,\frac{|X|+|Y|}{\sqrt{n}}$, so $\mathbb{E}R \le 2b$.

Let us give a simpler expression by noting that

$$\frac{|X|^4 + |Y|^4}{(|X|^2 + |Y|^2)^2} - \frac{1}{2} = \frac{(|X|^2 - |Y|^2)^2}{2\,(|X|^2 + |Y|^2)^2},$$

so that

$$R = \frac{1}{8\,n^{3/2}}\, \frac{(|X|^2 - |Y|^2)^2}{(|X|^2 + |Y|^2)^{3/2}} + \frac{(|X|^2 + |Y|^2)^{1/2}}{\sqrt{n}} \left(1 + \frac{1}{8n}\right) - \frac{|X - Y|}{\sqrt{n}} \left(1 + \frac{1}{4n}\right).$$

The first term here is actually of order $\sigma_4^2(X)/n^2$, which follows from the next general observation.

Lemma 15.5.2 *Let ξ be a non-negative random variable with finite second moment (not identically zero), and let η be its independent copy. Then*

$$\mathbb{E}\, \frac{(\xi - \eta)^2}{(\xi + \eta)^{3/2}}\, 1_{\{\xi+\eta>0\}} \le 12\, \frac{\mathrm{Var}(\xi)}{(\mathbb{E}\,\xi)^{3/2}}.$$

Proof By homogeneity, we may assume that $\mathbb{E}\xi = 1$. In particular, $\mathbb{E}\,|\xi - \eta| \le 2$.

We have

$$\begin{aligned}\mathbb{E}\, \frac{(\xi - \eta)^2}{(\xi + \eta)^{3/2}}\, 1_{\{\xi+\eta>1/2\}} &\le 2^{3/2}\, \mathbb{E}\,(\xi - \eta)^2\, 1_{\{\xi+\eta>1/2\}} \\ &\le 2^{3/2}\, \mathbb{E}\,(\xi - \eta)^2 = 4\sqrt{2}\, \mathrm{Var}(\xi).\end{aligned}$$

Also note that, by Chebyshev's inequality,

$$\mathbb{P}\,\{\xi \le 1/2\} = \mathbb{P}\,\{1 - \xi \ge 1/2\} \le 4\, \mathrm{Var}(\xi)^2,$$

so

$$\mathbb{P}\,\{\xi + \eta \le 1/2\} \le \mathbb{P}\,\{\xi \le 1/2\}\, \mathbb{P}\,\{\eta \le 1/2\} \le 16\, \mathrm{Var}(\xi)^2.$$

Hence, since $\frac{|\xi-\eta|}{\xi+\eta} \leq 1$ for $\xi + \eta > 0$, we have, by Cauchy's inequality,

$$\mathbb{E}\,\frac{(\xi-\eta)^2}{(\xi+\eta)^{3/2}}\,1_{\{0<\xi+\eta\leq 1/2\}} \leq \mathbb{E}\,\sqrt{|\xi-\eta|}\,1_{\{0<\xi+\eta\leq 1/2\}}$$
$$\leq \sqrt{\mathbb{E}\,|\xi-\eta|}\,\sqrt{\mathbb{P}\,\{\xi+\eta\leq 1/2\}} \leq 4\sqrt{2}\,\mathrm{Var}(\xi).$$

It remains to combine both inequalities, which yield

$$\mathbb{E}\,\frac{(\xi-\eta)^2}{(\xi+\eta)^{3/2}}\,1_{\{\xi+\eta>0\}} \leq 8\sqrt{2}\,\mathrm{Var}(\xi) \leq 12\,\mathrm{Var}(\xi).$$

The lemma is proved. □

Applying it with $\xi = |X|^2$, $\eta = |Y|^2$, and assuming that $\mathbb{E}\,|X|^2 = n$, we get that

$$\mathbb{E}\,\frac{(|X|^2-|Y|^2)^2}{(|X|^2+|Y|^2)^{3/2}} \leq 12\,\frac{\mathrm{Var}(|X|^2)}{(\mathbb{E}\,|X|^2)^{3/2}} = 12\,\frac{\mathrm{Var}(|X|^2)}{n^{3/2}} \leq 12\,\frac{\sigma_4^2(X)}{n^{1/2}}.$$

Hence, from Lemma 15.5.1 we obtain:

Proposition 15.5.3 *Let X be a random vector in $\mathbb{R}^n$ satisfying $\mathbb{E}\,|X|^2 = n$. Then, for some quantity c bounded by an absolute constant, we have*

$$\mathbb{E}_\theta\,\omega^2(F_\theta, F) = \frac{1}{\sqrt{2\pi}}\,\mathbb{E}R + c\,\frac{1+\sigma_4^2}{n^2},$$

where $\sigma_4^2 = \frac{1}{n}\,\mathrm{Var}(|X|^2)$ and

$$R = \frac{(|X|^2+|Y|^2)^{1/2}}{\sqrt{n}}\left(1+\frac{1}{8n}\right) - \frac{|X-Y|}{\sqrt{n}}\left(1+\frac{1}{4n}\right).$$

In this rather general setup (we only assume that $\mathbb{E}\,|X|^2 = n$), this formula may be effectively used in many specific situations. For example, when $|X| = \sqrt{n}$ a.s., then $\sigma_4 = 0$, and we return to the setting of the previous two sections. Moreover, all obtained results can be derived from Proposition 15.5.3 in this particular case.

Let us now see what simplification can be done in the symmetric case. Define the random variable

$$\xi = \frac{2\,\langle X, Y\rangle}{|X|^2+|Y|^2},$$

assigning to it the zero value, when $X = Y = 0$. Since $|X-Y| = (|X|^2+|Y|^2)^{1/2}\sqrt{1-\xi}$, we have

$$R = \frac{(|X|^2+|Y|^2)^{1/2}}{\sqrt{n}}\left[\left(1+\frac{1}{8n}\right) - \sqrt{1-\xi}\left(1+\frac{1}{4n}\right)\right]$$
$$= \frac{(|X|^2+|Y|^2)^{1/2}}{\sqrt{n}}\left[\left(1+\frac{1}{4n}\right)(1-\sqrt{1-\xi}) - \frac{1}{8n}\right].$$

One can now apply Lemma 15.4.4, which yields

$$R = \frac{(|X|^2 + |Y|^2)^{1/2}}{\sqrt{n}} \left[\left(1 + \frac{1}{4n}\right)\left(\frac{1}{2}\,\xi + \frac{1}{8}\,\xi^2 + \frac{1}{16}\,\xi^3 + c\xi^4\right) - \frac{1}{8n} \right]$$

with $0.01 \leq c \leq 3$. If X has a symmetric distribution (about the origin), then the expectation of the terms containing ξ and ξ^3 will vanish. As a result, we arrive at the following assertion.

Corollary 15.5.4 *Let X be a random vector in $\mathbb{R}^n$ with a symmetric distribution satisfying $\mathbb{E}\,|X|^2 = n$. For some quantities c_1, c_2 bounded by an absolute constant and such that c_1 is positive and bounded away from zero, we have*

$$\mathbb{E}_\theta\, \omega^2(F_\theta, F) = \frac{1}{\sqrt{2\pi}} \left(1 + \frac{1}{4n}\right) \mathbb{E}R + c_1\, \mathbb{E}\, \frac{(|X|^2 + |Y|^2)^{1/2}}{\sqrt{n}}\, \xi^4 + c_2\, \frac{1 + \sigma_4^2}{n^2},$$

where

$$R = \frac{(|X|^2 + |Y|^2)^{1/2}}{8\sqrt{n}} \left(\xi^2 - \frac{1}{n}\right), \quad \xi = \frac{2\,\langle X, Y\rangle}{|X|^2 + |Y|^2}.$$

15.6 General Lower Bounds

To derive effective general bounds on $\mathbb{E}_\theta\, \omega^2(F_\theta, F)$, let us return to the representation of Lemma 15.5.1. According to the remark after the lemma, if $\mathbb{E}\,|X| \leq b\sqrt{n}$ then we have

$$\mathbb{E}_\theta\, \omega^2(F_\theta, F) = \frac{1}{\sqrt{2\pi}}\, \mathbb{E}\,(R_0 + R_1) + \frac{cb}{n^2},$$

where

$$R_0 = \frac{1}{8\,n^{3/2}}\, \frac{(|X|^2 - |Y|^2)^2}{(|X|^2 + |Y|^2)^{3/2}},$$

$$R_1 = \frac{(|X|^2 + |Y|^2)^{1/2}}{\sqrt{n}} \left(1 + \frac{1}{8n}\right) - \frac{|X - Y|}{\sqrt{n}} \left(1 + \frac{1}{4n}\right)$$

with the assumption that $R_0 = 0$ when $X = Y = 0$. Here c is some quantity bounded by an absolute constant. (Although R_1 has already appeared as the expression R in Proposition 15.5.3, we prefer to keep the small summand R_0.) Recall that in terms of $\xi = \frac{2\langle X,Y\rangle}{|X|^2+|Y|^2}$, the expression R_1 admits the lower bound

$$R_1 \geq \frac{(|X|^2 + |Y|^2)^{1/2}}{\sqrt{n}} \left[\left(1 + \frac{1}{4n}\right)\left(\frac{1}{2}\,\xi + \frac{1}{8}\,\xi^2 + \frac{1}{16}\,\xi^3 + 0.01\,\xi^4\right) - \frac{1}{8n} \right].$$

The expectation of the terms on the right-hand side containing ξ and ξ^3 is non-negative. This follows from the first inequality of Proposition 1.4.4,

$$\mathbb{E}\,\langle X,Y\rangle^p\,(|X|^2+|Y|^2)^{-\alpha}\geq 0,$$

which holds true for any integer $p \geq 1$ and any real $0 \leq \alpha \leq p$. Hence, removing the unnecessary factor $1+\frac{1}{4n}$, up to some absolute constants $c_1, c_2 > 0$, we get

$$\begin{aligned}\mathbb{E}_\theta\,\omega^2(F_\theta,F) &\geq \frac{1}{\sqrt{2\pi}}\,\mathbb{E}R_0+\frac{1}{\sqrt{2\pi}}\,\mathbb{E}\,\frac{(|X|^2+|Y|^2)^{1/2}}{8\sqrt{n}}\Big(\xi^2-\frac{1}{n}\Big)\\ &\quad+c_1\,\mathbb{E}\,\frac{(|X|^2+|Y|^2)^{1/2}}{\sqrt{n}}\,\xi^4-c_2\frac{b}{n^2}.\end{aligned}\tag{15.10}$$

Now, by the second inequality of Proposition 1.4.4, applied with $\alpha = 3/2$,

$$\mathbb{E}\,(|X|^2+|Y|^2)^{1/2}\,\xi^2=4\,\mathbb{E}\,\frac{\langle X,Y\rangle^2}{(|X|^2+|Y|^2)^{3/2}}\geq\frac{4}{n}\,\mathbb{E}\,\frac{|X|^2\,|Y|^2}{(|X|^2+|Y|^2)^{3/2}}.$$

This gives

$$\begin{aligned}\mathbb{E}\,\frac{(|X|^2+|Y|^2)^{1/2}}{8\sqrt{n}}\Big(\xi^2-\frac{1}{n}\Big) &\geq \frac{1}{8n^{3/2}}\,\mathbb{E}\bigg[\frac{4\,|X|^2\,|Y|^2}{(|X|^2+|Y|^2)^{3/2}}-(|X|^2+|Y|^2)^{1/2}\bigg]\\ &= -\frac{1}{8n^{3/2}}\,\mathbb{E}\,\frac{(|X|^2-|Y|^2)^2}{(|X|^2+|Y|^2)^{3/2}} = -\mathbb{E}R_0.\end{aligned}$$

Thus, the summand $\mathbb{E}R_0$ in (15.6) neutralizes the term containing ξ^2, and we are left with the term containing ξ^4. That is, we arrive at:

Proposition 15.6.1 *Let X be a random vector in $\mathbb{R}^n$ satisfying $\mathbb{E}\,|X| \leq b\sqrt{n}$, and let Y be its independent copy. For some absolute constants $c_1, c_2 > 0$, we have*

$$\mathbb{E}_\theta\,\omega^2(F_\theta,F)\geq c_1\,\mathbb{E}\,\rho\,\xi^4-c_2\frac{b}{n^2},$$

where

$$\rho=\Big(\frac{|X|^2+|Y|^2}{2n}\Big)^{1/2},\qquad \xi=\frac{2\,\langle X,Y\rangle}{|X|^2+|Y|^2}.$$

If $|X|^2 = n$ a.s., this is exactly the lower bound of Corollary 15.4.3. In order to simplify the bound of Proposition 15.6.1 in the basic case where $\mathbb{E}\,|X|^2 = n$, note that $\mathbb{E}\rho^2 = 1$ and $\mathrm{Var}(\rho^2) = \frac{\sigma_4^2}{2n}$. Using

$$\xi=1-\frac{|X-Y|^2}{|X|^2+|Y|^2},$$

we have

$$\begin{aligned}\xi^4 &\geq (1-\alpha)^4\,1_{\{|X-Y|^2\,\leq\,\alpha\,(|X|^2+|Y|^2)\}}\\ &\geq (1-\alpha)^4\,1_{\{|X-Y|^2\,\leq\,\alpha\lambda n,\ |X|^2+|Y|^2\,\geq\,\lambda n\}},\qquad 0<\alpha,\lambda<1.\end{aligned}$$

The inequality $|X|^2 + |Y|^2 \geq \lambda n$ is equivalent to $\rho^2 \geq \frac{\lambda}{2}$, so,

$$\begin{aligned}\mathbb{E}\,\rho\,\xi^4 &\geq \frac{(1-\alpha)^4}{\sqrt{2}}\,\sqrt{\lambda}\,\mathbb{P}\Big\{|X-Y|^2 \leq \alpha\lambda n,\ |X|^2 + |Y|^2 \geq \lambda n\Big\} \\ &\geq \frac{(1-\alpha)^4}{\sqrt{2}}\,\sqrt{\lambda}\,\Big(\mathbb{P}\{|X-Y|^2 \leq \alpha\lambda n\} - \mathbb{P}\{|X|^2 + |Y|^2 \leq \lambda n\}\Big).\end{aligned}$$

But, by Proposition 1.6.1 with $p = 2$,

$$\mathbb{P}\Big\{|X|^2 + |Y|^2 \leq \lambda n\Big\} \leq \Big(\mathbb{P}\Big\{|X|^2 \leq \lambda n\Big\}\Big)^2 \leq \frac{1}{(1-\lambda)^4}\,\frac{\sigma_4^4}{n^2}.$$

Hence

$$\mathbb{E}\,\rho\,\xi^4 \geq \frac{(1-\alpha)^4}{\sqrt{2}}\,\sqrt{\lambda}\,\bigg(\mathbb{P}\{|X-Y|^2 \leq \alpha\lambda n\} - \frac{1}{(1-\lambda)^4}\,\frac{\sigma_4^4}{n^2}\bigg).$$

Choosing, for example, $\alpha = \lambda = \frac{1}{2}$, we get

$$\mathbb{E}\,\rho\,\xi^4 \geq \frac{1}{32}\,\mathbb{P}\Big\{|X-Y|^2 \leq \frac{1}{4}\,n\Big\} - \frac{\sigma_4^4}{2n^2}.$$

Hence, from Proposition 15.6.1 we obtain:

Proposition 15.6.2 *Let X be a random vector in $\mathbb{R}^n$ satisfying $\mathbb{E}\,|X|^2 = n$, and let Y be its independent copy. For some absolute constants $c_1, c_2 > 0$, we have*

$$\mathbb{E}_\theta\,\omega^2(F_\theta, F) \geq c_1\,\mathbb{P}\Big\{|X-Y| \leq \frac{1}{2}\sqrt{n}\Big\} - c_2\frac{1+\sigma_4^4}{n^2},$$

where $\sigma_4^2 = \frac{1}{n}\,\mathrm{Var}(|X|^2)$.

Chapter 16
Refinements for the Kolmogorov Distance

Let X be a random vector in $\mathbb{R}^n$ with finite 4-th moment and such that $\mathbb{E}\,|X|^2 = n$. According to Proposition 14.7.1, the distributions F_θ of the linear forms $\langle X, \theta\rangle$ satisfy the following bound on the Kolmogorov distance to the standard normal law Φ on average:

$$\mathbb{E}_\theta\, \rho(F_\theta, \Phi) \le C(m_4^{3/2} + \sigma_4^{3/2} + |a|)\, \frac{1}{\sqrt{n}}.$$

Here $a = \mathbb{E}X$, and as before

$$m_4^4 = \frac{1}{n^2}\, \mathbb{E}\,\langle X, Y\rangle^4, \quad \sigma_4^2 = \frac{1}{n}\, \mathrm{Var}(|X|^2),$$

with Y being an independent copy of X. While these quantities may be of the order 1 in many interesting situations, the dependence with respect to n in this bound does not seem to be quite satisfactory (although it cannot be improved in some other interesting examples). The aim of this chapter is to describe natural conditions which would ensure that the distance $\rho(F_\theta, \Phi)$ does not exceed $1/n$ up to a logarithmically growing factor. In particular, we would like to obtain a corresponding analog of Proposition 15.4.1 about the L^2-distance $\omega(F_\theta, \Phi)$. At the end of the chapter, we also discuss relations between these two distances which will allow us to develop lower bounds with a standard rate of normal approximation.

16.1 Preliminaries

Assume that $|X| = \sqrt{n}$ a.s., so that $\sigma_4^2 = 0$ and thus the characteristic function of the typical distribution is given by

$$f(t) = \mathbb{E}\, J_n(t|X|) = J_n(t\sqrt{n}).$$

S. Bobkov et al., *Concentration and Gaussian Approximation for Randomized Sums*, Probability Theory and Stochastic Modelling 104,
https://doi.org/10.1007/978-3-031-31149-9_16

In addition, in this case

$$\Delta_n(t) \equiv \mathbb{E}_\theta \, |f_\theta(t) - f(t)|^2 = \mathbb{E} J_n(t|X - Y|) - (J_n(t\sqrt{n}))^2.$$

Let us return to the Berry–Esseen-type bound (14.14) of Lemma 14.5.2: For all $T > T_0 \geq 1$, with some absolute constant $c > 0$, we have

$$\begin{aligned} c\,\mathbb{E}_\theta\, \rho^2(F_\theta, F) \leq &\int_0^1 \frac{\Delta_n(t)}{t^2}\,\mathrm{d}t + \log T \int_0^{T_0} \frac{\Delta_n(t)}{t}\,\mathrm{d}t \\ &+ \log T \int_{T_0}^T \frac{\mathbb{E}_\theta\, |f_\theta(t)|^2}{t}\,\mathrm{d}t + \frac{1}{T^2}\Big(\int_0^\infty |f(t)|\,\mathrm{d}t\Big)^2. \end{aligned}$$

The last integral is convergent and is bounded by an absolute constant uniformly over all $n \geq 4$ in view of the polynomial bound (11.8),

$$|f(t)| \leq 6\Big(1 + \frac{ct^2}{n}\Big)^{-\frac{n-1}{4}}, \quad t \in \mathbb{R}$$

(holding with $c = 1/(3\pi)^2$, Proposition 11.5.4). Hence, the above bound on the Kolmogorov distance simplifies to

$$\begin{aligned} c\,\mathbb{E}_\theta\, \rho^2(F_\theta, F) \leq &\int_0^1 \frac{\Delta_n(t)}{t^2}\,\mathrm{d}t + \log T \int_0^{T_0} \frac{\Delta_n(t)}{t}\,\mathrm{d}t \\ &+ \log T \int_{T_0}^T \frac{\mathbb{E}_\theta\, |f_\theta(t)|^2}{t}\,\mathrm{d}t + \frac{1}{T^2}. \end{aligned}$$

As we know from Lemma 13.1.2 with $p = 2$, for some absolute constant $c > 0$,

$$c\,\mathbb{E}_\theta\, |f_\theta(t)|^2 \leq \frac{m_4^4}{n^2} + e^{-t^2/8},$$

so,

$$c' \int_{T_0}^T \frac{\mathbb{E}_\theta\, |f_\theta(t)|^2}{t}\,\mathrm{d}t \leq \frac{m_4^4}{n^2}\,\log T + e^{-T_0^2/8}.$$

Let us choose here $T = 4n$ and $T_0 = 4\sqrt{\log n}$. Since $m_4 \geq 1$, we then arrive at the following preliminary general bound.

Lemma 16.1.1 *Let X be a random vector in $\mathbb{R}^n$ such that $|X| = \sqrt{n}$ a.s., and let Y be an independent copy of X. Then for some absolute constant $c > 0$,*

$$c\,\mathbb{E}_\theta\, \rho^2(F_\theta, F) \leq \int_0^1 \frac{\Delta_n(t)}{t^2}\,\mathrm{d}t + \log n \int_0^{4\sqrt{\log n}} \frac{\Delta_n(t)}{t}\,\mathrm{d}t + \frac{(\log n)^2}{n^4}\,\mathbb{E}\,\langle X, Y\rangle^4.$$

Since $\mathbb{E}\,\langle X, Y\rangle^4 \geq n^2$, this bound is fulfilled automatically for $n = 2$ and $n = 3$.

In order to study the integrals in Lemma 16.1.1, let us first focus on the first integral, additionally assuming that the random vector X in $\mathbb{R}^n$ is isotropic and has mean zero. We need to develop an asymptotic bound on $\Delta_n(t)$ for $t \in [0,1]$. Putting $\xi = \frac{\langle X,Y \rangle}{n}$, we have $|X-Y|^2 = 2n(1-\xi)$, so that

$$\Delta_n(t) = \mathbb{E} J_n\Big(t\sqrt{2n(1-\xi)}\Big) - (J_n(t\sqrt{n}))^2.$$

We use the asymptotic formula from Proposition 11.3.1,

$$J_n(t\sqrt{n}) = \Big(1 - \frac{t^4}{4n}\Big)\, \mathrm{e}^{-t^2/2} + \varepsilon_n(t), \quad t \in \mathbb{R},$$

where $\varepsilon_n(t)$ denotes a quantity of the form $O\big(n^{-2}\min(1,t^4)\big)$ with a universal constant in O (which may vary from place to place). It implies a similar bound

$$(J_n(t\sqrt{n}))^2 = \Big(1 - \frac{t^4}{2n}\Big)\, \mathrm{e}^{-t^2} + \varepsilon_n(t).$$

Since $|\xi| \leq 1$ a.s., we also have

$$J_n\Big(t\sqrt{2n(1-\xi)}\Big) = \Big(1 - \frac{t^4}{n}\,(1-\xi)^2\Big)\, \mathrm{e}^{-t^2(1-\xi)} + \varepsilon_n(t).$$

Hence, subtracting from $\mathrm{e}^{t^2\xi}$ the linear term $1 + t^2\xi$ and adding, one may write

$$\begin{aligned}\Delta_n(t) &= \mathrm{e}^{-t^2}\, \mathbb{E}\,\Big[\Big(1 - \frac{t^4}{n}\,(1-\xi)^2\Big)\, \mathrm{e}^{t^2\xi} - \Big(1 - \frac{t^4}{2n}\Big)\Big] + \varepsilon_n(t) \\ &= \mathrm{e}^{-t^2}\, \mathbb{E}\,(U+V) + \varepsilon_n(t)\end{aligned}$$

with

$$\begin{aligned}U &= \frac{t^4}{n}\Big(\frac{1}{2} - (1-\xi)^2\Big) + \Big(1 - \frac{t^4}{n}\,(1-\xi)^2\Big)\cdot t^2\xi, \\ V &= \Big(1 - \frac{t^4}{n}\,(1-\xi)^2\Big)(\mathrm{e}^{t^2\xi} - 1 - t^2\xi).\end{aligned}$$

Using $\mathbb{E}\xi = 0$, $\mathbb{E}\xi^2 = \frac{1}{n}$ and hence $\mathbb{E}\,|\xi|^3 \leq \mathbb{E}\xi^2 \leq \frac{1}{n}$, we find that in the interval $0 \leq t \leq 1$

$$\mathbb{E}U = -\frac{t^4}{2n} - \frac{t^4}{n^2} + \frac{2t^6}{n^2} - \frac{t^6}{n}\,\mathbb{E}\xi^3 = -\frac{t^4}{2n} + \varepsilon_n(t).$$

Next write

$$V = W - \frac{t^4}{n}\,(1-\xi)^2\, W, \quad W = \mathrm{e}^{t^2\xi} - 1 - t^2\xi.$$

Using $|\mathrm{e}^x - 1 - x| \leq 2x^2$ for $|x| \leq 1$, we have $|W| \leq 2t^4\xi^2$. Hence, the expected value of the second term in the representation for V does not exceed $8t^8/n^2$. Moreover, by

Taylor's expansion,

$$W = \frac{1}{2}\,t^4\xi^2 + \frac{1}{6}\,t^6\xi^3 + Rt^8\xi^4, \quad R = \sum_{k=4}^{\infty} \frac{t^{2k-8}}{k!}\,\xi^{k-4},$$

implying that

$$\mathbb{E}\,W = \frac{t^4}{2n} + \frac{t^6}{6}\,\mathbb{E}\xi^3 + ct^8\,\mathbb{E}\xi^4$$

with a quantity c bounded by an absolute constant. Summing the two expansions, we arrive at

$$\mathbb{E}\,(U+V) = \frac{t^6}{6}\,\mathbb{E}\xi^3 + ct^8\mathbb{E}\xi^4 + \varepsilon_n(t),$$

and therefore,

$$\int_0^1 \frac{\Delta_n(t)}{t^2}\,\mathrm{d}t \le c_1\,\mathbb{E}\xi^3 + c_2\,\mathbb{E}\xi^4 + O(n^{-2}),$$

for some absolute constants $c_1, c_2 > 0$. Here $\mathbb{E}\xi^4 \ge (\mathbb{E}\xi^2)^2 = n^{-2}$, so the term $O(n^{-2})$ may be absorbed by the 4-th moment of ξ. Since $\mathbb{E}\xi^3 \ge 0$, the bound of Lemma 16.1.1 may be simplified to

$$c\,\mathbb{E}_\theta\,\rho^2(F_\theta, F) \le \mathbb{E}\xi^3 + \mathbb{E}\xi^4 + \log n\int_0^{4\sqrt{\log n}} \frac{\Delta_n(t)}{t}\,\mathrm{d}t + \frac{(\log n)^2}{n^4}\,\mathbb{E}\,\langle X,Y\rangle^4.$$

Here again the last term dominates $\mathbb{E}\xi^4$, and we arrive at:

Lemma 16.1.2 *Let X be an isotropic random vector in $\mathbb{R}^n$ with mean zero and such that $|X| = \sqrt{n}$ a.s., and let Y be an independent copy of X. Then for some absolute constant $c > 0$,*

$$c\,\mathbb{E}_\theta\,\rho^2(F_\theta, F) \le \log n\int_0^{4\sqrt{\log n}} \frac{\Delta_n(t)}{t}\,\mathrm{d}t + \mathbb{E}\xi^3 + (\log n)^2\,\mathbb{E}\xi^4,$$

where $\xi = \frac{\langle X,Y\rangle}{n}$.

16.2 Large Interval. Final Upper Bound

Our next step (which is most important) is to show that the integral in Lemma 16.1.2 is dominated by the expression $\mathbb{E}\xi^3 + \mathbb{E}\xi^4$ up to logarithmic factors. This integral may be expressed in terms of the functions $g_n(t) = J_n(t\sqrt{2n})$ and

$$\psi(\alpha) = \int_0^T \frac{g_n(\alpha t) - g_n(t)}{t}\,\mathrm{d}t, \quad 0 \le \alpha \le \sqrt{2},$$

with $T = 4\sqrt{\log n}$. Indeed, in terms of $\xi = \frac{\langle X,Y\rangle}{n}$, we have

$$\int_0^T \frac{\Delta_n(t)}{t}\,dt = \int_0^T \frac{\mathbb{E}J_n(t\sqrt{2n(1-\xi)}) - (J_n(t\sqrt{n}))^2}{t}\,dt$$
$$= \mathbb{E}\,\psi\big(\sqrt{1-\xi}\big) + \int_0^T \frac{J_n(t\sqrt{2n}) - (J_n(t\sqrt{n}))^2}{t}\,dt.$$

To proceed, we need to develop a Taylor expansion for $\xi \to \psi\big(\sqrt{1-\xi}\big)$ around zero in powers of ξ. Recall that $g_n(t)$ represents the characteristic function of the random variable $\sqrt{2n}\,\theta_1$ on $(\mathbb{S}^{n-1}, \mathfrak{s}_{n-1})$. This already ensures that $|g_n(t)| \le 1$ and

$$|g_n'(t)| \le \sqrt{2n}\,\mathbb{E}\,|\theta_1| \le \sqrt{2n}\,(\mathbb{E}\,\theta_1^2)^{1/2} = \sqrt{2}$$

for all $t \in \mathbb{R}$. Hence

$$|g_n(\alpha t) - g_n(t)| \le \sqrt{2}\,(\alpha - 1)\,|t| \le |t|,$$

so that

$$|\psi(\alpha)| \le \int_0^1 \frac{|g_n(\alpha t) - g_n(t)|}{t}\,dt + \int_1^T \frac{|g_n(\alpha t) - g_n(t)|}{t}\,dt$$
$$\le 2 + 2\log T < 4\log T$$

(since $T > e$). In addition, $\psi(1) = 0$ and

$$\psi'(\alpha) = \int_0^T g_n'(\alpha t)\,dt = \frac{1}{\alpha}\,(g_n(\alpha T) - 1).$$

Therefore, we arrive at another expression

$$\psi(\alpha) = \int_1^\alpha \frac{g_n(Tx) - 1}{x}\,dx = \int_1^\alpha \frac{g_n(Tx)}{x}\,dx - \log\alpha.$$

For $|\varepsilon| \le 1$, let

$$v(\varepsilon) = \int_1^{(1-\varepsilon)^{1/2}} \frac{g_n(Tx)}{x}\,dx,$$
$$u(\varepsilon) = \psi\Big((1-\varepsilon)^{1/2}\Big) = v(\varepsilon) - \frac{1}{2}\,\log(1-\varepsilon),$$

so that $\mathbb{E}\,\psi\big(\sqrt{1-\xi}\big) = \mathbb{E}\,u(\xi)$. According to the bound of Proposition 11.4.1,

$$|J_n(t\sqrt{n})| \le 3\,(e^{-t^2/2} + e^{-n/12}),$$

which yields

$$|g_n(t)| \le 3\,(e^{-t^2} + e^{-n/12}).$$

Hence, for $-1 \le \varepsilon \le \frac{1}{2}$, up to an absolute constant C,

$$\begin{aligned} |v(\varepsilon)| &\le \sup_{\frac{1}{\sqrt{2}} \le x \le \sqrt{2}} |g_n(Tx)| \int_{\frac{1}{\sqrt{2}}}^{\sqrt{2}} \frac{1}{x}\,dx \\ &\le \sup_{z \ge T/\sqrt{2}} |g_n(z)| \le 3\,(e^{-T^2/2} + e^{-n/12}) \le \frac{C}{n^8}, \end{aligned}$$

where the last inequality is specialized to the choice $T = 4\sqrt{\log n}$. Using the Taylor expansion for the logarithmic function, we also have

$$-\log(1-\varepsilon) \le \varepsilon + \frac{1}{2}\,\varepsilon^2 + \frac{1}{3}\,\varepsilon^3 + 4\varepsilon^4, \quad \varepsilon \le \frac{1}{2}.$$

Combining the two inequalities, we get

$$u(\varepsilon) \le \frac{1}{2}\,\varepsilon + \frac{1}{4}\,\varepsilon^2 + \frac{1}{6}\,\varepsilon^3 + 2\varepsilon^4 + \frac{C}{n^8}, \quad -1 \le \varepsilon \le \frac{1}{2}.$$

In order to involve the remaining interval $\frac{1}{2} \le \varepsilon \le 1$ in an inequality of a similar type, recall that $|u(\varepsilon)| \le 3\log T$ for all $|\varepsilon| \le 1$. Hence, the above inequality will hold automatically by increasing the coefficient 2 in front of ε^4 to a multiple of $\log T$. As a result, we obtain the desired inequality on the whole segment, i.e.,

$$u(\varepsilon) \le \frac{1}{2}\,\varepsilon + \frac{1}{4}\,\varepsilon^2 + \frac{1}{6}\,\varepsilon^3 + c\log T\,\varepsilon^4 + \frac{c}{n^8}, \qquad -1 \le \varepsilon \le 1,$$

for some absolute constant $c > 0$.

Thus,

$$\psi\big(\sqrt{1-\xi}\big) \le \frac{1}{2}\,\xi + \frac{1}{4}\,\xi^2 + \frac{1}{6}\,\xi^3 + c\log T\,\xi^4 + \frac{c}{n^8},$$

and taking the expectation, we get

$$\mathbb{E}\,\psi\big(\sqrt{1-\xi}\big) \le \frac{1}{4n} + \frac{1}{6}\,\mathbb{E}\xi^3 + c\log T\,\mathbb{E}\xi^4,$$

where the term cn^{-8} was absorbed by the 4-th moment of ξ.

Now, let us turn to the integral

$$I_n = \int_0^T \frac{J_n(t\sqrt{2n}) - (J_n(t\sqrt{n}))^2}{t}\,dt$$

and recall the asymptotic formulas

$$J_n\big(t\sqrt{2n}\big) = \Big(1 - \frac{t^4}{n}\Big)\,e^{-t^2} + \varepsilon_n(t), \quad \Big(J_n\big(t\sqrt{n}\big)\Big)^2 = \Big(1 - \frac{t^4}{2n}\Big)\,e^{-t^2} + \varepsilon_n(t),$$

where $\varepsilon_n(t) = O\big(n^{-2}\min(1, t^4)\big)$. After integration, this remainder term will create an error of order at most $n^{-2}\log T$, up to which I_n is equal to

$$-\int_0^T \frac{t^4}{2n}\,\mathrm{e}^{-t^2}\,\frac{\mathrm{d}t}{t} = -\frac{1}{4n}\Big(1-(T^2+1)\,\mathrm{e}^{-T^2}\Big) = -\frac{1}{4n} + o(n^{-15}).$$

Thus,

$$I_n = -\frac{1}{4n} + O(n^{-2}\log T),$$

and therefore

$$\int_0^T \frac{\Delta_n(t)}{t}\,\mathrm{d}t = \mathbb{E}\,\psi\big(\sqrt{1-\xi}\big) + I_n \le \frac{1}{6}\,\mathbb{E}\xi^3 + c\log T\,\mathbb{E}\xi^4.$$

One can now apply this estimate in Lemma 16.1.2, and then we eventually obtain:

Proposition 16.2.1 *Let X be an isotropic random vector in $\mathbb{R}^n$ with mean zero and such that $|X| = \sqrt{n}$ a.s., and let Y be an independent copy of X. Then for some absolute constants $c_1, c_2 > 0$,*

$$\mathbb{E}_\theta\,\rho^2(F_\theta, F) \le c_1\,(\log n)\,\mathbb{E}\xi^3 + c_2\,(\log n)^2\,\mathbb{E}\xi^4,$$

where $\xi = \frac{\langle X,Y\rangle}{n}$. A similar inequality continues to hold for the standard normal distribution function Φ in place of F.

Although this assertion is mainly aimed at obtaining rates of order $1/n$ (modulo a logarithmic factor), it easily yields rates of order $1/\sqrt{n}$. Indeed, using $|\xi| \le 1$, we have that

$$\mathbb{E}\xi^4 \le \mathbb{E}\xi^2 = \frac{1}{n}, \quad \mathbb{E}\,|\xi|^3 \le \mathbb{E}\xi^2 = \frac{1}{n}.$$

Hence, from Proposition 16.2.1 we recover a particular case of the more general Proposition 14.6.1 (where $M_2 = 1$ and $\sigma_4 = 0$ under the current assumptions).

Corollary 16.2.2 *Let X be an isotropic random vector in $\mathbb{R}^n$ with mean zero and such that $|X| = \sqrt{n}$ a.s. Then for some absolute constant c,*

$$\mathbb{E}_\theta\,\rho^2(F_\theta, \Phi) \le \frac{c\,(\log n)^2}{n}.$$

16.3 Relations Between Kantorovich, L^2 and Kolmogorov distances

Let us now compare the L^2 and Kolmogorov distances on average, between the distributions F_θ of the weighted sums $\langle X, \theta\rangle$ and the typical distribution $F = \mathbb{E}_\theta F_\theta$. Such information will be needed to derive appropriate lower bounds on $\mathbb{E}_\theta\,\rho(F_\theta, F)$.

Proposition 16.3.1 *Let X be a random vector in $\mathbb{R}^n$ such that $|X| \le b\sqrt{n}$ a.s. Then, for any $\alpha \in [1,2]$,*

$$b^{-\alpha/2}\,\mathbb{E}_\theta\,\omega^\alpha(F_\theta,F) \le 14\,(\log n)^{\alpha/4}\,\mathbb{E}_\theta\,\rho^\alpha(F_\theta,F) + \frac{8}{n^4}. \tag{16.1}$$

As will be clear from the proof, at the expense of a larger coefficient in front of $\log n$, the last term n^{-4} can be shown to be of the form $n^{-\beta}$ for any prescribed value of β.

A relation similar to (16.1) is also true for the Kantorovich distance W in place of L^2. We state it for the case $\alpha = 1$.

Proposition 16.3.2 *Let X be a random vector in $\mathbb{R}^n$ such that $|X| \le b\sqrt{n}$ a.s. Then*

$$\mathbb{E}_\theta\,W(F_\theta,F) \le 10\,b\sqrt{\log n}\,\mathbb{E}_\theta\,\rho(F_\theta,F) + \frac{4b}{n^4}. \tag{16.2}$$

Proof Put $R_\theta(x) = F_\theta(-x) + (1 - F_\theta(x))$ for $x > 0$ and define R similarly on the basis of F. Using

$$\begin{aligned}(F_\theta(-x) - F(-x))^2 &\le 2F_\theta(-x)^2 + 2F(-x)^2,\\ (F_\theta(x) - F(x))^2 &\le 2\,(1 - F_\theta(x))^2 + 2\,(1 - F(x))^2,\end{aligned}$$

we have

$$(F_\theta(-x) - F(-x))^2 + (F_\theta(x) - F(x))^2 \le 2R_\theta(x)^2 + 2R(x)^2.$$

Hence, given $T > 0$ (to be specified later on), we have

$$\begin{aligned}\omega^2(F_\theta,F) &= \int_{-T}^{T} (F_\theta(x) - F(x))^2\,dx + \int_{|x|\ge T} (F_\theta(x) - F(x))^2\,dx\\ &\le 2T\rho^2(F_\theta,F) + 2\int_T^\infty R_\theta(x)^2\,dx + 2\int_T^\infty R(x)^2\,dx.\end{aligned}$$

It follows that, for any $\alpha \in [1,2]$,

$$\begin{aligned}\omega^\alpha(F_\theta,F) &\le (2T)^{\frac{\alpha}{2}}\,\rho^\alpha(F_\theta,F)\\ &\quad + \Big(2\int_T^\infty R_\theta(x)^2\,dx\Big)^{\frac{\alpha}{2}} + \Big(2\int_T^\infty R(x)^2\,dx\Big)^{\frac{\alpha}{2}},\end{aligned}$$

and therefore, by Jensen's inequality (since $\frac{\alpha}{2} \le 1$).

$$\begin{aligned}\mathbb{E}_\theta\,\omega^\alpha(F_\theta,F) &\le (2T)^{\frac{\alpha}{2}}\,\mathbb{E}_\theta\,\rho^\alpha(F_\theta,F)\\ &\quad + \Big(2\int_T^\infty \mathbb{E}_\theta\,R_\theta(x)^2\,dx\Big)^{\frac{\alpha}{2}} + \Big(2\int_T^\infty R(x)^2\,dx\Big)^{\frac{\alpha}{2}}.\end{aligned}$$

Next, by Markov's inequality, for any $x > 0$ and $p \geq 1$,

$$R_\theta(x)^2 \leq \left(\frac{\mathbb{E}\,|\,\langle X,\theta\rangle\,|^p}{x^p}\right)^2 \leq \frac{\mathbb{E}\,|\,\langle X,\theta\rangle\,|^{2p}}{x^{2p}}$$

and

$$\mathbb{E}_\theta R_\theta(x)^2 \leq \frac{\mathbb{E}_\theta\,\mathbb{E}\,|\,\langle X,\theta\rangle\,|^{2p}}{x^{2p}}.$$

Since $R = \mathbb{E}_\theta R_\theta$, a similar inequality holds true for R as well (by applying Cauchy's inequality). Hence

$$\mathbb{E}_\theta\,\omega^\alpha(F_\theta,F) \leq (2T)^{\frac{\alpha}{2}}\,\mathbb{E}_\theta\,\rho^\alpha(F_\theta,F) + 2\left(2\,\mathbb{E}_\theta\,\mathbb{E}\,|\,\langle X,\theta\rangle\,|^{2p}\int_T^\infty \frac{1}{x^{2p}}\,\mathrm{d}x\right)^{\frac{\alpha}{2}}.$$

Here, when θ is treated as a random vector with distribution $\mathfrak{s}_{n-1}$, which is independent of X, we have, by Corollary 10.3.3,

$$\mathbb{E}_\theta\,|\,\langle X,\theta\rangle\,|^{2p} \leq \left(\frac{2p}{n}\right)^p |X|^{2p}, \quad p \geq 1,$$

so that, by the assumption on X,

$$\mathbb{E}_\theta\,\mathbb{E}\,|\,\langle X,\theta\rangle\,|^{2p} \leq (2b^2p)^p.$$

From this,

$$2\left(2\,\mathbb{E}_\theta\int_T^\infty \frac{\mathbb{E}\,|\,\langle X,\theta\rangle\,|^{2p}}{x^{2p}}\,\mathrm{d}x\right)^{\frac{\alpha}{2}} \leq \frac{2^{\frac{\alpha}{2}+1}}{(2p-1)^{\frac{\alpha}{2}}}\,(2b^2p)^{\frac{\alpha p}{2}}\,T^{-\frac{\alpha(2p-1)}{2}}.$$

Thus,

$$\mathbb{E}_\theta\,\omega^\alpha(F_\theta,F) \leq (2T)^{\frac{\alpha}{2}}\,\mathbb{E}_\theta\,\rho^\alpha(F_\theta,F) + \frac{2^{\frac{\alpha}{2}+1}}{(2p-1)^{\frac{\alpha}{2}}}\,T^{\frac{\alpha}{2}}\left(\frac{2b^2p}{T^2}\right)^{\frac{\alpha p}{2}}.$$

Let us take $T = 2b\sqrt{p}$, in which case the above inequality becomes

$$\mathbb{E}_\theta\,\omega^\alpha(F_\theta,F) \leq (4b\sqrt{p})^{\frac{\alpha}{2}}\,\mathbb{E}_\theta\,\rho^\alpha(F_\theta,F) + \frac{2^{\alpha+1}}{(2p-1)^{\frac{\alpha}{2}}}\,(b\sqrt{p})^{\frac{\alpha}{2}}\,2^{-\frac{\alpha p}{2}}.$$

To simplify the last term, one can use $\sqrt{p} \leq 2p-1$ for $p \geq 1$ together with $2^{\alpha+1} \leq 8$ and $2^{-\frac{\alpha p}{2}} \leq 2^{\frac{p}{2}}$ (since $1 \leq \alpha \leq 2$), which leads to

$$\mathbb{E}_\theta\,\omega^\alpha(F_\theta,F) \leq (4b\sqrt{p})^{\frac{\alpha}{2}}\,\mathbb{E}_\theta\,\rho^\alpha(F_\theta,F) + 8b^{\frac{\alpha}{2}}\,2^{-p/2}.$$

Finally, choosing $p = 8\,\frac{\log n}{\log 2}$, we arrive at (16.1).

Now, turning to (16.2), we use the same functions R_θ and R as before and write

$$\begin{aligned} W(F_\theta, F) &= \int_{-T}^{T} |F_\theta(x) - F(x)| \, \mathrm{d}x + \int_{|x| \geq T} |F_\theta(x) - F(x)| \, \mathrm{d}x \\ &\leq 2T\rho(F_\theta, F) + \int_T^\infty R_\theta(x) \, \mathrm{d}x + \int_T^\infty R(x) \, \mathrm{d}x, \end{aligned}$$

which gives

$$\mathbb{E}_\theta \, W(F_\theta, F) \leq 2T \, \mathbb{E}_\theta \, \rho(F_\theta, F) + 2 \int_T^\infty R(x) \, \mathrm{d}x.$$

By Markov's inequality, for any $x > 0$ and $p \geq 1$,

$$R_\theta(x) \leq \frac{\mathbb{E} \, | \langle X, \theta \rangle |^p}{x^p}, \quad R(x) = \mathbb{E}_\theta R_\theta(x) \leq \frac{\mathbb{E}_\theta \, \mathbb{E} \, | \langle X, \theta \rangle |^p}{x^p}.$$

Hence

$$\mathbb{E}_\theta \, W(F_\theta, F) \leq 2T \, \mathbb{E}_\theta \, \rho(F_\theta, F) + 2 \, \mathbb{E}_\theta \, \mathbb{E} \, | \langle X, \theta \rangle |^p \int_T^\infty \frac{1}{x^p} \, \mathrm{d}x.$$

Here, one may use once more the bound $(\mathbb{E}_\theta \, | \langle X, \theta \rangle |^p)^{1/p} \leq \sqrt{\frac{p}{n}} \, |X|$ for $p \geq 2$, which yields

$$\mathbb{E}_\theta \, \mathbb{E} \, | \langle X, \theta \rangle |^p = \mathbb{E} \, |X|^p \, \mathbb{E}_\theta \, |\theta_1|^p \leq b^p n^{p/2} \, \mathbb{E}_\theta \, |\theta_1|^p \leq (b^2 p)^{p/2}$$

and

$$\mathbb{E}_\theta \, W(F_\theta, F) \leq 2T \, \mathbb{E}_\theta \, \rho(F_\theta, F) + \frac{2}{p-1} \, \frac{(b^2 p)^{p/2}}{T^{p-1}}.$$

Let us take $T = 2b\sqrt{p}$, in which case the above inequality becomes

$$\mathbb{E}_\theta \, W(F_\theta, F) \leq 4b\sqrt{p} \, \mathbb{E}_\theta \, \rho(F_\theta, F) + 4b \, \frac{\sqrt{p}}{p-1} \, 2^{-p}.$$

Here we choose $p = \frac{4 \log n}{\log 2} \geq 4$. Using $\frac{\sqrt{p}}{p-1} < 1$, we arrive at (16.2). $\square$

16.4 Lower Bounds

A lower bound on $\mathbb{E}_\theta \, \rho^2(F_\theta, F)$ which would be close to the upper bound of Proposition 16.2.1 (up to logarithmic terms) may be given with the help of the lower bound on $\mathbb{E}_\theta \, \omega^2(F_\theta, F)$. More precisely, this can be done in the case where the quantity $c_1 \, \mathbb{E}\xi^3 + c_2 \, \mathbb{E}\xi^4$ asymptotically dominates n^{-2}. As was shown in Section 15.4, in the isotropic mean zero case with $|X| = \sqrt{n}$ a.s., we have an asymptotic expansion

$$\sqrt{\pi}\,\mathbb{E}_\theta\,\omega^2(F_\theta,F) = \frac{1}{16}\,\mathbb{E}\xi^3 + c\,\mathbb{E}\xi^4 + O(n^{-2})$$

with $0.01 \le c \le 3$. As before, here we use the notation

$$\xi = \frac{1}{n}\,\langle X,Y\rangle\,,$$

where Y is an independent copy of X. Recall that $\mathbb{E}\xi^3 \ge 0$. Combining the above expansion with the bound (16.1) of Proposition 16.3.1 with $\alpha = 2$ and $b = 1$, we therefore obtain a lower bound on the Kolmogorov distance.

Proposition 16.4.1 *Let X be an isotropic random vector in $\mathbb{R}^n$ with mean zero and such that $|X| = \sqrt{n}$ a.s. For some absolute constants $c_j > 0$,*

$$\sqrt{\log n}\,\mathbb{E}_\theta\,\rho^2(F_\theta,F) \ge c_1\,\mathbb{E}\xi^3 + c_2\,\mathbb{E}\xi^4 - \frac{c_3}{n^2}. \tag{16.3}$$

A similar inequality continues to hold for the normal distribution function Φ in place of F.

The relation (16.2) in Proposition 16.3.1 for the Kantorovich distance $W(F_\theta,F)$ may be used to answer the following question: Is it possible to sharpen the lower bound (16.3) by replacing $\mathbb{E}_\theta\,\rho^2(F_\theta,F)$ with $\mathbb{E}_\theta\,\rho(F_\theta,F)$? To this end, we will need additional information about moments of $\omega(F_\theta,F)$ of order higher than 2.

Let us start with an elementary general inequality, connecting the three distances. Namely, given arbitrary distribution functions F and G, we have

$$\begin{aligned}\omega^2(F,G) &= \int_{-\infty}^{\infty} (F(x)-G(x))^2\,dx \\ &\le \int_{-\infty}^{\infty} \sup_x |F(x)-G(x)|\,|F(x)-G(x)|\,dx = \rho(F,G)\,W(F,G).\end{aligned}$$

Thus, putting

$$\omega = \omega(F_\theta,F), \quad W = W(F_\theta,F), \quad \rho = \rho(F_\theta,F),$$

we have $\omega^3 \le W^{3/2}\rho^{3/2}$ and, by Hölder's inequality with exponents $p = 4$ and $q = \frac{4}{3}$,

$$\Big(\mathbb{E}_\theta\,\omega^3\Big)^{1/3} \le \Big(\mathbb{E}_\theta\,W^6\Big)^{1/12}\Big(\mathbb{E}_\theta\,\rho^2\Big)^{1/4}.$$

By Proposition 14.2.1 with $p = 6$, and using $M_1 \le M_2$, we have

$$\Big(\mathbb{E}_\theta\,W^6\Big)^{1/6} \le \mathbb{E}_\theta\,W + M_2\,\frac{\sqrt{12}}{\sqrt{n}},$$

so that

$$\Big(\mathbb{E}_\theta\,\omega^3\Big)^{1/3} \le \Big(\mathbb{E}_\theta\,W + M_2\frac{\sqrt{12}}{\sqrt{n}}\Big)^{1/2}\Big(\mathbb{E}_\theta\,\rho^2\Big)^{1/4}.$$

If $|X| \leq b\sqrt{n}$ a.s., one may apply Proposition 16.3.2 so as to estimate $\mathbb{E}_\theta W$. This gives

$$\Big(\mathbb{E}_\theta\, \omega^3\Big)^{1/3} \leq C\Big(b\sqrt{\log n}\; \mathbb{E}_\theta\, \rho + \frac{b}{n^4} + \frac{M_2}{\sqrt{n}}\Big)^{1/2} \Big(\mathbb{E}_\theta\, \rho^2\Big)^{1/4} \tag{16.4}$$

for some absolute constant C.

In particular, if the parameters M_2 and b are of order 1, the L^3-norm of $\omega(F_\theta, F)$ is of the order of the L^2-norm of $\rho(F_\theta, F)$ modulo a logarithmic factor. As an illustration, let us return to Proposition 14.6.1, which under the assumption $\mathbb{E}\,|X|^2 = n$ provides the estimate

$$\mathbb{E}_\theta\, \rho \leq \Big(\mathbb{E}_\theta\, \rho^2\Big)^{1/2} \leq C(M_2^2 + \sigma_2)\, \frac{\log n}{\sqrt{n}},$$

for some absolute constant $C > 0$ in terms of our standard functionals

$$M_2^2 = \sup_{\theta \in \mathbb{S}^{n-1}} \mathbb{E}\,\langle X, \theta\rangle^2, \quad \sigma_2 = \frac{1}{\sqrt{n}}\, \mathbb{E}\,\big|\,|X|^2 - n\big|.$$

Since necessarily $b \geq 1$ and $M_2 \geq 1$, inserting the last bound in (16.4), we get:

Lemma 16.4.2 *Let X be a random vector in $\mathbb{R}^n$ such that $|X| \leq b\sqrt{n}$ a.s. and $\mathbb{E}\,|X|^2 = n$. Then for some absolute constant $c > 0$,*

$$c\,\Big(\mathbb{E}_\theta\, \omega^3(F_\theta, F)\Big)^{1/3} \leq (M_2^{3/2} + \sigma_2^{3/4})\sqrt{b}\, \frac{(\log n)^{5/4}}{\sqrt{n}}.$$

Let us now explain how this upper bound can be used to refine the lower bound (16.3). The argument is based on the following general elementary observation. Given a random variable η, introduce the L^p-norms $\|\eta\|_p = (\mathbb{E}\,|\eta|^p)^{1/p}$.

Lemma 16.4.3 *If $\eta \geq 0$ with $0 < \|\eta\|_3 < \infty$, then*

$$\mathbb{E}\,\eta \geq \frac{1}{\mathbb{E}\,\eta^3}\, (\mathbb{E}\,\eta^2)^2. \tag{16.5}$$

Moreover,

$$\mathbb{P}\Big\{\eta \geq \frac{1}{\sqrt{2}}\, \|\eta\|_2\Big\} \geq \frac{1}{8}\, \Big(\frac{\|\eta\|_2}{\|\eta\|_3}\Big)^6. \tag{16.6}$$

Thus, in the case where $\|\eta\|_2$ and $\|\eta\|_3$ are almost equal (up to not large factors), $\|\eta\|_1$ will be of a similar order. Moreover, η cannot be much smaller than its mean $\mathbb{E}\eta$ on a large part of the probability space (where it was defined).

Proof The relation (16.5) follows from the Cauchy inequality applied to the representation $\eta^2 = \eta^{1/2}\, \eta^{3/2}$. To prove (16.6), put $p = \mathbb{P}\{\eta \geq r\}$ for a given number $r > 0$. By Hölder's inequality with exponents $3/2$ and 3,

$$\mathbb{E}\,\eta^2\, 1_{\{\eta \geq r\}} \leq \big(\mathbb{E}\,\eta^3\big)^{2/3}\, p^{1/3}.$$

Hence, choosing $r = \frac{1}{\sqrt{2}} \|\eta\|_2$, we get

$$\begin{aligned} \mathbb{E}\,\eta^2 &= \mathbb{E}\,\eta^2\,1_{\{\eta\ge r\}} + \mathbb{E}\,\eta^2\,1_{\{\eta<r\}} \\ &\le \left(\mathbb{E}\,\eta^3\right)^{2/3} p^{1/3} + r^2 = \left(\mathbb{E}\,\eta^3\right)^{2/3} p^{1/3} + \frac{1}{2}\,\mathbb{E}\,\eta^2. \end{aligned}$$

Hence, $p^{1/3} \ge \frac{1}{2\,(\mathbb{E}\,\eta^3)^{2/3}}\,\mathbb{E}\,\eta^2$, which is the desired bound. □

We apply Lemma 16.4.3 on the unit sphere to $\eta(\theta) = \omega(F_\theta, F)$ viewed as a random variable on the probability space $(\mathbb{S}^{n-1}, \mathfrak{s}_{n-1})$. Recall that, in the case $\mathbb{E}\,|X|^2 = n$, necessarily $M_2 \ge 1$ and $b \ge 1$.

Proposition 16.4.4 *Let X be a random vector in $\mathbb{R}^n$ such that $|X| \le b\sqrt{n}$ a.s. with $\mathbb{E}\,|X|^2 = n$. Assume that*

$$\left(\mathbb{E}_\theta\,\omega^2(F_\theta, F)\right)^{1/2} \ge \frac{D}{\sqrt{n}}$$

for some $D > 0$. Then

$$\mathbb{E}_\theta\,\omega(F_\theta, F) \ge \frac{cb^{-3/2}}{M_2^{9/2} + \sigma_2^{9/4}}\,\frac{D^4}{(\log n)^{15/4}\sqrt{n}},$$

where $c > 0$ is an absolute constant. Moreover,

$$\mathfrak{s}_{n-1}\Big\{\omega(F_\theta, F) \ge \frac{1}{\sqrt{2n}}\,D\Big\} \ge \frac{cb^{-3}}{M_2^9 + \sigma_2^{9/2}}\,\frac{D^6}{(\log n)^{15/2}}.$$

The lower bound obtained for $\mathbb{E}_\theta\,\omega(F_\theta, F)$ implies a similar assertion for the Kolmogorov distance. Indeed, by Proposition 16.3.1 with $\alpha = 1$, we have

$$\frac{1}{\sqrt{b}}\,\mathbb{E}_\theta\,\omega(F_\theta, F) \le 14\,(\log n)^{1/4}\,\mathbb{E}_\theta\,\rho(F_\theta, F) + \frac{8}{n^4}.$$

This gives:

Corollary 16.4.5 *Under the assumptions of Proposition 16.4.4,*

$$\mathbb{E}_\theta\,\rho(F_\theta, F) \ge \frac{cb^{-2}}{M_2^{9/2} + \sigma_2^{9/4}}\,\frac{D^4}{(\log n)^4\,\sqrt{n}} - \frac{1}{4n^4},$$

where $c > 0$ is an absolute constant. In particular, if X is isotropic and $|X|^2 = n$ a.s., then

$$\mathbb{E}_\theta\,\rho(F_\theta, F) \ge \frac{cD^4}{(\log n)^4\,\sqrt{n}} - \frac{1}{4n^4}.$$

Furthermore, if D is of order 1, we obtain a lower bound on $\mathbb{E}_\theta\, \rho(F_\theta, F)$ with rate of the order $1/\sqrt{n}$ modulo a logarithmic factor. Hence, F may also be replaced with the standard normal distribution function Φ, since $\rho(F, \Phi) \leq \frac{c_1}{n}$, cf. Proposition 12.2.2. In this connection, let us emphasize that the rates for the normal approximation of F_θ that are better than $1/n$ (on average) cannot be obtained under the support assumption as above.

Proposition 16.4.6 *For any random vector X in $\mathbb{R}^n$ such that $|X|^2 = n$ a.s., we have*

$$\mathbb{E}_\theta\, \rho(F_\theta, \Phi) \geq \frac{c}{n},$$

where $c > 0$ is an absolute constant.

Proof Using the convexity of the distance function $H \to \rho(H, \Phi)$ and applying Jensen's inequality, we get $\mathbb{E}_\theta\, \rho(F_\theta, \Phi) \geq \rho(F, \Phi)$. To further estimate the latter distance, one may apply Proposition 3.4.1, which gives, for any $T > 0$,

$$\rho(F, \Phi) \geq \frac{1}{3T} \left| \int_0^T (f(t) - \mathrm{e}^{-t^2/2}) \Big(1 - \frac{t}{T}\Big)\, \mathrm{d}t \right|. \tag{16.7}$$

Since $|X|^2 = n$ a.s., and recalling Corollary 11.3.3, we have

$$f(t) = J_n\big(t\sqrt{n}\big) = \Big(1 - \frac{t^4}{4n}\Big)\, \mathrm{e}^{-t^2/2} + O\Big(n^{-2} \min\{1, t^4\}\Big),$$

which holds uniformly on the real line with a universal constant in O. With this approximation, choosing $T = 1$, it follows from (16.7) that $\rho(F, \Phi) \geq \frac{c}{n}$ with some absolute constant $c > 0$ for all $n \geq n_0$ (where n_0 is determined by c only). But, a similar bound also holds for $n < n_0$ since F is supported on the interval $[-\sqrt{n}, \sqrt{n}]$. □

16.5 Remarks

The material of Chapters 15–16 is based on [41].

Chapter 17
Applications of the Second Order Correlation Condition

We continue to use the notations of the previous chapters and denote by F_θ the distribution (function) of the linear form

$$\langle X, \theta \rangle = \theta_1 X_1 + \cdots + \theta_n X_n$$

of a given random vector $X = (X_1, \ldots, X_n)$ in $\mathbb{R}^n$, $n \geq 2$. The average value $F = \mathbb{E}_\theta F_\theta$ with integration on the unit sphere $\mathbb{S}^{n-1}$ over the uniform measure $\mathfrak{s}_{n-1}$ represents the typical distribution of the weighted sums of the components X_k of X.

The aim of this chapter is to derive almost $1/n$ bounds on the Kolmogorov distance $\rho(F_\theta, F)$ in the general case, that is, when the distribution of X is not necessarily supported on the sphere (of radius of $\sqrt{n}$). To this end we involve the second order concentration inequalities on $\mathbb{S}^{n-1}$, assuming throughout the chapter that X satisfies a second order correlation condition: With some (finite) constant $\Lambda = \Lambda(X)$,

$$\mathrm{Var}\bigg(\sum_{j,k=1}^{n} a_{jk} X_j X_k\bigg) \leq \Lambda \sum_{j,k=1}^{n} a_{jk}^2$$

for all $a_{jk} \in \mathbb{R}$. In addition, we always suppose that the distribution of X is isotropic, i.e., $\mathbb{E}\, X_j X_k = \delta_{jk}$ for all j, k. Recall that necessarily $\Lambda \geq \frac{1}{2}$ in the isotropic case.

17.1 Mean Value of $\rho(F_\theta, \Phi)$ Under the Symmetry Assumption

Like in Chapter 14, our basic tool will be the Berry–Esseen bounds. Hence, we need appropriate information about the behavior of the characteristic functions

$$f_\theta(t) = \mathbb{E}\, e^{it\langle X, \theta \rangle}, \quad \theta \in \mathbb{R}^n,\ t \in \mathbb{R},$$

treated as smooth functions on the Euclidean space (with t serving as a parameter). Preliminary results in this direction on the deviations of $\theta \to f_\theta(t)$ on $\mathbb{S}^{n-1}$ about

S. Bobkov et al., *Concentration and Gaussian Approximation for Randomized Sums*, Probability Theory and Stochastic Modelling 104,
https://doi.org/10.1007/978-3-031-31149-9_17

their means

$$f(t) = \mathbb{E}_\theta f_\theta(t)$$

have been already obtained in Section 13.4. In particular, it was shown that, if additionally the distribution of X is symmetric about the origin, then for the intervals $|t| \le An^{1/5}$ $(A > 0)$, we have

$$\mathbb{E}_\theta \, |f_\theta(t) - f(t)|^2 \le \frac{c}{n^2}\, \Lambda t^4,$$

where the constant $c > 0$ is determined by A only (Proposition 13.4.2). Hence

$$\mathbb{E}_\theta \, |f_\theta(t) - f(t)| \le \frac{c}{n} \sqrt{\Lambda}\, t^2. \tag{17.1}$$

This estimate is sufficient to control the distances $\rho(F_\theta, F)$ on average. Indeed, by Propositions 1.7.2 and 1.7.4, in the isotropic case we have $m_4^2 + \sigma_4^2 \le 1 + 2\Lambda$. Hence, the Berry–Esseen bound of Lemma 14.5.3 specialized to the case $p = 2$ yields, for all $T \ge T_0 \ge 1$,

$$c\, \mathbb{E}_\theta \, \rho(F_\theta, F) \le \int_0^{T_0} \frac{\mathbb{E}\, |f_\theta(t) - f(t)|}{t}\, dt + \frac{\Lambda}{n} \Big(1 + \log \frac{T}{T_0}\Big) + \frac{1}{T} + e^{-T_0^2/16}, \tag{17.2}$$

with some absolute constant $c > 0$. Let us choose here $T = 4n$ and $T_0 = 4\sqrt{\log n}$. Using (17.1), the integral in (17.2) produces the term $\frac{T_0^2}{n}\sqrt{\Lambda} = \frac{16 \log n}{n}\sqrt{\Lambda}$ (up to an absolute factor), the second term on the right-hand side produces $\frac{\log n}{n}\,\Lambda$, while the last two terms are of the order $1/n$. The condition $T_0 \le An^{1/5}$ in (17.1) is fulfilled for a suitable absolute constant A. Thus, we arrive at the following bound.

Proposition 17.1.1 *Given an isotropic random vector X in $\mathbb{R}^n$ with a symmetric distribution, we have*

$$\mathbb{E}_\theta \, \rho(F_\theta, F) \le \frac{c \log n}{n}\, \Lambda,$$

for some absolute constant $c > 0$. A similar inequality continues to hold for the standard normal distribution function Φ replacing F.

This bound is applicable to a large variety of probability distributions on $\mathbb{R}^n$, especially when the parameter Λ can effectively be bounded. For example, one can apply the bound $\Lambda \le 4/\lambda_1$ of Proposition 6.3.3, where $\lambda_1 = \lambda_1(X)$ is an optimal value in the Poincaré-type inequality

$$\lambda_1 \operatorname{Var}(u(X)) \le \mathbb{E}\, |\nabla u(X)|^2$$

(in the class of all bounded smooth functions u on $\mathbb{R}^n$). In this case, Proposition 17.1.1 leads to:

Corollary 17.1.2 *If X is isotropic and has a symmetric distribution satisfying a Poincaré-type inequality on $\mathbb{R}^n$ with $\lambda_1 > 0$, then for some absolute constant $c > 0$,*

$$\mathbb{E}_\theta\, \rho(F_\theta, \Phi) \le \frac{c \log n}{n}\, \lambda_1^{-1}.$$

In particular, suppose additionally that the distribution of X has density

$$p(x) = \mathrm{e}^{-V(x)}, \quad x \in K,$$

on the supporting (open convex) set $K \subset \mathbb{R}^n$, where the function V is twice continuously differentiable with Hessian satisfying $V''(x) \ge \lambda I_n$, $x \in K$ (in the matrix sense) for some $\lambda > 0$. Then, according to the Bakry–Emery criterion, $\lambda_1 \ge \lambda$ (cf. Proposition 7.2.7). Hence, the bound of Corollary 17.1.2 holds true with λ in place of λ_1. In fact, the symmetry assumption may be removed from Corollary 17.1.2. This will be discussed in Section 17.5.

17.2 Berry–Esseen Bounds Involving Λ

While Proposition 17.1.1 only gives a bound on $\rho(F_\theta, F)$ in the mean on the unit sphere, one may wonder whether this distance is at most of the same order $(\log n)/n$ on a large part of the sphere. Of course, by Markov's inequality,

$$\mathfrak{s}_{n-1}\Big\{\rho(F_\theta, F) \ge \frac{c\,\Lambda \log n}{n}\, r\Big\} \le \frac{1}{r} \quad (r > 0).$$

This already provides some information about deviations of $\rho(F_\theta, F)$ above its mean. However, the decay for probabilities such as $1/r$ is rather slow for growing r. One may try to sharpen this large deviation inequality under higher order moment hypotheses on X by applying the exponential inequality of Proposition 13.4.2 (in the symmetric case) or the more general ψ_1-bounds of Propositions 13.5.1/13.6.3.

With this aim, as a preliminary step, let us return to the Berry–Esseen-type inequality

$$c\,\rho(F_\theta, F) \le \int_0^T \frac{|f_\theta(t) - f(t)|}{t}\,\mathrm{d}t + \frac{1}{T}\int_0^T |f(t)|\,\mathrm{d}t,$$

which holds true for any $T > 0$ with some universal constant $c > 0$. By Lemma 13.1.3 with $p = 2$ and $T \ge n$, we have in the isotropic case (using $\sigma_4^2 \le \Lambda$, $\Lambda \ge \frac{1}{2}$)

$$\frac{c}{T}\int_0^T |f(t)|\,\mathrm{d}t \le \frac{1 + \sigma_4^2}{n} + \frac{1}{T} \le \frac{1 + \Lambda}{n} + \frac{1}{n} \le \frac{5\Lambda}{n}.$$

Moreover, the pointwise subgaussian bound of the same lemma yields

$$c\,|f(t)| \leq \frac{1+\sigma_4^2}{n} + \mathrm{e}^{-t^2/4} \leq \frac{3\Lambda}{n} + \mathrm{e}^{-t^2/4},$$

so that, for all $T \geq T_0 \geq 1$,

$$c \int_{T_0}^{T} \frac{|f(t)|}{t}\,\mathrm{d}t \leq \frac{\Lambda}{n} \log \frac{T}{T_0} + \mathrm{e}^{-T_0^2/4}.$$

Thus, using the Λ-functional, we have:

Lemma 17.2.1 *Let X be an isotropic random vector in $\mathbb{R}^n$. With some absolute constant $c > 0$, for all $\theta \in \mathbb{S}^{n-1}$,*

$$c\,\rho(F_\theta, F) \leq \int_0^{T_0} \frac{|f_\theta(t) - f(t)|}{t}\,\mathrm{d}t + \int_{T_0}^{T} \frac{|f_\theta(t)|}{t}\,\mathrm{d}t + \frac{\Lambda}{n} \log \frac{T}{T_0}, \tag{17.3}$$

provided that $T \geq 2T_0 \geq 4\sqrt{\log n}$ with $T \geq n$.

Our aim will be to show that the right-hand side of (17.3) is of the order Λ/n up to a logarithmically growing factor for most of $\theta \in \mathbb{S}^{n-1}$ rather than just on average.

First, let us consider the integrals of $|f_\theta(t) - f(t)|/t$ over moderate (that is, not too large) intervals $[0, T_0]$ with $T_0 \leq An^{1/6}$. If $0 < t \leq T_0$, and the random vector X has an isotropic symmetric distribution on $\mathbb{R}^n$, the exponential bound of Proposition 13.4.2 is applicable and may be stated in terms of the Orlicz norm generated by the Young function $\psi_1(s) = \mathrm{e}^{|s|} - 1$. On the space $(\mathbb{S}^{n-1}, \mathfrak{s}_{n-1})$, this norm is defined for any measurable function ξ on the unit sphere by

$$\|\xi\|_{\psi_1} = \inf\Big\{\lambda > 0 : \mathbb{E}_\theta\, \psi_1(|\xi(\theta)|/\lambda) \leq 1\Big\}.$$

Thus, by Proposition 13.4.2, in the symmetric case

$$c\,\|f_\theta(t) - f(t)\|_{\psi_1} \leq \frac{\Lambda t^2}{n}$$

for some absolute constant $c > 0$. To drop the symmetry assumption, we need to involve the linear part $l_t(\theta)$ of the function $\theta \to f_\theta(t) - f(t)$, that is, its orthogonal projection in $L^2(\mathbb{R}^n, \mathfrak{s}_{n-1})$ to the n-dimensional subspace of all linear functions on $\mathbb{R}^n$. By Proposition 13.5.1, we then have a more general bound

$$c\,\|f_\theta(t) - f(t)\|_{\psi_1} \leq \frac{\Lambda t^2}{n} + \sqrt{I(t)}, \quad I(t) = \|l_t\|_{L^2}^2.$$

An application of the triangle inequality in the Orlicz space yields

$$\begin{aligned} c\,\Big\| \int_0^{T_0} |f_\theta(t) - f(t)|\,\frac{\mathrm{d}t}{t} \Big\|_{\psi_1} &\le c \int_0^{T_0} \|f_\theta(t) - f(t)\|_{\psi_1}\,\frac{\mathrm{d}t}{t} \\ &\le \int_0^{T_0} \Big(\frac{\Lambda t^2}{n} + \sqrt{I(t)}\Big)\,\frac{\mathrm{d}t}{t} \\ &= \frac{\Lambda}{2n}\,T_0^2 + \int_0^{T_0} \sqrt{I(t)}\,\frac{\mathrm{d}t}{t}. \end{aligned}$$

By Markov's inequality, $\mathfrak{s}_{n-1}\{|\xi| \ge r\|\xi\|_{\psi_1}\} \le 2\,\mathrm{e}^{-r}$ $(r > 0)$. Hence, we get:

Lemma 17.2.2 *Let X be an isotropic random vector in $\mathbb{R}^n$. For all $0 < T_0 \le An^{1/6}$ and $r > 0$,*

$$\mathfrak{s}_{n-1}\Big\{c\int_0^{T_0} \frac{|f_\theta(t) - f(t)|}{t}\,\mathrm{d}t \ge \frac{r\Lambda}{n}\,T_0^2 + r\int_0^{T_0} \sqrt{I(t)}\,\frac{\mathrm{d}t}{t}\Big\} \le 2\,\mathrm{e}^{-r}$$

with a constant $c > 0$ depending on the parameter $A > 0$ only. Here, $I(t)$ is the squared L^2-norm of the linear part of the function $f_\theta(t) - f(t)$ in $L^2(\mathfrak{s}_{n-1})$.

In applications, the value $T_0 = T_0(n, r)$ may be chosen to be of the order $\sqrt{\log n}$ and depending on r, while the parameter T in (17.3) should be of the order n.

Outside the interval of t of moderate size, that is, on the long interval $[T_0, T]$, both $|f_\theta(t)|$ and $|f(t)|$ are expected to be small for most of θ, at least integrally. To study this property, let us consider the integral of the form

$$L(\theta) = \int_{T_0}^{T} \frac{|f_\theta(t)|}{t}\,\mathrm{d}t, \quad T \ge T_0 > 0. \tag{17.4}$$

By Hölder's inequality, for any integer $p \ge 1$,

$$L(\theta)^{2p} \le \log^{2p-1}\Big(\frac{T}{T_0}\Big) \int_{T_0}^{T} \frac{|f_\theta(t)|^{2p}}{t}\,\mathrm{d}t,$$

so that

$$\mathbb{E}_\theta\, L(\theta)^{2p} \le \log^{2p-1}\Big(\frac{T}{T_0}\Big) \int_{T_0}^{T} \frac{\mathbb{E}_\theta\,|f_\theta(t)|^{2p}}{t}\,\mathrm{d}t.$$

Since

$$|f_\theta(t)|^{2p} = \mathbb{E}\,\mathrm{e}^{it\langle \Sigma_p, \theta\rangle}, \quad \Sigma_p = \sum_{k=1}^{p} \Big(X^{(k)} - Y^{(k)}\Big),$$

where $X^{(k)}, Y^{(k)}$ $(1 \le k \le p)$ are independent copies of the random vector X, we may write

$$\mathbb{E}_\theta\,|f_\theta(t)|^{2p} = \mathbb{E} J_n(t\Sigma_p).$$

Thus,

$$\mathbb{E}_\theta\, L(\theta)^{2p} \le \log^{2p-1}\Big(\frac{T}{T_0}\Big) \int_{T_0}^{T} \mathbb{E}J_n(t\,\Sigma_p)\,\frac{\mathrm{d}t}{t}.$$

Next, given $\alpha > 0$, we split the above expectation into the events $A_\alpha = \big\{|\Sigma_p| \le \alpha\sqrt{n}\big\}$, $B_\alpha = \big\{|\Sigma_p| > \alpha\sqrt{n}\big\}$, and apply the upper bound of Proposition 11.4.1,

$$\big|J_n(t\sqrt{n})\big| \le 3\,\big(\mathrm{e}^{-t^2/2} + \mathrm{e}^{-n/12}\big).$$

This gives

$$\begin{aligned}\int_{T_0}^{T} \mathbb{E}J_n(t\,\Sigma_p)\,1_{B_\alpha}\,\frac{\mathrm{d}t}{t} &\le 3\int_{T_0}^{T} \frac{\mathrm{e}^{-\alpha^2 t^2/2} + \mathrm{e}^{-n/12}}{t}\,\mathrm{d}t \\ &\le 3\,\Big(\mathrm{e}^{-\alpha^2 T_0^2/2} + \mathrm{e}^{-n/12}\Big)\log\Big(\frac{T}{T_0}\Big).\end{aligned}$$

As for the complementary event, one may just use $|J_n| \le 1$ leading to the bound

$$\int_{T_0}^{T} \mathbb{E}J_n(t|\Sigma_p|)\,1_{A_\alpha}\,\frac{\mathrm{d}t}{t} \le \mathbb{P}(A_\alpha)\log\Big(\frac{T}{T_0}\Big).$$

Hence, we get

$$\mathbb{E}_\theta\, L(\theta)^{2p} \le 3\,\log^{2p}\Big(\frac{T}{T_0}\Big)\Big(\mathrm{e}^{-\alpha^2 T_0^2/2} + \mathrm{e}^{-n/12} + \mathbb{P}(A_\alpha)\Big).$$

More specifically, let us choose here $T_0 = \frac{2}{\alpha}\sqrt{p\log n}$ and $T = T_0 n$, which leads to

$$\mathbb{E}_\theta\, L(\theta)^{2p} \le 3\,(\log n)^{2p}\,\big(n^{-2p} + \mathrm{e}^{-n/12} + \mathbb{P}(A_\alpha)\big).$$

Using the inequality $x^{2p}\,\mathrm{e}^{-x} \le (\frac{2p}{e})^{2p}$ $(x \ge 0)$, we have $\mathrm{e}^{-n/12} \le (\frac{24\,p}{e})^{2p}\,n^{-2p}$, and the above bound is simplified. As a result, we get:

Lemma 17.2.3 *Let $X^{(k)}$, $Y^{(k)}$ $(1 \le k \le p)$ be independent copies of a random vector X in $\mathbb{R}^n$. Suppose that $L(\theta)$ is defined in* (17.4) *for limits of integration $T_0 = \frac{2}{\alpha}\sqrt{p\log n}$ and $T = T_0 n$ with $\alpha > 0$. Then, for some absolute constant $c > 0$,*

$$\mathbb{E}_\theta\, L(\theta)^{2p} \le (c\log n)^{2p}\,\Big(p^{2p}\,n^{-2p} + \mathbb{P}(A_\alpha)\Big), \tag{17.5}$$

where $A_\alpha = \big\{|\Sigma_p| \le \alpha\sqrt{n}\big\}$, $\Sigma_p = \sum_{k=1}^{p}\,\big(X^{(k)} - Y^{(k)}\big)$.

17.3 Deviations Under Moment Conditions

One may now develop several natural scenarios in Lemma 17.2.3, where the small ball probabilities $\mathbb{P}(A_\alpha)$ are used in various models of interest. For example, under the moment assumptions, using the variance functional $\sigma_4^2 = \frac{1}{n}\,\mathrm{Var}(|X|^2)$ and assuming that $\mathbb{E}\,|X|^2 = n$, Proposition 1.6.4 provides the bound

$$\mathbb{P}(A_{1/2}) \leq C_p\, n^{-2p}, \quad C_p = \big(16p^2 m_{4p}\big)^{4p} + \big(4\sigma_4^2\big)^{2p}, \tag{17.6}$$

where $m_{4p} = m_{4p}(X)$ and $\sigma_4^2 = \sigma_4^2(X)$. Hence, the growth of m_q for large q is responsible for large deviations of the integrals $L(\theta)$. To capture the behavior of the moments, one may start, for example, from the exponential moment condition

$$\mathbb{E}\,e^{|\langle X,\theta\rangle|/\lambda} \leq 2, \quad \theta \in \mathbb{S}^{n-1}, \tag{17.7}$$

with a parameter $\lambda > 0$. In analogy with the moment functionals $M_p = M_p(X)$, here the optimal value is described as the maximal Orlicz norm

$$\lambda = \sup_{\theta\in\mathbb{S}^{n-1}} \|\,\langle X,\theta\rangle\,\|_{\psi_1},$$

corresponding to the Young function $\psi_1(t) = e^{|t|} - 1$ $(t \in \mathbb{R})$. Then the previous results about the long and moderate intervals lead to:

Proposition 17.3.1 *Let X be an isotropic random vector in $\mathbb{R}^n$ with a symmetric distribution and a finite constant Λ. If X satisfies the moment condition* (17.7)*, then with some absolute constant $c > 0$, for all $r \geq 0$,*

$$\mathfrak{s}_{n-1}\Big\{\rho(F_\theta, F) \geq \frac{c\log n}{n}\,(\Lambda + \lambda^4)\,r\Big\} \leq 2\,\exp\{-r^{1/8}\}. \tag{17.8}$$

A similar inequality also holds for the normal distribution function Φ in place of F.

For the values $r = (\beta\log n)^9$, $\beta > 0$, this estimate provides a polynomial bound

$$\mathfrak{s}_{n-1}\Big\{\rho(F_\theta, F) \geq \frac{c\beta^8\,(\log n)^9}{n}\,(\Lambda + \lambda^4)\Big\} \leq 2\,n^{-\beta}$$

with respect to $1/n$. In other words, if a number A is sufficiently large, then with high $\mathfrak{s}_{n-1}$-probability

$$\rho(F_\theta, F) \leq \frac{A\,(\log n)^9}{n}\,(\Lambda + \lambda^4).$$

Technically, the power $1/8$ in (17.8) comes from the fact that the constant C_p in (17.6) grows at least like p^{8p} for large values of p.

Proof If $r < 1$, the right-hand side of (17.8) is greater than 1, while for $r > n$ and $c > 2$ the $\mathfrak{s}_{n-1}$-probability on the left-hand side is zero. Hence, one may assume that $1 \le r \le n$.

According to Propositions 1.4.2 and 1.3.3, we have $m_p \le M_p^2(X) \le \lambda^2 p^2$ for any $p \ge 2$. Hence, the quantity C_p in (17.6) admits the bound

$$C_p \le \left(c\lambda^2 p^4\right)^{4p} + \left(4\sigma_4^2\right)^{2p}.$$

Moreover, one may involve the second order correlation condition, which ensures that $\sigma_4^2 \le \Lambda$, according to Proposition 1.7.2. By (17.6), this gives

$$\mathbb{P}(A_{1/2}) \le c^p \left(\lambda^{8p} p^{16p} + \Lambda^{2p}\right) n^{-2p}$$

up to some absolute constant $c > 0$.

In view of the isotropy assumption, the parameter λ is bounded away from zero. Indeed, from (17.7) it follows that, for any $\theta \in \mathbb{S}^{n-1}$,

$$1 + \frac{1}{2}\,\mathbb{E}\left(\frac{\langle X, \theta\rangle}{\lambda}\right)^2 \le \mathbb{E}\, e^{|\langle X,\theta\rangle|/\lambda} \le 2,$$

and thus $\lambda^2 \ge \frac{1}{2}$. Hence, the term $\lambda^{8p} p^{16p}$ dominates the term p^{2p} which appears in (17.5). Therefore, being defined for the parameters $T_0 = 4\sqrt{p \log n}$ and $T = T_0 n$, the integrals $L(\theta)$ of Lemma 17.2.3 satisfy the moment bounds

$$\mathbb{E}_\theta\, L(\theta)^{2p} \le (c \log n)^{2p} \left(\lambda^{8p} p^{16p} + \Lambda^{2p}\right) n^{-2p}.$$

Applying Cauchy's inequality, it follows that

$$\mathbb{E}_\theta\, L(\theta)^p \le \left(\frac{c \log n}{n}\right)^p \left(\lambda^{4p} p^{8p} + \Lambda^p\right),$$

implying

$$\left(\mathbb{E}_\theta\, L(\theta)^p\right)^{1/p} \le \frac{c \log n}{n}\left(\Lambda + \lambda^4 p^8\right) \le \frac{c \log n}{n}\left(\Lambda + \lambda^4\right) p^8.$$

At this step (choosing a larger constant c, if necessary), one may assume that $p \ge 1$ is an arbitrary real number. One can now apply Markov's inequality which gives

$$\mathfrak{s}_{n-1}\Big\{L(\theta) \ge \frac{ce \log n}{n}\left(\Lambda + \lambda^4\right) s\Big\} \le \frac{p^{8p}}{(es)^p}, \quad s \ge 1.$$

Choosing here $p = s^{1/8}$ and replacing ce with c, we thus arrive at the inequality

$$\mathfrak{s}_{n-1}\left\{\int_{T_0}^{T} \frac{|f_\theta(t)-f(t)|}{t}\,dt \geq \frac{c\log n}{n}\,(\Lambda+\lambda^4)\,s\right\} \leq \exp\{-s^{1/8}\}.$$

Another change $s = x^{8/7}$ leads to

$$\mathfrak{s}_{n-1}\left\{\int_{T_0}^{T} \frac{|f_\theta(t)-f(t)|}{t}\,dt \geq \frac{c\log n}{n}\,(\Lambda+\lambda^4)\,x^{8/7}\right\} \leq \exp\{-x^{1/7}\}, \quad x \geq 1.$$

Note that, for this choice of p, we have

$$T_0 = 4s^{1/16}\sqrt{\log n} = 4x^{1/14}\sqrt{\log n}, \quad T_0^2 x = 16\,x^{8/7}\log n.$$

One may now apply Lemma 17.2.2 with x in place of r. Since $I(t) = 0$ for all $t \in \mathbb{R}$ (by the symmetry assumption), we then get

$$\mathfrak{s}_{n-1}\left\{\int_{0}^{T_0} \frac{|f_\theta(t)-f(t)|}{t}\,dt \geq \frac{c\log n}{n}\,\Lambda x^{8/7}\right\} \leq 2\,e^{-x}.$$

This bound holds as long as $T_0 \leq An^{1/6}$ with a constant $c > 0$ depending on the parameter $A > 0$.

Combining the two bounds obtained for the intervals $[0, T_0]$ and $[T_0, T]$, we obtain a similar bound for the whole interval $[0, T]$, namely

$$\mathfrak{s}_{n-1}\left\{\int_{0}^{T} \frac{|f_\theta(t)-f(t)|}{t}\,dt \geq \frac{c\log n}{n}\,(\Lambda+\lambda^4)\,x^{8/7}\right\} \leq 3\,\exp\{-x^{1/7}\}.$$

Replacing $x = r^{7/8}$, we arrive at

$$\mathfrak{s}_{n-1}\left\{\int_{0}^{T} \frac{|f_\theta(t)-f(t)|}{t}\,dt \geq \frac{c\log n}{n}\,(\Lambda+\lambda^4)\,r\right\} \leq 3\,\exp\{-r^{1/8}\}.$$

The required condition $T_0 = 4r^{1/16}\sqrt{\log n} \leq An^{1/6}$ is fulfilled with an absolute constant $A > 0$ in view of the assumption $r \leq n$.

It remains to apply Lemma 17.2.1, in which the condition $T_0 \geq 2\sqrt{\log n}$ is fulfilled due to the assumption $r \geq 1$. Thus, the bound (17.3) of this lemma is applicable and leads to

$$\mathfrak{s}_{n-1}\left\{\rho(F_\theta, F) \geq \frac{c\log n}{n}\,(\Lambda+\lambda^4)\,r\right\} \leq 3\,\exp\{-r^{1/8}\}$$

for some absolute constant $c > 0$. As explained before, this bound holds automatically for $0 \leq r \leq 1$ and $r \geq n$, hence, it holds true for all $r \geq 0$. Rescaling the variable r, we get the required inequality 17.8. Proposition 17.3.1 is proved. □

17.4 The Case of Non-symmetric Distributions

In order to extend the bound of Proposition 17.1.1,

$$\mathbb{E}_\theta\, \rho(F_\theta, F) \le \frac{c\, \log n}{n}\, \Lambda, \tag{17.9}$$

to the case where the distribution of X is not necessarily symmetric about the origin, we need to employ more sophisticated results reflecting the size of the linear part of the characteristic functions $f_\theta(t)$ in $L^2(\mathbb{S}^{n-1}, \mathfrak{s}_{n-1})$ as functions of the variable θ. This may be achieved at the expense of a certain term that has to be added to the right-hand side in (17.9). More precisely, we derive the following bound, denoting by c a positive absolute constant which may vary from place to place.

Proposition 17.4.1 *Let $X = (X_1, \dots, X_n)$ be an isotropic random vector in $\mathbb{R}^n$, and let Y be an independent copy of X. Then*

$$c\, \mathbb{E}_\theta\, \rho(F_\theta, F) \le \frac{\log n}{n}\, \Lambda + \Big(\frac{\log n}{n}\Big)^{1/4} \bigg(\mathbb{E}\, \frac{\langle X, Y\rangle}{\sqrt{|X|^2 + |Y|^2}}\bigg)^{1/2}. \tag{17.10}$$

A similar inequality remains to hold for the standard normal distribution function Φ in place of F.

Here, the ratio $\langle X, Y\rangle / \sqrt{|X|^2 + |Y|^2}$ is understood to be zero, if $X = Y = 0$. Moreover, the expectation on the right-hand side of (17.10) is non-negative, which follows from the representation

$$\mathbb{E}\, \frac{\langle X, Y\rangle}{\sqrt{|X|^2 + |Y|^2}} = \frac{1}{\sqrt{\pi}} \int_{-\infty}^{\infty} \sum_{k=1}^{n} \Big(\mathbb{E} X_k\, e^{-|X|^2 r^2}\Big)^2\, dr.$$

If the distribution of X is symmetric, this expectation is vanishing, and we return in (17.10) to the bound (17.9). Thus, Proposition 17.4.1 represents a generalization of this result.

To derive (17.10), we employ Proposition 13.6.3, which yields a second order concentration bound

$$c\, \mathbb{E}_\theta\, |f_\theta(t) - f(t)| \le \frac{1}{n} \sqrt{\Lambda}\, t^2 + \sqrt{I(t)}, \tag{17.11}$$

where $I(t)$ is the squared $L^2(\mathfrak{s}_{n-1})$-norm of the linear part of the function $\theta \to f_\theta(t)$. As we know,

$$I(t) = \frac{t^2}{n}\, \mathbb{E}\, \langle X, Y\rangle \Big(1 - \frac{(U^2 + V^2)\, t^4 - 8R^2 t^2}{4n}\Big)\, e^{-R^2 t^2} + O(t^2 n^{-5/2}) \tag{17.12}$$

with

$$U = \frac{|X|^2}{n}, \quad V = \frac{|Y|^2}{n}, \quad R^2 = \frac{U + V}{2} = \frac{|X|^2 + |Y|^2}{2n} \quad (R \ge 0).$$

As a preliminary step, first let us prove:

Lemma 17.4.2 *If the random vector X in $\mathbb{R}^n$ is isotropic, then*

$$c\int_0^{T_0} \frac{I(t)}{t^2}\,\mathrm{d}t \le \frac{1}{n}\,\mathbb{E}\,\frac{\langle X,Y\rangle}{R} + \frac{\Lambda^2}{n^2}, \tag{17.13}$$

where $T_0 = 4\sqrt{\log n}$.

Proof Introduce the events $A = \{R \le \frac{1}{2}\}$ and $B = \{R > \frac{1}{2}\}$. From the asymptotic representation (17.12), it follows that

$$\begin{aligned}\int_0^{T_0} \frac{I(t)}{t^2}\,\mathrm{d}t = {} & \frac{1}{n}\,\mathbb{E}\Big[\langle X,Y\rangle \int_0^{T_0} \mathrm{e}^{-R^2t^2}\,\mathrm{d}t\Big] \\ & + \frac{2}{n^2}\,\mathbb{E}\Big[\langle X,Y\rangle \int_0^{T_0} R^2t^2\mathrm{e}^{-R^2t^2}\,\mathrm{d}t\Big] \\ & - \frac{1}{4n^2}\,\mathbb{E}\Big[\langle X,Y\rangle \int_0^{T_0} (U^2+V^2)\,t^4\mathrm{e}^{-R^2t^2}\,\mathrm{d}t\Big] + O\Big(\frac{1}{n^2}\Big).\end{aligned}$$

After the change of the variable $Rt = s$ (assuming without loss of generality that $R > 0$) and putting $T_1 = RT_0$, the above is simplified to

$$\begin{aligned}\int_0^{T_0} \frac{I(t)}{t^2}\,\mathrm{d}t = {} & \frac{1}{n}\,\mathbb{E}\Big[\frac{\langle X,Y\rangle}{R} \int_0^{T_1} \mathrm{e}^{-s^2}\,\mathrm{d}s\Big] \\ & + \frac{2}{n^2}\,\mathbb{E}\Big[\frac{\langle X,Y\rangle}{R} \int_0^{T_1} s^2\mathrm{e}^{-s^2}\,\mathrm{d}s\Big] \\ & - \frac{1}{4n^2}\,\mathbb{E}\Big[\frac{\langle X,Y\rangle}{R}\,\frac{U^2+V^2}{R^4} \int_0^{T_1} s^4\mathrm{e}^{-s^2}\,\mathrm{d}s\Big] + O\Big(\frac{1}{n^2}\Big).\end{aligned}$$

At the expense of a small error, integration here may be extended from the interval $[0, T_1]$ to the whole half-axis $(0, \infty)$. To see this, one can use the estimate

$$\int_{T_1}^{\infty} (1+s^2+s^4)\,\mathrm{e}^{-s^2}\,\mathrm{d}s < c\mathrm{e}^{-T_1^2/2} \quad (T_1 \ge 0),$$

together with

$$\Big|\frac{\langle X,Y\rangle}{R}\Big| \le \frac{|X|\,|Y|}{R} \le \frac{|X|^2+|Y|^2}{2R} = Rn. \tag{17.14}$$

By Proposition 1.6.1 (with $p = 2$, $\lambda = \frac{1}{2}$), we have $\mathbb{P}\{|X|^2 \le \frac{n}{2}\} \le \frac{4\sigma_4^2}{n} \le \frac{4\Lambda}{n}$, so,

$$\mathbb{P}(A) = \mathbb{P}\Big\{|X|^2+|Y|^2 \le \frac{n}{2}\Big\} \le \mathbb{P}\Big\{|X|^2 \le \frac{n}{2}\Big\}\,\mathbb{P}\Big\{|Y|^2 \le \frac{n}{2}\Big\} \le \frac{16\Lambda^2}{n^2}. \tag{17.15}$$

On the other hand, on the complementary set B, we have $T_1^2 = 16R^2\log n > 4\log n$. Since $\mathbb{E}R^2 = 1$, it follows that

$$\mathbb{E}\, R\, e^{-T_1^2/2} = \mathbb{E}\, R\, e^{-T_1^2/2}\, 1_A + \mathbb{E}\, R\, e^{-T_1^2/2}\, 1_B \le \frac{1}{2}\, \mathbb{P}(A) + \frac{1}{n^2}\, \mathbb{E} R \le \frac{c\Lambda^2}{n^2},$$

where we used the universal lower bound $\Lambda \ge \frac{1}{2}$. Hence, using (17.14),

$$\mathbb{E}\Big[\frac{|\langle X, Y\rangle|}{R} \int_{T_1}^{\infty} e^{-s^2}\, ds\Big] \le cn\, \mathbb{E} R e^{-T_1^2} \le \frac{c\Lambda^2}{n}.$$

By a similar argument,

$$\mathbb{E}\Big[\frac{|\langle X, Y\rangle|}{R} \int_{T_1}^{\infty} s^2 e^{-s^2}\, ds\Big] \le cn\, \mathbb{E} R\, e^{-T_1^2/2} \le \frac{c\Lambda^2}{n}.$$

Using

$$\frac{U^2+V^2}{R^4} = \frac{4(U^2+V^2)}{(U+V)^2} \le 4,$$

we also have

$$\mathbb{E}\Big[\frac{|\langle X, Y\rangle|}{R}\, \frac{U^2+V^2}{R^4} \int_{T_1}^{\infty} s^4 e^{-s^2}\, ds\Big] \le cn\, \mathbb{E} R\, e^{-T_1^2/2} \le \frac{c\Lambda^2}{n}.$$

Thus, extending the integration to the whole positive half-axis, we get

$$\int_0^{T_0} \frac{I(t)}{t^2}\, dt = \frac{c_1}{n}\, \mathbb{E}\, \frac{\langle X, Y\rangle}{R} + \frac{c_2}{n^2}\, \mathbb{E}\, \frac{\langle X, Y\rangle}{R} - \frac{c_3}{n^2}\, \mathbb{E}\Big[\frac{\langle X, Y\rangle}{R}\, \frac{U^2+V^2}{R^4}\Big] + O(\Lambda^2 n^{-2})$$

for some positive absolute constants c_j. Moreover, using the identity

$$\frac{U^2+V^2}{R^4} = 2 + \frac{(U-V)^2}{2R^4} = 2 + 2\, \frac{(U-V)^2}{(U+V)^2},$$

and recalling that $\mathbb{E}\, \frac{\langle X, Y\rangle}{R} \ge 0$, it follows that for some other positive absolute constants

$$\int_0^{T_0} \frac{I(t)}{t^2}\, dt \le \frac{c_1}{n}\, \mathbb{E}\, \frac{\langle X, Y\rangle}{R} - \frac{c_2}{n^2}\, \mathbb{E}\, \frac{\langle X, Y\rangle}{R}\, \frac{(U-V)^2}{(U+V)^2} + O\Big(\frac{\Lambda^2}{n^2}\Big). \tag{17.16}$$

To get rid of the last expectation (by showing that it is bounded by a dimension free quantity), first note that, by (17.15), the expression under this expectation is bounded in absolute value by Rn. Hence, applying Cauchy's inequality and using $\mathbb{E} R^2 = 1$, from (17.14) we obtain that

$$\mathbb{E}\Big|\frac{\langle X, Y\rangle}{R}\Big|\, \frac{(U-V)^2}{(U+V)^2}\, 1_A \le \mathbb{E}\Big|\frac{\langle X, Y\rangle}{R}\Big|\, 1_A \le n\, \mathbb{E} R\, 1_A \le n\sqrt{\mathbb{P}(A)} \le 4\Lambda. \tag{17.17}$$

Turning to the complementary set, note that $|\frac{\langle X,Y\rangle}{R}| \leq 2\,|\,\langle X,Y\rangle\,|$ on B, while

$$\frac{(U-V)^2}{(U+V)^2} \leq \frac{|U-V|}{U+V} = \frac{|U-V|}{2R^2} \leq 2\,|U-V|.$$

Hence, by Cauchy's inequality, and using $\mathbb{E}\,\langle X,Y\rangle^2 = n$,

$$\begin{aligned}\mathbb{E}\Big|\frac{\langle X,Y\rangle}{R}\Big|\,\frac{(U-V)^2}{(U+V)^2}\,1_B &\leq 4\,\mathbb{E}\,|\,\langle X,Y\rangle\,|\,|U-V| \\ &\leq 4\sqrt{n}\,\sqrt{\mathbb{E}\,(U-V)^2} = 4\sqrt{2}\,\sigma_4 \leq 4\sqrt{2\Lambda}.\end{aligned}$$

Combining this bound with (17.17), we finally obtain that

$$\mathbb{E}\Big|\frac{\langle X,Y\rangle}{R}\Big|\,\frac{(U-V)^2}{(U+V)^2} \leq c\Lambda.$$

As a result, we arrive in (17.16) at the bound (17.13). Lemma 17.4.2 is thus proved.□

Proof (of Proposition 17.4.1) Inserting (17.11) into the Berry–Esseen-type bound (17.2), we get that with some absolute constant $c > 0$, for all $T \geq T_0 \geq 1$,

$$c\,\mathbb{E}_\theta\,\rho(F_\theta,F) \leq \frac{1}{n}\,T_0^2\sqrt{\Lambda} + \int_0^{T_0}\frac{\sqrt{I(t)}}{t}\,dt + \frac{\Lambda}{n}\Big(1+\log\frac{T}{T_0}\Big) + \frac{1}{T} + e^{-T_0^2/16}.$$

We choose here $T = 4n$, $T_0 = 4\sqrt{\log n}$, so that

$$c\,\mathbb{E}_\theta\,\rho(F_\theta,F) \leq \frac{\log n}{n}\,\Lambda + \int_0^{T_0}\frac{\sqrt{I(t)}}{t}\,dt.$$

Moreover, by Cauchy's inequality,

$$c\,\mathbb{E}_\theta\,\rho(F_\theta,F) \leq \frac{\log n}{n}\,\Lambda + \sqrt{T_0}\,\Big(\int_0^{T_0}\frac{I(t)}{t^2}\,dt\Big)^{1/2}.$$

Hence, an application of Lemma 17.4.2 yields

$$c\,\mathbb{E}_\theta\,\rho(F_\theta,F) \leq \frac{\log n}{n}\,\Lambda + \sqrt{T_0}\,\Big(\frac{1}{n}\,\mathbb{E}\,\frac{\langle X,Y\rangle}{R} + \frac{\Lambda^2}{n^2}\Big)^{1/2}.$$

Simplifying the expression on the right-hand side, we arrive at (17.10).
Proposition 17.4.1 is proved. □

17.5 The Mean Value of $\rho(F_\theta, \Phi)$ in the Presence of Poincaré Inequalities

Let us rewrite the bound (17.10) of Proposition 17.4.1 as

$$c\,\mathbb{E}_\theta\,\rho(F_\theta,\Phi) \le \frac{\log n}{n}\,\Lambda + \frac{(\log n)^{1/4}}{\sqrt{n}}\,\Big(\mathbb{E}\,\frac{\langle X,Y\rangle}{R}\Big)^{1/2}, \tag{17.18}$$

where $R^2 = \frac{|X|^2+|Y|^2}{2n}$ ($R \ge 0$), with Y being an independent copy of X. As a next step, we are going to simplify this bound by modifying the first term on the right-hand side and removing the second one for a large class of probability distributions μ of X on $\mathbb{R}^n$. As we know, this is possible for symmetric μ and also for the class of probability distributions that are supported on the sphere of radius $\sqrt{n}$ and have mean zero with $\mathbb{E}\,\langle X,Y\rangle^3 = 0$ (in this case, the first term should be modified to $\frac{\log n}{n^2}\,(\mathbb{E}\,\langle X,Y\rangle^4)^{1/2}$, cf. Proposition 16.2.1).

Note that, under our standard assumptions,

$$\mathbb{E}R^2 = 1, \quad \mathrm{Var}(R^2) = \frac{\sigma_4^2}{2n} \le \frac{\Lambda}{2n}.$$

Hence, with high probability the ratio $\frac{\langle X,Y\rangle}{R}$ is almost $\langle X,Y\rangle$, which in turn has zero expectation, as long as X has mean zero. However, in general it is not clear whether or not this approximation is sufficient to make further simplifications (having also in mind the factor $n^{-1/2}$). Nevertheless, the approximation $R^2 \sim 1$ is indeed sufficiently strong, for example, in the case where μ satisfies a Poincaré-type inequality

$$\lambda_1\,\mathrm{Var}_\mu(u) \le \int |\nabla u|^2\,d\mu$$

(in the class of all smooth functions u on $\mathbb{R}^n$). Namely, we have the following generalization of Corollary 17.1.2.

Proposition 17.5.1 *Let X be an isotropic random vector in $\mathbb{R}^n$ with mean zero and a positive Poincaré constant λ_1. Then for some absolute constant $c > 0$,*

$$c\,\mathbb{E}_\theta\,\rho(F_\theta,\Phi) \le \frac{\log n}{n}\,\lambda_1^{-1}. \tag{17.19}$$

The proof is based on the following statement of independent interest. We denote by c a positive absolute constant which may vary from place to place.

Lemma 17.5.2 *If the isotropic random vector $X = (X_1,\dots,X_n)$ has mean zero and a positive Poincaré constant λ_1, then*

$$\mathbb{E}\,\frac{\langle X,Y\rangle}{R} \le \frac{c}{\lambda_1^2\,n}. \tag{17.20}$$

As was emphasized in the previous section, the above expectation is always non-negative. Here, the ratio $\frac{\langle X,Y\rangle}{R}$ us understood to be zero in the case $X = Y = 0$.

Applying (17.20) in (17.18) and using $\Lambda \le 4/\lambda_1$ (Proposition 6.3.3), we get an estimate on average

$$c\,\mathbb{E}_\theta\, \rho(F_\theta, \Phi) \le \frac{\log n}{n}\,\frac{1}{\lambda_1} + \frac{(\log n)^{1/4}}{\sqrt{n}}\,\frac{1}{\lambda_1\sqrt{n}},$$

thus proving the relation (17.19).

Proof (of Lemma 17.5.2) Without loss of generality, assume that $R > 0$ a.s. Put

$$\delta_n = \frac{1}{\lambda_1 n}.$$

We apply the Poincaré-type inequality

$$\int\!\!\int |u(x,y)|^2\,\mathrm{d}\mu(x)\,\mathrm{d}\mu(y) \le \frac{1}{\lambda_1}\int\!\!\int |\nabla u(x,y)|^2\,\mathrm{d}\mu(x)\,\mathrm{d}\mu(y), \tag{17.21}$$

with respect to the product measure $\mu \otimes \mu$, which holds true with the same Poincaré constant as for μ in the class of all smooth functions u on $\mathbb{R}^n \times \mathbb{R}^n$ with $(\mu\otimes\mu)$-mean zero. Moreover, according to Proposition 6.2.1, in this class, for any $p \ge 2$,

$$\int\!\!\int |u(x,y)|^p\,\mathrm{d}\mu(x)\,\mathrm{d}\mu(y) \le \frac{p^p}{(2\lambda_1)^{p/2}}\int\!\!\int |\nabla u(x,y)|^p\,\mathrm{d}\mu(x)\,\mathrm{d}\mu(y). \tag{17.22}$$

Let us also recall that, by Proposition 6.3.4 applied to the $2n$-dimensional random vector (X,Y), the event $A = \{R \le \frac{1}{2}\}$ has probability

$$\mathbb{P}(A) \le 3\,\mathrm{e}^{-\sqrt{\lambda_1 n/2}}.$$

By (17.14), we have $\frac{|\langle X,Y\rangle|}{R} \le Rn$, so that $\frac{|\langle X,Y\rangle|}{R} \le \frac{1}{2}\,n$ on the set A. This implies

$$\mathbb{E}\,\frac{|\,\langle X,Y\rangle\,|}{R}\,1_A \le \frac{n}{2}\,\mathbb{P}(A) \le \frac{3n}{2}\,\mathrm{e}^{-\sqrt{\lambda_1 n/2}} \le \frac{c}{\lambda_1^2 n}. \tag{17.23}$$

Similarly,

$$\mathbb{E}\,|\,\langle X,Y\rangle\,|\,1_A \le \frac{n}{4}\,\mathbb{P}(A) \le \frac{c}{\lambda_1^2 n},$$

and since X has mean zero, for the complementary set $B = \{R > \frac{1}{2}\}$ we have the same bound

$$\big|\mathbb{E}\,\langle X,Y\rangle\,1_B\big| \le \frac{c}{\lambda_1^2 n}.$$

Using once more (17.14), on the set A we also have

$$\mathbb{E}\,|\langle X,Y\rangle\,|\,R^2\,1_A \le \frac{n}{16}\,\mathbb{P}(A) \le \frac{c}{\lambda_1^2 n},$$
$$\mathbb{E}\,|\langle X,Y\rangle\,|\,R^4\,1_A \le \frac{n}{64}\,\mathbb{P}(A) \le \frac{c}{\lambda_1^2 n}.$$

Now, consider the function $w(\varepsilon) = (1+\varepsilon)^{-1/2}$ on the half-axis $\varepsilon \ge -\frac{3}{4}$. By Taylor's formula, for some point ε_1 between $-\frac{3}{4}$ and ε,

$$w(\varepsilon) = 1 - \frac{1}{2}\,\varepsilon + \frac{3}{8}\,\varepsilon^2 - \frac{15}{16}\,(1+\varepsilon_1)^{-7/2}\,\varepsilon^3 = 1 - \frac{1}{2}\,\varepsilon + \frac{3}{8}\,\varepsilon^2 - \beta\varepsilon^3$$

with some $0 \le \beta \le 40$. Putting $\varepsilon = R^2 - 1$, we then get on the set B

$$\begin{aligned}\frac{\langle X,Y\rangle}{R} &= \langle X,Y\rangle - \frac{1}{2}\,\langle X,Y\rangle\,(R^2-1) + \frac{3}{8}\,\langle X,Y\rangle\,(R^2-1)^2 - \beta\,\langle X,Y\rangle\,(R^2-1)^3\\ &= \frac{15}{8}\,\langle X,Y\rangle - \frac{5}{4}\,\langle X,Y\rangle\,R^2 + \frac{3}{8}\,\langle X,Y\rangle\,R^4 - \beta\,\langle X,Y\rangle\,(R^2-1)^3.\end{aligned}$$

By the independence of X and Y, and due to the mean zero assumption, $\mathbb{E}\,\langle X,Y\rangle = \mathbb{E}\,\langle X,Y\rangle\,R^2 = 0$. Hence, writing $1_B = 1 - 1_A$, we have

$$\begin{aligned}\mathbb{E}\,\frac{\langle X,Y\rangle}{R} = &-\frac{15}{8}\,\mathbb{E}\,\langle X,Y\rangle\,1_A + \frac{5}{4}\,\mathbb{E}\,\langle X,Y\rangle\,R^2\,1_A - \frac{3}{8}\,\mathbb{E}\,\langle X,Y\rangle\,R^4\,1_A\\ &+ \frac{3}{8}\,\mathbb{E}\,\langle X,Y\rangle\,R^4 - \beta\,\mathbb{E}\,\langle X,Y\rangle\,(R^2-1)^3\,1_B.\end{aligned}$$

Here, the first three expectations on the right-hand side do not exceed in absolute value a multiple $\frac{1}{\lambda_1^2 n}$. Thus, using (17.23), we get

$$\mathbb{E}\,\frac{\langle X,Y\rangle}{R} \le \frac{c_1}{\lambda_1^2 n} + \frac{3}{8}\,\mathbb{E}\,\langle X,Y\rangle\,R^4 + c_2\,\mathbb{E}\,|\,\langle X,Y\rangle\,|\,|R^2-1|^3, \tag{17.24}$$

where $c_1 > 0$ and $c_2 > 0$ are absolute constants.

By Cauchy's inequality, the square of the last expectation does not exceed

$$\mathbb{E}\,\langle X,Y\rangle^2\;\mathbb{E}\,(R^2-1)^6 = n\,\mathbb{E}\,(R^2-1)^6.$$

In turn, the latter expectation may be bounded by virtue of the inequality (17.22) applied with $p = 6$ to the function $u(x,y) = \frac{1}{2n}\,(|x|^2 + |y|^2) - 1$. Since

$$|\nabla u(x,y)|^2 = |\nabla_x u(x,y)|^2 + |\nabla_y u(x,y)|^2 = \frac{|x|^2 + |y|^2}{n^2},$$

it gives

$$\mathbb{E}\,(R^2-1)^6 \le \frac{c}{\lambda_1^3\,n^3}\,\mathbb{E}\,R^6. \tag{17.25}$$

On the other hand, the Poincaré-type inequality easily yields the bound $\mathbb{E}R^6 \le c/\lambda_1^3$. However, in this step a more accurate estimation is required. Write

$$R^6 = (R^2-1)^3 + 3\,(R^2-1)^2 + 3\,(R^2-1) + 1,$$

so that

$$\mathbb{E}R^6 = \mathbb{E}\,(R^2-1)^3 + 3\,\mathbb{E}\,(R^2-1)^2 + 1. \tag{17.26}$$

By (17.21) with the same function u, we have

$$\mathbb{E}\,(R^2-1)^2 \le \frac{2}{\lambda_1 n}\,\mathbb{E}R^2 = 2\delta_n,$$

while (17.22) with $p = 3$ gives

$$\mathbb{E}\,|R^2-1|^3 \le 27\,\delta_n^{3/2}\,\mathbb{E}\,|R|^3.$$

Putting $x^2 = \mathbb{E}R^6$ $(x > 0)$ and using $\mathbb{E}\,|R|^3 \le x$ (by applying Cauchy's inequality), we therefore get from (17.26) that

$$x^2 \le 27\,\delta_n^{3/2}\,x + 6\delta_n + 1.$$

This quadratic inequality is easily solved to yield $x \le c\,(\delta_n+1)^{3/2}$. One can now apply this bound in (17.25) to conclude that

$$\mathbb{E}\,(R^2-1)^6 \le \frac{c}{\lambda_1^3 n^3}\,(\delta_n+1)^3.$$

This implies

$$\mathbb{E}\,\langle X,Y\rangle^2\,\mathbb{E}\,(R^2-1)^6 \le \frac{c}{\lambda_1^3 n^2}\,(\delta_n+1)^3,$$

which allows us to simplify the inequality (17.24) to the form

$$\mathbb{E}\,\frac{\langle X,Y\rangle}{R} \le \frac{c_1}{\lambda_1^2 n} + \frac{c_2}{\lambda_1^{3/2} n}\,(\delta_n+1)^{3/2} + \frac{3}{8}\,\mathbb{E}\,\langle X,Y\rangle\,R^4. \tag{17.27}$$

We are left with the estimation of the last expectation. Since

$$\mathbb{E}\,\langle X,Y\rangle\,|X|^4 = \mathbb{E}\,\langle X,Y\rangle\,|Y|^4 = 0,$$

it follows that

$$\mathbb{E}\,\langle X,Y\rangle\,R^4 = \frac{1}{2n^2}\,\mathbb{E}\,\langle X,Y\rangle\,|X|^2\,|Y|^2 = \frac{1}{2n^2}\,\big|\mathbb{E}\,|X|^2 X\big|^2,$$

where the latter expectation is understood in the usual vector sense. That is, in terms of the components in X, we have

$$\mathbb{E}\,|X|^2 X = (a_1,\dots,a_n), \quad a_k = \mathbb{E}\,|X|^2 X_k = \mathbb{E}\,(|X|^2 - n)\,X_k.$$

Since the collection $\{X_1, \dots, X_n\}$ appears as an orthonormal system in $L^2(\Omega, \mathcal{F}, \mathbb{P})$, the numbers a_k represent the (Fourier) coefficients for the projection of the random variable $|X|^2 - n$ onto the span of X_k's. Hence, by Bessel's inequality,

$$\left| \mathbb{E}\, |X|^2 X \right|^2 = \sum_{k=1}^{n} a_k^2 \le \big\| \, |X|^2 - n \big\|_{L^2(\mathbb{P})}^2 = \mathrm{Var}(|X|^2) = n\, \sigma_4^2(X) \le \frac{4n}{\lambda_1}.$$

Thus,

$$\mathbb{E}\,\langle X, Y\rangle\, R^4 \le \frac{2}{\lambda_1 n}.$$

In view of the upper bound $\lambda_1 \le 1$ (cf. Proposition 6.3.1), this expectation is dominated by the first term in (17.27), and we arrive at

$$\mathbb{E}\, \frac{\langle X, Y\rangle}{R} \le \frac{c}{\lambda_1^2 n} + \frac{c}{\lambda_1^{3/2} n} \Big(\frac{1}{\lambda_1 n} + 1\Big)^{3/2}.$$

If $\lambda_1 \ge n^{-1}$, the first term on the right-hand side dominates the second one, and we obtain the desired inequality (17.20). In the other case, we have $\frac{1}{\lambda_1^2 n} \ge n$, and then (17.20) holds true, since, by (17.14), $\mathbb{E}\, \frac{\langle X, Y\rangle}{R} \le n\,\mathbb{E}R \le n$. Lemma 17.5.2 is thus proved. □

17.6 Deviations of $\rho(F_\theta, \Phi)$ in the Presence of Poincaré Inequalities

In the next step we complement Proposition 17.5.1 with a large deviation bound in the presence of a Poincaré-type inequality for the distribution of the random vector X. As we know, for some absolute $c > 0$,

$$\mathbb{E}\, e^{c\sqrt{\lambda_1}\, |\langle X, \theta\rangle|} \le 2 \quad \text{for all } \theta \in \mathbb{S}^{n-1}$$

(cf. Proposition 6.1.1 and the Remarks after it). Hence, the parameter λ in Proposition 17.3.1 satisfies $\lambda \le \frac{1}{c\sqrt{\lambda_1}}$, and we are led to the bound

$$\mathfrak{s}_{n-1}\Big\{\rho(F_\theta, \Phi) \ge \frac{c \log n}{n}\, \lambda_1^{-2}\, r\Big\} \le 2\, \exp\{-r^{1/8}\}, \quad r \ge 0.$$

However, here the decay in r on the right-hand side can be improved by a more careful estimation of the probability of the events A_α from Lemma 17.2.3, while λ_1^{-2} on the left may be replaced with a smaller quantity λ_1^{-1}. More precisely, we establish the following refinement of Proposition 17.5.1.

Proposition 17.6.1 *Let X be an isotropic random vector in $\mathbb{R}^n$ with mean zero, satisfying the Poincaré-type inequality with $\lambda_1 > 0$. Then with some absolute constant $c > 0$, for all $r \geq 0$,*

$$\mathfrak{s}_{n-1}\Big\{\rho(F_\theta, \Phi) \geq \frac{c\log n}{n}\,\lambda_1^{-1}\,r\Big\} \leq 2\,\exp\big\{-\sqrt{r}\big\}. \tag{17.28}$$

This inequality readily implies a similar estimate on the mean given in Proposition 17.5.1. For the values $r = (\beta\log n)^2$, $\beta > 0$, (17.28) provides a polynomial bound

$$\mathfrak{s}_{n-1}\Big\{\rho(F_\theta, \Phi) \geq \frac{c\beta^2\,(\log n)^3}{n}\,\lambda_1^{-1}\Big\} \leq 2n^{-\beta}.$$

In other words, if a number A is large enough, then with high $\mathfrak{s}_{n-1}$-probability

$$\rho(F_\theta, \Phi) \leq \frac{A\,(\log n)^3}{n}\,\lambda_1^{-1}.$$

Proof (of Proposition 17.6.1) Let us return to the Berry–Esseen-type bound of Lemma 17.2.1. Using $\Lambda \leq 4/\lambda_1$ and assuming that $T = T_0 n$, $T_0 \geq 2\sqrt{\log n}$, the bound (7.3) gives, for all $\theta \in \mathbb{S}^{n-1}$,

$$c\,\rho(F_\theta, \Phi) \leq \frac{\log n}{n}\,\lambda_1^{-1} + \int_0^{T_0} \frac{|f_\theta(t) - f(t)|}{t}\,dt + L(\theta). \tag{17.29}$$

Here

$$L(\theta) = \int_{T_0}^{T} \frac{|f_\theta(t)|}{t}\,dt,$$

and $c > 0$ is an absolute constant. In order to estimate $\mathfrak{s}_{n-1}$-probabilities for large deviations of this term, let us recall that, by Proposition 6.3.5,

$$\mathbb{P}\Big\{\Big|\sum_{k=1}^{p} (X^{(k)} - Y^{(k)})\Big|^2 \leq \frac{np}{2}\Big\} \leq 3\,e^{-\frac{1}{3}\sqrt{\lambda_1 n}},$$

where $X^{(k)}$, $Y^{(k)}$ $(1 \leq k \leq p)$ are independent copies of X. One may therefore apply Lemma 17.2.3 with parameter $\alpha = \sqrt{p/2}$ and conclude that for some absolute constant $c > 0$,

$$\mathbb{E}_\theta\, L(\theta)^{2p} \leq (c\log n)^{2p}\,\Big(p^{2p}\,n^{-2p} + e^{-\frac{1}{3}\sqrt{\lambda_1 n}}\Big). \tag{17.30}$$

Note that the integrals $L(\theta)$ were defined according to (17.4) with limits of integration

$$T_0 = \frac{2}{\alpha}\sqrt{p\log n} = \sqrt{8\log n} \quad \text{and} \quad T = T_0 n,$$

which is consistent with the choice of the parameters T_0 and T in (17.29). Using the inequality $e^{-x} \leq (\frac{4p}{ex})^{4p}$ $(x > 0)$ and the property $\lambda_1 \leq 1$, (17.30) simplifies to

$$\mathbb{E}_\theta\, L(\theta)^{2p} \le \Big(\frac{c\log n}{n}\,\lambda_1^{-1}\Big)^{2p}\, p^{4p}$$

(for some other constant), i.e.,

$$\|L\|_{L^{2p}(\mathfrak{s}_{n-1})} \le \frac{c\log n}{n}\,\lambda_1^{-1}\, p^2.$$

This inequality is readily extended to all real $p \ge 1/2$. Thus, for any $p \ge 1$,

$$\mathbb{E}_\theta\, L(\theta)^{p} \le \Big(\frac{c\log n}{n}\,\lambda_1^{-1}\Big)^{p}\, p^{2p}.$$

One can now repeat the arguments which were used in the proof of Proposition 17.3.1. First, by Markov's inequality, for all $r > 0$,

$$\mathfrak{s}_{n-1}\Big\{L(\theta) \ge \frac{ce\log n}{n}\,\lambda_1^{-1}\, r\Big\} \le \frac{p^{2p}}{(er)^p}.$$

Assuming that $r \ge 1$ and choosing $p = \sqrt{r}$, we thus have for some constant $c > 0$

$$\mathfrak{s}_{n-1}\Big\{L(\theta) \ge \frac{c\log n}{n}\,\lambda_1^{-1} r\Big\} \le e^{-\sqrt{r}}. \tag{17.31}$$

This may be combined with a large deviation bound of Lemma 17.2.2. First, by Lemmas 17.4.2 and 17.5.2, and repeating the argument in the proof of Proposition 17.4.1 with $T_0' = 4\sqrt{\log n}$, we have

$$\begin{aligned}\int_0^{T_0'} \frac{\sqrt{I(t)}}{t}\,dt &\le \sqrt{T_0'}\,\Big(\int_0^{T_0'} \frac{I(t)}{t^2}\,dt\Big)^{1/2}\\ &\le c\sqrt{T_0'}\,\Big(\frac{1}{n}\,\mathbb{E}\,\frac{\langle X,Y\rangle}{R} + \frac{\Lambda^2}{n^2}\Big)^{1/2} \le c'\,(\log n)^{1/4}\,\frac{1}{\lambda_1 n}.\end{aligned}$$

Hence, using $T_0 \le T_0' \le An^{1/6}$ (where $A > 0$ is an absolute constant) and once more $\Lambda \le 4/\lambda_1$, Lemma 17.2.2 yields

$$\mathfrak{s}_{n-1}\Big\{\int_0^{T_0} \frac{|f_\theta(t) - f(t)|}{t}\,dt \ge \frac{c\log n}{\lambda_1 n}\,\lambda_1^{-1} r\Big\} \le 2\,e^{-r}. \tag{17.32}$$

Thus, being applied in (17.29), the two inequalities (17.31)–(17.32) yield

$$\mathfrak{s}_{n-1}\Big\{\rho(F_\theta,\Phi) \ge \frac{c\log n}{n}\,\lambda_1^{-1}\, r\Big\} \le 3\,e^{-\min\{r,\sqrt{r}\}}.$$

Rescaling the variable r, one can now easily obtain the desired inequality (17.28). Proposition 17.6.1 is proved. □

17.7 Relation to the Thin Shell Problem

Returning to Proposition 17.5.1 with its bound

$$\mathbb{E}_\theta\, \rho(F_\theta, \Phi) \le \frac{c\, \log n}{n}\, \lambda_1^{-1},$$

recall that $\lambda_1 = \lambda_1(X) > 0$ for any log-concave random vector X in $\mathbb{R}^n$. So, one may introduce the positive numbers

$$\lambda_{1,n} = \inf_X \lambda_1(X), \quad \sigma_{4,n}^2 = \sup_X \sigma_4^2(X),$$

by taking the infimum and the supremum in the entire class of all isotropic, log-concave probability distributions on $\mathbb{R}^n$ with mean zero. Since $\sigma_4^2(X) = \frac{1}{n}\operatorname{Var}(|X|)^2$ is related to λ_1 via $\sigma_4^2(X) \le 4/\lambda_1$ (cf. Proposition 6.3.3), we have $\lambda_{1,n}^{-1} \ge \frac{1}{4}\sigma_{4,n}^2$. As was shown by Eldan [84], this inequality may be reversed in the form

$$\frac{1}{\lambda_{1,n}} \le c \log n \sum_{k=1}^n \frac{\sigma_{4,k}^2}{k}$$

for some absolute constant $c > 0$. It may be simplified to

$$\frac{1}{\lambda_{1,n}} \le c\sigma_{4,n}^2\, (\log n)^2, \tag{17.33}$$

since the sequence $\sigma_{4,n}^2$ is increasing in some sense (cf. Remark 17.7.2 below). Assuming that the thin shell conjecture stating that $\sigma_{4,n}^2$ is bounded by an absolute constant is true, Proposition 17.5.1 then yields the upper bound

$$\mathbb{E}_\theta\, \rho(F_\theta, \Phi) \le \frac{c\, (\log n)^3}{n}$$

for any mean zero isotropic random vector X in $\mathbb{R}^n$ with a log-concave distribution.

To refine the relationship between the central limit theorem and the thin-shell problem, let us complement Proposition 17.5.1 with the following general statement involving the maximal ψ_1-norm $\lambda = \max_{\theta \in \mathbb{S}^{n-1}} \|\langle X, \theta\rangle\|_{\psi_1}$.

Proposition 17.7.1 *Let X be a random vector in $\mathbb{R}^n$ with $\mathbb{E}\,|X|^2 = n$, satisfying the exponential moment condition* (17.7). *Then for some absolute constant $c > 0$,*

$$c\,\sigma_4^2(X) \le n\,(\lambda \log n)^4\, \mathbb{E}_\theta\, \rho(F_\theta, \Phi) + \frac{\lambda^4}{n^4} + 1 \tag{17.34}$$

In the isotropic log-concave case, the linear functionals $\langle X, \theta\rangle$ have log-concave distributions on the real line with second moment 1 for all $\theta \in \mathbb{S}^{n-1}$. Hence the condition (17.7) is fulfilled for some absolute constant λ (cf. Corollary 2.5.2 or 2.6.4), which simplifies (17.34) to the relation

$$c\,\sigma_4^2(X) \le n\,(\log n)^4\,\mathbb{E}_\theta\,\rho(F_\theta,\Phi) + 1.$$

Therefore, the potential property

$$\mathbb{E}_\theta\,\rho(F_\theta,\Phi) \le \frac{c\,\log n}{n} \tag{17.35}$$

would imply that

$$\sigma_4^2(X) \le c\,(\log n)^5. \tag{17.36}$$

Note that in the thin shell problem one may assume additionally (without loss of generality) that the distribution of X is symmetric about zero. Indeed, define $X' = (X-Y)/\sqrt{2}$, where Y is an independent copy of a random vector X with an isotropic log-concave distribution on $\mathbb{R}^n$ (with mean zero). Then, the distribution of X' is isotropic, log-concave, and symmetric about zero. Moreover,

$$\begin{aligned}\sigma_4^2(X') &= \frac{1}{4n}\,\mathrm{Var}\Big(|X|^2 + |Y|^2 - 2\,\langle X,Y\rangle\Big)\\ &= \frac{1}{4n}\,\mathrm{Var}(|X|^2) + \frac{1}{4n}\,\mathrm{Var}(|Y|^2) + \frac{1}{n}\,\mathbb{E}\,\langle X,Y\rangle^2 \;=\; \frac{1}{2}\,\sigma_4^2(X) + 1.\end{aligned}$$

Hence, if (17.36) is true for the random vector X', then it holds for X as well (with constant $2c$ in place of c).

Note also that, starting from the normal approximation such as (17.35) and applying the inequalities (17.33)–(17.34), we get $\lambda_{1,n}^{-1} \le c\,(\log n)^7$. Thus, modulo n-dependent logarithmic factors, the following three assertions are equivalent up to positive constants c and β (perhaps different in different places):

(i) $\lambda_{1,n}^{-1} \le c\,(\log n)^\beta$;
(ii) $\sigma_{4,n}^2 \le c\,(\log n)^\beta$;
(iii) $\mathbb{E}_\theta\,\rho(F_\theta,\Phi) \le \frac{c}{n}\,(\log n)^\beta$ for any isotropic random vector X in $\mathbb{R}^n$ with a symmetric log-concave distribution.

Proof (of Proposition 17.7.1) In view of the triangle inequality

$$\rho(F,\Phi) \le \mathbb{E}_\theta\,\rho(F_\theta,\Phi),$$

it is sufficient to derive (17.34) for $\rho(F,\Phi)$ in place of $\mathbb{E}_\theta\,\rho(F_\theta,\Phi)$. As we know, the typical distribution function is represented as $F(x) = \mathbb{P}\{|X|\,\theta_1 \le x\}$, assuming that the random vector X and $\theta = (\theta_1,\dots,\theta_n) \in \mathbb{S}^{n-1}$ viewed as a random vector uniformly distributed on the unit sphere are independent. This description yields

$$\int_{-\infty}^{\infty} x^4\,\mathrm{d}F(x) = \mathbb{E}\,|X|^4\,\mathbb{E}_\theta\,\theta_1^4 = (n^2 + \sigma_4^2 n)\,\frac{3}{n(n+2)},$$

or equivalently

$$\int_{-\infty}^{\infty} x^4\,\mathrm{d}F(x) - \int_{-\infty}^{\infty} x^4\,\mathrm{d}\Phi(x) = \frac{3}{n+2}\,(\sigma_4^2 - 2), \tag{17.37}$$

where $\sigma_4^2 = \sigma_4^2(X)$. On the other hand, it follows from (17.7) that

$$\int_{-\infty}^{\infty} e^{|x|/\lambda}\, dF(x) \leq 2. \tag{17.38}$$

Using $t^2 \leq 4e^{-2}\, e^t$ $(t \geq 0)$ together with the property $\int_{-\infty}^{\infty} x^2\, dF(x) = \frac{1}{n}\, \mathbb{E}\, |X|^2 = 1$, (17.38) implies that $\lambda \geq e/\sqrt{8}$, which in turn implies

$$1 - \Phi(x) \leq \frac{1}{2}\, e^{-x^2/2} \leq e^{-x/\lambda}, \quad x \geq 0.$$

In addition, by Markov's inequality, $F(-x) + (1 - F(x)) \leq 2\, e^{-x/\lambda}$, so that

$$|F(-x) - \Phi(-x)| + |F(x) - \Phi(x)| \leq 4\, e^{-x/\lambda}$$

for all $x \geq 0$. Hence, integrating by parts, we see that, for any $T \geq 6\lambda$, the absolute value of the left-hand side of (17.37) does not exceed

$$4 \int_{-T}^{T} |x|^3\, |F(x) - \Phi(x)|\, dx + 16 \int_{T}^{\infty} x^3\, e^{-x/\lambda}\, dx \leq 2T^4\, \rho(F, \Phi) + 32\, \lambda T^3\, e^{-T/\lambda}.$$

Choosing $T = 9\lambda \log n$, we get

$$\sigma_4^2 - 2 \leq Cn \left[\lambda^4\, (\log n)^4\, \rho(F, \Phi) + \frac{\lambda^4}{n^9}\, (\log n)^3\right],$$

thus proving (17.34). □

Remark 17.7.2 Let us comment on the monotonicity of the sequence $\sigma_{4,n}^2$. Given two independent isotropic random vectors X and Y with values in $\mathbb{R}^n$ and $\mathbb{R}^k$, respectively, and having log-concave distributions, the random vector $Z = (X, Y)$ is isotropic in $\mathbb{R}^{n+k}$ and has a log-concave distribution as well. Since $|Z|^2 = |X|^2 + |Y|^2$, we have $\mathrm{Var}(|Z|^2) = \mathrm{Var}(|X|^2) + \mathrm{Var}(|Y|^2)$. Choosing X and Y such that $\mathrm{Var}(|X|^2) = n\sigma_{4,n}^2$ and $\mathrm{Var}(|Y|^2) = k\sigma_{4,k}^2$, it follows that

$$\sigma_{4,n+k}^2 \geq \frac{n}{n+k}\, \sigma_{4,n}^2 + \frac{k}{n+k}\, \sigma_{4,k}^2. \tag{17.39}$$

In particular, $\sigma_{4,2n}^2 \geq \sigma_{4,n}^2$ implying that the sequence $\sigma_{4,n}^2$ is non-decreasing along the powers $n = 2^m$. In addition, by (17.39), $\sigma_{4,n}^2 \geq \frac{1}{2}\, \sigma_{4,k}^2$ whenever $\frac{n}{2} \leq k \leq n$. One may now combine these two properties. Given $1 \leq k < n$, let $2^{m-1} < n \leq 2^m$ and $2^{l-1} \leq k \leq 2^l$ for some integers $m \geq l \geq 1$. In the case $l = m$, the previous bound $\sigma_{4,n}^2 \geq \frac{1}{2}\, \sigma_{4,k}^2$ holds true. If $l < m$, then similarly

$$\sigma_{4,n}^2 \geq \frac{1}{2}\, \sigma_{4,2^{m-1}}^2 \geq \frac{1}{2}\, \sigma_{4,2^l}^2 \geq \frac{1}{4}\, \sigma_{4,k}^2.$$

This shows that $\sigma_{4,n}^2 \geq \frac{1}{4}\,\sigma_{4,k}^2$ whenever $n > k \geq 1$. In particular,

$$\sum_{k=1}^{n} \frac{\sigma_{4,k}^2}{k} \leq 4\sigma_{4,n}^2 \sum_{k=1}^{n} \frac{1}{k} \leq 4\sigma_{4,n}^2\,(1 + \log n).$$

17.8 Remarks

The material of this chapter is based on the papers [39] and [40].

Part VI
Distributions and Coefficients of Special Type

Chapter 18
Special Systems and Examples

We will now illustrate some general results on the closeness of the distributions F_θ of the weighted sums $\langle X, \theta \rangle$, $\theta \in \mathbb{S}^{n-1}$, to the typical distribution $F = \mathbb{E}_\theta F_\theta$ on several specific classes of random vectors $X = (X_1, \dots, X_n)$ in $\mathbb{R}^n$. Most of them can be united in one scheme, where it is possible to develop a common approach to lower bounds on $\mathbb{E}_\theta \rho(F_\theta, F)$ and $\mathbb{E}_\theta\, \omega(F_\theta, \Phi)$ that are similar to upper bounds (modulo logarithmic terms with respect to the dimension n).

18.1 Systems with Lipschitz Condition

We start with the description of a general situation where a rate of order $n^{-1/2}$ is achievable, namely for the L^2-distance $\omega(F_\theta, F)$ on average with respect to θ and therefore for the Kolmogorov distances $\rho(F_\theta, F)$ and $\rho(F_\theta, \Phi)$ modulo a logarithmic factor. While the upper bounds of order $n^{-1/2}$ for these distances have been studied in detail, in this section we focus on conditions that provide similar lower bounds.

Definition. Let L be a fixed measurable function on the underlying probability space $(\Omega, \mathcal{F}, \mathbb{P})$. We will say that the system $X_1, \dots, X_n$ of random variables on $(\Omega, \mathcal{F}, \mathbb{P})$, or the random vector $X = (X_1, \dots, X_n)$ in $\mathbb{R}^n$ satisfies a Lipschitz condition with a parameter function L if

$$\max_{1 \le k \le n} |X_k(t) - X_k(s)| \le n\,|L(t) - L(s)|, \quad t, s \in \Omega. \tag{18.1}$$

When Ω is an interval of the real line (finite or not), and $L(t) = Lt$, $L > 0$, this condition means that every function X_k in the system has Lipschitz semi-norm at most Ln.

As before, we involve the variance functional $\sigma_4^2 = \frac{1}{n}\,\mathrm{Var}(|X|^2)$.

S. Bobkov et al., *Concentration and Gaussian Approximation for Randomized Sums*, Probability Theory and Stochastic Modelling 104,
https://doi.org/10.1007/978-3-031-31149-9_18

Proposition 18.1.1 *Suppose that $\mathbb{E}\,|X|^2 = n$. If the random vector X satisfies the Lipschitz condition with a parameter function L, then*

$$\mathbb{E}_\theta\, \omega^2(F_\theta, F) \geq \frac{c_L}{n} - \frac{c_0\,(1+\sigma_4^4)}{n^2} \tag{18.2}$$

for some absolute constant $c_0 > 0$ and with a constant $c_L > 0$ depending on the distribution of L only. Moreover, if L has a finite second moment, then

$$\mathbb{E}_\theta\, \omega^2(F_\theta, F) \geq \frac{c_1}{n\sqrt{\mathrm{Var}(L)}} - \frac{c_0\,(1+\sigma_4^4)}{n^2} \tag{18.3}$$

for some absolute constant $c_1 > 0$. We also have similar estimates when F is replaced with the standard normal distribution function Φ.

Note that, if $X_1, \dots, X_n$ form an orthonormal system in $L^2(\Omega, \mathcal{F}, \mathbb{P})$, and L has a finite second moment $\|L\|_2^2 = \mathbb{E}L^2$, then this moment has to be bounded from below by a multiple of $1/n^2$. Indeed, integrating the inequality $|X_k(t) - X_k(s)|^2 \leq n^2\,|L(t) - L(s)|^2$ over the product measure $\mathbb{P} \otimes \mathbb{P}$, we obtain a lower bound

$$n^2\,\mathrm{Var}(L) \geq \mathrm{Var}(X_k) = 1 - (\mathbb{E}X_k)^2.$$

Summing over all $k \leq n$ and using $\sum_{k=1}^n (\mathbb{E}X_k)^2 \leq 1$ (cf. Proposition 1.1.2), we get

$$\mathrm{Var}(L) \geq \frac{n-1}{n^3} \geq \frac{1}{2n^2} \quad (n \geq 2).$$

The Lipschitz condition (18.1) guarantees the validity of the following property.

Lemma 18.1.2 *Suppose that the random vector $X = (X_1, \dots, X_n)$ satisfies the Lipschitz condition with the parameter function L, and let Y be an independent copy of X. Then*

$$\mathbb{P}\Big\{|X-Y|^2 \leq \lambda n\Big\} \geq \frac{c\sqrt{\lambda}}{n}, \quad 0 \leq \lambda \leq 1,$$

where the constant $c > 0$ depends on the distribution of L only. Moreover, if L has a finite second moment, then

$$\mathbb{P}\Big\{|X-Y|^2 \leq \lambda n\Big\} \geq \frac{\sqrt{\lambda}}{6n\sqrt{\mathrm{Var}(L)}}, \quad 0 \leq \lambda \leq n^2\,\mathrm{Var}(L).$$

In turn, this lemma is based on the following general observation.

Lemma 18.1.3 *If η is an independent copy of a random variable ξ, then for any $\varepsilon_0 > 0$,*

$$\mathbb{P}\{|\xi - \eta| \le \varepsilon\} \ge c\varepsilon, \quad 0 \le \varepsilon \le \varepsilon_0,$$

with some constant $c > 0$ independent of ε. Moreover, if the standard deviation $\sigma = \sqrt{\mathrm{Var}(\xi)}$ is finite, then

$$\mathbb{P}\{|\xi - \eta| \le \varepsilon\} \ge \frac{1}{6\sigma}\,\varepsilon, \quad 0 \le \varepsilon \le \sigma.$$

Proof The difference $\xi - \eta$ has a non-negative characteristic function $h(t) = |\psi(t)|^2$, where ψ is the characteristic function of ξ. Denoting by H the distribution function of $\xi - \eta$, we start with a general identity

$$\int_{-\infty}^{\infty} \hat{p}(x)\,\mathrm{d}H(x) = \int_{-\infty}^{\infty} p(t)h(t)\,\mathrm{d}t, \tag{18.4}$$

which is valid for any integrable function $p(t)$ on the real line with Fourier transform

$$\hat{p}(x) = \int_{-\infty}^{\infty} \mathrm{e}^{itx}\,p(t)\,\mathrm{d}t, \quad x \in \mathbb{R}.$$

Given $\varepsilon > 0$, here we take a standard pair

$$p(t) = \frac{1}{2\pi}\left(\frac{\sin\frac{\varepsilon t}{2}}{\frac{\varepsilon t}{2}}\right)^2, \quad \hat{p}(x) = \frac{1}{\varepsilon}\left(1 - \frac{|x|}{\varepsilon}\right)^+.$$

In this case,

$$\int_{-\infty}^{\infty} \hat{p}(x)\,\mathrm{d}H(x) \le \frac{1}{\varepsilon}\int_{[-\varepsilon,\varepsilon]} \mathrm{d}H(x) = \frac{1}{\varepsilon}\,\mathbb{P}\{|\xi - \eta| \le \varepsilon\}.$$

On the other hand, since the function $\frac{\sin u}{u}$ is decreasing in $0 < u < \frac{\pi}{2}$, we have

$$\int_{-\infty}^{\infty} p(t)h(t)\,\mathrm{d}t \ge \frac{1}{2\pi}\left(2\sin(1/2)\right)^2 \int_{-1/\varepsilon}^{1/\varepsilon} h(t)\,\mathrm{d}t \ge \frac{1}{7}\int_{-1/\varepsilon}^{1/\varepsilon} h(t)\,\mathrm{d}t.$$

Hence, whenever $0 < \varepsilon \le \varepsilon_0$, by (18.4),

$$\mathbb{P}\{|\xi - \eta| \le \varepsilon\} \ge \frac{\varepsilon}{7}\int_{-1/\varepsilon}^{1/\varepsilon} h(t)\,\mathrm{d}t \ge \frac{\varepsilon}{7}\int_{-1/\varepsilon_0}^{1/\varepsilon_0} h(t)\,\mathrm{d}t.$$

Since $h(t)$ is bounded away from zero near the origin, the first assertion follows.

One may quantify this statement in terms of the variance $\sigma^2 = \mathrm{Var}(\xi)$ by using Taylor's expansion for $h(t)$ about zero. Indeed, it gives $1 - h(t) \le \sigma^2 t^2$, and thus for $\varepsilon \le \varepsilon_0 = \sigma$, we get

$$\int_{-1/\varepsilon}^{1/\varepsilon} h(t)\,\mathrm{d}t \geq \int_{-1/\sigma}^{1/\sigma} (1 - \sigma^2 t^2)\,\mathrm{d}t = \frac{4}{3\sigma}.$$

Since $\frac{\varepsilon}{7} \cdot \frac{4}{3\sigma} \geq \frac{1}{6\sigma}\,\varepsilon$, the lemma is proved. □

Proof (of Lemma 18.1.2) Let us equip the product space $\Omega^2 = \Omega \times \Omega$ with the product measure $\mathbb{P}^2 = \mathbb{P} \otimes \mathbb{P}$ and redefine the random vector X on this new probability space as $X(t,s) = X(t)$, $(t,s) \in \Omega^2$. Then one can realize an independent copy of X in the form $Y(t,s) = X(s)$. By the Lipschitz condition,

$$|X(t,s) - Y(t,s)|^2 = \sum_{k=1}^{n} |X_k(t) - X_k(s)|^2 \leq n^3\,|L(t) - L(s)|^2.$$

Hence, if η is an independent copy of the random variable $\xi = L$, then

$$\mathbb{P}\Big\{|X - Y|^2 \leq \lambda n\Big\} \geq \mathbb{P}\Big\{n^3\,|\xi - \eta|^2 \leq \lambda n\Big\} = \mathbb{P}\Big\{|\xi - \eta| \leq \frac{\sqrt{\lambda}}{n}\Big\}.$$

But by Lemma 18.1.3 with $\varepsilon_0 = 1$, the latter probability is at least $c\,\frac{\sqrt{\lambda}}{n}$, where the constant c depends on L only (via its distribution). An application of the second inequality of Lemma 18.1.3 yields the second assertion. □

To involve more examples, let us now give a slightly more general form of Lemma 18.1.2, assuming that $(\Omega, \mathbb{P}) = (\Omega_1 \times \Omega_2, \mathbb{P}_1 \otimes \mathbb{P}_2)$ is a product probability space.

Lemma 18.1.4 *Let $X = (X_1, \dots, X_n) : \Omega \to \mathbb{R}^n$ be a random vector such that, for some measurable functions L_1, L_2, defined on Ω_1 and Ω_2 respectively, we have*

$$\max_{1 \leq k \leq n} |X_k(t_1,t_2) - X_k(s_1,s_2)| \leq n\,|L_1(t_1) - L_1(s_1)| + |L_2(t_2) - L_2(s_2)| \tag{18.5}$$

for all $(t_1,t_2), (s_1,s_2) \in \Omega$. If Y is an independent copy of X, then

$$\mathbb{P}\Big\{|X - Y|^2 \leq \lambda n\Big\} \geq \frac{c\lambda}{n}, \quad 0 \leq \lambda \leq 1,$$

where the constant $c > 0$ depends on the distributions of L_1 and L_2 only.

Proof By the Lipschitz condition (18.5), for any $k \leq n$,

$$|X_k(t_1,t_2) - X_k(s_1,s_2)|^2 \leq 2n^2\,|L_1(t_1) - L_1(s_1)| + 2\,|L_2(t_2) - L_2(s_2)|^2,$$

so,

$$\begin{aligned} |X(t_1,t_2) - Y(s_1,s_2)|^2 &= \sum_{k=1}^{n} |X_k(t_1,t_2) - X_k(s_1,s_2)|^2 \\ &\leq 2n^3\,|L_1(t_1) - L_1(s_1)|^2 + 2n\,|L_2(t_2) - L_2(s_2)|^2. \end{aligned}$$

Putting $L_1(t_1,t_2) = L_1(t_1)$ and $L_2(t_1,t_2) = L_2(t_2)$, one may treat L_1 and L_2 as independent random variables. If L_1' is an independent copy of L_1, and L_2' is an independent copy of L_2 (and L_1, L_1', L_2, L_2' are independent), we obtain that

$$\begin{aligned}
\mathbb{P}\Big\{|X-Y|^2 \le \lambda n\Big\} &\ge \mathbb{P}\Big\{n^2\,|L_1-L_1'|^2 + |L_2-L_2'|^2 \le \frac{\lambda}{2}\Big\} \\
&\ge \mathbb{P}\Big\{n^2\,|L_1-L_1'|^2 \le \frac{\lambda}{4}\Big\}\,\mathbb{P}\Big\{|L_2-L_2'|^2 \le \frac{\lambda}{4}\Big\} \\
&= \mathbb{P}\Big\{|L_1-L_1'| \le \frac{1}{2n}\sqrt{\lambda}\Big\}\,\mathbb{P}\Big\{|L_2-L_2'| \le \frac{1}{2}\sqrt{\lambda}\Big\}.
\end{aligned}$$

It remains to apply Lemma 18.1.3. □

Let us now explain how to apply Lemma 18.1.2 and Lemma 18.1.4 to obtain a general lower bound on the L^2-distance

$$\omega(F_\theta, F) = \Big(\int_{-\infty}^{\infty} |F_\theta(x) - F(x)|^2\,dx\Big)^{1/2}.$$

Here we denote by $F_\theta(x)$ the distribution function of the linear form $\langle X, \theta\rangle$ with $\theta \in \mathbb{S}^{n-1}$ and by $F(x) = \mathbb{E}\,F_\theta(x)$, $x \in \mathbb{R}$, the typical distribution function. To this end, let us recall the bound of Proposition 15.6.2,

$$\mathbb{E}_\theta\,\omega^2(F_\theta, F) \ge c_0\,\mathbb{P}\Big\{|X-Y|^2 \le \frac{1}{4}\,n\Big\} - c_1\,\frac{1+\sigma_4^4}{n^2},$$

valid for some positive absolute constants c_0 and c_1 under the assumption that $\mathbb{E}\,|X|^2 = n$. Combining this bound with Lemma 18.1.2, we arrive at the lower bounds (18.2)–(18.3). But, using Lemma 18.1.4, we come to a similar conclusion:

Proposition 18.1.5 *The statement of Proposition 18.1.1 continues to hold under the Lipschitz condition* (18.5). *Thus, for some absolute constant $c_0 > 0$,*

$$\mathbb{E}_\theta\,\omega^2(F_\theta, F) \ge \frac{c}{n} - \frac{c_0\,(1+\sigma_4^4)}{n^2},$$

where the constant $c > 0$ depends on the distributions of L_1 and L_2 only. A similar estimate also holds when F is replaced with the normal distribution function Φ.

The assertion on the normal approximation follows from the inequality $\omega(F, \Phi) \le C\,(1+\sigma_4^2)/n$, cf. Proposition 12.5.1.

18.2 Trigonometric Systems

For an illustration of Propositions 18.1.1 and 18.1.5, let us first consider a few examples where the distribution of the random vector X is supported on the sphere of radius $\sqrt{n}$ (in which case $\sigma_4^2 = 0$). The first classical example is described by the trigonometric system $X = (X_1, \dots, X_n)$ with components

$$\begin{aligned} X_{2k-1}(t) &= \sqrt{2}\cos(kt), \\ X_{2k}(t) &= \sqrt{2}\sin(kt) \qquad (-\pi < t < \pi,\ \ k = 1, \dots, n/2), \end{aligned}$$

assuming that n is even. They are treated as random variables on the probability space $\Omega = (-\pi, \pi)$ equipped with the normalized Lebesgue measure $\mathbb{P}$ (the uniform distribution). Note that the linear forms

$$\langle X, \theta\rangle = \sqrt{2}\sum_{k=1}^{\frac{n}{2}} \Big(\theta_{2k-1}\cos(kt) + \theta_{2k}\sin(kt)\Big), \qquad \theta = (\theta_1, \dots, \theta_n) \in \mathbb{S}^{n-1},$$

represent trigonometric polynomials of degree at most $\frac{n}{2}$. The normalization $\sqrt{2}$ is chosen for the convenience only, since in this case X is isotropic and satisfies $|X| = \sqrt{n}$.

By Sudakov's theorem, the distributions F_θ are almost standard normal for most of θ. And as usual, the main question we study is to determine the rate of approximation of F_θ by Φ for canonical probability metrics.

Since for all $k \le \frac{n}{2}$, we have

$$|X_k(t) - X_k(s)| \le k\sqrt{2}\,|t-s| \le \frac{n}{\sqrt{2}}\,|t-s|, \qquad t, s \in \Omega,$$

the Lipschitz condition is fulfilled with the function $L(t) = t/\sqrt{2}$, in which case $\mathrm{Var}(L) = \pi^2/6$. Hence Proposition 18.1.1 yields the lower bound

$$\mathbb{E}_\theta\, \omega^2(F_\theta, \Phi) \ge \frac{c_1}{n} - \frac{c_2}{n^2}$$

for some absolute constants $c_j > 0$. Such an estimate may also be obtained by applying Corollary 15.4.3, which gives

$$\mathbb{E}_\theta\, \omega^2(F_\theta, \Phi) \ge \frac{c}{n^4}\,\mathbb{E}\,\langle X, Y\rangle^4 - O(n^{-2}),$$

where Y is an independent copy of X. On the other hand, according to Proposition 15.2.1, $\mathbb{E}_\theta\, \omega^2(F_\theta, \Phi) \le \frac{c}{n}$. Hence, we arrive at:

Proposition 18.2.1 *For the trigonometric system X we have for some positive absolute constants $c_1 > c_0 > 0$,*

$$\frac{c_0}{n} \le \mathbb{E}_\theta\, \omega^2(F_\theta, \Phi) \le \frac{c_1}{n}. \tag{18.6}$$

The inequality $\frac{c_1}{n} - \frac{c_2}{n^2} \geq \frac{c_1'}{n}$ holds true with some absolute constant $c_1' > 0$ for all n large enough. So, strictly speaking, the lower bound in (18.6) may only be stated for $n \geq n_0$ (on the basis of Proposition 18.1.1). It should however be clear that $\mathbb{E}_\theta\, \omega^2(F_\theta, \Phi)$ does not vanish (since trigonometric polynomials are bounded and therefore may not have a normal distribution). Hence, one may involve in (18.6) all values of n.

Applying Proposition 16.4.4 with $D = c_0/n$, there are similar bounds for the L^1-norm (modulo logarithmic factors). Namely, it gives

$$\frac{c_0}{(\log n)^{\frac{15}{4}} \sqrt{n}} \leq \mathbb{E}_\theta\, \omega(F_\theta, \Phi) \leq \frac{c_1}{\sqrt{n}}.$$

We also get an analogous pointwise lower bound on the "essential" part of the unit sphere. It seems however natural that the logarithmic factor could be removed from the left-hand side. A similar statement is also true for the Kolmogorov distance. Combining Proposition 14.6.1 with Corollary 16.4.5, we have:

Proposition 18.2.2 *For the trigonometric system X we have with some positive absolute constants $c_1 > c_0 > 0$,*

$$\frac{c_0}{(\log n)^4 \sqrt{n}} \leq \mathbb{E}_\theta\, \rho(F_\theta, \Phi) \leq \frac{c_1 \log n}{\sqrt{n}}. \tag{18.7}$$

Analogous results continue to hold for the cosine trigonometric system X with components

$$X_k(t) = \sqrt{2}\, \cos(kt) \quad (0 < t < \pi, \ \ k = 1, \dots, n)$$

on the probability space $\Omega = (0, \pi)$ equipped with the normalized Lebesgue measure $\mathbb{P}$. In that case, the linear forms

$$\langle X, \theta \rangle = \sqrt{2} \sum_{k=1}^{n} \theta_k \cos(kt), \qquad \theta = (\theta_1, \dots, \theta_n) \in \mathbb{S}^{n-1},$$

represent trigonometric polynomials of degree n. Due to the normalization $\sqrt{2}$, the distribution of X is isotropic in $\mathbb{R}^n$. The property $|X| = \sqrt{n}$ is not true anymore. However, there is a pointwise bound $|X| \leq \sqrt{2n}$. In addition, the variance functional $\sigma_4^2 = \frac{1}{n}\,\mathrm{Var}(|X|^2)$ is bounded by a universal constant for all $n \geq 1$ (and actually it does not depend on n). Indeed, write $X_k^2 = 2\cos^2(kt) = 1 + \frac{1}{2}\,(e^{2ikt} + e^{-2ikt})$, so that

$$2\,(|X|^2 - n) = \sum_{0<|k|\leq n} e^{2ikt}, \quad 4\,(|X|^2 - n)^2 = \sum_{0<|k|,|l|\leq n} e^{2i(k+l)t}.$$

It follows that

$$4\,\mathrm{Var}(|X|^2) = \sum_{0<|k|,|l|\leq n} \mathbb{E}\, e^{2i(k+l)t} = \sum_{0<|k|\leq n,\, l=-k} 1 = 2n.$$

Hence

$$\sigma_4^2 = \frac{1}{n}\,\mathrm{Var}(|X|^2) = \frac{1}{2}.$$

Now, the Lipschitz condition for the system X is fulfilled with the function $L(t) = \sqrt{2}\,t$. Hence Proposition 18.1.1 yields a similar lower bound $\mathbb{E}_\theta\, \omega^2(F_\theta, \Phi) \geq \frac{c_1}{n} - \frac{c_2}{n^2}$. On the other hand, by Proposition 15.2.1,

$$\mathbb{E}_\theta\, \omega^2(F_\theta, \Phi) \leq \frac{c\,(1+\sigma_4^2)}{n} = \frac{3c/2}{n}.$$

As for the Kolmogorov distance, one may appeal again to Proposition 14.6.1 and Corollary 16.4.5.

Proposition 18.2.3 *For the cosine trigonometric system X on the interval* $(0,\pi)$, *the bounds* (18.6)–(18.7) *are also true for some positive absolute constants.*

Let us also note that the sums $\sum_{k=1}^n \cos(kt)$ remain bounded for growing n (for any fixed $0 < t < \pi$). Hence the normalized sums

$$S_n(t) = \frac{\sqrt{2}}{\sqrt{n}} \sum_{k=1}^n \cos(kt),$$

which correspond to $\langle X, \theta\rangle$ with equal coefficients, are convergent to zero pointwise on Ω as $n \to \infty$. In particular, they fail to satisfy the central limit theorem.

18.3 Chebyshev Polynomials

An example closely related to the cosine trigonometric system is represented by the normalized Chebyshev polynomials. They are defined by

$$\begin{aligned} X_k(t) &= \sqrt{2}\,\cos(k \arccos t) \\ &= \sqrt{2}\left[t^n - \binom{n}{2} t^{n-2}(1-t^2) + \binom{n}{4} t^{n-4}(1-t^2)^2 - \dots\right], \end{aligned}$$

which we consider for $k = 1, 2, \dots, n$. These polynomials are orthonormal on the interval $\Omega = (-1, 1)$ with respect to the probability measure

$$\frac{d\mathbb{P}(t)}{dt} = \frac{1}{\pi\sqrt{1-t^2}}, \qquad -1 < t < 1,$$

cf. e.g. [114], [120] (for simplicity, we exclude the constant function $X_0(t) = 1$).

Similarly to the cosine trigonometric system, for the random vector $X = (X_1, \dots, X_n)$ we find that

$$4\,(|X|^2 - n)^2 = \sum_{0<|k|,|l|\leq n} \exp\{2i(k+l)\arccos t\}.$$

It follows that

$$4\,\mathrm{Var}(|X|^2) = \sum_{0<|k|,|l|\le n} \mathbb{E}\,\exp\{2i(k+l)\,\arccos t\} = \sum_{0<|k|\le n} 1 = 2n,$$

so that $\sigma_4^2 = \frac{1}{n}\,\mathrm{Var}(|X|^2) = \frac{1}{2}$. Now, for all $k \le n$,

$$|X_k(t) - X_k(s)| \le k\sqrt{2}\,|\arccos t - \arccos s|, \quad t,s \in \Omega,$$

which implies that the Lipschitz condition for the system X is fulfilled with the function $L(t) = \sqrt{2}\,\arccos t$. One may also notice that

$$\mathrm{Var}(L) = 2\int_{-1}^{1} (\arcsin t)^2\,\mathrm{d}\mathbb{P}(t) = \frac{\pi^2}{6}.$$

Hence Proposition 18.1.1 yields the lower bound

$$\mathbb{E}_\theta\,\omega^2(F_\theta,\Phi) \ge \frac{c_1}{n} - \frac{c_2}{n^2}$$

for some absolute constants $c > 0$. Together with Proposition 15.2.1, we obtain a full analogue of Proposition 18.2.1, while an application of Proposition 14.6.1 and Corollary 16.4.5 lead to an analogue of Proposition 18.2.2.

Proposition 18.3.1 *For the system $(X_1,\dots,X_n)$ of the first n Chebyshev polynomials on $(-1,1)$, we have with some positive absolute constants $c_1 > c_0 > 0$,*

$$\frac{c_0}{n} \le \mathbb{E}_\theta\,\omega^2(F_\theta,\Phi) \le \frac{c_1}{n},$$

$$\frac{c_0}{(\log n)^4\sqrt{n}} \le \mathbb{E}_\theta\,\rho(F_\theta,\Phi) \le \frac{c_1\log n}{\sqrt{n}}.$$

18.4 Functions of the Form $X_k(t,s) = f(kt+s)$

Let f be a real-valued, 1-periodic Borel measurable function on the real line such that $\int_0^1 f(x)\,\mathrm{d}x = 0$ and $\int_0^1 f(x)^2\,\mathrm{d}x = 1$. On the square $\Omega = (0,1)\times(0,1)$, which we equip with the Lebesgue measure $\mathbb{P}$, we consider the system of random variables

$$X_k(t,s) = f(kt+s) \quad (0 < t,s < 1).$$

As we know from Section 2.2, $\{X_k\}_{k=1}^\infty$ represents a strictly stationary sequence of pairwise independent random variables on Ω. Since

$$\mathbb{E}X_k = \int_0^1 f(x)\,\mathrm{d}x = 0, \quad \mathbb{E}X_k^2 = \int_0^1 f(x)^2\,\mathrm{d}x = 1,$$

the vector $X = (X_1, \dots, X_n)$ is isotropic, and its components are pairwise independent. The latter insures that, if f has finite 4-th moment on $(0, 1)$, the variance functional

$$\sigma_4^2 = \frac{1}{n}\,\mathrm{Var}(|X|^2) = \int_0^1 f(x)^4\,dx - 1$$

is finite and does not depend on n.

In addition, if the function f has finite Lipschitz semi-norm $\|f\|_{\mathrm{Lip}}$, then for all (t_1, t_2) and (s_1, s_2) in Ω,

$$|X_k(t_1, t_2) - X_k(s_1, s_2)| \le \|f\|_{\mathrm{Lip}}\,\big(k\,|t_1 - s_1| + |t_2 - s_2|\big).$$

This means that the Lipschitz condition (18.5) is fulfilled with linear functions L_1 and L_2. Hence, one may apply Proposition 18.1.5, giving the lower bound

$$\mathbb{E}_\theta\,\omega^2(F_\theta, F) \ge \frac{c_f}{n} - \frac{c\,(1 + \sigma_4^4)}{n^2}.$$

Hence $\mathbb{E}_\theta\,\omega^2(F_\theta, \Phi) \ge \frac{c_f'}{n}$ for all $n \ge n_0$, where the positive constants c_f and c_f', as well as the integer $n_0 \ge 1$, depend on the distribution of f only. Since the collection $\{F_\theta\}_{\theta \in \mathbb{S}^{n-1}}$ is separated from the normal distribution function Φ in the weak sense for $n < n_0$ (by the uniform boundedness of the X_k's), the latter lower bound holds true for all $n \ge 1$.

On the other hand, by Proposition 15.2.1, there is a similar upper bound under the 4-th moment condition, and then we obtain another full analogue of Proposition 18.2.1. Also, as was used above, Lipschitz functions on $(0, 1)$ are bounded, so that $|X| \le b\sqrt{n}$ with $b = \sup_x |f(x)|$, and one may apply Corollary 16.4.5. Using Proposition 14.6.1, we thus obtain the analogue of Proposition 18.2.2.

Proposition 18.4.1 *Assume that a* 1*-periodic function* f *has finite* 4*-th moment on* $(0, 1)$ *and is normalized so that* $\int_0^1 f(x)\,dx = 0$, $\int_0^1 f(x)^2\,dx = 1$. *Then*

$$\mathbb{E}_\theta\,\omega^2(F_\theta, \Phi) \le \frac{c_1}{n}, \quad \mathbb{E}_\theta\,\rho(F_\theta, \Phi) \le \frac{c_2 \log n}{\sqrt{n}}$$

for all $n \ge 2$ *with some positive constants* c_j *depending on* f *only. Moreover, if* f *has a finite Lipschitz semi-norm, then*

$$\mathbb{E}_\theta\,\omega^2(F_\theta, \Phi) \ge \frac{c_0}{n}, \quad \mathbb{E}_\theta\,\rho(F_\theta, \Phi) \ge \frac{c_0}{(\log n)^4 \sqrt{n}}.$$

Choosing $f(t) = \cos t$, we obtain the system $X = (X_1, \dots, X_n)$ with $X_k(t, s) = \cos(kt + s)$, which is closely related to the cosine trigonometric system. The main difference is however the property that the X_k's are now pairwise independent. Nevertheless, the normalized sums

$$\frac{1}{\sqrt{n}} \sum_{k=1}^{n} \cos(kt + s)$$

fail to satisfy the central limit theorem (since they converge pointwise to zero). One can construct other similar examples in the form $f(kt + s)$.

18.5 The Walsh System on the Discrete Cube

Given an integer $d \geq 1$, consider the functions on the discrete cube $\Omega = \{-1, 1\}^d$,

$$X_\tau(t) = \prod_{k \in \tau} t_k, \quad t = (t_1, \dots, t_d) \in \Omega, \quad \tau \subset \{1, \dots, d\},$$

where one puts $X_\emptyset = 1$. Equipping the discrete cube with the uniform (i.e., normalized counting) measure $\mathbb{P}$, we obtain 2^d random variables which form a complete orthonormal system in $L^2(\Omega, \mathbb{P})$, called the Walsh system on the discrete cube.

Note that each X_τ with $\tau \neq \emptyset$ is a symmetric Bernoulli random variable: It takes only two values, -1 and 1, each with probability $\frac{1}{2}$. For simplicity, let us exclude the constant random variable $X_\emptyset$ and consider the random vector

$$X = \{X_\tau\}_{\tau \neq \emptyset}$$

in $\mathbb{R}^n$ of dimension $n = 2^d - 1$ (the ordering of the components does not play any role). As before, we denote by F_θ the distribution function of the linear forms

$$\langle X, \theta \rangle = \sum_{\tau \neq \emptyset} \theta_\tau X_\tau, \quad \theta = \{\theta_\tau\}_{\tau \neq \emptyset} \in \mathbb{S}^{n-1}.$$

Since $|X_\tau| = 1$ and thus $|X| = \sqrt{n}$, for the study of the asymptotic behavior of the L^2-distance $\omega(F_\theta, F)$ and $\omega(F_\theta, \Phi)$ on average, one may apply the two-sided bounds of Proposition 15.4.1 or Corollaries 15.4.2–15.4.3.

Let Y be an independent copy of X, which we realize on the product space $\Omega^2 = \Omega \times \Omega$ with product measure $\mathbb{P}^2 = \mathbb{P} \times \mathbb{P}$ by

$$X_\tau(t, s) = \prod_{k \in \tau} t_k, \;\; Y_\tau(t, s) = \prod_{k \in \tau} s_k \quad t = (t_1, \dots, t_d), \; s = (s_1, \dots, s_d) \in \Omega.$$

Then the inner product

$$\langle X, Y \rangle = \sum_{\tau \neq \emptyset} X_\tau(t, s) Y_\tau(t, s) \; = \; -1 + \prod_{k=1}^{d} (1 + t_k s_k)$$

takes only two values, namely $2^d - 1$ in the case $t = s$, and -1 if $t \neq s$. Hence

$$\mathbb{E}\,\langle X,Y\rangle^4 = (2^d-1)^4\,\mathbb{P}^2\big\{(t,s)\in\Omega^2 : t=s\big\} + \mathbb{P}^2\big\{(t,s)\in\Omega^2 : t\neq s\big\}$$
$$= (2^d-1)^4\,2^{-d} + (1-2^{-d}) = \frac{n^4}{n+1} + \Big(1-\frac{1}{n+1}\Big) \sim n^3$$

and

$$\mathbb{E}\,\langle X,Y\rangle^3 = (2^d-1)^3\,2^{-d} + (1-2^{-d}) = \frac{n^3}{n+1} + \Big(1-\frac{1}{n+1}\Big) \sim n^2.$$

As a result, we obtain:

Proposition 18.5.1 *For the Walsh system X on the discrete cube $\{-1,1\}^d$ of size $n = 2^d - 1$, we have with some positive absolute constants $c_1 > c_0 > 0$,*

$$\frac{c_0}{n} \le \mathbb{E}_\theta\,\omega^2(F_\theta,\Phi) \le \frac{c_1}{n}.$$

By Proposition 14.6.1 and Corollary 16.4.5, we also have

$$\frac{c_0}{(\log n)^4\sqrt{n}} \le \mathbb{E}_\theta\,\rho(F_\theta,\Phi) \le \frac{c_1\log n}{\sqrt{n}}.$$

18.6 Empirical Measures

Here is another interesting example leading to a similar rate of normal approximation. Let $e_1,\dots,e_n$ denote the canonical basis in $\mathbb{R}^n$. Assuming that the random vector $X = (X_1,\dots,X_n)$ takes only n values,

$$\sqrt{n}\,e_1,\dots,\sqrt{n}\,e_n,$$

each with probability $1/n$, the linear form $\langle X,\theta\rangle$ also takes n values, namely, $\sqrt{n}\,\theta_1,\dots,\sqrt{n}\,\theta_n$, each with probability $1/n$, for any $\theta = (\theta_1,\dots,\theta_n) \in \mathbb{S}^{n-1}$. That is, as a measure, the distribution of $\langle X,\theta\rangle$ is described as

$$F_\theta = \frac{1}{n}\sum_{k=1}^n \delta_{\sqrt{n}\,\theta_k},$$

which represents an empirical measure based on the observations $\zeta_k = \sqrt{n}\,\theta_k$, $k = 1,\dots,n$. Each ζ_k is almost standard normal, while jointly they are nearly independent (we have already considered in detail its densities $\varphi_n(x)$ and characteristic functions $J_n(t\sqrt{n})$).

Just taking a short break, let us recall that when Z_k are standard normal and independent, it is well-known that the empirical measures

$$G_n = \frac{1}{n}\sum_{k=1}^n \delta_{Z_k}$$

approximate the standard normal law Φ with rate $1/\sqrt{n}$ with respect to the Kolmogorov distance. More precisely, $\mathbb{E}G_n = \Phi$, and there is a subgaussian deviation bound

$$\mathbb{P}\Big\{\sqrt{n}\,\rho(G_n,\Phi) \geq r\Big\} \leq 2e^{-2r^2}, \quad r \geq 0$$

(cf. [136]). In particular, $\mathbb{E}\,\rho(G_n,\Phi) \leq \frac{c}{\sqrt{n}}$ for some absolute constant c. Note that the characteristic function of G_n is given by

$$g_n(t) = \frac{1}{n}\sum_{k=1}^{n} e^{itZ_k}, \quad t \in \mathbb{R}.$$

It has mean $g(t) = e^{-t^2/2}$ and variance

$$\begin{aligned}\mathbb{E}\,\big|g_n(t) - g(t)\big|^2 &= \mathrm{Var}(g_n(t)) = \frac{1}{n^2}\sum_{k=1}^{n}\mathrm{Var}(e^{itZ_k}) \\ &= \frac{1}{n}\mathrm{Var}(e^{itZ_1}) = \frac{1}{n}\Big(1 - |\mathbb{E}\,e^{itZ_1}|^2\Big) = \frac{1}{n}\Big(1 - e^{-t^2}\Big)\end{aligned}$$

(in view of the independence of Z_k). Hence, applying Plancherel's theorem together with Lemma 12.5.2 with $\alpha = 0$ and $\alpha_0 = 1$, we also have

$$\mathbb{E}\,\omega^2(G_n,\Phi) = \frac{1}{2\pi}\int_{-\infty}^{\infty}\mathbb{E}\Big|\frac{g_n(t) - g(t)}{t}\Big|^2\,dt = \frac{1}{2\pi n}\int_{-\infty}^{\infty}\frac{1 - e^{-t^2}}{t^2}\,dt = \frac{1}{n\sqrt{\pi}}.$$

Thus, on average the L^2-distance $\omega(G_n,\Phi)$ is of order $1/\sqrt{n}$ as well.

Similar properties may be expected for the random variables $\zeta_k = \sqrt{n}\,\theta_k$ and hence for the random vector X. Note that $|X| = \sqrt{n}$, while

$$\mathbb{E}\,\langle X,\theta\rangle^2 = \frac{1}{n}\sum_{k=1}^{n}(\sqrt{n}\,\theta_k)^2 = 1,$$

so that X is isotropic. We now involve the asymptotic formula of Proposition 15.3.1,

$$\mathbb{E}_\theta\,\omega^2(F_\theta,\Phi) = \frac{1}{\sqrt{\pi}}\Big(1 + \frac{1}{4n}\Big)\,\mathbb{E}\Big(1 - (1-\xi)^{1/2}\Big) - \frac{1}{8n\sqrt{\pi}} + O(n^{-2}),$$

where $\xi = \frac{\langle X,Y\rangle}{n}$ with Y being an independent copy of X. By the definition, ξ takes only two values, 1 with probability $\frac{1}{n}$, and 0 with probability $1 - \frac{1}{n}$. Hence, the last expectation is equal to $\frac{1}{n}$, and we get

$$\mathbb{E}_\theta\,\omega^2(F_\theta,\Phi) = \frac{7/8}{n\sqrt{\pi}} + O(n^{-2}).$$

Note that in this particular case, the lower bound of Corollary 15.3.2 is useless.

As for the Kolmogorov distance, by Proposition 14.6.1 and Corollary 16.4.5,

$$\frac{c_0}{(\log n)^4\sqrt{n}} \leq \mathbb{E}_\theta\, \rho(F_\theta, \Phi) \leq \frac{c_1 \log n}{\sqrt{n}}$$

for some absolute constants $c_1 > c_0 > 0$. Apparently, both logarithmic terms can be removed. Their appearance here is explained by the use of the Fourier tools (in the form of Berry–Esseen bounds), while the proof of the Dvoretzky–Kiefer–Wolfowitz inequality on $\rho(G_n, \Phi)$, [81], is based on entirely different arguments.

18.7 Lacunary Systems

An orthonormal sequence of random variables $\{X_k\}_{k=1}^{\infty}$ in $L^2(\Omega, \mathcal{F}, \mathbb{P})$ is called a lacunary system of order $p > 2$ if, for any sequence (a_k) in ℓ^2, the series

$$\sum_{k=1}^{\infty} a_k X_k$$

converges in the L^p-norm to some element of $L^p(\Omega, \mathcal{F}, \mathbb{P})$. This property is equivalent to the Khinchine-type inequality

$$\Big(\mathbb{E}\,|a_1 X_1 + \dots + a_n X_n|^p\Big)^{1/p} \leq M_p\, (a_1^2 + \dots + a_n^2)^{1/2}, \tag{18.8}$$

which should be valid for all $a_k \in \mathbb{R}$ with some constant M_p independent of n and the choice of the coefficients a_k. For basic properties of such systems we refer the interested reader to [120].

Starting from an orthonormal lacunary system of order $p = 4$, consider the random vector $X = (X_1, \dots, X_n)$. According to Corollary 15.4.2, if $|X|^2 = n$ a.s. and $\mathbb{E}X = 0$, then, up to some absolute constant $c > 0$,

$$c\, \mathbb{E}_\theta\, \omega^2(F_\theta, F) \leq \frac{1}{n^3}\, \mathbb{E}\,\langle X, Y\rangle^3 + \frac{1}{n^4}\, \mathbb{E}\,\langle X, Y\rangle^4, \tag{18.9}$$

where Y is an independent copy of X. A similar bound

$$c\, \mathbb{E}_\theta\, \rho^2(F_\theta, F) \leq \frac{\log n}{n^3}\, \mathbb{E}\,\langle X, Y\rangle^3 + \frac{(\log n)^2}{n^4}\, \mathbb{E}\,\langle X, Y\rangle^4 \tag{18.10}$$

also holds for the Kolmogorov distance (Proposition 16.2.1).

Note that, in the inequality (18.8), the optimal constant is just the moment functional $M_p = M_p(X)$, while according to Proposition 1.4.2, we have general relations

$$\mathbb{E}\,|\langle X, Y\rangle|^3 \leq M_3^6\, n^{3/2}, \quad \mathbb{E}\,\langle X, Y\rangle^4 \leq M_4^8\, n^2.$$

Hence, the bounds (18.9)–(18.10) lead to the estimates

$$c\,\mathbb{E}_\theta\,\omega^2(F_\theta,F) \le \frac{1}{n^{3/2}}\,M_3^6 + \frac{1}{n^2}\,M_4^8,$$
$$c\,\mathbb{E}_\theta\,\rho^2(F_\theta,F) \le \frac{\log n}{n^{3/2}}\,M_3^6 + \frac{(\log n)^2}{n^2}\,M_4^8,$$

Thus, if M_4 is bounded, both distances are at most of order $n^{-3/4}$ on average (modulo a logarithmic factor). Moreover, if

$$\Sigma_3(n) \equiv \mathbb{E}\,\langle X,Y\rangle^3 = \sum_{1\le i_1,i_2,i_3\le n} \Big(\mathbb{E}X_{i_1}X_{i_2}X_{i_3}\Big)^2 \tag{18.11}$$

is bounded by a multiple of n, then these distances are on average at most $1/n$ (modulo a logarithmic factor in the case of ρ). Indeed, by the lacunary property, we have a Khinchine-type inequality

$$\Big(\mathbb{E}_Y\,\langle X,Y\rangle^4\Big)^{1/4} \le M_4\,|X| = M_4\sqrt{n},$$

so that $\mathbb{E}\,\langle X,Y\rangle^4 \le M_4 n^2$, and one may apply (18.9)–(18.10).

For an illustration, on the interval $\Omega = (-\pi,\pi)$ with the uniform measure $\mathrm{d}\mathbb{P}(t) = \frac{1}{2\pi}\,\mathrm{d}t$, consider a finite trigonometric system $X = (X_1,\dots,X_n)$ with components

$$X_{2k-1}(t) = \sqrt{2}\,\cos(m_k t),$$
$$X_{2k}(t) = \sqrt{2}\,\sin(m_k t), \quad k = 1,\dots,n/2,$$

where m_k are positive integers such that

$$\frac{m_{k+1}}{m_k} \ge q > 1,$$

assuming that n is even. Then X is an isotropic random vector satisfying $|X|^2 = n$ and $\mathbb{E}X = 0$, and with M_4 bounded by a function of q only. For evaluation of the moment $\Sigma_3(n)$ as in (18.11), one may use the identities

$$\cos t = \mathbb{E}_\varepsilon\,\mathrm{e}^{i\varepsilon t}, \quad \sin t = \frac{1}{i}\,\mathbb{E}_\varepsilon\,\varepsilon\,\mathrm{e}^{i\varepsilon t},$$

where ε is a Bernoulli random variable taking the values ± 1 with probability $\frac{1}{2}$. Let $\varepsilon_1,\varepsilon_2,\varepsilon_3$ be independent copies of ε which are also independent of X. Using the property that $\varepsilon_1' = -\varepsilon_1\varepsilon_3$ and $\varepsilon_2' = -\varepsilon_2\varepsilon_3$ are independent, the identity for the cosine function implies that, for all integers $1 \le n_1 \le n_2 \le n_3$,

$$\begin{aligned}\mathbb{E}\,\cos(n_1t)\cos(n_2t)\cos(n_3t) &= \mathbb{E}_\varepsilon\,\mathbb{E}\,\exp\{i(\varepsilon_1n_1+\varepsilon_2n_2+\varepsilon_3n_3)\,t\}\\ &= \mathbb{E}_\varepsilon\,I\{\varepsilon_1n_1+\varepsilon_2n_2+\varepsilon_3n_3=0\}\\ &= \mathbb{E}_\varepsilon\,I\{\varepsilon_1'n_1+\varepsilon_2'n_2=n_3\} = \frac{1}{4}\,I\{n_1+n_2=n_3\},\end{aligned}$$

where $\mathbb{E}_\varepsilon$ means the expectation over $(\varepsilon_1,\varepsilon_2,\varepsilon_3)$, and where $I\{A\}$ denotes the indicator of the event A. Similarly, involving also the identity for the sine function, we have

$$\begin{aligned}\mathbb{E}\,\sin(n_1t)\sin(n_2t)\cos(n_3t) &= -\mathbb{E}_\varepsilon\,\mathbb{E}\,\varepsilon_1\varepsilon_2\,\exp\{i(\varepsilon_1n_1+\varepsilon_2n_2+\varepsilon_3n_3)\,t\}\\ &= -\mathbb{E}_\varepsilon\,\varepsilon_1\varepsilon_2\,I\{\varepsilon_1n_1+\varepsilon_2n_2+\varepsilon_3n_3=0\}\\ &= -\mathbb{E}_\varepsilon\,\varepsilon_1'\varepsilon_2'\,I\{\varepsilon_1'n_1+\varepsilon_2'n_2=n_3\} = -\frac{1}{4}\,I\{n_1+n_2=n_3\},\end{aligned}$$

$$\begin{aligned}\mathbb{E}\,\sin(n_1t)\cos(n_2t)\sin(n_3t) &= -\mathbb{E}_\varepsilon\,\mathbb{E}\,\varepsilon_1\varepsilon_3\,\exp\{i(\varepsilon_1n_1+\varepsilon_2n_2+\varepsilon_3n_3)\,t\}\\ &= -\mathbb{E}_\varepsilon\,\varepsilon_1\varepsilon_3\,I\{\varepsilon_1n_1+\varepsilon_2n_2+\varepsilon_3n_3=0\}\\ &= \mathbb{E}_\varepsilon\,\varepsilon_1'\,I\{\varepsilon_1'n_1+\varepsilon_2'n_2=n_3\} = \frac{1}{4}\,I\{n_1+n_2=n_3\},\end{aligned}$$

$$\begin{aligned}\mathbb{E}\,\cos(n_1t)\sin(n_2t)\sin(n_3t) &= -\mathbb{E}_\varepsilon\,\mathbb{E}\,\varepsilon_2\varepsilon_3\,\exp\{i(\varepsilon_1n_1+\varepsilon_2n_2+\varepsilon_3n_3)\,t\}\\ &= -\mathbb{E}_\varepsilon\,\varepsilon_2\varepsilon_3\,I\{\varepsilon_1n_1+\varepsilon_2n_2+\varepsilon_3n_3=0\}\\ &= \mathbb{E}_\varepsilon\,\varepsilon_2'\,I\{\varepsilon_1'n_1+\varepsilon_2'n_2=n_3\} = \frac{1}{4}\,I\{n_1+n_2=n_3\}.\end{aligned}$$

On the other hand, if the sine function appears in the product once or three times, such expectations will be vanishing. They are thus vanishing in all cases where $n_1+n_2\neq n_3$, and do not exceed $\frac{1}{4}$ in absolute value for any combination of sine and cosine terms in all cases where $n_1+n_2=n_3$. Therefore, the moment $\Sigma_3(n)$ in (18.11) is bounded by a multiple of $T_3(n)$ where

$$T_3(n) = \mathrm{card}\big\{(i_1,i_2,i_3) : 1\le i_1\le i_2<i_3\le n,\ m_{i_1}+m_{i_2}=m_{i_3}\big\}.$$

One can now involve the lacunary assumption. If $q>2$, the property $i_1\le i_2<i_3$ implies $m_{i_1}+m_{i_2}<m_{i_3}$, so that $T_3(n)=\Sigma_3(n)=0$. In the case $1<q\le 2$, define A_q to be the (finite) collection of all couples (k_1,k_2) of positive integers such that $q^{-k_1}+q^{-k_2}\ge 1$. By the lacunary assumption, if $1\le i_1\le i_2<i_3\le n$, we have

$$m_{i_1}+m_{i_2}\le\Big(q^{-(i_3-i_1)}+q^{-(i_3-i_2)}\Big)\,m_{i_3}<m_{i_3},$$

as long as the couple (i_3-i_1,i_2-i_1) is not in A_q. Hence,

$$T_3(n) \le \text{card}\{(i_1,i_2,i_3) : 1 \le i_1 \le i_2 < i_3 \le n,\ (i_3 - i_1, i_3 - i_2) \in A_q\}$$
$$\le n\,\text{card}(A_q) \le c_q n$$

with constant depending on q only. Returning to (18.9)–(18.10), we then obtain:

Proposition 18.7.1 *For the lacunary trigonometric system X of an even length n and with parameter $q > 1$,*

$$\mathbb{E}_\theta\, \omega^2(F_\theta, F) \le \frac{c_q}{n^2}, \quad \mathbb{E}_\theta\, \rho^2(F_\theta, F) \le \frac{c_q\,(\log n)^2}{n^2},$$

where the constants c_q depend on q only. Similar inequalities also hold when F is replaced with the standard normal distribution function Φ.

In this connection one should mention a classical result of Salem and Zygmund concerning distributions of the lacunary sums

$$S_n = \sum_{k=1}^n (a_k \cos(m_k t) + b_k \sin(m_k t))$$

with an arbitrary prescribed sequence of the coefficients $(a_k)_{k\ge1}$ and $(b_k)_{k\ge1}$. Assume that $\frac{m_{k+1}}{m_k} \ge q > 1$ for all k and put

$$v_n^2 = \frac{1}{2} \sum_{k=1}^n (a_k^2 + b_k^2) \quad (v_n \ge 0),$$

so that the normalized sums $Z_n = S_n/v_n$ have mean zero and variance one under the measure $\mathbb{P}$. It was shown in [164] that the Z_n are weakly convergent to the standard normal law, i.e., their distributions F_n under $\mathbb{P}$ satisfy $\rho(F_n, \Phi) \to 0$ as $n \to \infty$, if and only if $\frac{a_n^2+b_n^2}{v_n^2} \to 0$ (in fact, the weak convergence was established on every subset of Ω of positive measure).

Restricting to the coefficients $\theta_{2k-1} = a_k/v_n$, $\theta_{2k} = b_k/v_n$, Salem–Zygmund's theorem may be stated as the assertion that $\rho(F_\theta, \Phi)$ is small if and only if $\|\theta\|_\infty = \max_{1\le k\le n} |\theta_k|$ is small. The latter condition naturally appears in the central limit theorem about the weighted sums of independent identically distributed random variables. Thus, Proposition 18.7.1 complements this result in terms of the rate of convergence in the mean on the unit sphere.

The result of [164] was generalized in [165]; as it turns out there is no need to assume that all m_k are integers, and the asymptotic normality is saved for real m_k such that $\inf_k \frac{m_{k+1}}{m_k} > 1$. However, in this more general situation, the rate $1/n$ as in Proposition 18.7.1 is no longer true (although the rate $1/\sqrt{n}$ is valid). The main reason is that the means

$$\mathbb{E}X_{2k-1} = \sqrt{2}\,\mathbb{E}\, \cos(m_k t) = \sqrt{2}\, \frac{\sin(\pi m_k)}{\pi m_k}$$

may be non-zero. For example, choosing $m_k = 2^k + \frac{1}{2}$, we obtain an orthonormal system with $\mathbb{E}X_{2k} = 0$, while

$$\mathbb{E}X_{2k-1} = \frac{2\sqrt{2}}{\pi\,(2^{k+1}+1)}.$$

Hence

$$\mathbb{E}\,\langle X,Y\rangle = |\mathbb{E}X|^2 = \frac{8}{\pi^2}\sum_{k=1}^{n}\frac{1}{(2^{k+1}+1)^2} \to c \quad (n\to\infty)$$

for some absolute constant $c > 0$, where Y is an independent copy of X. It can be seen that

$$\mathbb{E}\,\langle X,Y\rangle^2 = n + O(1), \quad \mathbb{E}\,\langle X,Y\rangle^3 = O(n), \quad \mathbb{E}\,\langle X,Y\rangle^4 = O(n^2).$$

Putting $\xi = \frac{\langle X,Y\rangle}{n}$ and applying Proposition 15.3.1, we find that

$$\begin{aligned}\sqrt{\pi}\,\mathbb{E}_\theta\,\omega^2(F_\theta,F) &= \Big(1+\frac{1}{4n}\Big)\,\mathbb{E}\,\Big(1-(1-\xi)^{1/2}\Big) - \frac{1}{8n} + O(n^{-2})\\ &= \Big(1+\frac{1}{4n}\Big)\Big(\frac{1}{2}\,\mathbb{E}\xi + \frac{1}{8}\,\mathbb{E}\xi^2\Big) - \frac{1}{8n} + O(n^{-2}) = \frac{c}{2n} + O(n^{-2}).\end{aligned}$$

A similar asymptotic is also true when F is replaced with Φ.

18.8 Remarks

The material of this chapter is based on [41]

For lacunary trigonometric cosine sequences

$$X_n(\omega) = \sqrt{2}\cos(n\,\omega), \quad -\pi < \omega < \pi,$$

Gaposhkin [93] established the following variant of the Berry–Esseen bound. Given a sequence m_k of natural numbers satisfying the lacunary condition $\inf_k \frac{m_{k+1}}{m_k} \geq q > 1$, and a family of real numbers $\{a_{n,k}\}_{n,k\geq 1}$ such that $\sum_{k=1}^{\infty} a_{n,k}^2 = 1$, consider the sums

$$S_n = \sum_{k=1}^{\infty} a_{n,k} X_{m_k}.$$

Thus, $\mathbb{E}S_n = 0$ and $\mathbb{E}S_n^2 = 1$ on the interval $\Omega = (-\pi,\pi)$ with the normalized Lebesgue measure $\mathbb{P}$. It was shown that the distribution functions $F_n(x) = \mathbb{P}\{S_n \leq x\}$ satisfy

$$\rho(F_n,\Phi) \leq C_q \sup_{k\geq 1} |a_{n,k}|^{1/4},$$

where the constants C_q depend on q only.

Chapter 19
Distributions With Symmetries

In this chapter we consider separately another natural example of probability distributions on $\mathbb{R}^n$ for which one may potentially obtain rates better than the standard $\frac{1}{\sqrt{n}}$-rate for approximations of F_θ by the standard normal law (in contrast with most of the examples of the previous chapter). Nevertheless, in the first section we focus on the standard rate and postpone refinements to the next sections. In the last section, we consider the important case of distributions which are coordinatewise symmetric and log-concave and give the proof of a theorem due to Klartag [123].

19.1 Coordinatewise Symmetric Distributions

Recall that a random vector $X = (X_1, \dots, X_n)$ in $\mathbb{R}^n$ is said to have a coordinatewise symmetric distribution if it is symmetric under all reflections $(x_1, \dots, x_n) \rightarrow (\varepsilon_1 x_1, \dots, \varepsilon_n x_n)$, $\varepsilon_k = \pm 1$, of the space $\mathbb{R}^n$ around the coordinate axes. In Section 2.3 we discussed a few properties of such distributions. First, we consider a scheme with explicitly prescribed weights.

Let $(\Omega, \mathcal{F}, \mathbb{P})$ be the underlying probability space on which all random variables X_k are defined, and let $F_n(x) = \mathbb{P}\{S_n \leq x\}$ be the distribution functions of the sums

$$S_n = X_1 + \cdots + X_n.$$

When X has a coordinatewise symmetric distribution with a finite second moment, then necessarily $\mathbb{E}X_k = 0$, so that $\mathbb{E}S_n = 0$, and

$$\mathbb{E}S_n^2 = \mathbb{E}\,|X|^2 = \sum_{k=1}^n \mathbb{E}X_k^2.$$

We shall estimate the Kolmogorov distance

$$\rho(F_n, \Phi) = \sup_x \left|F_n(x) - \Phi(x)\right|$$

S. Bobkov et al., *Concentration and Gaussian Approximation for Randomized Sums*, Probability Theory and Stochastic Modelling 104,
https://doi.org/10.1007/978-3-031-31149-9_19

in terms of moment-type functionals. The following general observation may be obtained based on the Berry–Esseen theorem.

Proposition 19.1.1 *If the random vector $X = (X_1, \dots, X_n)$ in $\mathbb{R}^n$ has a coordinatewise symmetric distribution, and $\mathbb{E}\,|X|^2 = 1$, then*

$$\rho(F_n, \Phi) \le 0.56\, \mathbb{E}\left[\frac{1}{|X|^3} \sum_{k=1}^n |X_k|^3\right] + 2.1\, \mathrm{Var}(|X|^2), \tag{19.1}$$

where the expression under the expectation sign is understood to be zero in the case $|X| = 0$. Moreover, for some absolute constant $c > 0$,

$$\rho(F_n, \Phi) \le 0.56\, \mathbb{E}\left[\frac{1}{|X|^3} \sum_{k=1}^n |X_k|^3\right] + c\, \mathrm{Var}(|X|). \tag{19.2}$$

Since $\mathrm{Var}(|X|) \le \mathrm{Var}(|X|^2)$ in view of the assumption $\mathbb{E}\,|X|^2 = 1$, the inequality (19.2) is stronger and requires only finiteness of the second moment M_2. On the other hand, the variance $\mathrm{Var}(|X|^2)$ is more tractable; in addition, (19.1) contains an explicit numerical constant.

Proof Let $\xi_1, \dots, \xi_n$ be independent random variables with mean zero and finite 3-rd absolute moments. Putting $r^2 = \mathrm{Var}(\xi_1 + \dots + \xi_n)$, $r \ge 0$, we have using the Berry–Esseen Theorem (cf. Proposition 4.4.1) for some absolute constant $c > 0$,

$$\sup_x \left|\mathbb{P}\{\xi_1 + \dots + \xi_n \le x\} - \Phi_r(x)\right| \le c L_3. \tag{19.3}$$

Here

$$L_3 = \frac{1}{r^3} \sum_{k=1}^n \mathbb{E}\,|\xi_k|^3$$

denotes the corresponding Lyapunov ratio, and $\Phi_r(x) = \Phi(x/r)$ stands for the distribution function of the normal law with mean zero and standard deviation r. In the case $r = 0$, put $L_3 = 0$ and $\Phi_r(x) = 1_{\{x \ge 0\}}$. As we mentioned before, (19.3) holds true with $c = 0.56$ according to [166].

We apply this theorem with $\xi_k = \varepsilon_k X_k$, where $\varepsilon_1, \dots, \varepsilon_n$ are independent Bernoulli random variables taking the values ± 1 with probability $1/2$, which are also independent of the X_k's. By the assumption, the random vector $(\varepsilon_1 X_1, \dots, \varepsilon_n X_n)$ has the same distribution as X. Hence, for any fixed values of $X_k = X_k(\omega)$, $\omega \in \Omega$, (19.3) may be applied to the sums $T_n = \varepsilon_1 X_1 + \dots + \varepsilon_n X_n$ to get

$$\sup_x \left|\mathbb{P}_\varepsilon\{T_n \le x\} - \Phi_{|X(\omega)|}(x)\right| \le 0.56\, \frac{1}{|X(\omega)|^3} \sum_{k=1}^n |X_k(\omega)|^3.$$

Let us take expectations of both sides with respect to the X_k's and then insert the expectation on the left-hand side inside the supremum. We thus obtain the distribution function $F_n(x) = \mathbb{E}\,\mathbb{P}_\varepsilon\{T_{n,X} \le x\}$, while $\Phi_{|X|} = \mathbb{E}\,\Phi_{|X(\omega)|}$ represents a mixture of

Gaussian measures which we considered in Chapter 12. Therefore, we get

$$\sup_x \left|F_n(x) - \Phi_{|X|}(x)\right| \le 0.56\,\mathbb{E}\left[\frac{1}{|X|^3}\sum_{k=1}^n \mathbb{E}\,|X_k|^3\right].$$

On the other hand, by Proposition 12.3.1, if $\mathbb{E}\,|X|^2 = 1$, then

$$\sup_x |\Phi_{|X|}(x) - \Phi(x)| \le 2.1\,\mathrm{Var}(|X|^2),$$

and it remains to combine the two bounds to get (19.1). Moreover, by Proposition 12.4.1,

$$\sup_x |\Phi_{|X|}(x) - \Phi(x)| \le c\,\mathrm{Var}(|X|),$$

and then we obtain (19.2). Proposition 19.1.1 is thus proved. □

To simplify the bound (19.1), let us derive a simple corollary. Given $0 < \lambda < 1$, by Chebyshev's inequality, the event $A = \{|X| \le \lambda\}$ has probability

$$\mathbb{P}(A) = \mathbb{P}\{1 - |X|^2 \ge 1 - \lambda^2\} \le \frac{1}{(1-\lambda^2)^2}\,\mathrm{Var}(|X|^2).$$

Applying the inequality $(x_1^2 + \cdots + x_n^2)^{-3/2} \sum_{k=1}^n |x_k|^3 \le 1$ $x_k \in \mathbb{R})$, we then have

$$\mathbb{E}\left(\frac{1}{|X|^3}\sum_{k=1}^n |X_k|^3\, 1_A\right) \le \mathbb{P}(A) \le \frac{1}{(1-\lambda^2)^2}\,\mathrm{Var}(|X|^2).$$

As for the complementary event $B = \{|X| > \lambda\}$, one may just write

$$\mathbb{E}\left(\frac{1}{|X|^3}\sum_{k=1}^n |X_k|^3\, 1_B\right) \le \frac{1}{\lambda^3}\sum_{k=1}^n \mathbb{E}\,|X_k|^3.$$

Combining these two estimates, from (19.1) we get

$$\rho(F_n, \Phi) \le \frac{0.56}{\lambda^3}\sum_{k=1}^n \mathbb{E}\,|X_k|^3 + \left(2.1 + \frac{1}{(1-\lambda^2)^2}\right)\mathrm{Var}(|X|^2).$$

Choosing $\lambda = 0.28^{1/3}$, we arrive at:

Corollary 19.1.2 *If the random vector X in $\mathbb{R}^n$ has a coordinatewise symmetric distribution with $\mathbb{E}\,|X|^2 = 1$, then*

$$\rho(F_n, \Phi) \le 2\sum_{k=1}^n \mathbb{E}\,|X_k|^3 + 6\,\mathrm{Var}(|X|^2). \tag{19.4}$$

19.2 Behavior On Average

Given a random vector $X = (X_1, \dots, X_n)$ in $\mathbb{R}^n$ with a coordinatewise symmetric distribution, let us return to the weighted sums

$$\langle X, \theta \rangle = \theta_1 X_1 + \dots + \theta_n X_n$$

and consider their distribution functions $F_\theta(x) = \mathbb{P}\{\langle X, \theta \rangle \le x\}$ ($\theta \in \mathbb{S}^{n-1}$, $n \ge 2$). If X is isotropic, which is equivalent to the condition $\mathbb{E}X_k^2 = 1$, $k = 1, \dots, n$, then the random vector $X_\theta = (\theta_1 X_1, \dots, \theta_n X_n)$ satisfies $\mathbb{E}\,|X_\theta|^2 = 1$. Hence, Corollary 19.1.2 is applicable to X_θ (in place of X), and (19.4) gives

$$\rho(F_\theta, \Phi) \le 2 \sum_{k=1}^n |\theta_k|^3\, \mathbb{E}\,|X_k|^3 + 6\,\mathrm{Var}(|X_\theta|^2). \tag{19.5}$$

This inequality can be used to control the distances $\rho(F_\theta, \Phi)$ on average. To estimate the expectations over θ on the right-hand side of (19.5), let us represent a standard normal random vector $Z = (Z_1, \dots, Z_n)$ in $\mathbb{R}^n$ in the form $Z = |Z|\,\theta$, where $\theta = (\theta_1, \dots, \theta_n)$ is regarded as a random vector on $\mathbb{S}^{n-1}$, which is uniformly distributed on the sphere and is independent of $|Z|$. This gives

$$\frac{4}{\sqrt{2\pi}} = \mathbb{E}\,|Z_k|^3 = \mathbb{E}\,|Z|^3\, \mathbb{E}_\theta\, |\theta_k|^3.$$

Since $\mathbb{E}\,|Z|^3 \ge (\mathbb{E}\,|Z|^2)^{3/2} = n^{3/2}$, we get $\mathbb{E}_\theta\, |\theta_k|^3 \le \frac{4}{\sqrt{2\pi}}\, n^{-3/2}$. Using $Z_i^2 Z_j^2 = |Z|^4\, \theta_i^2 \theta_j^2$ and $\mathbb{E}\,|Z|^4 = n^2 + 2n$, we also find that

$$\mathbb{E}_\theta\, \theta_i^2 \theta_j^2 = \frac{1}{n^2 + 2n} \ (i \ne j), \quad \mathbb{E}_\theta\, \theta_i^4 = \frac{3}{n^2 + 2n}$$

(recall the more general identities (10.15) and (10.16)). Hence

$$\begin{aligned} \mathbb{E}_\theta\, \mathrm{Var}(|X_\theta|^2) &= \mathbb{E}_\theta \sum_{i,j=1}^n \theta_i^2 \theta_j^2\, \mathrm{cov}(X_i^2, X_j^2) \\ &= \frac{1}{n^2 + 2n} \sum_{i,j=1}^n \mathrm{cov}(X_i^2, X_j^2) + \frac{2}{n^2 + 2n} \sum_{i=1}^n \mathrm{Var}(X_i^2) \\ &= \frac{1}{n^2 + 2n}\, \mathrm{Var}(|X|^2) + \frac{2}{n^2 + 2n} \sum_{i=1}^n \mathrm{Var}(X_i^2). \end{aligned}$$

Thus, from (19.5),

$$\mathbb{E}_\theta\, \rho(F_\theta, \Phi) \le \frac{8}{\sqrt{2\pi}\, n^{3/2}} \sum_{k=1}^n \mathbb{E}\,|X_k|^3 + \frac{6}{n^2+2n}\, \mathrm{Var}(|X|^2) + \frac{12}{n^2+2n} \sum_{k=1}^n \mathrm{Var}(X_k^2).$$

To further simplify, note that the assumption $\mathbb{E}X_k^2 = 1$ implies $\mathbb{E}\,|X_k|^3 \le \mathbb{E}X_k^4$. We also have $\mathrm{Var}(X_k^2) \le \mathbb{E}X_k^4$, so that

$$\mathbb{E}_\theta\, \rho(F_\theta, \Phi) \le \Big(\frac{8}{\sqrt{2\pi}\, n^{3/2}} + \frac{12}{n^2+2n}\Big) \sum_{k=1}^n \mathbb{E}\,|X_k|^4 + \frac{6}{n^2+2n}\, \mathrm{Var}(|X|^2).$$

Introduce the mean 4-th moment

$$\bar\beta_4 = \frac{1}{n} \sum_{k=1}^n \mathbb{E}\,|X_k|^4. \tag{19.6}$$

Using $\frac{1}{n^2+2n} \le \frac{\sqrt{2}-1}{n^{1/2}}$ for $n \ge 2$ and $12\,(\sqrt{2}-1) + \frac{8}{\sqrt{2\pi}} < 9$, we then arrive at:

Proposition 19.2.1 *If an isotropic random vector $X = (X_1, \dots, X_n)$ in $\mathbb{R}^n$ has a coordinatewise symmetric distribution, then*

$$\mathbb{E}_\theta\, \rho(F_\theta, \Phi) \le \frac{9}{\sqrt{n}}\, \bar\beta_4 + \frac{6}{n^2}\, \mathrm{Var}(|X|^2).$$

Thus, if $\mathbb{E}\,|X_k|^4 \le c$ and $\sigma_4^2(X) = \frac{1}{n}\, \mathrm{Var}(|X|^2) \le c\sqrt{n}$, we are led to a bound with standard rate, namely

$$\mathbb{E}_\theta\, \rho(F_\theta, \Phi) \le \frac{15c}{\sqrt{n}}.$$

Nevertheless, in many interesting cases these bounds can be improved to get a rate of the order $1/n$, up to a logarithmic factor. To this end, let us note that, by the coordinatewise symmetry assumption, we necessarily have $\mathbb{E}\,\langle X, Y\rangle^3 = 0$, where Y is an independent copy of X. Moreover, $\mathbb{E}\,\langle X, Y\rangle^4 \le 3n^2\bar\beta_4^2$, according to Proposition 2.3.2. Applying Corollary 15.4.2 and Proposition 16.2.1, we get:

Proposition 19.2.2 *Let $X = (X_1, \dots, X_n)$ be an isotropic random vector in $\mathbb{R}^n$ with a coordinatewise symmetric distribution such that $|X|^2 = n$ a.s. Then for some positive absolute constant c,*

$$\mathbb{E}_\theta\, \omega^2(F_\theta, \Phi) \le \frac{c}{n^2}\, \bar\beta_4^2,$$

where $\bar\beta_4$ is the mean 4-th moment defined in (19.6). *In addition,*

$$\mathbb{E}_\theta\, \rho^2(F_\theta, \Phi) \le \frac{c\,(\log n)^2}{n^2}\, \bar\beta_4^2.$$

In order to get rid of the support assumption on the distribution of X, we need to involve the variance functional $\sigma_4^2(X) = \frac{1}{n}\,\mathrm{Var}(|X|^2)$, or the functional

$$V_2(X) = \sup_{\theta \in \mathbb{S}^{n-1}} \mathrm{Var}\Big(\theta_1 X_1^2 + \dots + \theta_n X_n^2\Big),$$

which was introduced in Proposition 2.3.3 with the aim to bound $\Lambda = \Lambda(X)$. Combining this proposition with Proposition 17.1.1, we obtain a similar rate with respect to the dimension n.

Proposition 19.2.3 *Let X be an isotropic random vector in $\mathbb{R}^n$ with a coordinatewise symmetric distribution. Then for some absolute constant c,*

$$\mathbb{E}_\theta\, \rho(F_\theta, \Phi) \le \frac{c\,\log n}{n}\,\Big(\max_{1 \le k \le n} \mathbb{E} X_k^4 + V_2(X)\Big).$$

If additionally the distribution of X is invariant under permutations of coordinates,

$$\mathbb{E}_\theta\, \rho(F_\theta, \Phi) \le \frac{c\,\log n}{n}\,\Big(\mathbb{E} X_1^4 + \sigma_4^2(X)\Big),$$

where the last term $\sigma_4^2(X)$ may be removed in the case $\mathrm{cov}(X_1^2, X_2^2) \le 0$.

19.3 Coordinatewise Symmetry and Log-concavity

We now consider one important class of coordinatewise symmetric distributions on $\mathbb{R}^n$ where a $\frac{1}{n}$-rate of normal approximation for the distributions F_θ of the weighted sums $\langle X, \theta\rangle = \theta_1 X_1 + \dots + \theta_n X_n$ is achieved on a large part of the unit sphere. Keeping our standard notations, we have the following (non-randomized) variant of the central limit theorem due to Klartag [123].

Proposition 19.3.1 *Suppose that a random vector X in $\mathbb{R}^n$ has an isotropic, coordinatewise symmetric log-concave distribution. Then, for all $\theta = (\theta_1, \dots, \theta_n)$ in $\mathbb{S}^{n-1}$, with some universal constant $c > 0$ we have*

$$\rho(F_\theta, \Phi) \le c \sum_{k=1}^n \theta_k^4. \tag{19.7}$$

In particular, $\mathbb{E}_\theta\, \rho(F_\theta, \Phi) \le \frac{c}{n}$. Moreover, by Proposition 10.8.1 with $p = 4$,

$$\mathfrak{s}_{n-1}\big\{n\rho(F_\theta, \Phi) \ge cr\big\} \le e^{-\sqrt{rn}}, \quad r \ge 1.$$

Without the log-concavity assumption, the relation (19.7) is no longer true, even when the components X_k of X are independent and symmetric about the origin (as in the case of Bernoulli X_k's).

The proof requires some preparation. By assumption, the random vector X has a coordinatewise symmetric, log-concave density on $\mathbb{R}^n$. In view of the isotropy, we have $\mathbb{E}X_k^2 = 1$, and a Khinchine-type inequality for log-concave random variables yields the upper bound $\mathbb{E}X_k^4 \le \hat{C}$, $k = 1, \dots, n$, with some absolute constant.

Lemma 19.3.2 *There exist absolute constants $C, \tilde{C} \ge 1$ such that, for every $t \ge 0$,*

$$1 - \Phi\Big(t + 2\,(1 - \Phi(t))^{1/4}\Big) \ge \frac{1}{C}\,(1 - \Phi(t)),$$
$$\Phi\Big(t - 2\,(1 - \Phi(t))^{1/4}\Big) \ge \Phi(-2) \ge \frac{1}{C}\,(1 - \Phi(t)).$$

If $x > 0$ satisfies $|\frac{1}{x} - \frac{1}{\varphi(t)}| \le 4C\,(1 - \Phi(t))^{-3/4}$ for some $t \ge 0$, then $x \le \tilde{C}\varphi(t)$.

This lemma is easily obtained by applying the elementary inequalities $\frac{1}{1+x}\,\varphi(x) \le 1 - \Phi(x) \le \frac{1}{x}\,\varphi(x)$, $x \ge 0$. Here, the upper bound is well known. To prove the left one, consider the function $u(x) = \varphi(x) - (1+x)(1-\Phi(x))$ on the positive half-axis. We have $u(0) = \frac{1}{\sqrt{2\pi}} - \frac{1}{2} < 0$, while $u''(x) = (1-x)\varphi(x)$. Hence, u is convex on $[0, 1]$ and is convex on $[1, \infty)$. Since $u(\infty) = 0$, it follows that $u(x) \le 0$ for all $x \ge 0$.

Now, given a random variable X with an even, continuously differentiable density p, we denote by $S_\varepsilon = S_\varepsilon(X)$ the subset of the interval $[0, N]$ such that

$$p(x) = \varphi(x), \quad x \in S_\varepsilon,$$

where a real number N is defined by the relation $1 - \Phi(N) = 2CA\varepsilon^2$. Here $C > 0$ is an absolute constant, $A \ge 1$ and $0 < \varepsilon < 1$ (satisfying $2CA\varepsilon^2 < 1$).

Let Γ denote a random variable with density $q(x) = cx^{-8}\sin^8(x/8)$, where c is a normalizing constant. This density is proportional to the Fourier transform of the 8-th convolution power of the indicator function $1_{[-\frac{1}{8},\frac{1}{8}]}$, so, the characteristic function $\gamma(t)$ of Γ is vanishing for $|t| \ge 1$.

Lemma 19.3.3 *Let X be a random variable, independent of Γ, with an even, continuously differentiable density p. Given $0 < \varepsilon < 1$ and $A \ge 1$, assume that*

$$|\mathbb{P}(X + \varepsilon\Gamma \le x) - \Phi(x)| \le A\varepsilon^2, \quad x \in \mathbb{R}. \tag{19.8}$$

If there exists an absolute constant C_1 such that $|p'(x+u)| \le C_1$ for all $x \in S_\varepsilon$ and $|u| \le (1 - \Phi(x))^{1/4}$, then for some absolute constant C_2,

$$|\mathbb{P}(X \le x) - \Phi(x)| \le C_2 A\varepsilon^2, \quad x \in \mathbb{R}. \tag{19.9}$$

Proof By Markov's inequality, $\mathbb{P}\{|\Gamma| \ge (2\,\mathbb{E}\,\Gamma^6)^{1/6}\} \le \frac{1}{2}$. Therefore, for $x \ge N$,

$$\begin{aligned}\mathbb{P}\big\{X + \varepsilon\Gamma \ge x - (2\,\mathbb{E}\,\Gamma^6)^{1/6}\,\varepsilon\big\} &\ge \mathbb{P}\big\{X + \varepsilon\Gamma \ge x - (2\,\mathbb{E}\,\Gamma^6)^{1/6}\,\varepsilon,\ |\Gamma| \le (2\,\mathbb{E}\,\Gamma^6)^{1/6}\big\}\\ &\ge \mathbb{P}\big\{X \ge x,\ |\Gamma| \le (2\,\mathbb{E}\,\Gamma^6)^{1/6}\big\}\\ &\ge \frac{1}{2}\,\mathbb{P}\{X \ge x\}.\end{aligned}$$

Since $1 - \Phi(x - (2\,\mathbb{E}\,\Gamma^6)^{1/6}\,\varepsilon) \leq C\varepsilon^2$ for $x \geq N$ with an absolute constant $C > 0$, we obtain from (19.8) and the above lower bound that

$$\mathbb{P}(X \geq x) \ \leq\ 2(A + C)\varepsilon^2, \quad x \geq N,$$

which implies the required bound (19.9) for $x \geq N$, recalling that $1-\Phi(N) = 2CA\varepsilon^2$.

Now, on the interval $0 \leq x \leq N$, consider the function

$$\psi(x) = \mathbb{P}(X \leq x) - \Phi(x).$$

The inequality (19.9) is fulfilled at $x = 0$ and $x = N$ (since $\psi(0) = 0$). But, this function attains its minimum and maximum either at the endpoints, or at one of the inner points, where necessarily $\psi'(x) = p(x) - \varphi(x) = 0$ (if they exist). Hence, it is sufficient to prove (19.9) for $x \in S_\varepsilon$. Note that $1 - \Phi(x) \geq 2CA\varepsilon^2$ and thus

$$\frac{1}{\varepsilon}\int_{|s|\geq(1-\Phi(x))^{1/4}} q(s/\varepsilon)\,\mathrm{d}s = \mathbb{P}\Big\{\varepsilon|\Gamma| \geq (1-\Phi(x)^{1/4}\Big\} \leq \frac{\varepsilon^4\,\mathbb{E}\,\Gamma^4}{1-\Phi(x)} \leq C\varepsilon^2$$

for some absolute constant C. On the other hand, by Taylor's formula,

$$\mathbb{P}\{X \leq x+s\} = \mathbb{P}\{X \leq x\} + p(x)s + \frac{1}{2}\,p'(x+as)s^2, \quad |a| \leq 1,$$

while, by the assumption, we have $|p'(x+as)| \leq C_1$ for $|s| \leq (1-\Phi(x))^{1/4}$. Hence

$$\begin{aligned}
|\mathbb{P}(X+\varepsilon\Gamma \leq x) - \mathbb{P}(X \leq x)| &= \Big|\frac{1}{\varepsilon}\int_{-\infty}^{\infty}\Big(\mathbb{P}(X \leq x+s) - \mathbb{P}(X \leq x)\Big)\,q(s/\varepsilon)\,\mathrm{d}s\Big| \\
&\leq \Big|\frac{1}{\varepsilon}\int_{|s|\leq(1-\Phi(x))^{1/4}}\Big(p(x)\,s + \frac{1}{2}\,C_1 s^2\Big)\,q(s/\varepsilon)\,\mathrm{d}s\Big| \\
&\quad + \frac{2}{\varepsilon}\int_{|s|\geq(1-\Phi(x))^{1/4}} q(s/\varepsilon)\,\mathrm{d}s \\
&\leq \frac{C_1}{2\varepsilon}\int_{|s|\leq(1-\Phi(x))^{1/4}} s^2\,q(s/\varepsilon)\,\mathrm{d}s + 2C\varepsilon^2 \\
&\leq C_1\,\mathbb{E}\,(\varepsilon\Gamma)^2 + 2C\varepsilon^2 \leq \tilde{C}\varepsilon^2
\end{aligned}$$

for an absolute constant $\tilde{C}$. It remains to note that, for $x \in S_\varepsilon$, by (19.8),

$$\begin{aligned}
|\mathbb{P}(X \leq x) - \Phi(x)| &\leq \tilde{C}\varepsilon^2 + |\mathbb{P}(X+\varepsilon\Gamma \leq x) - \Phi(x)| \\
&\leq \tilde{C}\varepsilon^2 + A\varepsilon^2 \leq C_2 A\varepsilon^2.
\end{aligned}$$

The lemma is proved. □

Lemma 19.3.4 *Let X be a random variable with an even, positive, continuously differentiable log-concave density p. Assuming that* (19.8) *holds, there exists an absolute constant C_3 such that $|p'(x+u)| \le C_3$ for all $x \in S_\varepsilon$ and $|u| \le (1-\Phi(x))^{1/4}$.*

Proof Fix $x \in S_\varepsilon$ and recall that $1 - \Phi(x) \ge 2CA\varepsilon^2$. By Markov's inequality,

$$\begin{aligned}\mathbb{P}\Big\{\varepsilon\,|\Gamma| \ge (1-\Phi(x))^{1/4}\Big\} &\le \frac{\varepsilon^6\,\mathbb{E}\,\Gamma^6}{(1-\Phi(x))^{3/2}} \\ &\le \frac{1}{(2CA)^{5/2}}\,\varepsilon\,\mathbb{E}\,\Gamma^6\,(1-\Phi(x)). \qquad (19.10)\end{aligned}$$

By Lemma 19.3.2, $1 - \Phi(x + 2\,(1-\Phi(x))^{1/4}) \ge \frac{1}{C}(1-\Phi(x))$, and therefore, by (19.8),

$$\begin{aligned}\mathbb{P}\Big\{X + \varepsilon\Gamma \ge x + 2\,(1-\Phi(x))^{1/4}\Big\} &\ge 1 - \Phi(x + 2(1-\Phi(x))^{1/4}) - A\varepsilon^2 \\ &\ge \frac{1}{C}\,(1-\Phi(x)) - A\varepsilon^2 \\ &\ge \frac{1}{C}\,(1-\Phi(x)) - \frac{1}{2C}(1-\Phi(x)) = \frac{1}{2C}\,(1-\Phi(x)).\end{aligned}$$

Consequently, we get from (19.10) under the legitimate assumption that ε is smaller than a given absolute constant,

$$\begin{aligned}\mathbb{P}\Big\{X \ge x + (1-\Phi(x))^{1/4}\Big\} &\ge \mathbb{P}\Big\{X + \varepsilon\Gamma \ge x + 2\,(1-\Phi(x))^{1/4}\Big\} \\ &\quad - \mathbb{P}\Big\{\varepsilon|\Gamma| \ge (1-\Phi(x))^{1/4}\Big\} \ge \frac{1}{4C}(1-\Phi(x)\}.\end{aligned}$$

A similar argument, using the second inequality of Lemma 19.3.2, shows that

$$\begin{aligned}\mathbb{P}\Big\{X \le x - (1-\Phi(x))^{1/4}\Big\} &\ge \mathbb{P}\Big\{X + \varepsilon\Gamma \le x - 2\,(1-\Phi(x))^{1/4}\Big\} \\ &\quad - \mathbb{P}\Big\{\varepsilon|\Gamma| \ge (1-\Phi(x))^{1/4}\Big\} \ge \frac{1}{4C}\,(1-\Phi(x)).\end{aligned}$$

We conclude that, whenever $|u| \le (1-\Phi(x))^{1/4}$,

$$\min\Big\{\mathbb{P}(X \ge x+u),\ \mathbb{P}(X \le x+u)\Big\} \ge \frac{1}{4C}(1-\Phi(x)). \qquad (19.11)$$

Now, fix $u_0 \in \mathbb{R}$. Since $p > 0$ and $\log p$ is concave, one may write the inequality

$$p(u) \le p(u_0)\exp\Big\{\frac{p'(u_0)}{p(u_0)}(u-u_0)\Big\}, \qquad u \in \mathbb{R}.$$

Therefore, in the case $p'(u_0) \neq 0$, we have

$$\min\Big(\int_{u_0}^{\infty} p(u)\,\mathrm{d}u,\ \int_{-\infty}^{u_0} p(u)\,\mathrm{d}u\Big) \le \int_{u_0}^{\infty} p(u_0)\exp\Big\{-\frac{|p'(u_0)|(u-u_0)}{p(u_0)}\Big\}\,du = \frac{p(u_0)^2}{|p'(u_0)|}.$$

We may thus conclude from (19.11) that

$$|p'(x+u)| \le p^2(x+u)\,\Big(\min\{\mathbb{P}(X \ge x+u),\ \mathbb{P}(X \le x+u)\}\Big)^{-1} \le \frac{4C}{1-\Phi(x)}\,p^2(x+u). \tag{19.12}$$

Equivalently, $|(1/p(x+u))'| \le 4C/(1-\Phi(x))$, which yields

$$\Big|\frac{1}{p(x+u)} - \frac{1}{p(x)}\Big| \le \frac{4C}{1-\Phi(x)}\,(1-\Phi(x))^{1/4} = 4C\,(1-\Phi(x))^{-3/4}$$

for $|u| \le (1-\Phi(x))^{1/4}$. Recall that $p(x) = \varphi(x)$ for $x \in S_\varepsilon$. Applying the third inequality of Lemma 19.3.2, we obtain that $p(x+u) \le \tilde{C}\varphi(x)$, and returning to (19.12), we thus get

$$|p'(x+u)| \le \frac{4C}{1-\Phi(x)}\,\tilde{C}^2\varphi(x)^2 \le C_3$$

for some absolute constant C_3. The lemma is proved. □

Lemma 19.3.5 *Let $\Delta_1, \dots, \Delta_n$ be independent, symmetric Bernoulli random variables. Given $\sigma > 0$ and $\theta = (\theta_1, \dots, \theta_n) \in \mathbb{R}^n$, $\theta \neq 0$, assume that*

$$\sum_{k:\,|\theta_k| > \sigma} \theta_k^2 \le \frac{1}{2}\,|\theta|^2. \tag{19.13}$$

Then with some absolute constant $C > 0$, for all $x \in \mathbb{R}$,

$$\Big|\mathbb{P}\Big\{\sum_{k=1}^{n} \theta_k\Delta_k + \sigma\Gamma \le x\Big\} - \Phi\Big(\frac{x}{|\theta|}\Big)\Big| \le C\Big(\frac{\sigma^2}{|\theta|^2} + \sum_{k=1}^{n}\frac{\theta_k^4}{|\theta|^4}\Big). \tag{19.14}$$

Proof In view of the homogeneity of the inequality (19.14), we may assume that $|\theta| = 1$. Fix $x \in \mathbb{R}$ and put $\varepsilon = (\sum_{k=1}^n \theta_k^4)^{1/2}$. The characteristic function of the random variable $\xi = \sum_{k=1}^n \theta_k\Delta_k + \sigma\Gamma$ is given by $\mathbb{E}\,\mathrm{e}^{it\xi} = \gamma(\sigma t)\prod_{k=1}^n \cos(\theta_k t)$. Therefore, by the Fourier inversion formula,

$$\mathbb{P}\{\xi \le x\} - \Phi(x) = \frac{1}{2\pi}\int_{-\infty}^{\infty}\Big(\gamma(\sigma t)\prod_{k=1}^{n}\cos(\theta_k t) - \mathrm{e}^{-t^2/2}\Big)\frac{\mathrm{e}^{ixt}}{it}\,\mathrm{d}t.$$

We need to bound the absolute value of this integral, up to an absolute factor, by $\sigma^2 + \varepsilon^2$. First consider the main case $\sigma \le \varepsilon^{1/2}$. Splitting the integration over the regions $|t| \le \varepsilon^{-1/2}$, $\varepsilon^{-1/2} \le |t| \le \sigma^{-1}$ and $|t| \ge \sigma^{-1}$, we represent the above integral as the sum $I_1 + I_2 + I_3$.

In order to estimate I_1, one may use the simple equality $e^{s^2/2}\cos s = e^{as^4}$ for $|s| \le 1$ with $|a| \le 1$. Since $|\theta_k| \le \varepsilon^{1/2}$ for all k, we have, for $|t| \le \varepsilon^{-1/2}$,

$$\Big|\prod_{k=1}^n e^{\theta_k^2 t^2/2}\cos(\theta_k t) - 1\Big| = \Big|\exp\Big\{at^4\sum_{k=1}^n \theta_k^4\Big\} - 1\Big| \le 3\varepsilon^2 t^4.$$

The characteristic function γ of Γ satisfies $1 - ct^2 \le \gamma(t) \le 1$ for $0 \le t \le 1$ with an absolute constant $c > 0$. We may thus conclude that for $|t| \le \varepsilon^{-1/2}$

$$\gamma(\sigma t)\prod_{k=1}^n e^{\theta_k^2 t^2/2}\cos(\theta_k t) = \big(1 + a_1 c\,\sigma^2 t^2\big)\big(1 + 3a_2\,\varepsilon^2 t^4\big) = 1 + a_3\big(4c\sigma^2 t^2 + 3\,\varepsilon^2 t^4\big),$$

with some $|a_j| \le 1$. The latter formula yields

$$\begin{aligned} |I_1| &= \Big|\int_{-\varepsilon^{-1/2}}^{\varepsilon^{-1/2}} e^{-t^2/2}\Big(\gamma(\sigma t)\prod_{k=1}^n e^{\theta_k^2 t^2/2}\cos(\theta_k t) - 1\Big)\frac{e^{ixt}}{it}\,dt\Big| \\ &\le C\int_{-\infty}^{\infty} e^{-t^2/2}(\sigma^2 t^2 + \varepsilon^2 t^4)\,\frac{dt}{|t|} \le C'(\sigma^2 + \varepsilon^2) \end{aligned}$$

for some absolute constants C, C'.

To bound I_2, let $J = \{k \le n : |\theta_k| \le \sigma\}$, so that $\sum_{k\in J}\theta_k^2 \ge \frac{1}{2}$, by the hypothesis (19.13). Applying $\cos s \le e^{-s^2/2}$ ($|s| \le 1$), we get, for $|t| \le \sigma^{-1}$,

$$\Big|\prod_{k=1}^n \cos(\theta_k t)\Big| \le \prod_{k\in J}\cos(\theta_k t) \le \exp\Big\{-\frac{1}{2}t^2\sum_{k\in J}\theta_k^2\Big\} \le e^{-t^2/4}.$$

Using the bound $1 - \Phi(x) \le \frac{1}{2}e^{-x^2/2}$ ($x > 0$), we then obtain that

$$\begin{aligned} |I_2| &\le 2\int_{\varepsilon^{-1/2}}^{\sigma^{-1}} \Big(\Big|\prod_{k=1}^n \cos(\theta_k t)\Big| + e^{-t^2/2}\Big)\frac{dt}{t} \\ &\le 4\int_{\varepsilon^{-1/2}}^{\infty} e^{-t^2/4}\,dt < 8\,e^{-1/(4\varepsilon)} < 64\,\varepsilon^2. \end{aligned}$$

The bound for I_3 is easy as well. Since $\gamma(\sigma t) = 0$ for $|t| \ge \sigma^{-1}$, we have

$$|I_3| \le 2\int_{\sigma^{-1}}^{\infty} e^{-t^2/2}\,\frac{dt}{t} < 2e^{-1/(2\sigma^2)} < 4\sigma^2.$$

It remains to collect together all bounds for the integrals I_j, which leads to (19.14).

In the case $\sigma > \varepsilon^{1/2}$, the argument is simpler, since then we only need to consider the integrals I_0 and I_3 over the regions $|t| \le \sigma^{-1}$ and $|t| \ge \sigma^{-1}$ respectively. Since the first region is contained in $|t| \le \varepsilon^{-1/2}$, we have $|I_0| \le C'\,(\sigma^2 + \varepsilon^2)$ according to the previous step for the integral I_1. The bound $|I_3| < 4\sigma^2$ also holds. □

The lemma is thus proved, and we are ready to return to the setting of Proposition 19.3.1.

Proof (of Proposition 19.3.1) Assume that the random variable Γ with a special distribution as above is independent of the random vector $X = (X_1, \dots, X_n)$. In view of Lemmas 19.3.3–19.3.4, it is sufficient to derive the inequality

$$\sup_{y\in\mathbb{R}} \Big|\mathbb{P}\Big(\varepsilon\Gamma + \sum_{k=1}^{n} \theta_k X_k \le y\Big) - \Phi(y)\Big| \le C\varepsilon^2, \quad (\theta_1, \dots, \theta_n) \in \mathbb{S}^{n-1},$$

with $\varepsilon = 4\sqrt{\hat{C}\sum_{k=1}^n \theta_k^4}$ and for some positive absolute constant C.

Let $\Delta_1, \dots, \Delta_n$ be independent, symmetric Bernoulli random variables that are also independent of X and Γ. Fix $y \in \mathbb{R}$, $\theta \in \mathbb{S}^{n-1}$, and define the function

$$P_y(x) = \mathbb{P}\Big\{\varepsilon\Gamma + \sum_{k=1}^{n} \theta_k x_k \Delta_k \le y\Big\}, \quad x = (x_1, \dots, x_n) \in \mathbb{R}^n.$$

Since the density of X is coordinate-wise symmetric, the random variable $\sum_{k=1}^n \theta_k X_k$ has the same distribution as $\sum_{k=1}^n \theta_k X_k \Delta_k$, so that

$$\mathbb{P}\Big\{\varepsilon\Gamma + \sum_{k=1}^{n} \theta_k X_k \le y\Big\} = \mathbb{E}P_y(X). \tag{19.15}$$

Let A denote the collection of all points $x \in \mathbb{R}^n$ such that $\frac{1}{2} \le \sum_{k=1}^n \theta_k^2 x_k^2 \le \frac{3}{2}$. By Lemma 19.3.5 applied to the vector $(\theta_1 x_1, \dots, \theta_n x_n)$ and with $\sigma = \varepsilon$, we have

$$\Big|P_y(x) - \Phi\Big(y\Big/\Big(\sum_{k=1}^{n} \theta_k^2 x_k^2\Big)^{1/2}\Big)\Big| \le C\Big(\varepsilon^2 + \sum_{i=1}^{n} \theta_k^4 x_k^4\Big), \quad y \in A. \tag{19.16}$$

On the other hand, for the random variable $U = \sum_{k=1}^n \theta_k^2 X_k^2$, the weighted Poincaré-type inequality of Proposition 6.6.1 with $f(x_1, \dots, x_n) = \theta_1^2 x_1^2 + \dots + \theta_n^2 x_n^2$ (cf. also Proposition 6.6.2) gives

$$\mathrm{Var}(U) \le 16 \sum_{k=1}^{n} \theta_k^4\, \mathbb{E}X_k^4 \le 16\,\hat{C} \sum_{k=1}^{n} \theta_k^4 = \varepsilon^2.$$

Since $\mathbb{E}\,U = 1$, Chebyshev's inequality yields $\mathbb{P}\{|U - 1| \ge \frac{1}{2}\} \le 4\mathrm{Var}(U) \le 4\varepsilon^2$, i.e., $\mathbb{P}(X \notin A) \le 4\varepsilon^2$. From (19.16) we thus get

$$\left|\mathbb{E}P_y(X) - \mathbb{E}\,\Phi(y/\sqrt{U})\right| \le \mathbb{P}\{X \notin A\} + C\,\mathbb{E}\Big(\varepsilon^2 + \sum_{k=1}^n \theta_k^4 X_k^4\Big) \le C'\varepsilon^2,$$

where we used the bound $\mathbb{E}X_k^4 \le \hat{C}$. As a result, in view of (19.15), we only need to show that for some constant C

$$|\mathbb{E}\,\Phi(y/\sqrt{U}) - \Phi(y)| \le C\varepsilon^2.$$

And indeed, according to Proposition 12.3.1 with $\rho = \sqrt{U}$, the expression on the left does not exceed $3\,\mathrm{Var}(\rho^2) = 3\,\mathrm{Var}(U) \le 3\varepsilon^2$. The proposition is proved. □

19.4 Remarks

The Berry–Esseen bound of Corollary 19.1.2 in the form (19.5) was proved by Goldstein and Shao [97] (with different absolute constants), cf. also an earlier paper by M. Meckes and E. Meckes [140] for similar results where the closeness of F_θ to Φ is given explicitly in terms of θ. Multidimensional projections of log-concave distributions with an unconditional basis, that is, the distributions of random vectors

$$\big(\langle X,\theta_1\rangle,\dots,\langle X,\theta_k\rangle\big), \quad \theta_i \in \mathbb{S}^{n-1}, \ \ 1 \le k \le n,$$

were considered by M. Meckes [141].

As we have already mentioned, Proposition 19.3.1 is due to Klartag [123].

Chapter 20
Product Measures

In this chapter we shall discuss the classical scheme of sums of independent random variables, which will allow us to sharpen many of the previous results. In particular, the logarithmic factor appearing in the bounds for the Kolmogorov distance in Propositions 17.1.1 and 17.5.1 may be removed (as well as in the deviation bound of Proposition 17.6.1). This is shown using Fourier Analysis, more precisely – a third order Edgeworth expansion for characteristic functions under the 4-th moment condition (cf. Chapter 4), and applying several results from Chapter 10 about deviations of elementary polynomials on the unit sphere. Even better bounds hold when applying a fourth order Edgeworth expansion under the 5-th moment condition.

20.1 Edgeworth Expansion for Weighted Sums

Throughout this chapter, let $X_1, \dots, X_n$ $(n \geq 2)$ be independent random variables with mean zero, variance one, and finite absolute moments $\beta_{p,k} = \mathbb{E}\,|X_k|^p$ of an integer order $p \geq 3$. As before, we consider the weighted sums

$$S_\theta = \theta_1 X_1 + \dots + \theta_n X_n, \quad \theta = (\theta_1, \dots, \theta_n) \in \mathbb{S}^{n-1},$$

together with their distribution functions $F_\theta(x) = \mathbb{P}\{S_\theta \leq x\}$. As we know from Chapter 4, every F_θ is approximated by the standard normal law with density and distribution function

$$\varphi(x) = \frac{1}{\sqrt{2\pi}}\, \mathrm{e}^{-x^2/2}, \quad \Phi(x) = \int_{-\infty}^{x} \varphi(y)\,\mathrm{d}y \quad (x \in \mathbb{R}).$$

Moreover, the rate of this approximation may be controlled by means of the Berry–Esseen inequality

$$\rho(F_\theta, \Phi) \leq c \sum_{k=1}^{n} |\theta_k|^3\, \beta_{3,k},$$

S. Bobkov et al., *Concentration and Gaussian Approximation for Randomized Sums*, Probability Theory and Stochastic Modelling 104,
https://doi.org/10.1007/978-3-031-31149-9_20

where $c > 0$ is an absolute constant, cf. (4.15). Here, the sum represents the Lyapunov coefficient L_3 for the weighted sequence $\{\theta_k X_k\}_{k=1}^n$ and is greater than or equal to $1/\sqrt{n}$. To improve this standard rate of approximation (which is however not guaranteed in general), one should consider an Edgeworth correction of the normal law of order $p-1$ with some $p \geq 4$.

To describe these corrections explicitly, let us recall Definition 4.5.2 (cf. (4.23), Proposition 4.7.1 and Definition 4.7.2. We note that the cumulants for the weighted sum exist up to order p and are given by the formula

$$\gamma_l(\theta) \equiv \gamma_l(S_\theta) = \sum_{k=1}^n \gamma_l(X_k)\, \theta_k^l, \quad l = 1, 2, \dots, p,$$

in which every cumulant $\gamma_l(X_k)$ represents a certain polynomial in moments $\mathbb{E}X_k^m$, $m \leq l$. In particular, $\gamma_1(X_k) = 0$ and $\gamma_2(X_k) = 1$, by the assumption that $\mathbb{E}X_k = 0$ and $\mathrm{Var}(X_k) = 1$. Using these functionals, the Edgeworth correction of order $p-1$ for F_θ is defined as a signed Borel measure on the real line with total measure 1, whose Fourier–Stieltjes transform, density and "distribution" function are respectively given by

$$g_{p-1,\theta}(t) = g(t) + g(t) \sum \frac{1}{k_1! \cdots k_{p-3}!} \left(\frac{\gamma_3(\theta)}{3!}\right)^{k_1} \cdots \left(\frac{\gamma_{p-1}(\theta)}{(p-1)!}\right)^{k_{p-3}} (it)^k,$$

$$\varphi_{p-1,\theta}(x) = \varphi(x) + \varphi(x) \sum \frac{1}{k_1! \cdots k_{p-3}!} \left(\frac{\gamma_3(\theta)}{3!}\right)^{k_1} \cdots \left(\frac{\gamma_{p-1}(\theta)}{(p-1)!}\right)^{k_{p-3}} H_k(x),$$

$$\Phi_{p-1,\theta}(x) = \Phi(x) - \varphi(x) \sum \frac{1}{k_1! \cdots k_{p-3}!} \left(\frac{\gamma_3(\theta)}{3!}\right)^{k_1} \cdots \left(\frac{\gamma_{p-1}(\theta)}{(p-1)!}\right)^{k_{p-3}} H_{k-1}(x)$$

(for $t \in \mathbb{R}$ and $x \in \mathbb{R}$). Here $g(t) = \mathrm{e}^{-t^2/2}$ is the characteristic function of the standard normal law, and $H_k(x)$ is the Chebyshev–Hermite polynomial of degree

$$k = 3k_1 + \cdots + (p-1)k_{p-3}.$$

In each formula, the summation runs over all non-negative integers $k_1, \dots, k_{p-3}$ that are not all zero and such that $k_1 + 2k_2 + \cdots + (p-3)k_{p-3} \leq p-3$.

In particular,

$$g_{3,\theta}(t) = \mathrm{e}^{-t^2/2}\left(1 + \frac{\gamma_3(\theta)}{3!}\,(it)^3\right),$$

$$g_{4,\theta}(t) = \mathrm{e}^{-t^2/2}\left(1 + \frac{\gamma_3(\theta)}{3!}\,(it)^3 + \frac{\gamma_4(\theta)}{4!}\,(it)^4 + \frac{\gamma_3(\theta)^2}{2!\,3!^2}\,(it)^6\right), \tag{20.1}$$

and

$$\Phi_{3,\theta}(x) = \Phi(x) - \frac{\gamma_3(\theta)}{3!}\,H_2(x)\varphi(x),$$

$$\Phi_{4,\theta}(x) = \Phi(x) - \varphi(x)\left(\frac{\gamma_3(\theta)}{3!}\,H_2(x) + \frac{\gamma_4(\theta)}{4!}\,H_3(x) + \frac{\gamma_3(\theta)^2}{2!\,3!^2}\,H_5(x)\right). \tag{20.2}$$

A number of properties of the functions $\Phi_{p-1,\theta}$ which we discussed before in Chapter 4 are controlled by the Lyapunov coefficients. In the case of the weighted sequence $\theta_1 X_1, \dots, \theta_n X_n$, they are defined by

$$L_p(\theta) = \sum_{k=1}^n \beta_{p,k}\, |\theta_k|^p, \quad \theta \in \mathbb{S}^{n-1},$$

thus representing weighted ℓ^p-norms on the sphere. Although $L_p(\theta) \geq n^{-\frac{p-2}{2}}$ for all θ (Proposition 4.2.2), these functions are of the same order for most of θ on the sphere with respect to the uniform measure $\mathfrak{s}_{n-1}$. Therefore, we use as basic functionals

$$\beta_p = \max_{1 \leq k \leq n} \beta_{p,k} \quad \text{and} \quad \bar{\beta}_p = \frac{1}{n} \sum_{k=1}^n \beta_{p,k}.$$

Clearly,

$$\beta_p \geq \bar{\beta}_p \geq 1 \quad \text{and} \quad L_p(\theta) \leq \beta_p \leq n\bar{\beta}_p.$$

Since $\bar{\beta}_2 = 1$, the function $p \to (\bar{\beta}_p)^{\frac{1}{p-2}}$ is non-decreasing for $p > 2$ in analogy with L_p (Proposition 4.2.1). Since each $\beta_{2,k} = 1$, the function $p \to \beta_{p,k}^{\frac{1}{p-2}}$ is non-decreasing as well, and the same is thus true for $\beta_p^{\frac{1}{p-2}}$. In particular, for all $p \geq 4$,

$$\bar{\beta}_3 \leq (\bar{\beta}_p)^{\frac{1}{p-2}}, \quad \beta_3 \leq \beta_p^{\frac{1}{p-2}}, \quad \bar{\beta}_4 \leq (\bar{\beta}_p)^{\frac{2}{p-2}}, \quad \beta_4 \leq \beta_p^{\frac{2}{p-2}}.$$

In terms of these functionals, one may control deviations of the Lyapunov coefficients, as the following statement shows, with a rate depending on the type of the functional. Here and in the sequel, we denote by c a positive absolute constant and by c_p a positive constant which depends on p only.

Proposition 20.1.1 *For any integer $p \geq 3$,*

$$\mathbb{E}_\theta\, L_p(\theta) \leq p^{p/2} \bar{\beta}_p\, n^{-\frac{p-2}{2}}. \tag{20.3}$$

Moreover,

$$\mathfrak{s}_{n-1}\Big\{n^{\frac{p-2}{2}} L_p(\theta) \geq c_p \bar{\beta}_p r\Big\} \leq 2 \exp\big\{-r^{2/p}\big\}, \quad r > 0, \tag{20.4}$$

$$\mathfrak{s}_{n-1}\Big\{n^{\frac{p-2}{2}} L_p(\theta) \geq c_p \beta_p r\Big\} \leq \exp\big\{-(rn)^{2/p}\big\}, \quad r \geq 1. \tag{20.5}$$

Indeed, by the definition,

$$\mathbb{E}_\theta\, L_p(\theta) = n\bar{\beta}_p\, \mathbb{E}_\theta\, |\theta_1|^p.$$

As was shown in Section 10.8, we have $\mathbb{E}_\theta\, |\theta_1|^p \leq p^{p/2} n^{-\frac{p}{2}}$, and (20.3) follows. This bound is also implied by (20.4), although with a worse p-dependent factor. In turn, (20.4) is obtained from the large deviation bound (10.23) on the unit sphere applied with coefficients $a_k = \beta_{p,k}/\sum_{j=1}^n \beta_{p,j}$. Using

$$L_p(\theta) \le \beta_p \sum_{k=1}^n |\theta_k|^p,$$

(20.5) follows from the deviation bound (10.22), cf. Proposition 10.8.1.

20.2 Approximation of Characteristic Functions of Weighted Sums

By the independence hypothesis, the characteristic functions of the weighted sums S_θ have a product structure,

$$f_\theta(t) = \mathbb{E}\, e^{itS_\theta} = \prod_{k=1}^n \mathbb{E}\, e^{it\theta_k X_k} \qquad (\theta \in \mathbb{S}^{n-1},\ t \in \mathbb{R}).$$

Hence, one may apply the general Proposition 4.8.2 in order to explore the rate of approximation of $f_\theta(t)$ by the corrected "normal" characteristic functions $g_{p-1,\theta}(t)$ for most directions $\theta \in \mathbb{S}^{n-1}$. This can be done by virtue of Proposition 20.1.1 on the t-intervals of a moderate size such as $|t| \le \sqrt{n}/(c\beta_3)$, which also allows one to consider an approximation for deviations of $f_\theta(t)$ from the mean (typical) characteristic function

$$f(t) = \mathbb{E}_\theta f_\theta(t), \quad t \in \mathbb{R},$$

in terms of deviations of $g_{p-1,\theta}(t)$ from their means. Recall that, by Proposition 4.7.4,

$$|g_{p-1,\theta}(t)| \le c_p \max\{L_p(\theta), 1\}.$$

Moreover, we have the following statement.

Proposition 20.2.1 *If β_p is finite for $p \ge 3$, then for all $\theta \in \mathbb{S}^{n-1}$,*

$$|f_\theta(t) - g_{p-1,\theta}(t)| \le c_p L_p(\theta)\, \min\{1, |t|^p\}\, e^{-t^2/8}, \quad |t| \le \frac{1}{L_3(\theta)}. \tag{20.6}$$

As a consequence, for all $\theta \in \mathbb{S}^{n-1}$ except for a set of measure $\le c_p\beta_p\, \exp\{-n^{2/p}\}$, we have in the interval $|t| \le \sqrt{n}/(c\beta_3)$

$$f_\theta(t) - f(t) = g_{p-1,\theta}(t) - \mathbb{E}_\theta\, g_{p-1,\theta}(t) + \varepsilon \tag{20.7}$$

with

$$|\varepsilon| \le c_p\beta_p \Big(n^{-\frac{p-2}{2}}\, |t|^p\, e^{-t^2/8} + \exp\big\{ - n^{2/p}\big\}\Big).$$

Proof The inequality (20.6) is a partial case of Proposition 4.8.2.

Using (20.5) with $p = 3$ and $r = 1$, we obtain that the inequality

$$L_3(\theta) \leq \frac{c\beta_3}{\sqrt{n}} \equiv \frac{1}{T_n}$$

holds true with some absolute constant $c > 0$ for all θ from a set Ω_0 on the sphere of measure at least $1 - \exp\{-n^{2/3}\}$. Hence, by (20.6), for all $\theta \in \Omega_0$,

$$\sup_{|t|\leq T_n} \Big[\frac{1}{\min\{1,|t|^p\}\,\mathrm{e}^{-t^2/8}}\,|f_\theta(t) - g_{p-1,\theta}(t)|\Big] \leq c_p L_p(\theta).$$

Applying once more (20.5) with $r = 1$, we get that the inequality

$$\big|f_\theta(t) - g_{p-1,\theta}(t)\big| \leq c_p \beta_p\, n^{-\frac{p-2}{2}}\, |t|^p\, \mathrm{e}^{-t^2/8}, \quad |t| \leq T_n, \tag{20.8}$$

holds for all θ in a set Ω on the sphere of measure at least $1-2\exp\{-n^{2/p}\}$. One may replace $f_\theta(t)$ and $g_{p-1,\theta}(t)$ in this inequality by their mean values $f(t) = \mathbb{E}_\theta\, f_\theta(t)$ and $g_{p-1}(t) = \mathbb{E}_\theta\, g_{p-1,\theta}(t)$ at the expense of an error not exceeding

$$(1 - \mathfrak{s}_{n-1}(\Omega))\, \sup_t\, |f_\theta(t) - g_{p-1,\theta}(t)| \leq 4\, \exp\big\{-n^{2/p}\big\}.$$

Averaging over θ in (20.8), this inequality thus yields in the same t-interval

$$\big|f(t) - g_{p-1}(t)\big| \,\leq\, c_p \beta_p\, n^{-\frac{p-2}{2}}\, |t|^p\, \mathrm{e}^{-t^2/8} + 4\, \exp\big\{-n^{2/p}\big\}.$$

Finally, combining the latter with (20.8), one may bound the expression

$$(f_\theta(t) - g_{p-1,\theta}(t)) - (f(t) - g_{p-1}(t)) = \big(f_\theta(t) - f(t)\big) - \big(g_{p-1,\theta}(t) - \mathbb{E}_\theta\, g_{p-1,\theta}(t)\big)$$

by a similar quantity. As a result, we arrive at (20.7), and Proposition 20.2.1 is proved. □

Let us now express the representation (20.7) in a more explicit form in the particular cases $p = 4$ and $p = 5$, assuming that the random variables $X_1, \dots, X_n$ are identically distributed, with $\mathbb{E}X_1 = 0$ and $\mathbb{E}X_1^2 = 1$. Note that $\bar{\beta}_p = \beta_p = \mathbb{E}\,|X_1|^p$.

For $p = 4$, the representation (20.1) for the corrected normal characteristic function $g_{3,\theta}(t)$ contains the term

$$\gamma_3(\theta) = \alpha_3\, \alpha_3(\theta), \quad \alpha_3(\theta) = \sum_{k=1}^n \theta_k^3, \quad \theta = (\theta_1, \dots, \theta_n) \in \mathbb{S}^{n-1},$$

where $\alpha_3 = \mathbb{E}X_1^3 = \gamma_3(X_1)$. Since these functions have s_{n-1}-mean zero, from (20.7) we obtain:

Corollary 20.2.2 *If β_4 is finite, then for all $\theta \in \mathbb{S}^{n-1}$ except for a set on the sphere of measure at most $c\beta_4\,\mathrm{e}^{-\sqrt{n}}$, in the interval $|t| \le \sqrt{n}/(c\beta_3)$, we have*

$$f_\theta(t) - f(t) = \alpha_3\,\alpha_3(\theta)\,\frac{(it)^3}{3!}\,\mathrm{e}^{-t^2/2} + \varepsilon$$

with

$$|\varepsilon| \le c\beta_4\Big(\frac{1}{n}\,t^4\,\mathrm{e}^{-t^2/8} + \exp\big\{-\sqrt{n}\big\}\Big).$$

Turning to the next order of approximation with $p = 5$, let us recall that $\gamma_4(X_1) = \mathbb{E}X_1^4 - 3 = \beta_4 - 3$. Hence, we need to involve the cumulant function

$$\gamma_4(\theta) = (\beta_4 - 3)\,l_4(\theta), \quad l_4(\theta) = \sum_{k=1}^n \theta_k^4.$$

Moreover, assuming that $\alpha_3 = 0$, the expression in (20.1) for $g_{4,\theta}(t)$ is considerably simplified. Since $\mathbb{E}_\theta\,l_4(\theta) = \frac{3}{n+2}$, we then obtain:

Corollary 20.2.3 *If β_5 is finite and $\alpha_3 = 0$, then for all $\theta \in \mathbb{S}^{n-1}$ except for a set of measure at most $c\beta_5\,\exp\{-n^{2/5}\}$, in the interval $|t| \le \sqrt{n}/(c\beta_3)$, we have*

$$f_\theta(t) - f(t) = (\beta_4 - 3)\Big(l_4(\theta) - \frac{3}{n+2}\Big)\,\frac{t^4}{4!}\,\mathrm{e}^{-t^2/2} + \varepsilon$$

with

$$|\varepsilon| \le c\beta_5\Big(\frac{1}{n^{3/2}}\,|t|^5\,\mathrm{e}^{-t^2/8} + \exp\big\{-n^{2/5}\big\}\Big).$$

20.3 Integral Bounds on Characteristic Functions

Let us now return to the general (not necessarily i.i.d.) setting and consider the behavior of characteristic functions $f_\theta(t)$ on large t-intervals. More precisely, here we focus on integrals of the form

$$I_p(\theta) = 1_{\{\delta \le L_3(\theta) \le 1\}} \int_{1/L_3(\theta)}^{1/\delta} \frac{|f_\theta(t)|}{t}\,\mathrm{d}t, \tag{20.9}$$

which appear in the Berry–Esseen bound of Proposition 4.8.1. Here and in the sequel, we choose the parameter

$$\delta = \bar{\beta}_p\,n^{-\frac{p-2}{2}} = n^{-p/2}\sum_{k=1}^n \mathbb{E}\,|X_k|^p.$$

Proposition 20.3.1 *Let $p \geq 4$. For some absolute constants $c > 0$ and $C > 0$,*

$$\mathbb{E}_\theta\, I_p(\theta) \leq Cp\, 1_{\{\delta \leq 1\}} \exp\Big\{ -cn^{1/4}\, \bar\beta_p^{-\frac{1}{2(p-2)}} \Big\}. \tag{20.10}$$

Moreover,

$$\mathbb{E}_\theta\, I_p(\theta) \leq Cp\, 1_{\{\delta \leq 1\}} \Big(\exp\Big\{ -2n^{2/3} \Big\} + \exp\Big\{ -cn\beta_p^{-\frac{2}{p-2}} \Big\} \Big). \tag{20.11}$$

Proof If $\delta > 1$, then $I_p(\theta) = 0$, and there is nothing to prove. So, let $\delta \leq 1$.

On the sphere $\mathbb{S}^{n-1}$ we consider the sets

$$\Omega_0 = \{\delta \leq L_3(\theta) \leq 1\}, \quad \Omega_1 = \{L_3(\theta) < \varepsilon\}, \quad \Omega_2 = \{L_3(\theta) \geq \varepsilon\}$$

with parameter $\varepsilon > 0$ to be chosen later on. To estimate the measure of the last set, note that the bound $\bar\beta_3 \leq (\bar\beta_p)^{\frac{1}{p-2}}$ implies

$$\frac{\sqrt{n}}{\bar\beta_3} \geq \sqrt{n}\, (\bar\beta_p)^{-\frac{1}{p-2}} = \delta^{-\frac{1}{p-2}}.$$

Hence, applying the bound (20.4) of Proposition 20.1.1 with $p = 3$, we have

$$\mathfrak{s}_{n-1}(\Omega_2) \leq 2\, \exp\Big\{ -c\Big(\frac{\sqrt{n}}{\bar\beta_3}\,\varepsilon\Big)^{2/3} \Big\} \leq 2\, \exp\Big\{ -c\,\varepsilon^{2/3}\, \delta^{-\frac{2}{3(p-2)}} \Big\}$$

with an absolute constant $c > 0$. Since $|f_\theta(t)| \leq 1$, while the lower limit of integration in (20.9) is necessarily greater than or equal to 1, we have $I_p(\theta) \leq \log(1/\delta)$ for all $\theta \in \mathbb{S}^{n-1}$. This gives

$$\mathbb{E}_\theta\, \big[I_p(\theta)\, 1_{\Omega_2}\big] \leq 2\log(1/\delta)\, \exp\Big\{ -c\,\varepsilon^{2/3}\, \delta^{-\frac{2}{3(p-2)}} \Big\}.$$

Turning to the integral over the complementary set and putting $\varepsilon' = \min\{\varepsilon, 1\}$, note that, for all $\theta \in \Omega_1$,

$$I_p(\theta) \leq 1_{\{\delta \leq \varepsilon'\}} \int_{1/\varepsilon'}^{1/\delta} \frac{|f_\theta(t)|}{t}\, dt,$$

so that

$$\mathbb{E}_\theta\, I_p(\theta)\, 1_{\Omega_1} \leq 1_{\{\delta \leq \varepsilon'\}} \int_{1/\varepsilon'}^{1/\delta} \frac{\mathbb{E}\,|f_\theta(t)|}{t}\, dt.$$

In this step we involve Lemma 13.1.4. Since $\bar\beta_4 \leq (\bar\beta_p)^{\frac{2}{p-2}}$, this gives

$$\begin{aligned}\mathbb{E}_\theta\, |f_\theta(t)| &\leq \sqrt{3}\, e^{-t^2/4} + \sqrt{6}\, \exp\Big\{ -\frac{n}{32}\, (\bar\beta_p)^{-\frac{2}{p-2}} \Big\} \\ &= \sqrt{3}\, e^{-t^2/4} + \sqrt{6}\, \exp\Big\{ -\frac{1}{32}\, \delta^{-\frac{2}{p-2}} \Big\}.\end{aligned}$$

Hence

$$\mathbb{E}_\theta\, I_p(\theta)\, 1_{\Omega_1} \le 1_{\{\delta \le \varepsilon'\}} \int_{1/\varepsilon'}^{1/\delta} \Big(2\, e^{-t^2/4} + 3\, \exp\Big\{-\frac{1}{32}\, \delta^{-\frac{2}{p-2}}\Big\}\Big)\, \frac{dt}{t}$$
$$\le 4\, \exp\Big\{-\frac{1}{4\varepsilon^2}\Big\} + 3\log(1/\delta)\, \exp\Big\{-\frac{1}{32}\, \delta^{-\frac{2}{p-2}}\Big\}. \qquad (20.12)$$

One can now combine the two estimates, which leads to

$$\mathbb{E}_\theta\, I_p(\theta) \le 4\, \exp\Big\{-\frac{1}{4\varepsilon^2}\Big\} + 5\log(1/\delta)\, \exp\Big\{-c\, \varepsilon^{2/3}\, \delta^{-\frac{2}{3(p-2)}}\Big\}.$$

In order to (approximately) optimize the right-hand side over ε, it is sufficient to equalize the two exponential terms on the right-hand side. Namely, choosing $\varepsilon = \delta^{\frac{1}{4(p-2)}} \le 1$, the resulting bound simplifies to

$$\mathbb{E}_\theta\, I_p(\theta) \le (4 + 5\log(1/\delta))\, \exp\Big\{-c\, \delta^{-\frac{1}{2(p-2)}}\Big\}.$$

Choosing a smaller value of the constant $c > 0$, one may replace the logarithmic term with a quantity not exceeding Cp, since

$$\sup_{0<\delta\le 1} \Big[\log(1/\delta)\, \exp\Big\{-\frac{c}{2}\, \delta^{-\frac{1}{2(p-2)}}\Big\}\Big] \le Cp.$$

As a result, we arrive at (20.10).

Turning to the second bound of the lemma, define the sets Ω_j in a similar way with $\varepsilon = cb_n$, $b_n = \beta_3/\sqrt{n}$, where the constant $c > 0$ is chosen according to the bound (20.5) of Proposition 20.1.1, so that

$$\mathfrak{s}_{n-1}(\Omega_2) \le \exp\big\{-3n^{2/3}\big\}.$$

Note that $\bar{\beta}_p \ge 1$, which implies $\log(1/\delta) \le \frac{p-2}{2}\log n$. Since $|f_\theta(t)| \le 1$, we get

$$\mathbb{E}_\theta\, \big[I_p(\theta)\, 1_{\Omega_2}\big] \le \frac{p-2}{2}\, \mathfrak{s}_{n-1}(\Omega_2)\, \log n \le Cp\, \exp\big\{-2n^{2/3}\big\}$$

for some constant C. Using (20.12) once more, we also have

$$\mathbb{E}_\theta\, I_p(\theta)\, 1_{\Omega_1} \le 4\, \exp\Big\{-\frac{1}{4\,(cb_n)^2}\Big\} + 3\log(1/\delta)\exp\Big\{-\frac{1}{32}\, \delta^{-\frac{2}{p-2}}\Big\}$$
$$\le 4\, \exp\Big\{-\frac{n}{4\beta_3^2}\Big\} + Cp\, \exp\Big\{-\frac{1}{64}\, \delta^{-\frac{2}{p-2}}\Big\}.$$

It remains to combine the two bounds and use $\beta_3 \le \beta_p^{\frac{1}{p-2}}$ together with $\delta \le \beta_p\, n^{-\frac{p-2}{2}}$.

Proposition 20.3.1 is proved. □

20.4 Approximation in the Kolmogorov Distance

The approximation of F_θ by means of the corrected normal "distribution" function $\Phi_{p-1,\theta}$ on average and actually for most of the weight vectors $\theta \in \mathbb{S}^{n-1}$ can be given in terms of the M_p-functional, which is equivalent in the case of independent components to the maximal L^p-norm $\beta_p = \max_k (\mathbb{E}\,|X_k|^p)^{1/p}$ (within p-dependent constants, cf. Proposition 2.1.3). One may also use a more flexible functional $\bar{\beta}_p$.

Recall that, by Proposition 4.7.4, for all $\theta \in \mathbb{S}^{n-1}$,

$$\rho(\Phi_{p-1,\theta}, \Phi) \le c_p \max\{L_p(\theta), 1\}$$

with some constant $c_p \ge 1$ depending on p only. Since $\rho(\Phi, F_\theta) \le 1$, a similar inequality also holds if we replace Φ with F_θ so that

$$\rho(F_\theta, \Phi_{p-1,\theta}) \le c_p \max\{L_p(\theta), 1\}.$$

As a consequence, applying (20.4), we obtain that, for any $r \ge 1/\bar{\beta}_p$,

$$\begin{aligned} \mathfrak{s}_{n-1}\big\{\rho(F_\theta, \Phi_{p-1,\theta}) \ge c_p^2 \bar{\beta}_p r\big\} &\le \mathfrak{s}_{n-1}\big\{\max\{L_p(\theta), 1\} \ge c_p \bar{\beta}_p r\big\} \\ &= \mathfrak{s}_{n-1}\big\{L_p(\theta) \ge c_p \bar{\beta}_p r\big\} \\ &\le 2\,\exp\big\{-r^{2/p}\, n^{\frac{p-2}{p}}\big\}. \end{aligned} \tag{20.13}$$

In other words, the distance from F_θ to $\Phi_{p-1,\theta}$ is essentially bounded by $1/\bar{\beta}_p$. We now derive much stronger inequalities. In particular, one may remove the requirement that the variable r is bounded away from zero in the inequality (20.13).

Proposition 20.4.1 *If $p \ge 4$, we have*

$$\mathbb{E}_\theta\, \rho(F_\theta, \Phi_{p-1,\theta}) \le c_p \bar{\beta}_p\, n^{-\frac{p-2}{2}}. \tag{20.14}$$

Moreover, for all $r > 0$,

$$\mathfrak{s}_{n-1}\big\{n^{\frac{p-2}{2}} \rho(F_\theta, \Phi_{p-1,\theta}) \ge c_p \bar{\beta}_p\, r\big\} \le 2\,\exp\big\{-r^{2/p}\big\}. \tag{20.15}$$

These large deviation bounds may be further sharpened in terms of the moment functional β_p (which is often of the same order as $\bar{\beta}_p$).

Proposition 20.4.2 *If $p \ge 4$, we have, for all $r \ge 1$,*

$$\mathfrak{s}_{n-1}\big\{n^{\frac{p-2}{2}} \rho(F_\theta, \Phi_{p-1,\theta}) \ge c_p \beta_p\, r\big\} \le \varepsilon_p(n, r), \tag{20.16}$$

where

$$\varepsilon_p(n,r) = \exp\big\{-(rn)^{2/p}\big\} + 2\,I(\bar{\beta}_p \le n^{\frac{p-2}{2}})\,\exp\big\{-\min\big(n^{2/3}, cn\beta_p^{-\frac{2}{p-2}}\big)\big\}.$$

Here, the last term should be removed when $\bar{\beta}_p > n^{\frac{p-2}{2}}$.

For example, choosing $r = 1$ and assuming that n is large enough, this bound gives

$$\mathfrak{s}_{n-1}\big\{n^{\frac{p-2}{2}}\rho(F_\theta,\Phi_{p-1,\theta}) \geq c_p\beta_p\big\} \leq \exp\big\{-n^{2/p}\big\}.$$

Our basic tool is Proposition 4.8.1, which relates the Kolmogorov distance to the Lyapunov coefficients $L_p(\theta)$ and the characteristic functions $f_\theta(t)$ via the upper bound

$$c_p\,\rho(F_\theta,\Phi_{p-1,\theta}) \leq L_p(\theta) + I_p(\theta) + \delta,$$

where $I_p(\theta)$ is the integral defined in (20.9) with arbitrary $\delta \geq 0$. As before, we choose the value $\delta = \bar{\beta}_p\, n^{-\frac{p-2}{2}}$, so that

$$c_p\,\rho(F_\theta,\Phi_{p-1,\theta}) \leq L_p(\theta) + I_p(\theta) + \bar{\beta}_p\, n^{-\frac{p-2}{2}}. \tag{20.17}$$

With this approach we are reduced to the study of the distributions of these two functionals on the sphere, $L_p(\theta)$ and $I_p(\theta)$, considered in the previous two sections.

Proof (of Proposition 20.4.1) Using $e^{-x} \leq p^{2p}\,x^{-2(p-2)}$ $(x > 0)$, from the bound (20.10) of Proposition 20.3.1 it follows that

$$\mathbb{E}_\theta\, I_p(\theta) \leq c_p\bar{\beta}_p\, n^{-\frac{p-2}{2}}.$$

Applying this in (20.17) together with a similar bound (20.3) on $\mathbb{E}_\theta\, L_p(\theta)$, we arrive at the first inequality (20.14).

By Markov's inequality, (20.10) also implies that

$$\mathfrak{s}_{n-1}\big\{n^{\frac{p-2}{2}} I_p(\theta) \geq c_p\bar{\beta}_p\big\} \leq 1_{\{\delta\leq 1\}}\,\frac{Cp}{c_p\,\delta}\,\exp\Big\{-c\delta^{-\frac{1}{2(p-2)}}\Big\}. \tag{20.18}$$

Since

$$\sup_{\delta>0}\Big[\frac{1}{\delta}\,\exp\Big\{-\frac{c}{2}\,\delta^{-\frac{1}{2(p-2)}}\Big\}\Big] \leq \Big(\frac{2}{c}\Big)^{2p},$$

one can choose suitable values of $c > 0$ and $c_p > 0$ in order to simplify (20.18) to the form

$$\mathfrak{s}_{n-1}\big\{n^{\frac{p-2}{2}} I_p(\theta) \geq c_p\bar{\beta}_p\big\} \leq 1_{\{\delta\leq 1\}}\,\exp\Big\{-c\delta^{-\frac{1}{2(p-2)}}\Big\}.$$

Let us now combine this with the deviation bound (20.4) on $L_p(\theta)$, and then from (20.17) we get that

$$\begin{aligned}\mathfrak{s}_{n-1}\big\{n^{\frac{p-2}{2}}\rho(F_\theta,\Phi_{p-1,\theta}) \geq 2c_p\bar{\beta}_p r\big\} &\leq 2\,\exp\big\{-r^{2/p}\big\}\\ &\quad + 1_{\{\delta\leq 1\}}\,\exp\Big\{-c\delta^{-\frac{1}{2(p-2)}}\Big\}.\end{aligned}$$

If $\delta > 1$, the last term on the right-hand side is vanishing, and we obtain the large deviation bound (20.15). But, if $\delta \leq 1$, we only have (replacing $2c_p$ with c_p)

$$\mathfrak{s}_{n-1}\big\{n^{\frac{p-2}{2}}\rho(F_\theta,\Phi_{p-1,\theta}) \geq c_p\bar{\beta}_p r\big\} \leq 3\,\exp\Big\{-\min\big\{r^{2/p}, c\delta^{-\frac{1}{2(p-2)}}\big\}\Big\}.$$

In the range $r^{2/p} \le \delta^{-\frac{1}{2(p-2)}}$, that is, for $r \le r_0 = \delta^{-\frac{p}{4(p-2)}}$, the above also leads to the required bound

$$\mathfrak{s}_{n-1}\{n^{\frac{p-2}{2}}\rho(F_\theta, \Phi_{p-1,\theta}) \ge c_p\bar{\beta}_p r\} \le 3\,\exp\{-cr^{2/p}\}. \tag{20.19}$$

Let now $r > r_0$, in which case

$$\bar{\beta}_p\, rn^{-\frac{p-2}{2}} = \delta r > \delta r_0 = \delta^{-\frac{3p-8}{4(p-2)}} \ge 1.$$

Hence, we may apply the inequality (20.13) with $rn^{-\frac{p-2}{2}}$ in place of r, which gives

$$\begin{aligned}\mathfrak{s}_{n-1}\{n^{\frac{p-2}{2}}\rho(F_\theta, \Phi_{p-1,\theta}) \ge c_p\bar{\beta}_p r\} &\le 2\,\exp\Big\{-(rn^{-\frac{p-2}{2}})^{\frac{2}{p}}\,n^{\frac{p-2}{p}}\Big\}\\ &= 2\,\exp\{-r^{2/p}\}.\end{aligned}$$

Thus, we have obtained (20.19) in all cases with arbitrary $r > 0$. Rescaling the variable r, this inequality may be modified to the form (20.15). Proposition 20.4.1 is proved. □

Proof (of Proposition 20.4.2) We now apply Markov's inequality on the basis of (20.11). Since $\beta_p n^{\frac{p-2}{2}} \ge \delta$, it gives

$$\begin{aligned}\mathfrak{s}_{n-1}\{n^{\frac{p-2}{2}}I_p(\theta) \ge c_p\beta_p\} &\le 1_{\{\delta\le 1\}}\,\frac{C p\, n^{\frac{p-2}{2}}}{c_p\beta_p}\,\exp\{-2n^{2/3}\}\\ &\quad + 1_{\{\delta\le 1\}}\,\frac{Cp}{c_p\delta}\,\exp\Big\{-c\delta^{-\frac{2}{p-2}}\Big\}.\end{aligned}$$

As before, choosing a smaller value of c and a larger value of c_p, the coefficient $\frac{Cp}{c_p\delta}$ in the last term may be removed. Similarly, using $\beta_p \ge 1$, the first term may be replaced with $\exp\{-n^{2/3}\}$. Hence, the above bound simplifies to

$$\mathfrak{s}_{n-1}\{n^{\frac{p-2}{2}}I_p(\theta) \ge c_p\beta_p\} \le 1_{\{\delta\le 1\}}\,\exp\{-n^{2/3}\} + 1_{\{\delta\le 1\}}\exp\Big\{-cn\beta_p^{-\frac{2}{p-2}}\Big\}.$$

It remains to combine this with (20.5). Recalling (20.17), we then get that with some constant $c_p > 0$, for all $r \ge 1$,

$$\begin{aligned}\mathfrak{s}_{n-1}\{n^{\frac{p-2}{2}}\rho(F_\theta, \Phi_{p-1,\theta}) \ge c_p\beta_p r\} &\le \exp\{-(rn)^{2/p}\} + 1_{\{\delta\le 1\}}\,\exp\{-n^{2/3}\}\\ &\quad + 1_{\{\delta\le 1\}}\exp\Big\{-cn\beta_p^{-\frac{2}{p-2}}\Big\}.\end{aligned}$$

Here, the condition $\delta \le 1$ is the same as $\bar{\beta}_p \le n^{\frac{p-2}{2}}$, which leads to the inequality (20.16). □

20.5 Normal Approximation Under the 4-th Moment Condition

As we have seen so far, the approximation of the distribution functions F_θ by the corrected normal "distribution" functions $\Phi_{p-1,\theta}$ can be made as good as we wish for most θ under natural moment-type conditions (with respect to the growing parameter n). However, these approximating functions still depend on θ, and one may wonder if one can use a fixed distribution (or measure) to approximate F_θ, for example, by the average corrected normal distribution function

$$G_{p-1}(x) = \mathbb{E}_\theta\, \Phi_{p-1,\theta}(x), \quad x \in \mathbb{R}.$$

This seems to be a natural approach; however, in general the concentration of $\Phi_{p-1,\theta}$ around G_{p-1} may not be as strong as the concentration of F_θ around $\Phi_{p-1,\theta}$.

In the case $p = 4$, the approximating functions are described in (20.2); let us rewrite them in a more explicit form as

$$\Phi_{3,\theta}(x) = \Phi(x) - \frac{\alpha_3(\theta)}{6}\,(x^2 - 1)\varphi(x), \quad \alpha_3(\theta) = \sum_{k=1}^n \alpha_{3,k}\theta_k^3,$$

where $\alpha_{3,k} = \mathbb{E}X_k^3$. Hence $G_3 = \Phi$ is the standard normal distribution function. Our basic functionals quantifying the strength of concentration of F_θ about $\Phi_{3,\theta}$ have been

$$\beta_4 = \max_{1 \le k \le n} \beta_{4,k} \quad \text{and} \quad \bar\beta_4 = \frac{1}{n}\sum_{k=1}^n \beta_{4,k} \quad (\beta_{4,k} = \mathbb{E}X_k^4).$$

They can be used to quantify the strength of concentration of $\Phi_{3,\theta}$ around Φ as well. This leads to the following observation due to Klartag and Sodin [125].

Proposition 20.5.1 *We have*

$$\mathbb{E}_\theta\, \rho(F_\theta, \Phi) \le \frac{c}{n}\,\bar\beta_4. \tag{20.20}$$

Moreover, for any $r > 0$,

$$\mathfrak{s}_{n-1}\big\{n\rho(F_\theta, \Phi) \ge c\bar\beta_4 r\big\} \le 2\,\exp\big\{-r^{1/2}\big\}. \tag{20.21}$$

The strength of approximation may actually be sharpened in terms of β_4.

Proposition 20.5.2 *With some absolute constants $c, c_0 > 0$, for all $r > 0$,*

$$\mathfrak{s}_{n-1}\big\{n\rho(F_\theta, \Phi) \ge cr\beta_4\big\} \le 2\,\exp\big\{-r^{2/3}\big\}. \tag{20.22}$$

Moreover, if $\mathbb{E}X_k^3 = 0$ $(k \le n)$ and $n \ge \beta_4^2$, then

$$\mathfrak{s}_{n-1}\big\{n\rho(F_\theta, \Phi) \ge c\beta_4\big\} \le 2\,\exp\big\{-c_0\sqrt{n}\big\}. \tag{20.23}$$

Proof (of Proposition 20.5.1) By Proposition 20.4.1, $\mathbb{E}_\theta\, \rho(F_\theta, \Phi_{3,\theta}) \le \frac{c}{n}\bar{\beta}_4$, so that, by the triangle inequality,

$$\mathbb{E}_\theta\, \rho(F_\theta, \Phi) \le \frac{c}{n}\bar{\beta}_4 + \mathbb{E}_\theta\, \rho(\Phi_{3,\theta}, \Phi). \tag{20.24}$$

To get a similar bound on the last integral, we use $|x^2 - 1|\, \varphi(x) \le 1, x \in \mathbb{R}$. Hence, by the very definition of the 3-rd order corrected normal distribution function,

$$\rho(\Phi_{3,\theta}, \Phi) \le \frac{1}{6}\, |\alpha_3(\theta)|. \tag{20.25}$$

Put $a_k = \mathbb{E}X_k^3$ and note that $a_k^2 \le \mathbb{E}X_k^4$, which is due to the assumption $\mathbb{E}X_k^2 = 1$. Recalling the identity (10.18) with $Q_3 = \alpha_3$, we obtain that

$$\mathbb{E}_\theta\, \rho^2(\Phi_{3,\theta}, \Phi) \le \frac{1}{36}\, \mathbb{E}_\theta \Big|\sum_{k=1}^n a_k \theta_k^3\Big|^2 = \frac{5/12}{n(n+2)(n+4)} \sum_{k=1}^n a_k^2.$$

The last sum does not exceed $n\bar{\beta}_4$, so that $\mathbb{E}_\theta\, \rho^2(\Phi_{3,\theta}, \Phi) \le \frac{1}{n^2}\, \bar{\beta}_4$ and hence

$$\mathbb{E}_\theta\, \rho(\Phi_{3,\theta}, \Phi) \le \frac{1}{n}\sqrt{\bar{\beta}_4} \le \frac{1}{n}\, \bar{\beta}_4$$

(since $\bar{\beta}_4 \ge 1$). Applying this estimate in (20.24), we arrive at (20.20).

Next, we apply Proposition 10.6.3 to Q_3 with coefficients $a_k = \mathbb{E}X_k^3/(\bar{\beta}_4)^{1/2}$ to obtain from (10.18) that, for any $r > 0$,

$$\mathfrak{s}_{n-1}\big\{n\rho(\Phi_{3,\theta}, \Phi) \ge 6(\bar{\beta}_4)^{1/2}\, r\big\} \le 2\, \exp\Big\{ -\frac{1}{23}\, r^{2/3}\Big\}.$$

After the change of variable, and using $\bar{\beta}_4 \ge 1$, one may rewrite this as

$$\mathfrak{s}_{n-1}\big\{n\rho(\Phi_{3,\theta}, \Phi) \ge c\bar{\beta}_4\, r\big\} \le 2\, \exp\big\{ -r^{2/3}\big\}. \tag{20.26}$$

On the other hand, by the deviation bound (20.15) of Proposition 20.4.1 with $p = 4$,

$$\mathfrak{s}_{n-1}\big\{n\rho(F_\theta, \Phi_{3,\theta}) \ge c\bar{\beta}_4 r\big\} \le 2\, \exp\big\{ -r^{1/2}\big\}.$$

Combining these two bounds, we thus get

$$\mathfrak{s}_{n-1}\big\{n\rho(F_\theta, \Phi) \ge 2c\bar{\beta}_4 r\big\} \le 4\, \exp\Big\{ -\min\big(r^{2/3}, r^{1/2}\big)\Big\},$$

which can easily be simplified to the form (20.21) by changing the variable r and choosing a suitable constant $c > 0$. Proposition 20.5.1 is now proved. □

Proof (of Proposition 20.5.2) Turning to Proposition 20.5.2, we employ the inequality (20.16), which in the case $p = 4$ gives, for all $r \ge 1$,

$$\mathfrak{s}_{n-1}\big\{n\rho(F_\theta, \Phi_{3,\theta}) \ge c\beta_4 r\big\} \le 3\, \exp\Big\{-\min\big((rn)^{1/2}, n^{2/3}, c_0 n\beta_4^{-1}\big)\Big\}. \tag{20.27}$$

In the second assertion of the proposition, we necessarily have $\Phi_{3,\theta} = \Phi$, and, by the assumption, $n\beta_4^{-1} \ge \sqrt{n}$. Hence, choosing here $r = 1$, we arrive at a concentration bound which is equivalent to (20.23).

As for the first assertion, first assume that $r \ge 1$. We combine (20.27) with (20.26) and use $\bar{\beta}_4 \le \beta_4$, to get that

$$\mathfrak{s}_{n-1}\big\{n\rho(F_\theta, \Phi) \ge c\beta_4 r\big\} \le 5\varepsilon(n, r) \tag{20.28}$$

with

$$\varepsilon(n, r) = \exp\Big\{-\min\big((rn)^{1/2}, n^{2/3}, c_0 n\beta_4^{-1}, r^{2/3}\big)\Big\}.$$

Since $\rho(F_\theta, \Phi) \le 1$, one may restrict (20.28) to the region $1 \le r \le \frac{n}{c\beta_4}$ (in particular, $\beta_4 \le n/c$). But then, $r^{2/3}$ is dominated by all other terms in the definition of $\varepsilon(n, r)$, so that (20.28) simplifies to

$$\mathfrak{s}_{n-1}\big\{n\rho(F_\theta, \Phi) \ge cr\beta_4\big\} \le 5\, \exp\big\{-r^{2/3}\big\}, \quad r \ge 1.$$

This inequality also holds for $0 \le r < 1$, since its right-hand side is greater than 1 in this case. Hence it holds for all $r \ge 0$. Finally, rescaling the variable r and choosing a larger constant $c > 0$, the above inequality yields (20.22). Proposition 20.5.2 is proved. □

20.6 Approximation With Rate $n^{-3/2}$

In order to improve the $\frac{1}{n}$-rate of normal approximation for most of the distribution functions F_θ, we need to involve the next term in the Edgeworth expansion. For simplicity, let us consider the i.i.d. case so that $X_1, \dots, X_n$ have equal distributions with

$$\mathbb{E}X_1 = 0, \quad \mathbb{E}X_2^2 = 1, \quad \bar{\beta}_p = \beta_p = \mathbb{E}\,|X_1|^p.$$

Recall that for $p = 5$, the representation (20.2) for the corrected normal distribution function $\Phi_{4,\theta}$ contains the term

$$\gamma_3(\theta) = \alpha_3\,(\theta_1^3 + \cdots + \theta_n^3), \quad \theta = (\theta_1, \dots, \theta_n) \in \mathbb{S}^{n-1},$$

where $\alpha_3 = \mathbb{E}X_1^3$. It has fluctuations on the sphere of order $1/n$, which follows from the formula (10.18), implying that

$$\mathbb{E}_\theta\, \gamma_3(\theta)^2 = \frac{15}{(n+2)(n+4)}\, \alpha_3^2.$$

Since the fluctuations of the other terms around their means in (20.2) are of a smaller order, the distances $\rho(\Phi_{4,\theta}, G_4)$ with $G_4 = \mathbb{E}_\theta\, \Phi_{4,\theta}$ are going to be of the same order $1/n$, as long as $\alpha_3 \neq 0$. This means that the improvement in comparison with the normal approximation obtained in the previous section can only be achieved when $\alpha_3 = 0$. This is not a strict condition. It is fulfilled, for example, when the underlying distribution (of X_1) is symmetric about the origin.

So, let us assume that $\alpha_3 = 0$. Since

$$\gamma_4(\theta) = (\beta_4 - 3)\, l_4(\theta), \quad l_4(\theta) = \sum_{k=1}^n \theta_k^4,$$

the representation (20.2) is simplified in this case to

$$\Phi_{4,\theta}(x) = \Phi(x) - \frac{\beta_4 - 3}{4!} H_3(x)\varphi(x)\, l_4(\theta), \quad H_3(x) = x^3 - 3x. \tag{20.29}$$

These functions have the spherical mean

$$G_4(x) = \Phi(x) - \frac{\beta_4 - 3}{8\,(n+2)} H_3(x)\varphi(x), \quad x \in \mathbb{R}. \tag{20.30}$$

We shall verify below that the distances $\rho(\Phi_{4,\theta}, G_4)$ are of the order $n^{-3/2}$, which leads to the following statement.

Proposition 20.6.1 *If* $\alpha_3 = 0$ *and* $\beta_5 < \infty$, *then*

$$\mathbb{E}_\theta\, \rho(F_\theta, G_4) \leq \frac{c}{n^{3/2}}\, \beta_5. \tag{20.31}$$

Proof Using (20.30), let us rewrite (20.29) as

$$\Phi_{4,\theta}(x) = G_4(x) - \frac{\beta_4 - 3}{24} H_3(x)\,\varphi(x)\,\Big(l_4(\theta) - \frac{3}{n+2}\Big).$$

It is easy to check that $|H_3(x)|\,\varphi(x) < 1$ for all $x \in \mathbb{R}$. Since $\beta_4 \geq 1$ so that $|\beta_4 - 3| \leq 2\beta_4$, it follows that

$$\rho(\Phi_{4,\theta}, G_4) \leq \frac{\beta_4}{12}\Big|l_4(\theta) - \frac{3}{n+2}\Big|. \tag{20.32}$$

Now, let us recall that, by (10.20) with $a_k = 1$,

$$\mathbb{E}_\theta\Big|l_4(\theta) - \frac{3}{n+2}\Big| \leq \sqrt{\mathrm{Var}_\theta(l_4(\theta))} \leq \sqrt{96}\, n^{-3/2},$$

which implies that

$$\mathbb{E}_\theta\, \rho(\Phi_{4,\theta}, G_4) \leq \beta_4\, n^{-3/2}.$$

Also, according to Proposition 20.4.1 with $p = 5$,

$$\mathbb{E}_\theta\, \rho(F_\theta, \Phi_{4,\theta}) \le c\beta_5\, n^{-3/2}.$$

Applying the triangle inequality for the Kolmogorov distance ρ together with $\beta_4 \le \beta_5$, the two bounds yield the required inequality (20.31). □

It is possible to strengthen this proposition in terms of deviation bounds.

Proposition 20.6.2 *If $\alpha_3 = 0$ and $\beta_5 < \infty$, then, for all $r \ge 0$,*

$$\mathfrak{s}_{n-1}\big\{n^{3/2}\rho(F_\theta, G_4) \ge c\beta_5\, r\big\} \le 2\,\exp\big\{-r^{2/5}\big\}. \tag{20.33}$$

Moreover, if $n \ge \beta_5$,

$$\mathfrak{s}_{n-1}\big\{n^{3/2}\rho(F_\theta, G_4) \ge c\beta_5\, r\big\} \le 2\,\exp\big\{-r^{1/2}\big\}. \tag{20.34}$$

Proof Rescaling the variable r (if necessary), we may assume that $r \ge 1$.
Applying Corollary 10.6.2, we have

$$\mathfrak{s}_{n-1}\Big\{n^{3/2}\Big|l_4(\theta) - \frac{3}{n+2}\Big| \ge r\Big\} \le 2\,\exp\Big\{-\frac{1}{38}\,r^{1/2}\Big\}, \quad r \ge 0.$$

Hence, from (20.32) it also follows that

$$\mathfrak{s}_{n-1}\Big\{\rho(\Phi_{4,\theta}, G_4) \ge \frac{\beta_4 r}{12}\, n^{-3/2}\Big\} \le 2\,\exp\Big\{-\frac{1}{38}\,r^{1/2}\Big\},$$

which implies

$$\mathfrak{s}_{n-1}\big\{\rho(\Phi_{4,\theta}, G_4) \ge C\beta_5 r\, n^{-3/2}\big\} \le 2\,\exp\big\{-r^{1/2}\big\} \tag{20.35}$$

for some absolute constant $C > 0$. On the other hand, by the bound (20.15) of Proposition 20.4.1 with $p = 5$,

$$\mathfrak{s}_{n-1}\big\{\rho(F_\theta, \Phi_{4,\theta}) \ge C\beta_5\, r\, n^{-3/2}\big\} \le 2\,\exp\big\{-r^{2/5}\big\}.$$

In combination with (20.35), the latter yields an inequality which is equivalent to (20.33).

To sharpen this bound, we employ Proposition 20.4.2. The inequality (20.16) specialized to the case $p = 5$ provides the bound

$$\mathfrak{s}_{n-1}\big\{\rho(F_\theta, \Phi_{4,\theta}) \ge C\beta_5\, r\, n^{-3/2}\big\} \le 3\,\exp\Big\{-\min\big((rn)^{2/5}, n^{2/3}, cn\beta_5^{-2/3}\big)\Big\}$$

for some absolute constants $c > 0$ and $C > 0$. Combining this with (20.35), we get that

$$\mathfrak{s}_{n-1}\big\{\rho(F_\theta, G_4) \ge C\beta_5 r\, n^{-3/2}\big\} \le 5\,\exp\Big\{-\min\big(r^{1/2}, (rn)^{2/5}, n^{2/3}, cn\beta_5^{-2/3}\big)\Big\}.$$

If

$$1 \leq r \leq r_0 = \min\left\{n^{4/3},\, (n\beta_5^{-2/3})^2\right\},$$

both $(rn)^{2/5}$ and $n^{2/3}$ will dominate $r^{1/2}$, and the above bound gives the desired relation

$$\mathfrak{s}_{n-1}\{\rho(F_\theta, G_4) \geq C\beta_5 r\, n^{-3/2}\} \leq 5\, \exp\{-cr^{1/2}\}. \tag{20.36}$$

As for larger values of r, the usual Berry–Esseen bound

$$\rho(F_\theta, \Phi) \leq CL_3(\theta) = C\beta_3 \sum_{k=1}^{n} |\theta_k|^3$$

with a purely Gaussian approximation is more accurate (here one may take $C = 1$). Indeed, applying the inequality (20.5) of Proposition 20.1.1 with $p = 3$, we get that, for all $r \geq n$,

$$\begin{aligned}\mathfrak{s}_{n-1}\{\rho(F_\theta, \Phi) \geq C\beta_3 r\, n^{-3/2}\} &\leq \mathfrak{s}_{n-1}\left\{n^{1/2}\, L_3(\theta) \geq C\beta_3\, rn^{-1}\right\}\\ &\leq \exp\{-r^{2/3}\}\end{aligned}$$

with some absolute constant C. Since $\beta_3 \leq \beta_5$, we obtain that

$$\mathfrak{s}_{n-1}\{\rho(F_\theta, \Phi) \geq C\beta_5 r\, n^{-3/2}\} \leq \exp\{-r^{2/3}\}. \tag{20.37}$$

But this inequality also holds for $r > (n\beta_5^{-2/3})^2$. Indeed, in this case

$$\beta_5 r\, n^{-3/2} > \beta_5 n^{-3/2}\, (n\beta_5^{-2/3})^2 = \Big(\frac{n^{3/2}}{\beta_5}\Big)^{1/3} > 1.$$

Hence, choosing $C > 1$ in (20.37), the left $\mathfrak{s}_{n-1}$-probability in this inequality is vanishing due to the property $\rho(F_\theta, \Phi) \leq 1$. Thus, the inequality (20.37) holds true for all $r > r_0$.

For sufficiently large values of r, one may replace Φ with G_4 in (20.37). Indeed, from the representation (20.30) it follows that

$$\rho(G_4, \Phi) \leq \frac{\beta_4}{n} \leq \frac{\beta_5}{n}.$$

Here, the last ratio does not exceed $\beta_5 r\, n^{-3/2}$ for all $r > r_0$ if and only if $r_0 \geq \sqrt{n}$. By the definition of r_0, the latter is equivalent to $n \geq \beta_5^{8/9}$ (which is fulfilled as long as $n \geq \beta$). Thus, (20.37) holds true with G_4 in place of Φ for all $r > r_0$, and therefore the inequality (20.36) is fulfilled for some constants $c > 0$ and $C > 0$ without any constraint on the range of r. It remains to rescale the variable r in (20.36) to obtain the bound (20.34). Proposition 20.4.1 is now proved. □

20.7 Lower Bounds

Keeping the notations and the basic assumptions in the i.i.d. case, let us now examine the sharpness of the bounds of Propositions 20.5.1 and 20.6.1,

$$\mathbb{E}_\theta\, \rho(F_\theta, \Phi) \leq \frac{c}{n}\, \beta_4, \quad \mathbb{E}_\theta\, \rho(F_\theta, G_4) \leq \frac{c}{n^{3/2}}\, \beta_5,$$

where $\beta_4 = \mathbb{E} X_1^4$ and $\beta_5 = \mathbb{E}\,|X_1|^5$. The second bound is optimal in the sense that it can be reversed in a typical situation, where the 4-th moment of X_1 is different from the 4-th moment of the standard normal law. The same observation applies to the first inequality, when the 3-rd moment $\alpha_3 = \mathbb{E} X_1^3$ is not zero.

Denote by $\mathcal{G}$ the collection of all functions G of bounded variation on the real line such that $G(-\infty) = 0$ and $G(\infty) = 1$.

Proposition 20.7.1 *If $\alpha_3 \neq 0$, $\beta_4 < \infty$, then the inequality*

$$\inf_{G \in \mathcal{G}} \mathbb{E}_\theta\, \rho(F_\theta, G) \geq \frac{c}{n} \tag{20.38}$$

holds for all n with a constant $c > 0$ depending on α_3 and β_4 only. Moreover, if $\alpha_3 = 0$, $\beta_4 \neq 3$, $\beta_5 < \infty$, then

$$\inf_{G \in \mathcal{G}} \mathbb{E}_\theta\, \rho(F_\theta, G) \geq \frac{c}{n^{3/2}}, \tag{20.39}$$

where the constant $c > 0$ depends on β_4 and β_5.

Proof Our basic tool will be Lemma 14.5.4. According to the lower bound of this lemma, for any $T > 0$,

$$\mathbb{E}_\theta\, \rho(F_\theta, G) \geq \frac{1}{6T}\, \mathbb{E}_\theta \bigg| \int_0^T (f_\theta(t) - f(t)) \Big(1 - \frac{t}{T}\Big)\, dt \bigg|. \tag{20.40}$$

As before, here $f_\theta(t)$ and $f(t) = \mathbb{E}_\theta f_\theta(t)$ denote the characteristic functions of F_θ and the typical distribution $F = \mathbb{E}_\theta\, F_\theta$. We use c, c', c'' to denote positive absolute constants which may vary from place to place.

First, let us apply the representation from Corollary 20.2.2, which holds for all θ from a set $\Omega \subset \mathbb{S}^{n-1}$ of $\mathfrak{s}_{n-1}$-measure at least $1 - c\beta_4 \exp\{-n^{1/2}\}$ in the interval $|t| \leq \sqrt{n}/(c\beta_3)$. Given $0 < T \leq 1$, we have

$$\int_0^T t^3\, e^{-t^2/2} \Big(1 - \frac{t}{T}\Big)\, dt > \frac{1}{4} \int_0^{T/2} t^3\, dt = \frac{1}{256}\, T^4.$$

On the other hand,

$$\int_0^T t^4\, e^{-t^2/8} \Big(1 - \frac{t}{T}\Big)\, dt < \frac{1}{5}\, T^5.$$

Therefore, for all $\theta \in \Omega$ and $n \geq (c\beta_3)^2$,

$$\left|\int_0^T (f_\theta(t) - f(t)) \left(1 - \frac{t}{T}\right) dt\right| \geq c\,|\alpha_3|\,|\alpha_3(\theta)|\,T^4 - \frac{c'}{n}\,\beta_4 T^5 - c'\beta_4 \exp\big\{-n^{1/2}\big\}\,T. \qquad (20.41)$$

To explore the behavior of the right-hand side on average for the growing parameter n, note that

$$\mathbb{E}_\theta\,|\alpha_3(\theta)| = \mathbb{E}_\theta\,|\theta_1^3 + \dots + \theta_n^3| \geq \frac{c}{n}. \qquad (20.42)$$

To see this, write a standard normal random vector $Z = (Z_1, \dots, Z_n)$ in $\mathbb{R}^n$ in the form $Z = r\theta$, where the random vector θ is uniformly distributed on $\mathbb{S}^{n-1}$ and $r = |Z|$ is independent of θ. This gives $\mathbb{E}_\theta\,|\alpha_3(\theta)|\,\mathbb{E}\,r^3 = \mathbb{E}\,|S_n|$ where $S_n = Z_1^3 + \dots + Z_n^3$. Since $\mathbb{E}\,r^4 = n(n+2) \leq 3n$, we have, by Hölder's inequality, $\mathbb{E}\,r^3 < 3n^{3/2}$, so,

$$\mathbb{E}_\theta\,|\alpha_3(\theta)| \geq \frac{1}{3n^{3/2}}\,\mathbb{E}\,|S_n|.$$

By the central limit theorem, $\frac{1}{\sqrt{n}}\,\mathbb{E}\,|S_n| \to \sqrt{15}$ as $n \to \infty$, and we arrive at (20.42).

Moreover, since $|\alpha_3(\theta)| \leq 1$ for all $\theta \in \mathbb{S}^{n-1}$, we have $\mathbb{E}_\theta\,|\alpha_3(\theta)|\,1_\Omega(\theta) \geq \frac{c}{n}$ for all n large enough. One can now integrate the inequality (20.41) over the set Ω, to conclude that

$$\frac{1}{T}\,\mathbb{E}_\theta \left|\int_0^T (f_\theta(t) - f(t)) \left(1 - \frac{t}{T}\right) dt\right| \geq \frac{T^3}{n}\,\big(c\,|\alpha_3| - c'\beta_4 T\big) - c''\beta_4 \exp\big\{-n^{1/2}\big\}$$

for all $n \geq n_0$. Choosing an appropriate value of $T \sim |\alpha_3|/\beta_4$ and applying (20.40), we get a lower bound of the form

$$\mathbb{E}_\theta\,\rho(F_\theta, G) \geq c\,\frac{|\alpha_3|^4}{\beta_4^3\,n} - c''\beta_4 \exp\big\{-n^{1/2}\big\}$$

with an arbitrary function G of bounded variation such that $G(-\infty) = 0$, $G(\infty) = 1$. The latter immediately yields the required relation (20.37) for the range $n \geq n_0$ with constant $c = c_0\,|\alpha_3|^4/\beta_4^3$ and for a sufficiently large n_0 depending α_3 and β_4.

In order to treat the remaining values $2 \leq n < n_0$, let us note that the infimum in (20.38) is positive. Indeed, assuming the opposite for a fixed n, there would exist a $G \in \mathcal{G}$ such that $\mathbb{E}_\theta\,\rho(F_\theta, G) = 0$ and hence $F_\theta(x) = G(x)$ for all $\theta \in \mathbb{S}^{n-1}$ and $x \in \mathbb{R}$. In particular, all the weighted sums S_θ would be equidistributed, which is only possible when all the random variables X_k have a standard normal distribution, according to the Pólya characterization theorem [158], cf. also [115]. But this contradicts the assumption $\alpha_3 \neq 0$.

The second assertion, where $\alpha_3 = 0$ and $\beta_4 \neq 3$, is similar. We now apply the representation of Corollary 20.2.3, which holds for all θ in a set $\Omega \subset \mathbb{S}^{n-1}$ of measure at least $1 - c\beta_5 \exp\{-n^{2/5}\}$ in the same interval $|t| \leq \sqrt{n}/(c\beta_3)$. Given $0 < T \leq 1$, we have

$$\int_0^T t^4 \, e^{-t^2/2} \Big(1 - \frac{t}{T}\Big) dt > \frac{1}{4} \int_0^{T/2} t^4 \, dt = \frac{1}{640} T^5.$$

On the other hand,

$$\int_0^T t^5 \, e^{-t^2/8} \Big(1 - \frac{t}{T}\Big) dt < \frac{1}{6} T^6.$$

Therefore, for all $\theta \in \Omega$ and $n \geq (c\beta_3)^2$,

$$\begin{aligned} \bigg|\int_0^T (f_\theta(t) - f(t)) \Big(1 - \frac{t}{T}\Big) dt\bigg| &\geq c\,|\beta_4 - 3| \, \Big|l_4(\theta) - \frac{3}{n+2}\Big| \, T^5 \\ &\quad - \frac{c'}{n^{3/2}} \, \beta_5 \, T^6 - c'\beta_5 \exp\big\{-n^{2/5}\big\}. \qquad (20.43) \end{aligned}$$

Using the arguments based on the application of the central limit theorem, we similarly have

$$\mathbb{E}_\theta \, \Big|l_4(\theta) - \frac{3}{n+2}\Big| \geq \frac{c}{n^{3/2}}. \qquad (20.44)$$

Indeed, the representation $Z = r\theta$ yields

$$S_n \equiv \sum_{k=1}^n (Z_k^4 - 3) = r^4 \Big(l_4(\theta) - \frac{3}{n+2}\Big) + \frac{3}{n+2} \, (r^4 - n(n+2)),$$

implying that

$$\mathbb{E}\,|S_n| \leq n(n+2)\, \mathbb{E}\, \Big|l_4(\theta) - \frac{3}{n+2}\Big| + \frac{3}{n+2}\, \mathbb{E}\,|r^4 - n(n+2)|.$$

To estimate the last expectation, recall that $\mathbb{E}\,|Z|^{2p} = n(n+2)\dots(n+2p-2)$ for $p = 1, 2, \dots$ In particular, $\mathbb{E}\, r^4 = n(n+2)$ and

$$\begin{aligned} \mathrm{Var}(r^4) &= \mathbb{E}\,|Z|^8 - (\mathbb{E}\,|Z|^4)^2 \\ &= n(n+2)(n+4)(n+6) - n^2(n+2)^2 = 8\,n(n+2)(n+3). \end{aligned}$$

Hence

$$\mathbb{E}\,|r^4 - n(n+2)| \leq 2\sqrt{2\,n(n+2)(n+3)},$$

so that

$$\sqrt{n}\,(n+2)\,\mathbb{E}\,\Big|l_4(\theta) - \frac{3}{n+2}\Big| \geq \frac{1}{\sqrt{n}}\,\mathbb{E}\,|S_n| - 6\sqrt{2}\sqrt{\frac{n+3}{n+2}}. \qquad (20.45)$$

Here, by the central limit theorem, $\frac{1}{\sqrt{n}}\,\mathbb{E}\,|S_n| \to \sqrt{\mathrm{Var}(Z_1^4)} = 4\sqrt{6}$ as $n \to \infty$, which is greater than $6\sqrt{2}$. This shows that the right-hand side of (20.45) is bounded away from zero, and therefore the lower bound (20.44) holds true for all $n \ge 2$ with some absolute constant $c > 0$. Moreover, since $|l_4(\theta) - \frac{3}{n+2}| \le 1$ on $\mathbb{S}^{n-1}$, necessarily

$$\mathbb{E}_\theta \left| l_4(\theta) - \frac{3}{n+2} \right| 1_\Omega(\theta) \ge \frac{c}{n}$$

with some absolute constant $c > 0$ for all n large enough. One can now integrate the inequality (20.43) over the set Ω, to conclude that

$$\frac{1}{T}\,\mathbb{E}_\theta \left| \int_0^T (f_\theta(t) - f(t)) \left(1 - \frac{t}{T}\right) \mathrm{d}t \right| \ge \frac{T^4}{n^{3/2}} \left(c\,|\beta_4 - 3| - c'\beta_5 T\right) - c''\beta_5 \exp\left\{ - n^{2/5} \right\}.$$

Choosing an appropriate value of $T \sim |\beta_4 - 3|/\beta_5$ and applying (20.40), we get

$$\mathbb{E}_\theta\, \rho(F_\theta, G) \ge c\, \frac{|\beta_4 - 3|^5}{\beta_5^4\, n} - c''\beta_5 \exp\left\{ - n^{2/5} \right\}.$$

The latter yields the required relation (20.39) for the range $n \ge n_0$ with constant $c = c_0\,|\beta_4 - 3|^5/\beta_5^4$ and with a sufficiently large n_0 depending on β_5.

A similar argument based on an application of the Pólya theorem allows us to involve the remaining values $2 \le n < n_0$. Proposition 20.7.1 is proved. □

20.8 Remarks

As mentioned before, Proposition 20.5.1 is due to Klartag and Sodin [125]. There it has also been shown that, in the i.i.d. case, a $\frac{1}{n}$-bound such as

$$\rho(F_\theta, \Phi) \le \frac{c}{n}$$

holds for some universal collections of coefficients. For example, when n is divisible by 4, one may take

$$\theta = \sqrt{\frac{2}{3n}}\,(1, \sqrt{2}, -1, -\sqrt{2}, 1, \sqrt{2}, -1, -\sqrt{2}, \dots)$$

The rate of convergence is closely related to Diophantine approximations, cf. e.g. [34].

Most of the other material of the chapter is contained in [35].

Chapter 21
Coefficients of Special Type

There are a number of results on distribution functions F_θ of the weighted sums

$$\langle X, \theta \rangle = \theta_1 X_1 + \cdots + \theta_n X_n$$

for coefficients θ_k which have a special structure. In this chapter we briefly discuss some of these results while skipping proofs.

21.1 Bernoulli Coefficients

When all $\theta_k = \pm\frac{1}{\sqrt{n}}$, we are dealing with distribution functions $F_\varepsilon(x) = \mathbb{P}\{S_\varepsilon \le x\}$ of the randomized sums

$$S_\varepsilon = \frac{1}{\sqrt{n}}\,(\varepsilon_1 X_1 + \cdots + \varepsilon_n X_n), \quad \varepsilon = (\varepsilon_1, \dots, \varepsilon_n) \in \{-1, 1\}^n.$$

In this scheme, the uniform measure $\mathfrak{s}_{n-1}$ on the unit sphere is replaced with the normalized counting measure μ_n on the discrete cube $\Omega_n = \{-1, 1\}^n$, and then the typical distribution is understood as the average

$$F(x) = \int_{\Omega_n} F_\varepsilon(x)\, \mathrm{d}\mu_n(\varepsilon) = \frac{1}{2^n} \sum_\varepsilon F_\varepsilon(x), \quad x \in \mathbb{R}.$$

The following observation was made in [26].

Proposition 21.1.1 *Given an isotropic random vector X in $\mathbb{R}^n$, for any $r > 0$,*

$$\mu_n\big\{L(F_\varepsilon, F) \ge r\big\} \le Cn^{1/4} \exp\big\{-cnr^8\big\} \tag{21.1}$$

with some absolute constants $C > 0$ and $c > 0$.

S. Bobkov et al., *Concentration and Gaussian Approximation for Randomized Sums*, Probability Theory and Stochastic Modelling 104,
https://doi.org/10.1007/978-3-031-31149-9_21

This statement may be viewed as a discrete analog of Propositions 14.4.1–14.4.2 for general spherical coefficients. It implies that

$$L(F_\varepsilon, F) \le c \left(\frac{\log n}{n}\right)^{1/8}$$

for most ± 1 sequences ε.

The property that the right-hand side of (21.1) has an exponential decay with respect to n allows one to study the asymptotic normality of the weighted sums along arrays of random variables $\{X_{n,i}\}$, $1 \le i \le n$. Let μ denote the uniform (infinite product Bernoulli) measure on $\{-1,1\}^\infty$, and assume that

a) $\mathbb{E} X_{n,i} X_{n,j} = \delta_{i,j}$ for all $n \ge 1$ (isotropy);

b) $\frac{1}{\sqrt{n}} \max_{i \le n} |X_{n,i}| \to 0$ in probability as $n \to \infty$;

c) $\frac{1}{n} \sum_{i=1}^n X_{n,i}^2 \to 1$ in probability as $n \to \infty$ (law of large numbers for squares).

Corollary 21.1.2 *Under the assumptions a)–c), for μ-almost all ± 1-sequences $(\varepsilon_1, \varepsilon_2, \dots)$, we have that, as $n \to \infty$,*

$$\mathbb{P}\Big\{\frac{1}{\sqrt{n}} \sum_{i=1}^n \varepsilon_i X_{n,i} \le x\Big\} \to \Phi(x), \quad x \in \mathbb{R},$$

We refer to [26] for details. Let us stress that this variant of the central limit theorem does not hold any longer for prescribed coefficients, say for $\varepsilon_i = 1$, even if we assume that all $\{X_{n,i}\}_{1 \le i \le n}$ are uniformly bounded, symmetrically distributed and pairwise independent. For example, for the values $n = d(d-1)/2$, $d = 1, 2, \dots$, consider the family of random variables

$$X_{n,(k,j)} = \xi_k \xi_j, \qquad 1 \le k < j \le d,$$

where $\xi_1, \dots, \xi_d$ are independent and have a symmetric Bernoulli distribution on $\{-1,1\}$. In this case, the conditions *b)*–*c)* are fulfilled automatically, since $|X_{n,(k,j)}| = 1$, while the isotropy condition *a)* follows from the pairwise independence of $X_{n,(k,j)}$. On the other hand, by the central limit theorem for the sums of ξ_j's, weakly in distribution

$$\frac{1}{\sqrt{n}} \sum_{1 \le k < j \le d} X_{n,(k,j)} = \frac{1}{2\sqrt{n}} \Big(\sum_{k=1}^d \xi_k\Big)^2 - \frac{d}{2\sqrt{n}} \Rightarrow \frac{Z^2 - 1}{\sqrt{2}} \quad (d \to \infty),$$

where $Z \sim N(0,1)$. Thus, the limit law for the normalized sums exists, but it is not normal.

21.2 Random Sums

Similar concentration-type observations can be made for the distribution functions F_τ of the sums

$$S_\tau = \frac{1}{\sqrt{k}}\,(X_{i_1} + \cdots + X_{i_k}),$$

where $\tau = \{i_1, \dots, i_k\}$, $1 \le i_1 < \cdots < i_k \le n$. In this case, the average distribution function is defined to be

$$F(x) = \int_{\mathcal{G}_{n,k}} F_\tau(x)\, \mathrm{d}\mu_{n,k}(\tau) = \frac{1}{C_n^k} \sum_\tau F_\tau(x), \quad x \in \mathbb{R}, \tag{21.2}$$

where $\mu_{n,k}$ $(k = 1, \dots, n)$ denotes the normalized counting measure on the collection $\mathcal{G}_{n,k}$ of all subsets $\tau \subset \{1, \dots, n\}$ of cardinality $|\tau| = k$. Note that $|\mathcal{G}_{n,k}|$ has cardinality $C_n^k = \frac{n!}{k!(n-k)!}$.

Proposition 21.2.1 *Given an isotropic random vector X in $\mathbb{R}^n$, for any $r > 0$,*

$$\mu_{n,k}\big\{L(F_\tau, F) \ge r\big\} \le Ck^{3/4} \exp\big\{-ckr^8\big\}$$

with some absolute constants $C > 0$ and $c > 0$.

Here, the concentration property is telling us that the distribution of the resulting sum S_τ does not essentially depend on the concrete times $i_1, \dots, i_k$ when the observations X_j are made, see [28].

The question of whether or not the average distribution F in (21.2) is close to the normal distribution or, more flexibly, to an appropriate Gaussian mixture, is connected with certain elementary polynomials in n complex variables of degree $1 \le k \le n$, namely

$$P_{n,k}(z) = \frac{1}{C_n^k} \sum_{i_1 < \cdots < i_k} z_{i_1} \dots z_{i_k}, \quad z = (z_1, \dots, z_n) \in \mathbb{C}^n.$$

One can show that, if $|z_i| \le 1$ for all $i \le n$, then

$$|P_{n,k}(z) - \bar{z}^k| \le 6\,\frac{k-1}{n-1},$$

where $\bar{z} = \frac{1}{n}\,(z_1 + \cdots + z_n)$. This leads to the following consequence of Proposition 21.2.1 in analogy with Corollary 21.1.2 for an infinite sequence of random variables $\{X_n\}_{n \ge 1}$. Below, they are assumed to satisfy the moment conditions

$$\mathbb{E}X_i = 0,\ \mathbb{E}X_iX_j = \delta_{i,j} \quad \text{and} \quad \sup_i \mathbb{E}\,|X_i|^3 < \infty$$

for all $i, j \ge 1$. As before, we denote by Φ_ρ the distribution of the random variable ρZ, assuming that $Z \sim N(0,1)$ is independent of ρ.

Corollary 21.2.2 *Assume that, for some random variable $\rho \geq 0$, weakly in distribution*

$$\frac{1}{n}\sum_{i=1}^{n} X_i^2 \Rightarrow \rho^2 \quad (n \to \infty).$$

Then for all (k,n) in the range $1 \ll k \ll n$, for any $r > 0$ and for all $\tau \in \mathcal{G}_{n,k}$ except for a set of $\mu_{n,k}$-measure at most $Ck^{3/4}\,\mathrm{e}^{-ckr^8}$, we have

$$L(F_\tau, \Phi_\rho) < r + o(1).$$

21.3 Existence of Infinite Subsequences of Indexes

For a closely related type of randomization, one may start from an infinite sequence of random variables $\{X_n\}_{n\geq 1}$ and consider the sums

$$S_n = \frac{1}{\sqrt{n}}\sum_{k=1}^{n} X_{i_k}$$

along increasing subsequences of indexes $\{i_n\}_{n\geq 1}$. Similarly to the Bernoulli case, one should pose the following natural conditions:

$a)$ $\mathbb{E}\, X_i X_j = \delta_{i,j}$ for all i, j;

$b)$ $\max_{1\leq i\leq n} |X_i| = o(\sqrt{n})$ in probability as $n \to \infty$;

$c)$ $\frac{1}{n}\sum_{i=1}^n X_i^2 \Rightarrow \rho^2$ weakly in distribution as $n \to \infty$ for some random $\rho \geq 0$.

Note that the condition $b)$ is fulfilled under standard moment assumptions such as

$$C = \sup_{n\geq 1}\, \mathbb{E}\, X_n^2 \, \log(1 + |X_n|) < \infty,$$

for example. Indeed, in this case, by Markov's inequality, for any $\varepsilon > 0$,

$$\begin{aligned}
\mathbb{P}\Big\{\max_{1\leq i\leq n} |X_i| \geq \varepsilon\sqrt{n}\Big\} &\leq \sum_{i=1}^{n} \mathbb{P}\big\{|X_i| \geq \varepsilon\sqrt{n}\big\} \\
&\leq \sum_{i=1}^{n} \frac{\mathbb{E}\, X_i^2 \log(1 + |X_i|)}{\varepsilon^2 n \log(1 + \varepsilon\sqrt{n})} \leq \frac{C}{\varepsilon^2 \log(1 + \varepsilon\sqrt{n})} \to 0
\end{aligned}$$

as $n \to \infty$.

Recall that $N(0,\rho^2)$ denotes the law of the random variable of ρZ, assuming that ρ is independent of $Z \sim N(0,1)$. We have the following statement proved in [45].

Proposition 21.3.1 *Under the conditions a)–c), weakly in distribution*

$$S_n \Rightarrow N(0, \rho^2) \quad (n \to \infty) \tag{21.3}$$

for at least one sequence i_n. Moreover, for any prescribed sequence j_n such that $j_n/n \to \infty$, we may have

$$i_n \leq j_n \quad \text{for all } n \text{ large enough.} \tag{21.4}$$

The first observations of this kind with a purely normal limit are apparently due to Kac in his 1937 paper [111], and somewhat later there is a similar result by Fortet [90], cf. also [112]. On the unit interval $\Omega = (0, 1)$ equipped with Lebesgue measure $\mathbb{P}$, they considered subsystems of $X_n(\omega) = f(n\omega)$, where f is a fixed 1-periodic function on the real line. The existence of a sequence satisfying (21.3) without the tightness property (21.4) for i_n is actually known to hold in more general situations and for several schemes of weighted sums. In particular, for sequences of real numbers $a = (a_n)$ such that

$$A_n = (a_1^2 + \cdots + a_n^2)^{1/2} \to \infty, \quad a_n = o(A_n)$$

(which we call an admissible sequence), one may consider the weighted sums

$$S_n(a) = \frac{1}{A_n} \sum_{k=1}^{n} a_k X_{i_k}.$$

In 1955 Morgenthaler [147] considered an arbitrary uniformly bounded orthonormal system X_n on the unit interval $((0, 1), \mathbb{P})$ (thus, satisfying a)–c)). He proved that there exists an increasing sequence i_n and a measurable function $\rho \geq 0$ on Ω with $\|\rho\|_2 = 1$ and $\|\rho\|_\infty \leq \sup_n \|X_n\|_\infty$ such that (21.3) holds true for any admissible $a = (a_n)$. Moreover, this convergence is stable in the sense that (21.3) holds on every measurable set $B \subset \Omega$ of positive measure with respect to the normalized restriction of $\mathbb{P}$ to B.

More general statements including necessary and sufficient conditions were obtained in the mid 60s – early 70s by Gaposhkin in a series of papers, and here we mention two results. Let $(X_n)_{n\geq 1}$ and $\rho \geq 0$ be random variables such that $\mathbb{E}X_n^2 = \mathbb{E}\rho^2 = 1$.

Proposition 21.3.2 *The following properties are equivalent:*

1) *There exists an increasing sequence i_n such that the central limit theorem* (21.3) *holds true for any admissible $a = (a_n)$;*
2) *There exists an increasing sequence i_n such that $X_{i_n} \to 0$ weakly in L^2 and $X_{i_n}^2 \to \rho_0^2$ weakly in L^1 as $n \to \infty$ for some $\rho_0 \geq 0$ on Ω equidistributed with ρ.*

We refer to [91], Theorem 5, and [94], Theorem 6.

Here, the case where $\rho = 1$ when the limit distribution is standard normal is of special interest. The existence of a random variable ρ with $\|\rho\|_2 = 1$ such that (21.3) holds for some increasing sequence i_n is equivalent to the weak convergence $X_{i_n} \to 0$ together with the uniform integrability of the sequence $X_{i_n}^2$ (cf. [91], Theorem 8, or [92], Theorem 1.5.3).

Proposition 21.3.3 *If $X_n \to 0$ weakly in L^2, then there exists an increasing sequence i_n such that* (21.3) *holds true for any admissible $a = (a_n)$ and some random variable $\rho \geq 0$.*

We refer to [92], Theorem 1.5.2, and [94], Theorem 5.

In [94], Gaposhkin introduced an "equivalence lemma" which made it possible to reduce many problems on subsequences of X_n to martingale differences (such as convergence of series, the central limit theorem, the law of the iterated logarithm) and eventually to extend the corresponding statements from Lebesgue measure and the space $\Omega = (0, 1)$ to arbitrary probability spaces.

The above Proposition 21.3.3 was rediscovered by Chatterji [75] with a similar martingale approach. Chatterji introduced an informal statement known as the principle of subsequences; it states that any limit theorem about independent, identically distributed random variables continues to hold under proper moment assumptions for a certain subsequence of a given sequence of random variables. This general observation was made precise and developed by Aldous [3], and later by Berkes and Péter [13].

21.4 Selection of Indexes from Integer Intervals

One should however note that not much is known about the speed of increase of a subsequence chosen to satisfy a central limit theorem like in Propositions 21.3.1 and 21.3.2. For example, sharpening results of Salem and Zygmund [164], [165] on lacunary trigonometric subsequences, Erdös [85] proved that any sequence

$$X_{i_n}(\omega) = \cos(2\pi i_n \omega), \quad \omega \in (0, 1),$$

with $\frac{i_{n+1}}{i_n} \geq 1 + \frac{c_n}{\sqrt{n}}$, $c_n \to \infty$, satisfies

$$S_n = \frac{1}{\sqrt{n}} (X_{i_1} + \cdots + X_{i_n}) \Rightarrow N(0, 1/2) \tag{21.5}$$

with respect to the uniform measure $\mathbb{P}$ on $(0, 1)$; cf. also [14]. Note that the indexes i_n in Erdös' theorem must grow faster than $e^{\sqrt{n}}$. Using a randomization of indexes in a trigonometric orthonormal system, Berkes [15] showed that a sequence i_n can actually be chosen to satisfy $i_{n+1} - i_n = O(j_n)$ for any prescribed $j_n \to \infty$. Proposition 21.3.1 above describes a similar property, with possible applications to other orthonormal systems.

To describe Berkes' construction, suppose that the set of all natural numbers is partitioned into non-empty consecutive intervals Δ_k, $k \geq 1$, of respective lengths (cardinalities) $|\Delta_k| \to \infty$. Berkes [14] proved that, if we select each i_k from Δ_k independently and at random according to the discrete uniform distribution on that interval, then for almost all choices of indexes, (21.5) still holds true. Hence, the gaps $i_{n+1} - i_n$ may grow as slow as we wish.

Moreover, as was shown in [46], the Berkes construction is also applicable to general systems of random variables $\{X_n\}$ satisfying the following properties:

a) $\mathbb{E}\, X_i X_j = \delta_{i,j}$ for all i, j;

b) $\max_{1 \leq k \leq n} \max_{i \in \Delta_k} |X_i| = o(\sqrt{n})$ in probability as $n \to \infty$;

c) $\frac{1}{n} \sum_{i=1}^n X_i^2 \Rightarrow \rho^2$ weakly in distribution as $n \to \infty$, for some random $\rho \geq 0$.

Proposition 21.4.1 *Under the conditions a)–c), for almost all indexes $\{i_k\}_{k\geq 1}$ selected independently and uniformly from Δ_k, weakly in distribution*

$$S_n \Rightarrow N(0, \rho^2) \quad (n \to \infty).$$

In fact, here the condition *a*) may be weakened to

$$M_2 = \sup_{n\geq 1} \sup_{\theta \in \mathbb{S}^{n-1}} \mathbb{E}\, |\theta_1 X_1 + \dots + \theta_n X_n|^2 < \infty,$$

which means that the spectral norms of the correlation operators of the random vectors $(X_1, \dots, X_n)$ remain bounded for growing n.

This statement is based on a strong concentration property of the distribution functions F_τ of $S_n = S_n(\tau)$ with respect to finite selections $\tau = (i_1, \dots, i_n)$ from the space $\Delta_1 \times \dots \times \Delta_n$, which is equipped with the product $\mu_n = \lambda_1 \otimes \dots \otimes \lambda_n$ of the uniform measures λ_k on Δ_k. The conditions *b*)–*c*) are not needed, while *a*) is replaced with $M_2 < \infty$.

Proposition 21.4.2 *For any $r > 0$,*

$$\mu_n\big\{\tau : L(F_\tau, F) \geq r\big\} \leq C e^{-cn},$$

where $F = \int F_\tau \, d\mu_n(\tau)$ is the average distribution, and where C, c denote positive constants depending on M_2 and r, only.

Returning to the cosine trigonometric system, Berkes raised the natural question of whether or not it is possible to find i_n with bounded gaps $i_{n+1} - i_n$, still satisfying (21.5). As was also shown in [46], the answer is negative: It turns out that a non-trivial limit $S_n \Rightarrow \xi$ is possible in the case of bounded gaps, but then necessarily $\mathbb{E}\xi^2 < \frac{1}{2}$ (so, part of the second moment must be vanishing in the limit distribution). To see what may happen in the typical situation, the next observation from [46] gives a hint.

Proposition 21.4.3 *In the case of the cosine trigonometric system, for almost all indexes* $\{i_k\}_{k\geq 1}$ *selected independently and uniformly from the two point integer sets* $\Delta_k = \{2k-1, 2k\}$, *weakly in distribution we have*

$$\sqrt{2}\, S_n \Rightarrow N(0, \rho^2) \quad (n \to \infty),$$

where ρ *is distributed according to the arcsine law.*

More precisely, ρ has the distribution function $F(x) = \frac{2}{\pi} \arccos(x)$, $0 < x < 1$. Hence $\mathbb{E}\rho^2 = \frac{1}{2}$, while $\mathbb{E}\,(\sqrt{2}\, S_n)^2 = 1$. Other typical distributions may appear in the limit for intervals Δ_k of a larger length.

References

1. Aida, S.; Masuda, T.; Shigekawa, I. Logarithmic Sobolev inequalities and exponential integrability. *J. Func. Anal.* **126**, 83–101 (1994).
2. Aida, S.; Stroock, D. Moment estimates derived from Poincaré and logarithmic Sobolev inequalities. *Math. Res. Lett.* **1**, no. 1, 75–86 (1994).
3. Aldous, D. J. Limit theorems for subsequences of arbitrarily-dependent sequences of random variables. *Z. Wahrscheinlichkeitstheorie und Verw. Gebiete* **40**, no. 1, 59–82 (1977).
4. Antilla, M.; Ball, K.; Perissinaki, I. The central limit problem for convex bodies. *Trans. Amer. Math. Soc.* **355**, no. 12, 4723–4735 (2003).
5. Alonso-Gutiérrez, D.; Bastero, J. *Approaching the Kannan–Lovász-Simonovits and Variance Conjectures*. Lecture Notes in Mathematics, **2131**. Springer, Cham, (2015). x+148 pp.
6. Artstein-Avidan, S.; Giannopoulos, A.; Milman, V. D. *Asymptotic Geometric Analysis. Part I.* Mathematical Surveys and Monographs, **202**. American Mathematical Society, Providence, RI (2015). xx+451 pp.
7. Bakry, D.; Emery, M. Diffusions hypercontractives. In: *Sém. Probab., XIX*, Lecture Notes in Math. **1123**, pp. 177–206, Springer-Verlag, New York, Berlin (1985).
8. Bakry, D.; Gentil, I.; Ledoux, M. *Analysis and Geometry of Markov Diffusion Operators*. Springer, Berlin Heidelberg New York (2014).
9. Bakry, D.; Ledoux, M. Lévy–Gromov's isoperimetric inequality for an infinite-dimensional diffusion generator. *Invent. Math.* **123**, no. 2, 259–281 (1996).
10. Ball, K. Logarithmically concave functions and sections of convex sets in R^n. *Studia Math.* **88**, no. 1, 69–84 (1988).
11. Bateman, H. *Higher Transcendental Functions. Vol. II*, McGraw-Hill Book Company, Inc., 1953, 396 pp.
12. Bentkus, V.; Götze, F. Optimal rates of convergence in the CLT for quadratic forms. *Ann. Probab.* **24**, no. 1, 466–490 (1996).
13. Berkes, I.; Péter, E. Exchangeable random variables and the subsequence principle. *Probab. Theory Related Fields* **73**, no. 3, 395–413(1986).
14. Berkes, I. On the central limit theorem for lacunary trigonometric series. *Anal. Math.* **4**, no. 3, 159–180 (1978).
15. Berkes, I. A central limit theorem for trigonometric series with small gaps. *Z. Wahrsch. Verw. Gebiete* **47**, no. 2, 157–161 (1979).
16. Bhattacharya, R. N.; Ranga Rao, R. *Normal Approximation and Asymptotic Expansions*. John Wiley & Sons, Inc. (1976). Also: Soc. for Industrial and Appl. Math., Philadelphia (2010).
17. Bikjalis, A. Remainder terms in asymptotic expansions for characteristic functions and their derivatives. (Russian) *Litovsk. Mat. Sb.* **7** (1967) 571–582 (1968). *Selected Transl. in Math. Statistics and Probability* **11**, 149–162 (1973).

S. Bobkov et al., *Concentration and Gaussian Approximation for Randomized Sums*, Probability Theory and Stochastic Modelling 104,
https://doi.org/10.1007/978-3-031-31149-9

18. Bobkov, S. G. Extremal properties of half-spaces for log-concave distributions. *Ann. Probab.* **24**, no. 1, 35–48 (1996).
19. Bobkov, S. G. An isoperimetric inequality on the discrete cube, and an elementary proof of the isoperimetric inequality in Gauss space. *Ann. Probab.* **25**, no. 1, 206–214 (1997).
20. Bobkov, S. G. Isoperimetric and analytic inequalities for log-concave probability measures. *Ann. Probab.* **27**, no. 4, 1903–1921 (1999).
21. Bobkov, S. G. Remarks on the Gromov–Milman inequality. (Russian) *Vestn. Syktyvkar. Univ., Ser. 1 Mat. Mekh. Inform.* **3**, 15–22 (1999).
22. Bobkov, S. G. Remarks on the growth of L^p-norms of polynomials. In: *Geometric Aspects of Functional Analysis*, pp. 27–35, Lecture Notes in Math., **1745**, Springer, Berlin (2000).
23. Bobkov, S. G. Some generalizations of Prokhorov's results on Khinchin-type inequalities for polynomials. (Russian) *Teor. Veroyatnost. i Primenen.* **45**, no. 4, 745–748 (2000). Translation in: *Theory Probab. Appl.* **45**, no. 4, 644–647 (2002).
24. Bobkov, S. G. A localized proof of the isoperimetric Bakry–Ledoux inequality and some applications. (Russian) *Teor. Veroyatnost. i Primenen.* **47**, no. 2, 340–346 (2002). Translation in: *Theory Probab. Appl.* **47**, no. 2, 308–314 (2003).
25. Bobkov, S. G. On concentration of distributions of random weighted sums. *Ann. Probab.* **31**, no. 1, 195–215 (2003).
26. Bobkov, S. G. Concentration of distributions of the weighted sums with Bernoullian coefficients. In: *Geometric Aspects of Functional Analysis*, Lecture Notes in Math. **1807**, pp. 27–36, Springer, Berlin (2003).
27. Bobkov, S. G. Spectral gap and concentration for some spherically symmetric probability measures. In: *Geometric Aspects of Functional Analysis*, Lecture Notes in Math. **1807**, pp. 37–43, Springer, Berlin, 2003.
28. Bobkov, S. G. Concentration of normalized sums and a central limit theorem for noncorrelated random variables. *Ann. Probab.* **32**, no. 4, 2884–2907 (2004).
29. Bobkov, S. G. Large deviations and isoperimetry over convex probability measures with heavy tails. *Electr. J. Probab.* **12**, 1072–1100 (2007).
30. Bobkov, S. G. On isoperimetric constants for log-concave probability distributions. In: *Geometric Aspects of Functional Analysis*, Lecture Notes in Math. **1910**, pp. 81–88, Springer, Berlin (2007).
31. Bobkov, S. G. On a theorem of V. N. Sudakov on typical distributions. (Russian) *Zap. Nauchn. Sem. S.-Peterburg. Otdel. Mat. Inst. Steklov. (POMI)* **368** (2009), *Veroyatnost i Statistika.* **15**, 59–74. Translation in: J. Math. Sciences (N. Y.), **167**, no. 4, 464–473 (2010).
32. Bobkov, S. G. Closeness of probability distributions in terms of Fourier–Stieltjes transforms. (Russian) *Uspekhi. Matemat. Nauk*, vol. 71, issue 6(432), 37–98 (2016). Translation in: *Russian Math. Surveys* **71**:6, 1021–1079 (2016).
33. Bobkov, S. G. Asymptotic expansions for products of characteristic functions under moment assumptions of non-integer orders. In: *Convexity and Concentration*, pp. 297–357, IMA Vol. Math. Appl., **161**, Springer, New York (2017).
34. Bobkov, S. G. Central limit theorem and Diophantine approximations. *J. Theoret. Probab.* **31**, no. 4, 2390–2411 (2018).
35. Bobkov, S. G. Edgeworth corrections in randomized central limit theorems. In: *Geometric Aspects of Functional Analysis*, Lecture Notes in Math. **2256**, 71–97 (2020).
36. Bobkov, S. G.; Chistyakov G. P.; Götze, F. Second order concentration on the sphere. *Communications in Contemp. Math.* **19**, no. 5, 1650058 (2017). 20 pp.
37. Bobkov, S. G.; Chistyakov G. P.; Götze, F. Gaussian mixtures and normal approximation for V. N. Sudakov's typical distributions. *Zap. Nauchn. Sem. S.-Peterburg. Otdel. Mat. Inst. Steklov. (POMI)* **457**, *Veroyatnost i Statistika.* **25**, 37–52 (2017). Also in: *J. Math. Sciences* **238**, no. 4, 366–376 (2019).
38. Bobkov, S. G.; Chistyakov G. P.; Götze, F. Berry–Esseen bounds for typical weighted sums. *Electronic J. of Probability* **23**, no. 92, 1–22 (2018).

39. Bobkov, S. G.; Chistyakov G. P.; Götze, F. Normal approximation for weighted sums under a second order correlation condition. *Ann. Probab.* **48**, no. 3, 1202–1219 (2020).
40. Bobkov, S. G.; Chistyakov G. P.; Götze, F. Poincaré inequalities and normal approximation for weighted sums. *Electron. J. of Probab.* **25** (2020), Paper no. 155 (2020). 31 pp.
41. Bobkov, S. G.; Chistyakov G. P.; Götze, F. Asymptotic expansions and two-sided bounds in randomized central limit theorems. *Preprint* (2020).
42. Bobkov, S. G.; Cordero-Erausquin, D. K-L-S-type isoperimetric bounds for log-concave probability measures. *Ann. Mat. Pura Appl.* (4) 195, no. 3, 681–695 (2016).
43. Bobkov, S. G.; Gentil, I.; Ledoux, M. Hypercontractivity of Hamilton–Jacobi equations. *J. Math. Pures Appl. 80* 7, 669–696 (2001).
44. Bobkov, S. G.; Götze, F. Exponential integrability and transportation cost related to logarithmic Sobolev inequalities. *J. Func. Anal.* **1**, 1–28 (1999).
45. Bobkov, S. G.; Götze, F. On the central limit theorem along subsequences of noncorrelated observations. *Teor. Veroyatnost. i Primenen.* **48**, no. 4, 745–765 (2003). Translation in: *Theory Probab. Appl.* **48**, no. 4, 604–621 (2004).
46. Bobkov, S. G.; Götze, F. Concentration inequalities and limit theorems for randomized sums. *Probab. Theory Related Fields* **137**, no. 1–2, 49–81 (2007).
47. Bobkov, S. G.; Götze, F. Hardy-type inequalities via Riccati and Sturm–Liouville equations. In: *Sobolev Spaces in Mathematics, Sobolev Type Inequalities, Vol. I*, Intern. Math. Series, Vol. 8, ed. by V. Maz'ya, Springer (2008).
48. Bobkov, S. G.; Götze, F. Concentration of empirical distribution functions with applications to non-i.i.d. models. *Bernoulli* **16**, no. 4, 1385–1414 (2010).
49. Bobkov, S. G.; Houdré, C. *Some Connections Between Isoperimetric and Sobolev-type Inequalities*. Mem. Amer. Math. Soc. **129**, no. 616 (1997). viii+111 pp.
50. Bobkov, S. G.; Houdré, C. Isoperimetric constants for product probability measures. *Ann. Probab.* **25**, no. 1, 184–205 (1997).
51. Bobkov, S. G.; Koldobsky, A. On the central limit property of convex bodies. In: *Geometric Aspects of Functional Analysis*, Lecture Notes in Math. **1807**, pp. 44–52, Springer, Berlin (2003).
52. Bobkov, S. G.; Ledoux, M. Poincaré's inequalities and Talagrand's concentration phenomenon for the exponential distribution. *Probab. Theory Rel. Fields* **107**, 383–400 (1997).
53. Bobkov, S. G.; Ledoux, M. From Brunn–Minkowski to Brascamp–Lieb and to logarithmic Sobolev inequalities. *Geom. Funct. Anal.* **10**, no. 5, 1028—1052 (2000).
54. Bobkov, S. G.; Ledoux, M. From Brunn–Minkowski to sharp Sobolev inequalities. *Ann. Mat. Pura Appl.* (4) 187, no. 3, 369–384 (2008).
55. Bobkov, S. G.; Ledoux, M. Weighted Poincaré-type inequalities for Cauchy and other convex measures. *Ann. Probab.* **37**, 403–427 (2009).
56. Bobkov, S. G.; Ledoux, M. On weighted isoperimetric and Poincaré-type inequalities. In: *High Dimensional Probability V: The Luminy Volume. Vol. 5*, pp. 1–29, IMS Collections (2009).
57. Bobkov, S. G.; Ledoux, M. *One-dimensional Empirical Measures, Order Statistics and Kantorovich Transport Distances*. Memoirs of the AMS, **261**, no. 1259 (2019). 126 pp.
58. Bobkov, S. G.; Nazarov, F. L. On convex bodies and log-concave probability measures with unconditional basis. In: *Geometric Aspects of Functional Analysis*, pp. 53–69, Lecture Notes in Math. **1807**, Springer, Berlin, 2003.
59. Bogachev, V. I. *Measure theory. Vol. I, II.* Springer-Verlag, Berlin (2007). Vol. I: xviii+500 pp., Vol. II: xiv+575 pp.
60. Bogachev, V. I.; Kolesnikov, A. V. The Monge–Kantorovich problem: achievements, connections, and prospects. (Russian) *Uspekhi Mat. Nauk* **67**, no. 5 (407), 3–110 (2012); translation in: *Russian Math. Surveys* **67**, no. 5, 785–890 (2012).
61. Bonnefont, M.; Joulin, A.; Ma, Y. Spectral gap for spherically symmetric log-concave probability measures, and beyond. *J. Funct. Anal.* **270**, 2456–2482 (2016).

62. Borell, C. Complements of Lyapunov's inequality. *Math. Ann.* **205**, 323–331 (1973).
63. Borell, C. Convex measures on locally convex spaces. *Ark. Math.* **12**, 239–252 (1974).
64. Borell, C. Convex set functions in d-space. *Period. Math. Hungar.* **6**, no. 2, 111–136 (1975).
65. Borell, C. The Brunn–Minkowski inequality in Gauss space. *Invent. Math.* **30** (1975), no. 2, 207–216.
66. Borovkov A. A.; Utev S. A. On an inequality and a characterization of the normal distribution connected with it. *Probab. Theory Appl.* **28**, 209–218 (1983).
67. Bourgain, J. On the distribution of polynomials on high dimensional convex bodies. In: *Geometric Aspects of Functional Analysis*, pp. 127–137, Lecture Notes in Math., **1469**, Springer, Berlin (1991).
68. Brascamp, H. J.; Lieb, E. H. On extensions of the Brunn–Minkowski and Prékopa–Leindler theorems, including inequalities for log-concave functions, and with an application to the diffusion equation. *J. Funct. Anal.* **22**, 366–389 (1976).
69. Brazitikos, S.; Giannopoulos, A.; Valettas, P.; Vritsiou, B-H. *Geometry of Isotropic Convex Bodies*. Mathematical Surveys and Monographs, 196. American Mathematical Society, Providence, RI (2014). xx+594 pp.
70. Brehm, U.; Hinow, P.; Vogt, H.; Voigt, J. Moment inequalities and central limit properties of isotropic convex bodies. *Math. Z.* **240**, no. 1, 37–51 (2002).
71. Brehm, U.; Voigt, J. Asymptotics of cross sections for convex bodies. *Beiträge Algebra Geom.* **41**, no. 2, 437–454 (2000).
72. Bulinskii, A. V. Estimates for mixed semi-invariants and higher covariances of bounded random variables. (Russian) *Teor. Verojatnost. i Premenen.* **19**, 869–873 (1974).
73. Burago, Yu. D.; Zalgaller, V. A. *Geometric Inequalities*. Translated from the Russian by A. B. Sosinskii. Fundamental Principles of Mathematical Sciences, **285**. Springer Series in Soviet Mathematics. Springer-Verlag, Berlin (1988). xiv+331 pp.
74. Caffarelli, L. A. Monotonicity properties of optimal transportation and the FKG and related inequalities. *Comm. Math. Phys.* **214**, no. 3, 547–563 (2000).
75. Chatterji, S. D. A principle of subsequences in probability theory: the central limit theorem. *Advances in Math.* **13**, 31–54 (1974); correction, ibid. **14**, 266–269 (1974).
76. Cheeger, J. A lower bound for the smallest eigenvalue of the Laplacian. In: *Problems in Analysis (Papers Dedicated to Salomon Bochner, 1969)*, pp. 195–199. Princeton Univ. Press, Princeton, N. J. (1970).
77. Chen, Y. An almost constant lower bound of the isoperimetric coefficient in the KLS conjecture. *Geom. Funct. Anal.* **31**, no. 1, 34–61 (2021).
78. Davidovich, Ju. S.; Korenbljum, B. I.; Khacet, B. I. A certain property of logarithmically concave functions. (Russian) *Dokl. Akad. Nauk SSSR* **185**, 1215–1218 (1969). English translation in: *Soviet Math. Dokl.* **10**, 477–480 (1969).
79. Diaconis, P.; Freedman, D. Asymptotics of graphical projection pursuit. *Ann. Stat.* **12**, no. 3, 793–815 (1984).
80. Dudley, R. M. *Real Analysis and Probability*. The Wadsworth & Brooks/Cole Mathematics Series. Wadsworth & Brooks/Cole Advanced Books & Software, Pacific Grove, CA (1989).
81. Dvoretzky, A.; Kiefer, J.; Wolfowitz, J. Asymptotic minimax character of the sample distribution function and of the classical multinomial estimator. *Ann. Math. Statist.* **27**, 642–669(1956).
82. Ehrhard, A. Isoperimetric inequalities and Gaussian Dirichlet integrals. (French) *Ann. Sci. École Norm. Sup.* (4) 17, no. 2, 317–332 (1984).
83. Eldan, R.; Klartag, B. Approximately Gaussian marginals and the hyperplane conjecture. In: *Concentration, Functional Inequalities and Isoperimetry*, pp. 55–68, Contemp. Math. 545, Amer. Math. Soc., Providence, RI (2011).
84. Eldan, R. Thin shell implies spectral gap up to polylog via a stochastic localization scheme. *Geom. Funct. Anal.* **23**, no. 2, 532–569 (2013).
85. Erdös, P. On trigonometric sums with gaps. *Magyar Tud. Akad. Mat. Kutató Int. Közl* **7** 37–42 (1962).

86. Esseen, C.-G. Fourier analysis of distribution functions. A mathematical study of the Laplace–Gaussian law. *Acta Math.* **77**, 1–125 (1945).
87. Evans, L. C. *Partial Differential Equations*. Graduate Studies in Math. vol. **19**, Amer. Math. Soc., Providence, RI (1998). xviii+662 pp.
88. Fainleib, A. S. A generalization of Esseen's inequality and its application in probabilistic number theory. (Russian) *Izv. Akad. Nauk SSSR Ser. Mat.* **32**, 859–879 (1968).
89. Figiel, T.; Lindenstrauss, J.; Milman, V. D. The dimension of almost spherical sections of convex bodies. *Acta Math.* **139**, no. 1–2, 53–94 (1977).
90. Fortet, R. Sur une suite egalement répartie. (French) *Studia Math.* **9**, 54–70 (1940).
91. Gaposhkin, V. F. Sequences of functions and the central limit theorem. (Russian) *Mat. Sbornik (N.S.)* **70** (112), 145–171 (1966), *Select. Translat. Math. Statist. Probab.* **9**, 89–118 (1971).
92. Gaposhkin, V. F. Lacunary series and independent functions. (Russian) *Uspekhi Mat. Nauk* **21**, no. **6** (132), 3–82 (1966). Translation in: *Russian. Math. Surveys* **21**, 1–82 (1966).
93. Gaposhkin, V. F. The rate of approximation to the normal law of the distributions of weighted sums of lacunary series. (Russian) *Teor. Verojatnost. i Primenen.* **13**, 445–461 (1968).
94. Gaposhkin, V. F. Convergence and limit theorems for subsequences of random variables. (Russian) *Teor. Verojatnost. i Primenen.* **17**, 401–423 (1972).
95. Götze, F. Asymptotic expansions in functional limit theorems. *J. Multivariate Anal.* **16**, no. 1, 1–20 (1985).
96. Götze, F.; Naumov, A.; Ulyanov, V. Asymptotic analysis of symmetric functions. *J. Theoret. Probab.* **30**, no. 3, 876–897 (2017).
97. Goldstein, L.; Shao, Q.-M. Berry–Esseen bounds for projections of coordinate symmetric random vectors. *Electron. Commun. Probab.* **14**, 474–485 (2009).
98. Gozlan, G.; Léonard, C. Transport inequalities. A survey. *Markov Process. Rel. Fields* **16** 635–736 (2010).
99. Gozlan, N.; Roberto, C.; Samson, P.-M. Hamilton–Jacobi equations on metric spaces and transport-entropy inequalities. *Rev. Mat. Iberoam.* **30**, no. 1, 133–163 (2014).
100. Gromov, M.; Milman V. D. A topological application of the isoperimetric inequality. *Amer. J. Math.* **105**, 843–854 (1983).
101. Guédon, O.; Milman, E. Interpolating thin-shell and sharp large-deviation estimates for isotropic log-concave measures. *Geom. Funct. Anal.* **21**, no. 5, 1043–1068 (2011).
102. Guédon, O. Kahane–Khinchine type inequalities for negative exponent. *Mathematika* **46**, no. 1, 165–173 (1999).
103. Gross, L. Logarithmic Sobolev inequalities. *Amer. J. Math.* **97**, no. 4, 1061–1083 (1975).
104. Grünbaum, B. Partitions of mass-distributions and of convex bodies by hyperplanes. *Pacific J. Math.* **10**, 1257–1261 (1960).
105. Holley, R., Stroock, D. Logarithmic Sobolev inequalities and stochastic Ising models. *J. Statist. Phys.* **46**, no. 5–6, 1159–1194 (1987).
106. Ibragimov, I. A.; Linnik, Yu. V. *Independent and Stationary Sequences of Random vVariables*. With a supplementary chapter by I. A. Ibragimov and V. V. Petrov. Translation from the Russian edited by J. F. C. Kingman. Wolters-Noordhoff Publishing, Groningen (1971). 443 pp.
107. Ibragimov, R.; Sharakhmetov, Sh. On an exact constant for the Rosenthal inequality. (Russian) *Teor. Veroyatnost. i Primenen.* **42**, no. 2, 341–350 (1997); translation in *Theory Probab. Appl.* **42** (1997), no. 2, 294–302 (1998).
108. Jambulapati, A.; Lee, Y. T.; Vempala, S. S. A slightly improved bound for the KLS constant. *Preprint* (2022), arXiv:2208.11644.
109. Johnson, W. B.; Schechtman, G.; Zinn, J. Best constants in moment inequalities for linear combinations of independent and exchangeable random variables. *Ann. Probab.* **13**, no. 1, 234–253 (1985).
110. Joffe, A. On a sequence of almost deterministic pairwise independent random variables. *Proc. Amer. Math. Soc.* **29**, no. 2, 381–382 (1971).

111. Kac, M. Sur les fonctions indépendantes. V. *Studia Math.* **7**, 96–100 (1937).
112. Kac, M. On the distribution of values of sums of the type $\sum f(2^k t)$. *Ann. Math. (2)* **47**, 33–49 (1946).
113. Kac, I. S.; Krein, M. G. Criteria for the discreteness of the spectrum of a singular string. (Russian) *Izv. Vysš. Učebn. Zaved. Matematika*, no. 2 (3), 136–153 (1958).
114. Kaczmarz, S.; Steinhaus, H. *Theory of Orthogonal Series*. Warszawa, Lwow (1935); Russian ed.: Izdat. Fiz.-Mat. Lit., Moscow (1958. 507 pp.
115. Kagan, A. M.; Linnik, Yu. V.; Rao, C. R. *Characterization Problems in Mathematical Statistics*. Translated from the Russian by B. Ramachandran. Wiley Series in Probability and Mathematical Statistics. John Wiley & Sons, New York-London-Sydney (1973). xii+499 pp.
116. Kannan, R.; Lovász, L.; Simonovits, M. Isoperimetric problems for convex bodies and a localization lemma. *Discrete and Comput. Geom.* **13**, 541–559 (1995).
117. Kantorovitch, L. On mass transportation. (Russian) Reprinted from *C. R. (Doklady) Acad. Sci. URSS (N.S.)* **37**, no. 7–8 (1942). *Zap. Nauchn. Sem. S.-Peterburg. Otdel. Mat. Inst. Steklov. (POMI)* **312** (2004), *Teor. Predst. Din. Sist. Komb. i Algoritm. Metody.* **11**, 11–14; translation in: *J. Math. Sci. (N.Y.)* **133**, no. **4**, 1381–1382 (2006).
118. Kantorovitch, L. On a problem of Monge. (Russian) Reprinted from *C. R. (Doklady) Acad. Sci. URSS (N.S.)* **3**, no. 2 (1948). *Zap. Nauchn. Sem. S.-Peterburg. Otdel. Mat. Inst. Steklov. (POMI)* **312** (2004), *Teor. Predst. Din. Sist. Komb. i Algoritm. Metody.* **11**, 15–16; translation in *J. Math. Sci. (N.Y.)* **133**, no. 4, 1383 (2006).
119. Karlin, S.; Proschan, F.; Barlow, R. E. Moment inequalities of Pólya frequency functions. *Pacific J. Math.* **11**, 1023–1033 (1961).
120. Kashin, B. S.; Saakyan, A. A. *Orthogonal Series*. Translated from the Russian by Ralph P. Boas. Translation edited by Ben Silver. Translations of Mathematical Monographs **75**. American Mathematical Society, Providence, RI (1989). xii+451 pp.
121. Klartag, B. A central limit theorem for convex sets. *Invent. Math.* **168**, no. 1, 91–131 (2007).
122. Klartag, B. Power-law estimates for the central limit theorem for convex sets. *J. Funct. Anal.* **245**, no. 1, 284–310 (2007).
123. Klartag, B. A Berry–Esseen type inequality for convex bodies with an unconditional basis. *Probab. Theory Related Fields* **145**, no. 1-2, 1–33 (2009).
124. Klartag, B.; Lehec, J. Bourgain's slicing problem and KLS isoperimetry up to polylog. *Geom. Funct. Anal.* **32**, no. 5, 1134–1159 (2022).
125. Klartag, B.; Sodin, S. Variations on the Berry–Esseen theorem. *Teor. Veroyatn. Primen.* **56**, no. 3, 514–533 (2011); reprinted in *Theory Probab. Appl.* **56**, no. 3, 403–419 (2012).
126. Koldobsky, A.; Lifshits, M. Average volume of sections of star bodies. In: *Geometric Aspects of Functional Analysis*, pp. 119–146, Lecture Notes in Math. **1745**, Springer, Berlin (2000).
127. Latala, R. On the equivalence between geometric and arithmetic means for log-concave measures. In: *Convex Geometric Gnalysis (Berkeley, CA, 1996)*, pp. 123–127, Math. Sci. Res. Inst. Publ. **34**, Cambridge Univ. Press, Cambridge (1999).
128. Ledoux, M. Remarks on logarithmic Sobolev constants, exponential integrability and bounds on the diameter. *J. Math. Kyoto Univ.* **35**, no. 2, 211–220 (1995).
129. Ledoux, M. Isoperimetry and Gaussian analysis. In: *Lectures on Probability Theory and Statistics (Saint-Flour, 1994)*, pp. 165–294, Lecture Notes in Math., **1648**, Springer, Berlin (1996).
130. Ledoux, M. Concentration of measure and logarithmic Sobolev inequalities. In: *Séminaire de Probabilités XXXIII*. Lect. Notes in Math. **1709**, pp. 120–216, Springer (1999).
131. Ledoux, M. *The Concentration of Measure Phenomenon. Math. Surveys and monographs*, vol. **89**, AMS (2001).
132. Ledoux, M. Spectral gap, logarithmic Sobolev constant, and geometric bounds. In: *Surveys in Differential Geometry. Vol. IX*, pp. 219-–240, Surv. Differ. Geom., IX, Int. Press, Somerville, MA (2004).

133. Lee, Y. T.; Vempala, S. S. Eldan's stochastic localization and the KLS hyperplane conjecture: an improved lower bound for expansion. In: *58th Annual IEEE Symposium on Foundations of Computer Science – FOCS 2017*, pp. 998–1007, IEEE Computer Soc., Los Alamitos, CA (2017).
134. Lee, Y. T.; Vempala, S. S. The Kannan–Lovász–Simonovits Conjecture. In: *Current Developments in Mathematics 2017*, pp. 1–36, Int. Press, Somerville, MA (2019).
135. Matskyavichyus, V. K. A lower bound for the rate of convergence in the central limit theorem. (Russian) *Theory Probab. Appl.* **28**, no. 3, 565–569 (1983).
136. Massart, P. The tight constant in the Dvoretzky–Kiefer–Wolfowitz inequality. *Ann. Probab.* **18**, no. 3, 1269–1283 (1990).
137. Maurey, B. Some deviation inequalities. *Geom. and Funct. Anal.* **1**, 188–197 (1991).
138. Maz'ya, V. G. Classes of domains and embedding theorems for functions spaces. *Dokl. Akad. Nauk SSSR*, 527–530 (1960) (Russian); translated in *Soviet Math. Dokl.* **1**, 882–885 (1960).
139. Maz'ya, V. G. *Sobolev Spaces*. Springer-Verlag, Berlin - New York (1985).
140. Meckes, M. W.; Meckes, E. The central limit problem for random vectors with symmetries. *J. Theoret. Probab.* **20**, no. 4, 697–720 (2007).
141. Meckes, M. W. Gaussian marginals of convex bodies with symmetries. *Beiträge Algebra Geom.* **50**, no. 1, 101–118 (2009).
142. Milman, V. D. A new proof of A. Dvoretzky's theorem on cross-sections of convex bodies. *Func. Anal. Appl.* **5**, 28–37 (1971).
143. Milman, V. D. Asymptotic properties of functions of several variables that are defined on homogeneous spaces. *Soviet Math. Dokl.* **12**, 1277–1281 (1971). Translated from: *Dokl. AN SSSR* **199**, no. 6, 1247–1250 (1971) (Russian).
144. Milman, V. D.; Pajor, A. Isotropic position and inertia ellipsoids and zonoids of the unit ball of a normed n-dimensional space. In: *Geometric Aspects of Functional Analysis (1987–88)*, pp. 64–104, Lecture Notes in Math. **1376**, Springer, Berlin (1989).
145. Milman, V. D.; Schechtman, G. *Asymptotic Theory of Finite-Dimensional Normed Spaces.* With an appendix by M. Gromov. Lecture Notes in Mathematics **1200**, Springer-Verlag, Berlin (1986). viii+156 pp.
146. Milman, E. On the role of convexity in isoperimetry, spectral gap and concentration. *Invent. Math.* **177**, no. 1, 1–43 (2009).
147. Morgenthaler, G. W. A central limit theorem for uniformly bounded orthonormal systems. *Trans. Amer. Math. Soc.* **79**, 281–311 (1955).
148. Muckenhoupt, B. Hardy's inequality with weights. *Studia Math.* XLIV, 31–38 (1972).
149. Mueller, C. E.; Weissler, F. B. Hypercontractivity for the heat semigroup for ultraspherical polynomials and on the n-sphere. *J. Funct. Anal.* **48**, no. 2, 252–283 (1982).
150. Otto, F.; Villani, C. Generalization of an inequality by Talagrand and links with the logarithmic Sobolev inequality. *J. Funct. Anal.* **173**, no. 2, 361–400 (2000).
151. Paouris, G. Concentration of mass on convex bodies. *Geom. Funct. Anal.* **16**, no. 5, 1021–1049 (2006).
152. Paouris, G. Small ball probability estimates for log-concave measures. *Trans. Amer. Math. Soc.* **364**, no. 1, 287–308 (2012).
153. Payne, L. E.; Weinberger, H. F. An optimal Poincaré inequality for convex domains. *Arch. Rational Mech. Anal.* **5**, 286–292 (1960).
154. Petrov, V. V. *Sums of Independent Random Variables*. Translated from the Russian by A. A. Brown. Ergebnisse der Mathematik und ihrer Grenzgebiete, Band 82. Springer-Verlag, New York-Heidelberg (1975). x+346 pp. Russian ed.: Moscow, Nauka (1972). 414 pp.
155. Petrov, V. V. *Limit Theorems for Sums of Independent Random Variables*. Moscow, Nauka (1987) (in Russian).
156. Pinelis, I. Exact Rosenthal-type bounds. *Ann. Probab.* **43**, no. 5, 2511–2544 (2015).
157. Pisier, G. *The Volume of Convex Bodies and Banach Space Geometry*. Cambridge Tracts in Mathematics **94**. Cambridge University Press, Cambridge (1989). xvi+250 pp.

158. Pólya, G. Herleitung des Gaußschen Fehlergesetzes aus einer Funktionalgleichung. (German) *Math. Z.* **18**, no. 1, 96–108 (1923).
159. Prawitz, H. Limits for a distribution, if the characteristic function is given in a finite domain. *Skand. Aktuarietidskr.* 1972, 138–154 (1973).
160. Prékopa, A. On logarithmic concave measures and functions. *Acta Sci. Math. (Szeged)* **34**, 335–343 (1973).
161. Rokhlin, V. A. On the fundamental ideas of measure theory. (Russian) M*at. Sbornik N.S.* **25** (67), 107–150 (1949). Also in *Selected Works*. Edited and with a preface by Vershik. Moskovskii Tsentr Nepreryvnogo Matematicheskogo Obrazovaniya, Moscow (1999). 496 pp.
162. Rosenthal, H. P. On the subspaces of L_p $(p > 2)$ spanned by sequences of independent random variables. *Israel J. Math.* **8**, 273–303 (1970).
163. Rothaus, O. S. Analytic inequalities, isoperimetric inequalities and logarithmic Sobolev inequalities. *J. Funct. Anal.* **64**, no. 2, 296–313 (1985).
164. Salem, R.; Zygmund, A. On lacunary trigonometric systems. *Proc. Nat. Acad. Sci. USA* **33**, 333–338 (1947).
165. Salem, R.; Zygmund, A. On lacunary trigonometric series. II. *Proc. Nat. Acad. Sci. USA* **34**, 54–62 (1948).
166. Shevtsova, I. G. Refinement of estimates for the rate of convergence in Lyapunov's theorem. (Russian) *Dokl. Akad. Nauk* **435**, no. 1, 26–28 (2010); translation in *Dokl. Math.* **82**, no. 3, 862–864 (2010).
167. Sodin, S. Tail-sensitive Gaussian asymptotics for marginals of concentrated measures in high dimension. In: *Geometric Aspects of Functional Analysis*, pp. 271–295, Lecture Notes in Math. **1910**, Springer, Berlin (2007).
168. Stein, E. M.; Weiss, G. *Introduction to Fourier Analysis on Euclidean Spaces*. Princeton Mathematical Series, No. 32. Princeton University Press, Princeton, N.J. (1971). x+297 pp.
169. Sudakov, V. N. Typical distributions of linear functionals in finite-dimensional spaces of higher dimension. *Soviet Math. Dokl.* **19**, no. 6, 1578–1582 (1978). Translated from: *Dokl. Akad. Nauk SSSR* **243**, no. 6 (1978).
170. Sudakov, V. N.; Tsirelson, B. S. Extremal properties of half-spaces for spherically invariant measures: Problems in the theory of probability distributions, II. (Russian) *Zap. Nauchn. Sem. Leningrad. Otdel. Mat. Inst. Steklov. (LOMI)* **41**, 14–24, 165 (1974).
171. Talagrand, M. A new isoperimetric inequality and the concentration of measure phenomenon. In: *Geometric Aspects of Functional Analysis*, Israel Seminar, pp. 94–124 Springer Verlag, Lect. Notes Math. **1469** (1991).
172. Talagrand, M. Concentration of measure and isoperimetric inequalities in product spaces. *Publications Mathématiques de l'I.H.E.S.* **81**, 73–205 (1995).
173. Talagrand, M. Transportation cost for Gaussian and other product measures. *Geom. Funct. Anal.* **6**, 587–600 (1996).
174. Villani, C. *Topics in Optimal Transportation*. Graduate Studies in Mathematics **58**. American Mathematical Society, Providence, RI (2003). xvi+370 pp.
175. Wang, F.-Y. Logarithmic Sobolev inequalities on noncompact Riemannian manifolds. *Probab. Theory Related Fields* **109**, 417–424 (1997).
176. von Weizsäcker, H. Sudakov's typical marginals, random linear functionals and a conditional central limit theorem. *Probab. Theory Rel. Fields* **107**, 313–324 (1997).
177. Zolotarev, V. M. Estimates of the difference between distributions in the Lévy metric. (Russian) In: *Collection of Articles Dedicated to Academician Ivan Matveevich Vinogradov on his 80th Birthday*, I. Trudy Mat. Inst. Steklov. **112**, pp. 224–231 (1971).

Glossary

Algebra

$\mathbb{R}^n = \mathbb{R} \times \cdots \times \mathbb{R}$	Euclidean n-dimensional space
$\langle x, y \rangle = x_1 y_1 + \cdots + x_n y_n$	inner product for vectors (points) $x = (x_1, \dots, x_n)$, $y = (y_1, \dots, y_n) \in \mathbb{R}^n$
$\lvert x \rvert = (x_1^2 + \cdots + x_n^2)^{1/2}$	Euclidean norm, length of vector $x \in \mathbb{R}^n$
$\mathbb{C}^n = \mathbb{C} \times \cdots \times \mathbb{C}$	complex Euclidean n-dimensional space
$\langle x, y \rangle = x_1 \bar{y}_1 + \cdots + x_n \bar{y}_n$	inner product for vectors (points) $x = (x_1, \dots, x_n)$, $y = (y_1, \dots, y_n) \in \mathbb{C}^n$
$\lvert x \rvert = (\lvert x_1 \rvert^2 + \cdots + \lvert x_n \rvert^2)^{1/2}$	Euclidean norm of vector $x \in \mathbb{C}^n$
$\lVert A \rVert_{\mathrm{HS}} = \left(\sum_{i,j=1}^n \lvert a_i \rvert^2 \right)^{1/2}$	Hilbert–Schmidt norm of $n \times n$ matrix $A = (a_{ij})_{i,j=1}^n$ with complex entries
$\lVert A \rVert = \sup\{\lvert Ax \rvert : x \in \mathbb{C}^{n-1}, \lvert x \rvert = 1\}$	operator norm of complex matrix A

Probability

$(\Omega, \mathcal{F}, \mathbb{P})$	underlying probability space
$\mathbb{E}\xi = \int \xi \, d\mathbb{P}$	expectation of random variable (or vector) ξ
$\mathrm{Var}(\xi) = \mathbb{E}\, \lvert \xi - \mathbb{E}\xi \rvert^2$	variance of complex-valued random variable ξ
$\lVert \xi \rVert_p = (\mathbb{E}\, \lvert \xi \rvert^p)^{1/p}$	L^p-norm of random variable or vector ξ (may be used for $p \geq 0$)
$\lVert \xi \rVert_\psi = \inf\{\lambda > 0 : \mathbb{E}\, \psi(\xi/\lambda) \leq 1\}$	Orlicz norm of ξ generated by the Young function ψ

S. Bobkov et al., *Concentration and Gaussian Approximation for Randomized Sums*, Probability Theory and Stochastic Modelling 104,
https://doi.org/10.1007/978-3-031-31149-9

$\|\|\xi\|\|_{\psi_p}$	Orlicz norm of ξ generated by the Young function $\psi_p(r) = \exp\{\|r\|^p\} - 1$
$\gamma_p(\xi) = \frac{d^p}{i^p\, dt^p}\, \mathbb{E}\, \exp\{it\xi\}\big\|_{t=0}$	cumulant of random variable ξ of order $p = 1, 2, \dots$

Moment- and variance-type functionals

$X = (X_1, \dots, X_n)$	random vector in $\mathbb{R}^n$
$Y = (Y_1, \dots, Y_n)$	independent copy of X
$M_p = M_p(X)$ $= \sup_{\theta \in \mathbb{S}^{n-1}} \|\| \langle X, \theta \rangle \|\|_p$	moment of random vector X in $\mathbb{R}^n$ of order $p \geq 1$, 6
$m_p = m_p(X) = \frac{1}{\sqrt{n}} \|\| \langle X, Y \rangle \|\|_p$	moment of X of order p, using independent copy Y of X, 8
$\sigma_4^2 = \sigma_4^2(X) = \frac{1}{n} \mathrm{Var}(\|X\|^2)$	variance-type functional, 11
$\sigma_{2p} = \sigma_{2p}(X) = \sqrt{n} \left(\mathbb{E} \left\| \frac{\|X\|^2}{n} - 1 \right\|^p \right)^{1/p}$	variance-type functional of order $2p$, assuming that $\mathbb{E}\,\|X\|^2 = n$, 15
$\Lambda = \sup_{\sum a_{ij}^2 = 1} \mathrm{Var}\Big(\sum_{i,j=1}^n a_{ij} X_i X_j \Big)$	functional responsible for second order correlation condition

Metric spaces

(M, d)	(abstract) metric space M with metric d, without isolated points
$\|\nabla f(x)\| = \limsup_{d(x,y) \to 0} \frac{\|f(x) - f(y)\|}{d(x,y)}$	generalized modulus of gradient, 92 of a function $f : \mathbb{M} \to \mathbb{R}$.
$\|\nabla^2 f(x)\| = \|\nabla \|\nabla f(x)\|\,\|$	second order modulus of gradient, 93
(M, d, μ)	metric space with probability measure μ
$\mu^+(A)$	μ-perimeter of set $A \subset M$, 94 (outer Minkowski content)
$\lambda_1 = \lambda_1(\mu)$	spectral gap (Poincaré constant), 96
$h = h(\mu)$	Cheeger constant, 102
$\rho = \rho(\mu)$	logarithmic Sobolev constant, 134

Analysis on the sphere

$\mathbb{S}^{n-1} = \{x \in \mathbb{R}^n : \|x\| = 1\}$	unit sphere in $\mathbb{R}^n$
$f'(x) = \nabla f(x) = \left(\frac{\partial f(x)}{\partial x_i} \right)_{i=1}^n$	Euclidean gradient (first derivative) of function $f = f(x)$ at the point x
$f''(x) = \left(\frac{\partial^2 f(x)}{\partial x_i \partial x_j} \right)_{i,j=1}^n$	Euclidean Hessian (second derivative)

$\nabla_{\mathbb{S}} f(x)$	spherical gradient (first derivative) of f at the point $x \in \mathbb{S}^{n-1}$, 169
$f''_{\mathbb{S}}(x)$	spherical second derivative of f at x, 170
$\lvert\nabla^2_{\mathbb{S}} f(x)\rvert = \lvert\nabla_{\mathbb{S}} \lvert\nabla_S f(x)\rvert\,\rvert$	spherical second order modulus of gradient, 172
$\Delta_{\mathbb{S}} f(x)$	spherical Laplacian of f at x, 175
$\mathfrak{s}_{n-1}$	uniform distribution on $\mathbb{S}^{n-1}$ (normalized Lebesgue measure)
$\mathbb{E}_\theta f(\theta) = \int f \, \mathrm{d}\mathfrak{s}_{n-1}$	expectation (integral) with respect to $\mathfrak{s}_{n-1}$
$J_n(t) = \int \exp\{it\theta_1\} \, \mathrm{d}\mathfrak{s}_{n-1}(\theta)$	characteristic function of the first coordinate of a point θ under $\mathfrak{s}_{n-1}$, 212

Weighted sums and probability metrics

$S_\theta = \langle X, \theta\rangle = \theta_1 X_1 + \cdots + \theta_n X_n$	weighted sum of coordinates of random vector $X = (X_, \dots, X_n)$ with weights θ_i
$F_\theta(x) = \mathbb{P}\{S_\theta \le x\}$	distribution function of S_θ, 223
$F(x) = \mathbb{E}_\theta F_\theta(x)$	typical distribution function, 224
$f_\theta(t) = \mathbb{E}\, e^{itS_\theta}$	characteristic function of S_θ, 241
$f(t) = \mathbb{E}_\theta f_\theta(t)$	characteristic function of F, 241
$W(F_\theta, F) = \int_{-\infty}^{\infty} \lvert F_\theta(x) - F(x)\rvert \, \mathrm{d}x$	L^1-distance between F_θ and F, 259 (Kantorovich distance)
$L(F_\theta, F)$	Lévy distance between F_θ and F, 269
$\rho(F_\theta, F) = \sup_x \lvert F_\theta(x) - F(x)\rvert$	L^∞-distance between F_θ and F, 274 (Kolmogorov distance)
$\omega(F_\theta, F) = \left(\int_{-\infty}^{\infty} (F_\theta(x) - F(x))^2 \, \mathrm{d}x\right)^{1/2}$	L^2-distance between F_θ and F, 299

Standard notations/abbreviations

1_A, $I(A)$	indicator of set (event) A
$x^+ = \max\{x, 0\}$	positive part of a real number x
$x^- = \max\{-x, 0\}$	negative part of x
a.e., a.s.	almost everywhere, almost surely

Index

S. Bobkov et al., *Concentration and Gaussian Approximation for Randomized Sums*, Probability Theory and Stochastic Modelling 104,
https://doi.org/10.1007/978-3-031-31149-9

GPSR Compliance
The European Union's (EU) General Product Safety Regulation (GPSR) is a set of rules that requires consumer products to be safe and our obligations to ensure this.

If you have any concerns about our products, you can contact us on

ProductSafety@springernature.com

In case Publisher is established outside the EU, the EU authorized representative is:

Springer Nature Customer Service Center GmbH
Europaplatz 3
69115 Heidelberg, Germany

www.ingramcontent.com/pod-product-compliance
Lightning Source LLC
Chambersburg PA
CBHW071611300625
28940CB00004B/355
* 9 7 8 3 0 3 1 3 1 1 4 8 2 *